Modern Zoology

Modern Zoology

Edited by Simon Benson

SYRAWOOD
PUBLISHING HOUSE

New York

Published by Syrawood Publishing House,
750 Third Avenue, 9th Floor,
New York, NY 10017, USA
www.syrawoodpublishinghouse.com

Modern Zoology
Edited by Simon Benson

International Standard Book Number: 978-1-68286-742-6 (Hardback)

Cataloging-in-Publication Data

Modern zoology / edited by Simon Benson.
    p. cm.
Includes bibliographical references and index.
ISBN 978-1-68286-742-6
1. Zoology. 2. Animals. I. Benson, Simon.
QL45.2 .M64 2019
590--dc23

TABLE OF CONTENTS

PREFACE

The purpose of the book is to provide a glimpse into the dynamics and to present opinions and studies of some of the scientists engaged in the development of new ideas in the field from very different standpoints. This book will prove useful to students and researchers owing to its high content quality.

Zoology is the branch of biology that is concerned with the scientific study of animals. All aspects of animal structure, classification, embryology and distribution are explored in the domain of zoology. The fundamental theories of cell biology, anatomy, physiology, evolutionary biology, ethology, etc. aid in developing a comprehensive understanding of all animals. Depending on taxonomy, zoology can be divided into the fields of mammalogy, ornithology, entomology and herpetology. The other sub-disciplines include zoography, behavioral ecology, animal physiology, vertebrate and invertebrate zoology, etc. This book traces the progress of this discipline and highlights some of its key concepts and applications. It attempts to understand the multiple branches that fall under the domain of zoology. This book includes contributions of experts and scientists, which will provide innovative insights into this field.

At the end, I would like to appreciate all the efforts made by the authors in completing their chapters professionally. I express my deepest gratitude to all of them for contributing to this book by sharing their valuable works. A special thanks to my family and friends for their constant support in this journey.

Editor

Partitioning of seed dispersal services between birds and bats in a fragment of the Brazilian Atlantic Forest

Raissa Sarmento[1,4], Cecília P. Alves-Costa[2], Adriana Ayub[2] & Marco A.R. Mello[3]

[1] Laboratório de Bioecologia e Conservação de Aves Neotropicais, Setor de Biodiversidade e Ecologia, Universidade Federal de Alagoas. Avenida Lourival Melo Mota, Tabuleiro do Martins, 57072-900 Maceió, AL, Brazil.
[2] Laboratório de Ecologia e Restauração da Biodiversidade (LERBIO), Departamento de Botânica, Universidade Federal de Pernambuco. Avenida Prof. Moraes Rego 1235, Cidade Universitária, 50670-901 Recife, PE, Brazil.
[3] Departamento de Biologia Geral, Instituto de Ciências Biológicas, Universidade Federal de Minas Gerais. Avenida Antônio Carlos 6627, Pampulha, Caixa Postal 486, 31270-910 Belo Horizonte, MG, Brazil.
[4] Corresponding author. E-mail: raissa.pereira@gmail.com

ABSTRACT. Community-level network studies suggest that seed dispersal networks may share some universal properties with other complex systems. However, most of the datasets used so far in those studies have been strongly biased towards temperate birds, including not only dispersers, but also seed predators. Recent evidence from multi-taxon networks suggests that seed dispersal networks are not all alike and may be more complex than previously thought. Here, we used network theory to evaluate seed dispersal in a strongly impacted Atlantic Forest fragment in northeastern Brazil, where bats and birds are the only extant dispersers. We hypothesized that the seed dispersal network should be more modular then nested, and that the dispersers should segregate their services according to dispersal syndromes. Furthermore, we predicted that bat and bird species that are more specialized in frugivory would be more important for maintaining the network structure. The mixed network contained 56 plant species, 12 bat species, and eight bird species, and its structure was more modular (M = 0.58) then nested (NODF = 0.21) compared with another multi-taxon network and 21 single-taxon networks (with either bats or birds). All dispersed fruits had seeds smaller than 9 mm. Bats dispersed mainly green fruits, whereas birds dispersed fruits of various colors. The network contained eight modules: five with birds only, two with bats only, and one mixed. Most dispersers were peripheral, and only specialized frugivores acted as hubs or connectors. Our results strongly support recent studies, suggesting that seed dispersal networks are complex mosaics, where different taxa form separate modules with different properties, which in turn play complementary roles in the maintenance of the associated ecosystem functions and services.

KEY WORDS. Diet overlap; frugivory; modularity; nestedness; networks.

Seed dispersal is a vital component of ecosystem functioning in tropical forests, where most plants depend on animals for the successful completion of their reproductive cycle (FLEMING & KRESS 2011). Although considerable knowledge has been accumulated on seed dispersal at the organism and population levels, relatively few studies have focused on the community level (MELLO et al. 2011a) – a pattern of research repeated for mutualisms in general (BRONSTEIN 1994). More recently, network theory has helped to fill this gap (BASCOMPTE & JORDANO 2007).

A network approach can be used to identify common properties of different kinds of mutualisms at the community level, such as nestedness (i.e., species with fewer interactions are connected to a subset of the mutualistic partners of species with more interactions, BASCOMPTE et al. 2003) and power law degree distribution (i.e., only a few species have a large number of interactions, JORDANO et al. 2003), and to identify underlying mechanisms (KRISHNA et al. 2008). Despite the assumed universality of nestedness, it has been recently discovered that seed dispersal networks may also be highly modular (i.e., these networks are composed of subsets of densely connected species, usually phylogenetically close to each other, MELLO et al. 2011a), and that modularity and nestedness are not mutually exclusive (FORTUNA et al. 2010).

In seed-dispersal networks, the balance between modularity and nestedness may largely depend on the diversity of frugivores available in the local community (MELLO et al. 2011a), as different frugivore groups focus on different plants (VAN DER PIJL 1972), but also on connectance (i.e., the proportion of realized interactions in the network, FORTUNA et al. 2010). Furthermore, different frugivore species contribute differently to the structure of the whole network, varying from peripherals with few interactions, to hubs with several interactions, and to connectors that bind different guilds (MELLO et al. 2011b). Unfortunately, most of the accumulated knowledge on seed dispersal networks came from studies based on datasets that contained only birds, mainly temperate or subtemperate species, and they included even several seed predators, which causes a strong bias in the biological interpretation and weakens the inferences (MELLO *et al.* 2011a). Due to differences in dietary preferences (MUSCARELLA & FLEMING 2007), frugivorous bats and birds, the main seed dispersers in the Neotropics (GALINDO-GONZÁLES *et al.* 2000), may form separated modules with different structure within multi-taxon seed dispersal networks (MELLO *et al.* 2011a). In our study site, a fragment of the northeastern Brazilian Atlantic Forest (see below), the importance of the remaining bats and birds is even greater, as other large frugivores (including some large-bodied birds) have become locally extinct or are so rare that they no longer make an significant ecological contribution to the dispersal system (PONTES *et al.* 2006, RODA 2006). Many of these locally extinct frugivores, such as medium- and large-bodied mammals, birds and reptiles, are generalists that dispersed the fruit of several species (TERBORGH *et al.* 2002, ALVES-COSTA & ETEROVICK 2007).

In the absence of other dispersers, it is important to understand how the remaining fauna maintains the seed dispersal function. Thus, in our study we used network theory to evaluate how bats and birds share the local seed dispersal service. Our main hypothesis was that bats and birds should form separate guilds in the community, reflected as modules within the seed dispersal network, as they have different diets (MUSCARELLA & FLEMING 2007) and other frugivores which might have had overlapping niches with bats and birds are locally extinct. Based on the theory of seed dispersal syndromes (VAN DER PIJL 1972) and on more recent studies (e.g., KORINE & KALKO 2005), we expected that bats would disperse mainly seeds of green fruits, while birds would disperse mainly seeds of other colors, and also that bats would disperse smaller seeds on average. Finally, we would expect bat and bird species that are more specialized for frugivory to be more important to the whole network structure (as in MELLO *et al.* 2011b), i.e., to be hubs or connectors, as they depend on fruits and thus probably participate more actively in frugivory interactions than occasional frugivores, regardless of abundance.

MATERIAL AND METHODS

Study area

The Atlantic Forest of the northern reach of São Francisco River, northeastern Brazil, is one of the most threatened ecosystems in the world, with less than five percent of its original forest cover remaining (TABARELLI & RODA 2005). In this region, known as the Pernambuco Center of Endemism (PCE), about one third of all tree species may be endangered due to the loss of their seed dispersers (SILVA & TABARELLI 2000). Our study was carried out in the Coimbra Forest, on land owned by Usina Serra Grande (a sugar mill). The site is located in Ibateguara, state of Alagoas (8°58'S, 36°3'W), at medium altitude of 500-600 m. The area covers 24,000 ha, and its land-

scape comprises fragments of Atlantic Forest (a total of 8,000 ha), embedded in a matrix of sugar-cane plantations. Coimbra Forest has an area of 3,500 ha and is the largest primary forest fragment in PCE (GRILLO *et al.* 2006). A detailed description of its vegetation is given by OLIVEIRA (2005).

Field data

Mist nets (12 x 2 m, 36 mm mesh) were set up to capture birds and bats in the forest understory (0-2 m above ground). Mist nets were distributed along trails inside the fragment and its edges, at three sampling stations, and were kept open for at least one day in each station every month. Stations were located at least 2 km from each other. As bats and birds were captured at the same sites and dates, differences between the diets of both taxa are unlikely to be related to spatial and temporal differences in fruit availability.

Bird captures were made from July 2007 to December 2008. Nets remained open from 05:30 a.m. to 05:00 p.m., totaling a capture effort of 4.4×10^4 h.m^2 (the area of one mist net multiplied by the total number of nets and the total number of hours worked, sensu STRAUBE & BIANCONI 2002). Birds were kept inside cloth bags in order to obtain fecal samples, which were stored in individual plastic vials. Birds were identified using a field guide (SIGRIST 2006), with identities confirmed by Sônia Roda (CEPAN). Birds were photographed and then released at their capture sites.

Bats were captured from August 2007 to July 2008. Nets remained open from 6:00 a.m. to 0:00 a.m., totaling an effort to capture of 1.8×10^4 h.m^2. Bats were also kept inside cloth bags to defecate, and feces were stored in plastic vials. The bats were released after being examined. A combination of keys was used to identify bats (VIZOTTO & TADDEI 1973, EMMONS & FEER 1997, GARDNER 2008) and taxonomy followed GARDNER (2008). The bats that could not be identified in the field were deposited in the Mammal Collection of Universidade Federal de Pernambuco in order to confirm identification (Table I). For the network analysis, we considered only data from August 2007 to July 2008, in order to ensure that bird and bat data were comparable.

Table I. Bat species and voucher number of individuals deposited in the Mammal Collection of the Universidade Federal de Pernambuco.

Species	Voucher number	Sex
Carollia perspicillata (Linnaeus, 1758)	1696	M
	1699	M
	1702	M
Rhinophylla pumilio W. Peters, 1865	1685	F
	1703	F
Artibeus obscurus (Schinz, 1821)	1698	M
Trinycteris nicefori Sanborn, 1949	1704	M
Tonatia saurophila Koopman and Williams, 1951	1701	F
Micronycteris sp.	1700	F
Lasiurus blossevillii (Lesson, 1826)	1694	M

Guild structure of the network

In order to describe the structure of the seed dispersal network, we built a binary (presence/absence) adjacency matrix with animals (bats and birds) as i rows and plants as j columns. Cells were filled with value 1 when seeds of a j plant were found in the feces of an i animal, and with value 0 when there was no record of interactions. We focused on binary metrics of network structure, as this approach is very useful to search for general patterns (JORDANO et al. 2003) and the literature is richer for this approach, providing a better benchmark for comparison. As the sampling period was longer for birds than for bats, we discarded bird data from August 2008 and built the total matrix only with simultaneously collected bat and bird data. We also analyzed the bat-fruit and bird-fruit sub-networks separately, in order to evaluate the combined topology of the whole network.

To analyze the overlap between seed-dispersal services by bats and birds we used a simulated annealing analysis (GUIMERÀ & AMARAL 2005), based on the concept of modularity. Here, a module is defined as a subgroup of species that are more densely connected to each other than to other species in the same network. The degree of modularity in the network was measured with the index M, in the Netcarto software (GUIMERÀ & AMARAL 2005), which varies from 0 (no subgroups) to 1 (totally separated subgroups). As the Monte Carlo analysis in Netcarto was designed for unipartite networks, we used a custom-made Monte Carlo procedure (first used by MELLO et al. 2011a): 1) we generated 1,000 randomized networks from the original network, using the null model 2 with marginal totals fixed (as in BASCOMPTE et al. 2003) and a MatLab script written by Paulo R. Guimarães Jr; 2) we measured M for each of these networks with a version of Netcarto that was modified by Flavia M. D. Marquitti, which allows analyzing multiple networks at a time and annotating M-values in a text file; 3) we carried out a Z-test to estimate the significance of M. This analysis allowed an estimate of the number of modules in the network and the distribution of bats, birds, and plants into different modules.

We estimated the degree of nestedness in the studied network with the index NODF (ALMEIDA-NETO et al. 2008), which varies from 0 (no nestedness) to 1 (perfect nestedness); values were normalized. The significance of NODF was estimated with a Monte Carlo procedure (1,000 randomizations) in the software Aninhado 3.0 (GUIMARÃES & GUIMARÃES 2006), using the null model Ce, in which marginal totals are fixed (null model 2 of BASCOMPTE et al. 2003). The index NODF measures nestedness better than the widely used index N (derived from T, ATMAR & PATTERSON 1993) because it considers the pairwise nesting of rows and columns, and more closely reflects the original concept of nestedness (ULRICH et al. 2009).

We used Z tests to test whether the values of NODF and M observed in the overall network differed from the values measured for other 21 networks (compiled by MELLO et al. 2011a)

that contained either birds ($N = 10$) or bats ($N = 11$), and from another mixed network with both disperser groups from a protected rainforest in Peru (the GORCHOV et al. 1995 dataset analyzed by MELLO et al. 2011b) (Appendix S1*).

Seed size and fruit color

Fecal samples from bats and birds were taken to the laboratory, washed in water, passed through a 1-mm sieve, sun-dried, measured, photographed, enumerated, and separated into morphospecies (according to color, size and texture). We also recorded whether seeds were damaged or intact. Seeds were identified to the lowest possible taxonomic level and deposited in the LERBIO seed bank. Fruit color and seed size of the plants dispersed were determined, to test for differences among the species subsets dispersed by each group. We determined fruit color by direct observation of plants in the field, by checking herbarium specimens, and by searching in the literature. The diameter of each seed was measured in the program Image Tool 3.0.

Relative importance of bat and bird species

A species that either makes a disproportionally large number of interactions (hub) or binds together different modules of the network (connector) was considered as relatively more important to the entire seed dispersal network. Thus, the relative importance of each frugivore and plant species for the overall network structure was measured with three surrogate metrics: degree centrality (kr), betweenness centrality (BC), and network functional role (FR). Degree centrality is calculated as the number of interactions made by a species (NOOY et al. 2005), and can be expressed as a proportion in relation to the total number of interactions that each vertex could make in its network; this metric is used as a surrogate for the ecological concept of niche breadth (as the 'normalized degree' used in MELLO et al. 2011b). Betweenness centrality is calculated as the proportion of small paths (the shortest path between two species, measured in number of interactions) in the network, which the species of interest crosses (NOOY et al. 2005). It is a measure of the role of a species in binding together different guilds within the mutualistic community (DUPONT & OLESEN 2009).

The concept of network functional role comes from the (previously described) modularity analysis (GUIMERÀ & AMARAL 2005). Based on the within module degree (z – how many interactions a species makes within its module) and the participation coefficient (P – the percentage of interactions made by a species that link it to species outside its module) measured for each species in the modularity analysis, a functional role from R1 to R7 was attributed to the species. The most important roles are considered connectors (R3, i.e., species that bind different modules) or a mixture of hubs (i.e., species with a disproportionally high number of interactions) and connectors (R6 and R7). For details on the analysis of functional roles applied to mutualistic networks, see MELLO et al. (2013).

In order to test the hypothesis that more specialized frugivores establish interactions with a larger proportion of the plants available in the network (and have higher importance for the whole network structure), we classified each bat and bird species according to its dietary specialization. Here, specialization is related to 'level of frugivory', i.e., how dependent a species is on fruits. When a frugivore feeds exclusively or mostly on fruits, it was classified as 'specialized'; when fruits are an important part of the diet, but the species does not depend exclusively on them, it was classified as 'secondary'; and when the species feeds on fruits only occasionally, it was classified as 'opportunistic'.

A general linear mixed-effects multivariate model (GLMM) was used to test whether the variation in centrality (k_r, BC, and FR – dependent variables) was affected by level of frugivory (specialized, secondary or opportunistic) and disperser group (bats or birds) (fixed factors); as the values of centrality could be also biased by sampling effort, we considered the number of captures (a measure of relative abundance) and the number of fecal samples analyzed of each disperser species as cofactors in our model. Each of our variables followed a different statistical distribution, so we estimated significance by bootstrapping (1,000 randomizations). As there are differences among specialized frugivores in the number and diversity of fruit species consumed (e.g., among Carollia bats, THIES & KALKO 2004), we expected specialized frugivores to play a wider variety of functional roles, and secondary and occasional frugivores to be more peripheral. Thus, we tested also for differences in the variance of k_r, BC, and FR among species with different levels of frugivory with Levene tests. We also searched for extreme values and outliers in our data, in order to identify the most important species according to each of the centrality metrics analyzed. Percentage data were arcsine transformed prior to the analysis. Finally, we tested for differences between the seed sizes ingested by birds and bats using a Mann-Whitney U test, as the data were not normally distributed. All analyses were run in SPSS 20 for Mac and followed ZAR (1996) and MANLY (2007).

RESULTS

Species and interaction records

We captured 284 birds of 43 species and 14 families, from which we obtained 177 fecal samples (Appendix S2*). Only 10 bird species dispersed seeds; 45.7 percent of their feces (N = 81) contained seeds of 39 plant species. We considered only eight of these bird species in the network analysis, as we discarded data collected after August 2008. The three most abundant bird species belonged to Pipridae: *Ceratopipra rubrocapilla* (Temminck, 1821) (19% of all individuals), *Chiroxiphia pareola* (Linnaeus, 1766) (18.3%) and *Manacus manacus* (Linnaeus, 1766) (7.7%). These species were also the main seed dispersers,

contributing with 54.8 percent of the fecal samples, of which 82.8 percent contained seeds. Capture success was higher for bats (bat = 535 x 10^4 ind/h.m² and bird = 64 x 10^4 ind/h.m²), as we recorded 964 individuals from 27 species and two families, from which we obtained 629 fecal samples (Appendix S2*). Twelve bat species dispersed seeds, and 57.1 percent (N = 359) of their fecal samples contained the seeds of 33 species. *Carollia perspicillata* (Linnaeus, 1758) (Phyllostomidae) was the most abundant bat (73.2% of all individuals), and was responsible for 80 percent of all fecal samples and 85.2 percent of all samples with seeds. Almost half of the captured bat species but only one quarter of the bird species dispersed seeds. Bats dispersed almost five times more seeds than the birds and also more seeds per individual (birds: 5,415 seeds, 66.8 seeds/individual; bats: 26,670 seeds, 75.1 seeds/individual). Although the number of bird fecal samples was five times lower than the number for bats, on average each bird species dispersed more plant species: the plant/animal ratio was 3.9 for birds and 2.7 for bats. However, on average, birds (k_r = 9 ± 7%) and bats (k_r = 9 ± 14%) interacted with a similar proportion of the fruit species available in the overall network (d.f. = 20, t = -0.26, P = 0.80).

Seed size and fruit color

Seeds ingested by birds (N = 61, median = 2.00 mm, range = 0.3-8.5) were larger than seeds consumed by bats (N = 63, median = 1.00 mm, range = 0.3-6.0) (U = 1526.0, p = 0.048). Birds dispersed seeds from fruits of six different colors; purple (50%), blue (21%), green (14%), brown (7%), red (5%) and yellow (3%), whereas bats dispersed mainly seeds of green fruits (91%), with a lower percentage of brown (6%) and purple (3%) fruits.

Guild structure of the network

The overall seed dispersal network comprised of eight bird and 12 bat species, which dispersed seeds from 56 plant species (Fig. 1, Appendix S3*). Seeds of 28 plant species were dispersed exclusively by bats, 23 exclusively by birds, and 5 by both taxa, in a total of 124 interactions. The overall network was more modular (M = 0.58, p < 0.001) than the 21 single-taxon networks with either bats or birds used for comparison (M = 0.35 ± 0.07, d.f. = 20, t = 11.4, p < 0.001) and another mixed network from Peru (M = 0.45), but exhibited lower nestedness (NODF = 0.21, p < 0.001) than the same 21 single-taxon networks (NODF = 0.48 ± 0.10, d.f. = 20, t = 15.0, p < 0.001) and mixed network (NODF = 0.31). The bat sub-network (NODF = 0.48, p < 0.001) had a higher degree of nestedness than the overall network, whereas the bird sub-network was not significantly nested (NODF = 0.21, p = 0.59).

Eight modules were detected in the overall network. Five modules were composed exclusively of birds, two exclusively of bats, and one module contained species of both groups (Fig. 1).

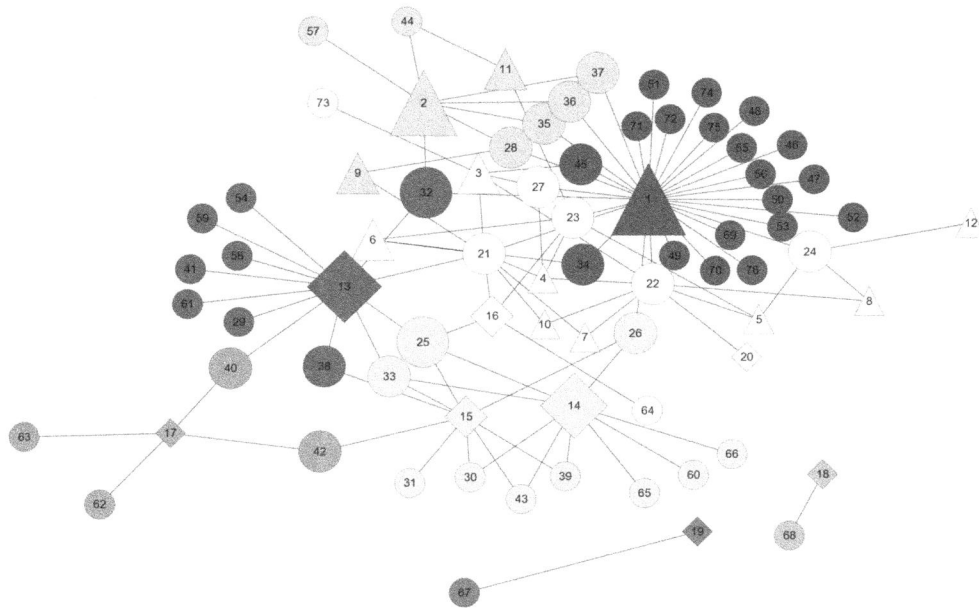

Figure 1. The structure of the mixed bipartite seed dispersal network from Coimbra Forest, northeastern Brazil, was more modular than nested, with a strong separation between bats (triangles) and birds (diamonds), and their food-plants (ellipses) in different modules (identified by grey tones). The most important frugivore and plant species (larger symbols) were located in the center of some modules (hubs) or bound together different modules (connectors). Symbol size is proportional to network functional role. Species codes are presented in Table II.

Relative importance of bat and bird species

The most important species, identified as connector hubs (R6), were the bat *Carollia perspicillata* and the bird *Chiroxiphia pareola*. Modules were centered on these hubs. These were followed by two provincial hubs (R5), the bat *Rhinophylla pumilio* Peters, 1865 (Phyllostomidae) and the bird *Ceratopipra rubrocapilla*. Two plant species functioned as non-hub connectors (R3), binding together different modules, although having fewer interactions than a hub: *Piper caldense* C.DC 1872 (Piperaceae) and *Philodendron* sp. Schott.1829 (Araceae). There was large variation in all centrality metrics among the frugivore species in the seed dispersal network of Coimbra Forest (Table II). Variances did not differ among categories of level of frugivory for network functional role (FR: F = 2.53, p = 0.08), degree centrality (k_r: F = 1.85, p = 0.17), and betweenness centrality (BC: F = 1.35, p = 0.30). The GLMM was significant for degree centrality (N = 20, F = 8.01, p = 0.001, Power = 0.99) and betweenness centrality (N = 20, F = 6.35, p = 0.003, Power = 0.97), but not for network functional role (N = 20, d.f. = 7, F = 1.81, p = 0.18, Power = 0.48). There were no significant effects of level of frugivory (level of frugivory – FR: F = 0.78, p = 0.48, partial eta² = 0.12; k_r: F = 2.50, p = 0.12, partial eta² = 0.29; BC: F = 1.44, p = 0.28, partial eta² = 0.19) and disperser group (FR: F = 0.005, p = 0.95, partial eta² = 0.00; k_r: F = 0.27, p = 0.62, partial eta² = 0.02; BC: F = 1.37, p = 0.26, partial eta² = 0.10) on the centrality metrics

analyzed (Figs 2-7). The number of captures (p-values – FR: 0.80, k_r = 0.97, BC = 0.76) and the number of fecal samples analyzed (P-values – FR: 0.72, k_r = 0.79, BC = 0.60) for each species did also not explain the variation in the centrality metrics. Despite the lack of differences in averages and variances among levels of frugivory in all three centrality metrics, there was an interesting pattern of outliers and extreme values. Specialized frugivores, such as the bat *Carollia perspicillata* and the bird *Chiroxiphia pareola*, were the only species to reach the highest values of all centrality metrics (Table III).

DISCUSSION

Our findings support the hypothesis that the seed dispersal network of Coimbra Forest is highly modular, with bats and birds forming separated guilds and playing different functional roles. This separation of guilds may be explained by interaction syndromes. Also, as anticipated, the most important species for the maintenance of the seed dispersal system were specialized frugivores. In summary, our results suggest that the maintenance of seed dispersal within our highly impacted area largely depends on the complementarity among modules and on the survival of specialized frugivorous bats and birds.

The studied network had different properties compared to other seed-dispersal networks reported in the literature –

Table II. The studied species of bats, birds, and plants showed a large variation in their ecological functional roles in the mixed seed dispersal network form the Coimbra Forest, northeastern Brazil. Specialized frugivores reached the highest values of network functional role (FR), degree centrality (kr), and betweenness centrality (BC). The codes presented after the species' names are the same used in Fig. 1.

Species	Code	FR	kr	BC	Frug
Bats					
Carollia perspicillata (Linnaeus, 1758)	1	6	0.54	0.58	specialized
Rhinophylla pumilio W. Peters, 1865	2	5	0.13	0.05	specialized
Artibeus fimbriatus Gray, 1838	3	2	0.09	0.03	specialized
Glossophaga soricina (Pallas, 1766)	9	2	0.04	0.00	secondary
Trachops cirrhosus (Spix, 1823)	6	2	0.05	0.01	opportunistic
Trinycteris nicefori Sanborn, 1949	11	2	0.04	0.01	opportunistic
Artibeus cinereus (P. Gervais, 1856)	5	1	0.05	0.00	specialized
Artibeus lituratus (Olfers, 1818)	4	1	0.07	0.01	specialized
Artibeus obscurus (Schinz, 1821)	8	1	0.04	0.00	specialized
Artibeus planirostris (Spix, 1823)	7	1	0.04	0.00	specialized
Phyllostomus elongatus (É. Geoffroy, 1810)	12	1	0.02	0.00	opportunistic
Platyrrhinus lineatus (É. Geoffroy, 1810)	10	1	0.04	0.00	specialized
Birds					
Chiroxiphia pareola (Linnaeus, 1766)	13	6	0.21	0.26	specialized
Ceratopipra rubrocapilla (Temminck, 1821)	14	5	0.16	0.11	specialized
Manacus manacus (Linnaeus, 1766)	15	2	0.16	0.11	specialized
Saltator maximus (Statius Muller, 1776)	16	2	0.07	0.05	secondary
Dysithamnus mentalis (Temminck, 1823)	18	1	0.02	0.00	opportunistic
Euphonia violacea (Linnaeus, 1758)	19	1	0.02	0.00	specialized
Tangara cayana (Linnaeus, 1766)	20	1	0.02	0.00	secondary
Turdus albicollis Vieillot, 1818	17	1	0.07	0.05	secondary
Plants					
Philodendron sp. Schott.1829	25	3	0.20	0.05	
Piper caldense C.DC. 1872	32	3	0.15	0.09	
Fabaceae sp. 1	37	2	0.10	0.01	
Fabaceae sp. 2	38	2	0.10	0.01	
M42	23	2	0.30	0.06	
M46	27	2	0.15	0.01	
M47	28	2	0.15	0.02	
M52	33	2	0.15	0.02	
M53	34	2	0.10	0.00	
M55	36	2	0.10	0.01	
M59	40	2	0.10	0.05	
M64	45	2	0.10	0.01	

Continues

Table II. Continued.

Species	Code	FR	kr	BC	Frug
Piper arboreum Aubl., 1775	22	2	0.40	0.07	
Piper marginatum Jacq., 1791	21	2	0.45	0.19	
Piper sp. 2	35	2	0.10	0.01	
Piper sp. 5	26	2	0.15	0.16	
Solanum sp.	42	2	0.10	0.02	
Vismia guianensis (Aubl.) Pers., 1807	24	2	0.20	0.04	
Cecropia pachystachya Trécul., 1847	72	1	0.05	0.00	
Clidemia debilis Crueg., 1847	49	1	0.05	0.00	
Clidemia hirta (L.) D. Don., 1823	75	1	0.05	0.00	
Fabaceae sp.3	62	1	0.05	0.00	
Ficus gomelleira Kunth & Bouché, 1847	73	1	0.05	0.00	
M10	63	1	0.05	0.00	
M11	59	1	0.05	0.00	
M15	64	1	0.05	0.00	
M17	65	1	0.05	0.00	
M18	66	1	0.05	0.00	
M2	54	1	0.05	0.00	
M21	68	1	0.05	0.00	
M24	52	1	0.05	0.00	
M25	69	1	0.05	0.00	
M3	61	1	0.05	0.00	
M36	70	1	0.05	0.00	
M39	71	1	0.05	0.00	
M5	60	1	0.05	0.00	
M58	39	1	0.10	0.00	
M6	50	1	0.05	0.00	
M66	47	1	0.05	0.00	
M67	48	1	0.05	0.00	
Melastomataceae sp. 1	76	1	0.05	0.00	
Melastomataceae sp. 2	67	1	0.05	0.00	
Miconia prasina (Sw.) DC., 1828	57	1	0.05	0.00	
Miconia sp. 1	56	1	0.05	0.00	
Miconia sp. 2	51	1	0.05	0.00	
Passifloraceae	44	1	0.10	0.00	
Piper aduncum L., 1753	74	1	0.05	0.00	
Piper sp. 1	43	1	0.10	0.00	
Piper sp. 3	41	1	0.05	0.00	
Piper sp. 4	29	1	0.05	0.00	
Poaceae	58	1	0.05	0.00	
Schefflera morototoni (Aubl.)	55	1	0.05	0.00	
Solanaceae	46	1	0.05	0.00	
Solanum americanum Mill., 1768	31	1	0.05	0.00	
Solanum rugosum Dunal. 1852	30	1	0.10	0.00	
Vismia sp.	53	1	0.05	0.00	

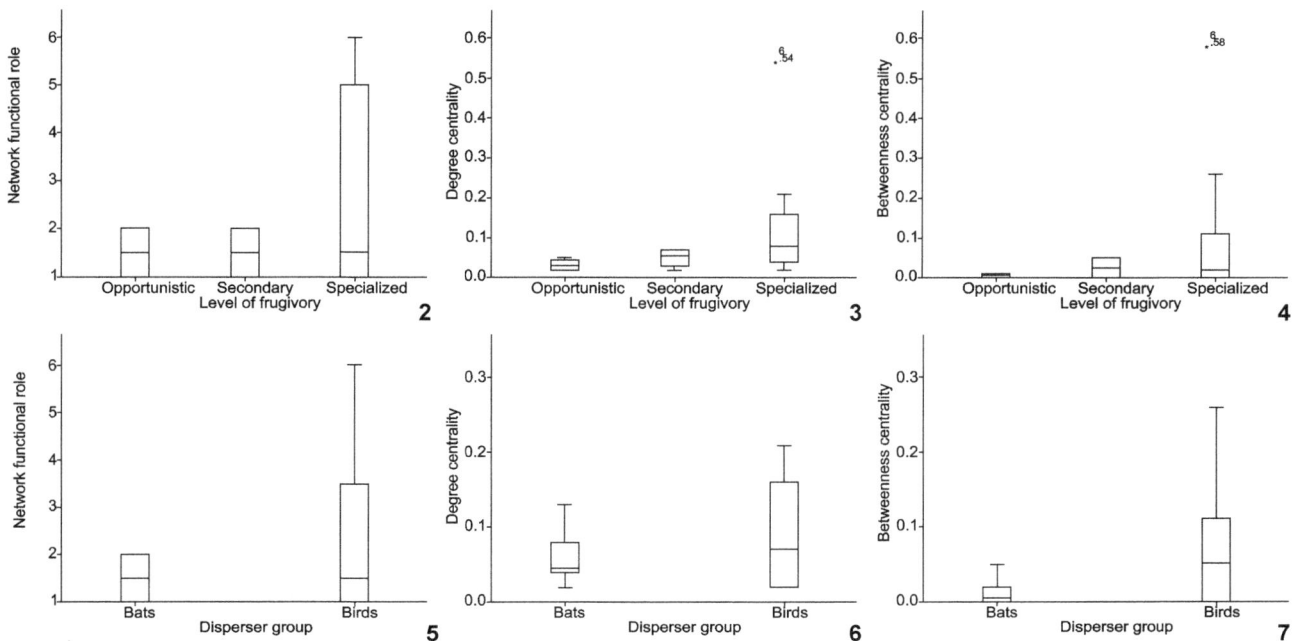

Figures 2-7. On average, there was no effect of level of frugivory (2-4) and disperser group (5-7) on the three centrality metrics studied (network functional role, degree centrality, and betweenness centrality), and variances were also similar among categories. However, we noticed that all outliers and extreme values occurred only in the category 'specialized'; furthermore, there was a slight, but non-significant, trend towards higher variance in this category. In the plot, the central line represents the median, the boxes represent the quartiles, the whiskers represent the 95% interval, and * represents outliers.

which comprised either birds or bats. First, as observed in other seed dispersal networks (MELLO *et al.* 2011a) and oil-flower pollination networks (BEZERRA *et al.* 2009), mixed networks are less nested than single-taxon networks. Although we should be careful when interpreting the topological metrics of the bat or bird subnetworks – as they are very small – these results are consistent with the hypothesis that mutualistic modules formed by phylogenetically related species are tiny worlds (BEZERRA *et al.* 2009) within small worlds (OLESEN *et al.* 2006), i.e., subnetworks with higher cohesiveness than whole networks. They also corroborate the hypothesis that plant-animal mutualisms at the community level are mosaics of subsystems with different structure and dynamics (JORDANO (1987).

This conclusion is also supported by the much higher modularity observed in the mixed seed-dispersal network as compared with single-taxon networks. In the Coimbra Forest network there was strong separation between bird and bat modules, as from the eight modules detected only one was mixed. Therefore, although species within a module play somewhat redundant roles by dispersing particular subsets of plants, modules in the network are complementary. This information is relevant for conservation, as efficient strategies for maintenance of ecosystem services should take into account ecological redundancy (WALKER 1992, ALVES-COSTA & ETEROVICK 2007).

Our results show that bats and birds probably do not replace each other in the seed-dispersal service; so both groups need to be preserved within local systems – especially when other dispersers have already been lost.

The biological features that explain the observed modular and hierarchical structure within the disperser network are especially interesting. The separation between birds and bats in terms of dispersal services is explained at least in part by the theory of dispersal syndromes (VAN DER PIJL 1972), as in the area fruit color was related to disperser choice. Although this theory has been criticized (e.g., OLLERTON *et al.* 2009), some of its predictions are useful to understand fruit selection by bats and birds. Specifically, there is growing evidence that these disperser groups focus on different plant species (MUSCARELLA & FLEMING 2007, LOBOVA *et al.* 2009), and that their choices are largely related to fruit characteristics (KALKO *et al.* 1996, KALKO & CONDON 1998, KORINE & KALKO 2005). Results were also consistent with syndromes, as bats dispersed smaller seeds. However, as our study was based only on fecal samples, it remains unknown to what extent the remaining birds and bats can disperse large-seed plants. Evidence from a rainforest in Mexico suggests that bats can partly replace larger dispersers (MELO *et al.* 2009).

Similar differences in nestedness and modularity were observed between the study network and a mixed seed-dispersal

Table III. Some specialized frugivores, such as the bat *Carollia perspicillata* and the bird *Chiroxiphia pareola*, were the only species to reach extreme values of all centrality metrics, although there were no differences on average in centrality among categories of dietary specialization.

Centrality metric	Level of frugivory			Species	Value
Network functional role	Opportunistic	Highest	1	*Trachops cirrhosus*	2
			2	*Trinycteris nicefori*	2
		Lowest	1	*Dysithamnus mentalis*	1
			2	*Phyllostomus elongatus*	1
	Secondary	Highest	1	*Glossophaga soricina*	2
			2	*Saltator maximus*	2
		Lowest	1	*Turdus albicollis*	1
			2	*Tangara cayana*	1
	Specialized	Highest	1	*Carollia perspicillata*	6
			2	*Chiroxiphia pareola*	6
			3	*Rhinophylla pumilio*	5
			4	*Ceratopipra rubrocapilla*	5
			5	*Artibeus fimbriatus*	2
		Lowest	1	*Euphonia violacea*	1
			2	*Platyrrhinus lineatus*	1
			3	*Artibeus planirostris*	1
			4	*Artibeus obscurus*	1
			5	*Artibeus lituratus*	1
Degree centrality	Opportunistic	Highest	1	*Trachops cirrhosus*	0.05
			2	*Trinycteris nicefori*	0.04
		Lowest	1	*Dysithamnus mentalis*	0.02
			2	*Phyllostomus elongatus*	0.02
	Secondary	Highest	1	*Saltator maximus*	0.07
			2	*Turdus albicollis*	0.07
		Lowest	1	*Tangara cayana*	0.02
			2	*Glossophaga soricina*	0.04
	Specialized	Highest	1	*Carollia perspicillata*	0.54
			2	*Chiroxiphia pareola*	0.21
			3	*Manacus manacus*	0.16
			4	*Ceratopipra rubrocapilla*	0.16
			5	*Rhinophylla pumilio*	0.13
		Lowest	1	*Euphonia violacea*	0.02
			2	*Platyrrhinus lineatus*	0.04
			3	*Artibeus planirostris*	0.04
			4	*Artibeus obscurus*	0.04
			5	*Artibeus cinereus*	0.05
Betweenness centrality	Opportunistic	Highest	1	*Trachops cirrhosus*	0.01
			2	*Trinycteris nicefori*	0.01
		Lowest	1	*Dysithamnus mentalis*	0.00
			2	*Phyllostomus elongatus*	0.00
	Secondary	Highest	1	*Saltator maximus*	0.05
			2	*Turdus albicollis*	0.05
		Lowest	1	*Tangara cayana*	0.00
			2	*Glossophaga soricina*	0.00
	Specialized	Highest	1	*Carollia perspicillata*	0.58
			2	*Chiroxiphia pareola*	0.26
			3	*Manacus manacus*	0.11
			4	*Ceratopipra rubrocapilla*	0.11
			5	*Rhinophylla pumilio*	0.05
		Lowest	1	*Euphonia violacea*	0.00
			2	*Platyrrhinus lineatus*	0.00
			3	*Artibeus planirostris*	0.00
			4	*Artibeus obscurus*	0.00
			5	*Artibeus cinereus*	0.00

network from Peru (analyzed by MELLO *et al.* 2011a). The higher modularity and lower nestedness of the study network compared with the Peruvian network are probably an effect of the loss of other frugivores in Coimbra Forest. Medium- and large-bodied mammals, birds, and reptiles are frequently generalistic frugivores (ALVES-COSTA & ETEROVICK 2007), binding different parts of the network and increasing its cohesiveness. These animals are also among those preferred by poachers and hunters (REDFORD 1992), and are among the first to be lost due to fragmentation effects (FAHRIG 2003). Therefore, species loss due to human influence may significantly increase modularity, i.e., guild segregation, in seed-dispersal networks. This hypothesis remains to be tested in future studies.

Our mixed model detected no significant effect of level of frugivory on network functional role, degree centrality, and betweenness centrality (contrary to the observations by MELLO *et al.* 2011b). However, in our study only specialized frugivores attained the highest functional roles (i.e., extremes and outliers) in the mixed network: hubs and connectors. This is an important finding, consistent with results obtained in studies on pollination (BEZERRA *et al.* 2009, DUPONT & OLESEN 2009, GONZALEZ *et al.* 2010) and seed dispersal (MELLO *et al.* 2011b), which indicate that, regardless on their level of specialization, the keystones in each network are dietary specialists (frugivores that feed exclusively or mainly on fruits). Interestingly, there were some differences in centrality between specialized frugivores, especially bats of the subfamilies Stenodermatinae and Carolliinae. Among specialized frugivorous bats, the first are considered more specialized than the latter (LOBOVA *et al.* 2009). However, those subtle differences did not result in stenodermatines being more central than carolliines, as what matters in our context in how dependent on fruits an animal is, no matter on how many fruit species it specializes.

It should be noted that the concept of specialization is a source of disagreement in the literature (BLÜTHGEN 2010, MELLO *et al.* 2011b). There are several ecological concepts of specialization, which take into account not only the number of interactions made by a species, but also different aspects of the phylogenetic signal in those interactions (DEVICTOR *et al.* 2010). However, in the network literature, specialization is in most cases referred to as simply the number or strength of the interactions made by a species (BASCOMPTE *et al.* 2006) or the uniqueness of those interactions when compared to the patterns observed in other species in the same network (BLÜTHGEN *et al.* 2006). We propose that, in network studies, specialization should be defined on biological grounds, while centrality metrics should be used as surrogates for relative importance in the community as a network. Centrality metrics can be also used as surrogates for ecological specialization, although care must be taken not to mistake the former for the latter. For instance, in the studied network, a highly specialized frugivore, such as the phyllostomid bat *Carollia perspicillata*, could have been called a 'generalist' (following the terminology of most network studies), which does not make any sense from a biological perspective.

Two plant species functioned as connectors in the network (*Piper caldense* and *Philodendron* sp.), as they bound together different modules. These two genera are core components of the diet of some numerically important bats (Carolliinae and Stenodermatinae) genera (Henry & Kalko 2007). Therefore, these plants are good candidates to be used in forest restoration programs, as they may accelerate the process of regeneration of the seed-dispersal network in degraded areas.

In summary, the seed-dispersal service in the Coimbra Forest appears to be structured as a network with a combined topology. Future conservation plans should take into account the modular structure of seed-dispersal networks. Moreover, guild diversity should be prioritized since the loss of some disperser groups seems to cause a more modular system in which interactions are fragmented and the system as a whole is less resilient.

ACKNOWLEDGEMENTS

We thank many colleagues who helped us during this study. Gleice P. Silva assisted us in the field. Sônia Roda and Thaís Lira helped us identify frugivore species. Andrej Mrvar, Carsten Dormann, Flávia Marquitti, Mário Almeida-Neto, Roger Guimerà, and Vladmir Batagelj provided us with their network software. Elisabeth Kalko and Nico Blüthgen discussed with us the biological meaning of network metrics, what was extremely helpful in trying to strengthen the link between natural history, ecological theory, and network ecology. Scott Heald and Richard Ladle proofread our manuscript. Usina Serra Grande allowed us to carry out research in Mata de Coimbra. CEPAN made this project possible by establishing a partnership between Usina Serra Grande and Universidade Federal de Pernambuco. This project was funded by the Brazilian Research Council (CNPq, 485309/2006-8). Marco Mello was sponsored by the Alexander von Humboldt Foundation (AvH, 1134644) and the Federal University of Minas Gerais (UFMG, edital 2013/1). This work is dedicated to the memory of Elisabeth Kalko.

REFERENCES

Almeida-Neto, M.; P.R. Guimarães; P.R. Guimarães Jr; R.D. Loyola & W. Ulrich. 2008. A consistent metric for nestedness analysis in ecological systems: reconciling concept and measurement. **Oikos 117**: 1227-1239. doi: 10.1111/j.0030-1299.2008.16644.x

Alves-Costa, C.P. & P.C. Eterovick. 2007. Seed dispersal services by coatis (*Nasua nasua*, Procyonidae) and their redundancy with other frugivores in southeastern Brazil. **Acta Oecologica 32**: 77-92. doi:10.1016/j.actao.2007.03.001

Atmar, W. & B.D. Patterson. 1993. The measure of order and disorder in the distribution of species in fragmented habitat.

Oecologia 96: 373-382. doi:10.1007/BF00317508

Bascompte, J. & P. Jordano. 2007. Plant-animal mutualistic networks: the architecture of biodiversity. **Annual Review of Ecology Evolution and Systematics 38**: 567-593. doi: 10.1146/annurev.ecolsys.38.091206.095818

Bascompte, J.; P. Jordano; C.J. Melian & J.M. Olesen. 2003. The nested assembly of plant-animal mutualistic networks. **Proceedings of the National Academy of Sciences of the United States of America 100**: 9383-9387. doi: 10.1073/pnas.1633576100

Bascompte, J.; P. Jordano & J.M. Olesen. 2006. Asymmetric coevolutionary networks facilitate biodiversity maintenance. **Science 312**: 431-433. doi: 10.1126/science.1123412

Bezerra, E.L.S.; I.C.S. Machado & M.A.R. Mello. 2009. Pollination networks of oil-flowers: a tiny world within the smallest of all worlds. **Journal of Animal Ecology 78**: 1096-1101. doi: 10.1111/j.1365-2656.2009.01567.x

Blüthgen, N. 2010. Why network analysis is often disconnected from community ecology: a critique and an ecologist's guide. **Basic and Applied Ecology 11**: 185-195. doi:10.1016/j.baae.2010.01.001

Blüthgen, N.; F. Menzel & N. Blüthgen. 2006. Measuring specialization in species interaction networks. **BMC Ecology 6**: 1-12. doi:10.1186/1472-6785-6-9

Bronstein, J.L. 1994. Our current understanding of mutualism. **The Quarterly Review of Biology 69**: 31-51.

Devictor, V.; J. Clavel; R. Julliard; S. Lavergne; D. Mouillot; W. Thuiller; P. Venail; S. Villéger & N. Mouquet. 2010. Defining and measuring ecological specialization. **Journal of Applied Ecology 47**: 15-25. doi: 10.1111/j.1365-2664.2009.01744.x

Dupont, Y.L. & J.M. Olesen. 2009. Ecological modules and roles of species in heathland plant-insect flower visitor networks. **Journal of Animal Ecology 78**: 346-353. doi: 10.1111/j.1365-2656.2008.01501.x

Emmons, L.H. & F. Feer. 1997. **Neotropical rainforest mammals: a field guide.** Chicago, University of Chicago Press.

Fahrig, L. 2003. Effects of habitat fragmentation on biodiversity. **Annual Review of Ecology, Evolution, and Systematics 34**: 487-515. doi: 10.1146/annurev.ecolsys.34011802.132419

Fleming, T.H. & W.J. Kress. 2011. A brief history of fruits and frugivores. **Acta Oecologica 37**: 521-530. doi:10.1016/j.actao.2011.01.016

Fortuna, M.A.; D.B. Stouffer; J.M. Olesen; P. Jordano; D. Mouillot; B.R. Krasnov; R. Poulin & J. Bascompte. 2010. Nestedness versus modularity in ecological networks: two sides of the same coin? **Journal of Animal Ecology 79**: 811-817. doi: 10.1111/j.1365-2656.2010.01688.x

Galindo-Gonzáles, J.; S. Guevara & V.J. Sosa. 2000. Bat and bird-generated seed rains at isolated trees in pastures in a tropical rainforest. **Conservation Biology 14**: 1693-1703. doi: 10.1111/j.1523-1739.2000.99072.x

Gardner, A.L. 2008. **Mammals of South America.** Chicago, University of Chicago Press, vol. 1.

GONZALEZ, A.M.M.; B. DALSGAARD & J.M. OLESEN. 2010. Centrality measures and the importance of generalist species in pollination networks. **Ecological Complexity 7**: 36-43. doi: 10.1016/j.ecocom.2009.03.008.

GORCHOV, D.L.; F. CORNEJO; C.F. ASCORRA & M. JARAMILLO. 1995. Dietary overlap between frugivorous birds and bats in the Peruvian Amazon. **Oikos 74**: 235-250.

GRILLO, A.S.; A.A. OLIVEIRA & M. TABARELLI. 2006. Árvores, p. 190-216. *In*: K.C. PÔRTO; J.S. ALMEIDA-CORTEZ & M. TABARELLI (Eds). **Diversidade biológica e conservação da floresta atlântica ao norte do Rio São Francisco**. Brasília, Ministério do Meio Ambiente, Série Biodiversidade 14.

GUIMARÃES, P.R. & P. GUIMARÃES. 2006. Improving the analyses of nestedness for large sets of matrices. **Environmental Modelling and Software 21**: 1512-1513. doi:10.1016/j.envsoft.2006.04.002.

GUIMERÀ, R. & L.A.N. AMARAL. 2005. Functional cartography of complex metabolic networks. **Nature 433**: 895-900. doi:10.1038/nature03288

HENRY, M. & E.K.V. KALKO. 2007. Foraging strategy and breeding constraints of Rhinophylla pumilio (Phyllostomidae) in the Amazon lowlands. **Journal of Mammalogy 88**: 81-93. **doi**: http://dx.doi.org/10.1644/06-MAMM-A-001R1.1

JORDANO, P. 1987. Patterns of mutualistic interactions in pollination and seed dispersal – connectance, dependence asymmetries, and coevolution. **The American Naturalist 129**: 657-677.

JORDANO, P.; J. BASCOMPTE & J.M. OLESEN. 2003. Invariant properties in coevolutionary networks of plant-animal interactions. **Ecology Letters 6**: 69-81. doi: 10.1046/j.1461-0248.2003.00403.x

KALKO, E.K.V. & M.A. CONDON. 1998. Echolocation, olfaction and fruit display: how bats find fruit of flagellichorous cucurbits. **Functional Ecology 12**: 364-372. doi: 10.1046/j.1365-2435.1998.00198.x

KALKO, E.K.V.; E.A. HERRE & C.O. HANDLEY. 1996. The relation of fig fruit syndromes to fruit-eating bats in the New and Old World tropics. **Journal of Biogeography 23**: 565-576. doi: 10.1111/j.1365-2699.1996.tb00018.x

KORINE, C. & E.K.V. KALKO. 2005. Fruit detection and discrimination by small fruit-eating bats (Phyllostomidae): echolocation call design and olfaction. **Behavioral Ecology and Sociobiology 59**: 12-23. doi: 10.1007/s00265-005-0003-1

LOBOVA, T.A.; C.K. GEISELMAN & S.A. MORI. 2009. **Seed dispersal by bats in the Neotropics**. New York, New York Botanical Garden Press.

KRISHNA, A.;P.R. GUIMARAES; P. JORDANO & J. BASCOMPTE. 2008. A neutral-niche theory of nestedness in mutualistic networks. **Oikos 117**: 1609-1618. doi: 10.1111/j.1600-0706.2008.16540.x

MANLY, B.F.J. 2007. **Randomization, bootstrap and Monte Carlo methods in biology**. Boca Raton, Chapman & Hall/CRC, 3rd ed.

MELLO, M.A.R.; E.L.S. BEZERRA & I.C. MACHADO. 2013. Functional roles of Centridini oil bees and Malpighiaceae oil flowers in biome-wide pollination networks. **Biotropica 45**: 45-

53.doi: 10.1111/j.1744-7429.2012.00899.x

MELLO, M.A.R.; F. MARQUITTI; P. GUIMARÃES; E. KALKO; P. JORDANO & M. DE AGUIAR. 2011a. The modularity of seed dispersal: differences in structure and robustness between bat– and bird–fruit networks. **Oecologia 167**: 131-140. doi:10.1007/s00442-011-1984-2

MELLO, M.A.R.; F.M.D. MARQUITTI; P.R. GUIMARÃES Jr; E.K.V. KALKO; P. JORDANO & M.A.M. DE AGUIAR. 2011b. The missing part of seed dispersal networks: structure and robustness of bat-fruit interactions. **PLOS One 6**: e17395. doi:10.1371/journal.pone.0017395

MELO, F.P.L.; B. RODRIGUEZ-HERRERA; R.L. CHAZDON; R.A. MEDELLIN & G.G. CEBALLOS. 2009. Small tent-roosting bats promote dispersal of large-seeded plants in a Neotropical forest. **Biotropica 41**: 737-743. doi: 10.1111/j.1744-7429.2009.00528.x

MUSCARELLA, R. & T.H. FLEMING. 2007. The role of frugivorous bats in tropical forest succession. **Biological Reviews 82**: 573-590. doi: 10.1111/j.1469-185X.2007.00026.x

NOOY, W.; A. MRVAR & V. BATAGELJ. 2005. **Exploratory social network analysis with Pajek**. New York, Cambridge University Press.

OLESEN, J.M.; J. BASCOMPTE; Y.L. DUPONT & P. JORDANO. 2006. The smallest of all worlds: Pollination networks. **Journal of Theoretical Biology 240**: 270-276. doi:10.1016/j.jtbi.2005.09.014

OLIVEIRA, M.; A. GRILLO & M. TABARELLI. 2005. **Caracterização da flora dos remanescentes da Usina Serra Grande, Alagoas**. Recife, Centro de Pesquisas Ambientais do Nordeste.

OLLERTON, J.; R. ALARCON; N.M. WASER; M.V. PRICE; S. WATTS; L. CRANMER; A. HINGSTON; C.I. PETER & J. ROTENBERRY. 2009. A global test of the pollination syndrome hypothesis. **Annals of Botany 103**: 1471-1480. doi:10.1093/aob/mcp03.

PONTES, A.R.M.; P.H.A.L. PERES; I.C. NORMANDE & C.M. BRAZIL. 2006. Mamíferos, p. 303-324. *In*: K.C. PÔRTO; J.S. ALMEIDA-CORTEZ & M. TABARELLI (Eds). **Diversidade biológica e conservação da floresta atlântica ao norte do Rio São Francisco**. Brasília, Ministério do Meio Ambiente, Série Biodiversidade 14.

REDFORD, K. 1992. The empty forest. **BioScience 42**: 412-422.

RODA, S.A. 2006. Aves, p. 279-302. *In*: K.C. PÔRTO; J.S. ALMEIDA-CORTEZ & M. TABARELLI (Eds). **Diversidade biológica e conservação da floresta atlântica ao norte do Rio São Francisco**. Brasília, Ministério do Meio Ambiente, Série Biodiversidade 14.

SIGRIST, T. 2006. **Aves do Brasil: uma visão artística**. São Paulo, Auk Paper, 2nd ed.

SILVA, J.M.C. & M. TABARELLI. 2000. Tree species impoverishment and the future flora of the Atlantic Forest of northeast Brazil. **Nature 404**: 72-74. doi:10.1038/35003563

STRAUBE, F. & G. BIANCONI. 2002. Sobre a grandeza e a unidade utilizada para estimar esforço de captura com utilização de redes-de-neblina. **Chiroptera Neotropical 8**: 150-152.

TABARELLI, M. & S.A. RODA. 2005. Uma oportunidade para o Centro de Endemismo Pernambuco. **Natureza & Conservação 3**: 22-28.

TERBORGH, J.; N. PITMAN; M. SILMAN; H. SCHICHTER & V.P. NÚÑES. 2002. Maintenance of tree diversity in tropical forests, p. 1-17. *In*: D.J. LEVEY; W.R. SILVA & M. GALETTI (Eds). **Seed dispersal and frugivory: ecology, evolution, and conservation.** New York, CABI Publishing.

THIES, W. & E.K.V. KALKO. 2004. Phenology of Neotropical pepper plants (Piperaceae) and their association with their main dispersers, two short-tailed fruit bats, *Carollia perspicillata* and *C. castanea* (Phyllostomidae). **Oikos 104**: 362-376. doi: 10.1111/j.0030-1299.2004.12747.x

ULRICH, W.; M. ALMEIDA-NETO & N.J. GOTELLI. 2009. A consumer's guide to nestedness analysis **Oikos 118**: 3-17. doi: 10.1111/j.1600-0706.2008.17053.x

VAN DER PIJL, L. 1972. **Principles of dispersal in higher plants.** Berlin, Springer Verlag.

VIZOTTO, L.D. & V.A. TADDEI. 1973. **Chave para determinação de quirópteros brasileiros.** São José do Rio Preto, EDUSP.

WALKER, B.H. 1992. Biodiversity and ecological redundancy. **Conservation Biology 6**: 18-23. doi: 10.1046/j.1523-1739.1992.610018.x

ZAR, J.H. 1996. **Biostatistical analysis.** New Jersey, Prentice-Hall, 3rd ed.

2

Evolution of bill size in relation to body size in toucans and hornbills (Aves: Piciformes and Bucerotiformes)

Austin L. Hughes

Department of Biological Sciences, University of South Carolina, Columbia SC 29205 USA. E-mail: austin@biol.sc.edu

ABSTRACT. Evidence that the bill of the Toco Toucan, *Ramphastos toco* Statius Muller, 1776, has a specialized role in heat dissipation suggests a new function for the large and light-weight bill of the toucan family (Piciformes: Ramphastidae). A prediction of this hypothesis is that bill length in toucans will increase with body mass at a rate greater than the isometric expectation. This hypothesis was tested in a phylogenetic context with measurements of skeletal elements in adult males of 21 toucan species. In these species, 64.3% of variance in relative skeletal measurements was accounted for by the contrast between bill and body size. Maxilla length and depth increased with body mass at a greater than isometric rate relative to both body mass and other linear skeletal measures. By contrast, no such trend was seen in a parallel analysis of 24 hornbill species (Bucerotiformes), sometimes considered ecological equivalents of toucans. The unique relationship between bill size and body mass in toucans supports the hypothesis that the evolution of a heat dissipation function has been a persistent theme of bill evolution in toucans.

KEY WORDS. Allometry; heat dissipation; Ramphastidae.

The adaptive significance of the large and remarkably light-weight bill of members of the Neotropical toucan family (Piciformes: Ramphastidae) has been the subject of much speculation (SHORT & HORNE 2001, TATTERSALL *et al.* 2009). Although VAN TYNE (1929: 39) suggested that the toucan's bill has no "especial adaptive function", a number of adaptive hypotheses have been proposed. BÜHLER (1995) proposed that the bill's large size and serrated edges originally evolved primarily as an adaptation for reaching and grasping fruit; later "tooth-like" markings on the bill may have evolved as adaptations to minimize mobbing by other birds when toucans prey on their nests (SICK 1993, BÜHLER 1995). SHORT & HORNE (2001) suggested a similar evolutionary sequence, while emphasizing the likely importance of species-specific bill markings in species recognition and courtship. Toucan bills are often brightly colored, and a few species show sexual dimorphism in bill coloration (SHORT & HORNE 2001). When bill color dimorphism occurs it is usually not very marked (SHORT & HORNE 2001), but its presence suggests that sexual selection may be another evolutionary force acting on toucan bills, at least in some species.

A further contribution to understanding the function of the toucan's bill was provided by evidence that the bill of the Toco Toucan, *Ramphastos toco* Statius Muller, 1776, serves as a key surface area for heat dissipation (TATTERSALL *et al.* 2009), which the bird can use to regulate body temperature by controlling blood flow. There is evidence that bills of a variety of avian taxa can function in heat dissipation (HAGAN & HEATH 1980, SCOTT *et al.* 2008, GREENBERG *et al.* 2012a, b, GREENBERG & DANNER 2013), suggesting that heat dissipation may be a plesiomorphic function of the avian bill. In the Ramphastidae, it might be hypothesized that the ancestral heat-dissipation function has become elaborated by the evolution of a highly modifiable vascular radiator (TATTERSALL *et al.* 2009). On this hypothesis, the emergence of this vascular adaptation has been an additional factor favoring the evolution of large bill size in toucans, in conjunction with other selective pressures such as frugivory and signaling. Relatively little is known of toucans' thermal biology in nature, but the family is entirely Neotropical in distribution, and most species inhabit tropical lowland forests (SHORT & HORNE 2001), where high daily maximum temperatures occur year-round (GRUBB & WHITMORE 1996).

Consistent with a role for the toucan bill in heat-dissipation, TATTERSALL *et al.* (2009) presented evidence that bill length in juvenile and adult Toco Toucan increases as a function of body mass at a rate greater than the isometric expectation; i.e., greater than an exponent of 1/3 expected for a linear dimension (ALEXANDER 1971). Likewise, SYMONS & TATTERSALL (2010) provided evidence that across toucan species bill length increases as a function of body mass at a rate greater than linear expectation, using published data on 34 species of Ramphastidae. Such a relationship is expected if the bill plays a role in dissipating body heat, since metabolic rate increases with body mass with an exponent between 2/3 and 1.0, depending on activity level (GLAZIER 2008).

Here I analyze the evolution of bill size in relation both to the size of other major skeletal elements and to body mass across the family Ramphastidae in order to test the hypothesis of isometry against an alternative consistent with the bill's proposed role in heat dissipation, using statistical methods that control for phylogenetic relationships. A phylogenetic approach makes it possible to test the hypothesis that there has been a trend toward bill sizes greater than the isometric expectation throughout the evolution of this family. The hornbills (Bucerotiformes) are considered Old World ecological equivalents of the toucans, filling similar ecological niches in their respective ecosystems; most members of both families are cavity-nesting frugivores of tropical forests, and the two families have convergently evolved large slightly downcurved bills (KEMP 1995, KINNAIRD & O'BRIEN 2007). Because of these ecological parallels, I conduct a similar analysis with hornbills to compare the patterns of bill evolution relative to body size in the two groups of large-billed tropical birds. By comparing pattern of bill allometry in these two families, I test for a distinctive pattern of bill evolution in toucans, which would be consistent with the hypothesis that the toucan's bill plays an exceptionally highly developed role in heat dissipation.

MATERIAL AND METHODS

Measurements were made on complete skeletal specimens of adult males belonging to 21 species of toucans (Piciformes: Ramphastidae) and on complete skeletons of adult males of 24 species of hornbills (Bucerotiformes: Bucorvidae and Bucerotidae) from the U.S. National Museum of Natural History. The same measurements were also made on adult females of 16 of the toucan species and 14 of the hornbill species; since the patterns were similar for males and females, only the results for males are reported here. As an outgroup to root the phylogenetic tree of toucans, two species of New World barbets (Piciformes: Capitonidae) were used, *Capito aurovirens* (Cuvier, 1829) and *Semnornis ramphastinus* (Jardine, 1855) (Fig. 1). As an outgroup to root the phylogenetic tree of hornbills, the Eurasian Hoopoe, *Upupa epops* Linnaeus, 1758, (Upupiformes: Upupidae) was used (Fig. 2). Species were included based on available specimens but sampled all major lineages of both toucans and hornbills (Figs 1 and 2).

Because no comprehensive molecular phylogeny of toucans has been published, the phylogeny of toucans (Fig. 1) was derived from a combination of published DNA sequence-based phylogenies. The relationships among the ramphastid genera were based on NAHUM *et al.* (2003); see also PATANÉ *et al.* (2009). Relationships within the genus *Ramphastos* were based on PATANÉ *et al.* (2009); see also WECKSTEIN (2005). Relationships within *Pteroglossus* and *Baillonius* were based on EBERHARD & BERMINGHAM (2005) and PATEL *et al.* (2011). Relationships within *Aulacorhynchus* were based on BONACCORSO *et al.* (2011); and those within *Andigena* and *Selenidera* were based on LUTZ *et al.* (2013). The phylogeny of hornbills (Fig. 2) was based on the DNA sequence phylogeny of GONZALEZ *et al.* (2013). Most branching patterns indicated in Figs 1 and 2 were strongly supported in the original phylogenetic analyses by bootstrap prob-

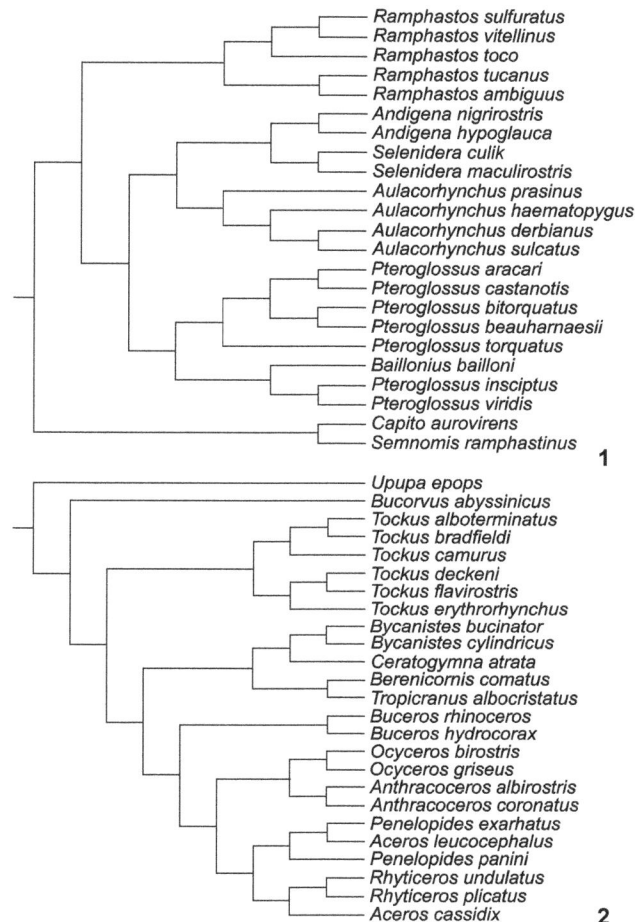

Figures 1-2. Phylogenies of species used in analyses: (1) phylogeny of toucans (Ramphastidae), rooted with two species of barbet, *Capito aurovirens* and *Semnornis ramphastinus*; (2) phylogeny of hornbills (Bucerotiformes), rooted with *Upupa epops*.

abilities, Bayesian posterior probabilities, or both. Preliminary analyses using the phylogeny of *Ramphastos* from HAFFER (1974, 1997) showed essentially identical results to those based on the phylogeny of PATANÉ *et al.* (2009); only the latter results are reported here.

In the case of toucans and barbets, the following nine skeletal measurements were made by digital caliper (BAUMEL 1993): 1) maxilla length, measured from the dorsal junction of the maxilla with the cranium to the tip of the bill (*Rostrum maxillare*); 2) maxilla depth, measured at the point of widest dorsal to ventral depth; 3) maxilla width, measured at the point of greatest lateral width; 4) cranium length, measured from the dorsal junction of the maxilla with the cranium to the posterior end (*Proeminentia cerebellaris*) of the cranium; 5) cranium width, measured at the point of greatest lateral width; 6) sternum, measured from *Apex carinae* to *Margo caudalis*; 7) synsacrum, measured from the anterior edge of *Ala preacetabularis* to the posterior edge of *Ala ischii*; 8) *femur*, measured from the proximal point of *Crista trochanteris* to the distal point of *Condylus lateralis*; and 9) tibiotarsus, measured from the proximal point of *Facies gastrocnemialis* to *Incisura intercondylaris*. In the case of the hornbill sample, Maxilla depth and cranium length were not included in analyses because the presence of the casque prevented comparable measurements in most species. Mean body mass values (in grams) for each species were obtained from DUNNING (2008). In most species, values for males and females were given separately (DUNNING 2008); and in those cases values for males were used. Data for male and female toucans are available in Appendix S1*, while male and female hornbills are available in Appendix S2*.

To test the sensitivity of these measurements to within-species variation, the same nine measurements were made on 10 adult males of *Ramphastos sulfuratus* Lesson, 1830 and 11 adult males of *R. toco*. Analysis of variance applied to log-transformed measurements was used to test for the relative magnitude of within-species and between-species components of variance in each of the nine measurements. In the case of all measurements, between species variance was significantly greater than within-species variance (p < 0.001 in every case except for cranium length, where p = 0.017, F-tests). The less pronounced between-species difference in cranium length than in the other measures was consistent with previous reports of low variance in similar measures (HÖFLING 1991).

Size-corrected transformations ("Mosimann transformations" – MOSIMANN 1970) were computed for the 9 skeletal measurements on toucans. Where the x_i are the individual measurements, let $z_i = \ln [x_i/G(x)]$, where $G(x)$ is the geometric mean of the nine measurements within each species (MOSIMANN 1970). Principal components (PCs) were extracted from the correlation matrix of the z_is; the PC scores were used to provide size-independent indices of body shape for each toucan species (DARROCH & MOSIMANN 1985, JUNGERS *et al.* 1995, HUGHES 2013). The values used in these computations are shown in Appendix S1*. Principal components extracted Mosimann-transformed variables are preferable to principal components extracted from raw data, because the former are more effective in correctly identifying similarities in shape independent of body size (JUNGERS *et al.* 1995). Because maxilla depth and cranium length could not be accurately measured in the case of hornbills, in order to compare the two families, Mosimann transformations were computed for the remaining seven variables separately for each family; and principal components were extracted from these transformed variables.

To test hypotheses regarding isometric relationships among skeletal measures and between skeletal measures and body mass, all measurements were first log-transformed. The isometric expectation for the slope (b) of a log-log regression (i.e., the allometric exponent) of any linear skeletal measure on any other linear skeletal measure is 1.0. The isometric expectation for b in a log-log regression of a linear measure on body mass (predicted to be proportional to body volume) is 1/3 (ALEXANDER 1971). Because the toucan's maxilla is approximately triangular in cross-section (SHORT & HORNE 2001), the external surface area of the bill consists largely of the area on the two lateral bill surfaces. Assuming that each of these surfaces has the approximate shape of an elongated triangle, the surface area of the maxilla can be roughly approximated by the product maxilla length times maxilla depth. The isometric expectation for b in a log-log regression of the product of two linear measures on body mass is 2/3 (ALEXANDER 1971).

Isometric expectations were tested in two ways: 1) traditional analyses, in which phylogeny was not taken into account but rather each species was treated an independent unit of analysis; and 2) phylogenetically independent contrasts. In traditional analyses, the outgroup species were not included in the regressions. On the basis of the phylogenetic trees (Figs 1 and 2), phylogenetically independent contrasts were constructed using the PDAP (GARLAND *et al.* 1993) contrasts plug-in within Mesquite version 2.75 (MADDISON & MADDISON 2011). Regressions between phylogenetically independent contrasts were conducted without fitting an intercept (GARLAND *et al.* 1992). PCs extracted from the correlation matrix of the z_is were mapped on the toucan phylogeny by maximum parsimony using the "Map Continuous" function in Mesquite with default settings.

Following the recommendation of SMITH (2009) for testing the null hypothesis of isometry, reduced major axis (RMA) was used rather than ordinary least squares (OLS) to estimate regression coefficients (SOKAL & ROHLF 1995). The results with OLS (not shown) were very similar to those of RMA in the present case because correlations between variables were high. For all allometric regressions reported here (N = 58), the linear correlation coefficient ranged from 0.735 to 0.988 (mean = 0.887 ± 0.008 S.E., median = 0.893). OLS was used to estimate regression lines (SMITH 2009). All reported significance levels are corrected for multiple testing by the Bonferroni method (SOKAL & ROHLF 1995). Statistical analyses were conducted in Minitab (http://www.minitab.com).

RESULTS

Relative length of skeletal elements

The first principle component (PC1) extracted from the correlation matrix of size-corrected transformations of nine linear skeletal measures of toucans accounted for 64.3% of the variance and represented a contrast between two sets of variables: 1) maxilla length and maxilla depth; and 2) the other variables except for sternum (Table I). Thus PC1 could be interpreted as a size-corrected measure of the contrast between bill size and body size. PC2, accounting for 17.7% of the variance, seemed to mainly consist of a contrast between sternum and maxilla width (Table I). In order to provide a visual image of how the contrast between bill and body size has evolved across the Ramphastidae, PC1 values were mapped across the phylogeny of toucans. The highest values (indicating greatest bill size relative to body size) were seen in *Ramphastos* (Fig. 3). The phylogeny also supported the hypothesis of a parallel increase in bill size relative to body size in the *Pteroglossus/Baillonius* lineage (Fig. 3).

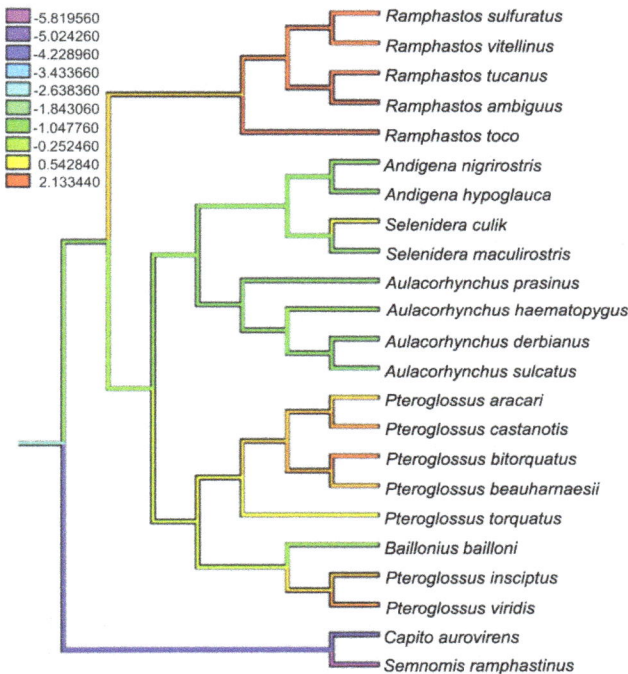

Figure 3. Parsimony-reconstructed PC1 scores across the phylogeny of toucans.

Table I. Variable loadings on the first two principal components (PC1 and PC2) derived from transformed skeletal measurements of 21 species of Ramphastidae.

Variable	PC1	PC2
Maxilla Length	0.407	-0.065
Maxilla Depth	0.391	0.069
Maxilla Width	-0.030	-0.691
Cranium Length	-0.375	-0.106
Cranium Width	-0.379	-0.059
Sternum	0.008	0.663
Synsacrum	-0.368	0.239
Femur	-0.327	0.003
Tibiotarsus	-0.391	0.002
% variance	64.300	17.700

Because maxilla depth and cranium length could not be accurately measured in the case of hornbills, principal were extracted from the correlation matrix of size-corrected transformations of remaining seven linear skeletal measures of in each family (Table II). Even excluding maxilla depth and cranium length, PC1 (accounting for 60.8% of the variance) in the toucan data again appeared mainly to represent a contrast between bill size and body size (Table II). By contrast, in hornbills, PC1 accounted for only 30.8% of the variance and appeared to reflect mainly a contrast between body size and the width of both bill and cranium (Table II). The loading of maxilla length on PC1 in hornbills (-0.063) differed strikingly from that in toucans (0.481, Table II). Thus hornbills appeared to differ from toucans in that bill length relative to body size was not a major factor in cross-species comparisons of major skeletal elements.

Table II. Variable loadings on the principal components (PC1) derived from transformed skeletal measurements of 21 species of Ramphastidae and 24 species of Bucerotidae.

Variable	Ramphastidae	Bucerotidae
Maxilla Length	0.481	-0.063
Maxilla Width	0.481	0.583
Cranium Width	0.039	0.342
Sternum	-0.421	-0.377
Synsacrum	0.001	-0.432
Femur	-0.451	-0.145
Tibiotarsus	-0.394	-0.436
% variance	60.800	38.000

Allometric relationships

In traditional analyses, not accounting for the phylogeny, the 9 log-transformed skeletal measures were regressed against log body mass (Table III). Likewise, phylogenetically independent contrasts in the same 9 log-transformed skeletal measures were regressed against phylogenetically independent contrasts in log body mass (Table III). The results were broadly similar in the two types of analysis (Table III). In both cases, the allometric exponent (b) for maxilla length and maxilla depth were significantly greater than the isometric expectation (1/3; Table III). In the case of phylogenetically independent contrasts, b for maxilla width was also significantly greater than the isometric expectation (Table III). In traditional analyses, but not in phylogenetically independent contrasts, b for femur was significantly greater than the isometric expectation (Table III). No other linear measure showed b significantly greater than the linear expectation in either type of analysis, but in the traditional analyses b for cranium length was significantly less than the isometric expectation (Table III). When the log of the product of maxilla length and maxilla depth,

Table III. Allometric exponents (b) of regression of skeletal measures on body mass of 21 species of Ramphastidae in traditional analyses and in phylogenetically independent contrasts.

Dependent variable	Null hypothesis[a]	Traditional analyses (non-phylogenetic)			Phylogenetically independent contrasts		
		b	t	p b	b	t	p[b]
Maxilla Length	1/3	0.647	3.44	< 0.05	0.884	3.55	< 0.05
Maxilla Depth	1/3	0.607	4.35	< 0.01	0.678	5.50	< 0.001
Maxilla Length x Maxilla Depth	2/3	1.239	3.96	< 0.01	1.527	4.88	< 0.001
Maxilla Width	1/3	0.339	0.28	N.S.	0.469	3.52	< 0.05
Cranium Length	1/3	0.189	-7.35	< 0.001	0.269	-2.51	N.S.
Cranium Width	1/3	0.305	-1.17	N.S.	0.348	0.41	N.S.
Sternum	1/3	0.430	1.58	N.S.	0.438	1.55	N.S.
Synsacrum	1/3	0.378	1.26	N.S.	0.417	1.30	N.S.
Femur	1/3	0.424	3.69	< 0.05	0.433	2.54	N.S.
Tibiotarsus	1/3	0.366	1.51	N.S.	0.371	1.26	N.S.

[a] The value shown is the isometric expectation of the exponent (b) under the relevant null hypothesis.
[b] All P-values shown have been corrected by the Bonferroni procedure.

was regressed against log body mass, in both types of analyses, b was significantly greater than the isometric expectation (2/3; Table III). These results imply that in the Ramphastidae both bill size and bill surface area increase with body mass at a greater rate than expected under isometry.

The hypothesis that maxilla length and maxilla depth show a distinctive pattern of evolution in the Ramphastidae was further tested by regressing logarithms of these measures on those of linear measures of non-maxillary structures (Table IV). In both traditional and phylogenetically based analyses, b exceeded the isometric expectation (1.0) in every case (Table IV). In both kinds of analyses, the regressions with maxilla length as the dependent variable, the b was significantly greater than the isometric expectation with cranium length, cranium width, and sternum as dependent variables (Table III). In both kinds of analyses, in regressions with maxilla depth as the dependent variable, b was significantly greater than the isometric expectation with cranium length, cranium width, sternum, synsacrum and tibiotarsus as dependent variables (Table IV). In the phylogenetically based analysis, maxilla length showed b greater than the isometric expectation when regressed on tibiotarsus, and maxilla depth also showed b greater than the isometric expectation when regressed on femur (Table IV).

When log-transformed skeletal measures of hornbills were regressed against log body mass, a very different pattern was seen from that seen in toucans (Table V). In hornbills, b for maxilla length did not differ significantly from the isometric expectation (Table V), resulting in distinct patterns in toucans and hornbills (Fig. 4). In traditional analyses, the only measure for which the slope significantly exceeded the isometric expectation was synsacrum, while the slope for cranium width was significantly less than the isometric expectation (Fig. 4). Likewise, in phylogenetically based analyses, the slope of the relationship for contrasts in log maxilla length did not

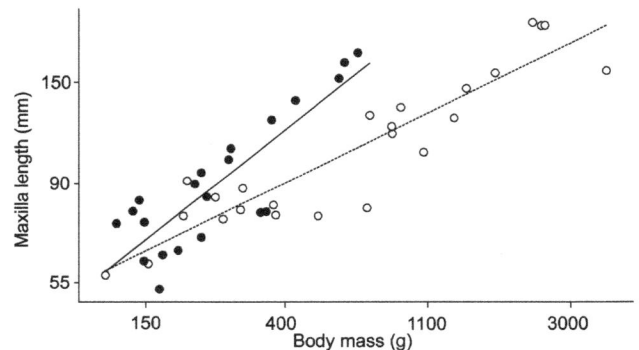

Figure 4. Maxilla length vs. body mass for toucans (solid circles) and hornbills (open circles) on a log scale, with OLS linear regression lines: for toucans, Y = 1.460 + 0.551X (solid line; adj. R^2 = 71.0%, p < 0.001) and for hornbills: Y = 2.445 + 0.342X (dotted line; adj. R^2 = 84.0%, p < 0.001).

differ significantly from the isometric expectation, and the only measure for which the slope exceeded the isometric expectation was synsacrum (Table V).

The Northern Ground Hornbill *Bucorvus abyssinicus* (Boddaert, 1783) (Fig. 2) had a relatively large synsacrum (129.9 mm) in comparison to the 23 other hornbill species of (mean = 62.6 ± 5.0 mm, range = 30.5 to 105.mm, Appendix S2*). A relatively large is consistent with the terrestrial habits and relatively large legs of the Northern Ground Hornbill (KEMP 1995). However, even when the Northern Ground Hornbill was excluded from the data set, a similar relationship was seen in the traditional analysis of the relationship between log synsacrum and log body mass (b = 0.409; test of equality to isometric expectation, p < 0.001). Likewise, in phylogenetically independent contrasts, when both the ancestral node and the node

Table IV. Allometric exponents (b) of regression of maxilla length and maxilla depth on other skeletal measures of 21 species of Ramphastidae in traditional analyses and phlogenetically independent contrasts.

| Independent Variable | Traditional analyses (non-phylogenetic) | | | | | | Phylogenetically Independent Contrasts | | | | | |
| | Maxilla Length | | | Maxilla Depth | | | Maxilla Length | | | Maxilla Depth | | |
	b	t	p [a]	b	t	p [a]	b	t	p [a]	b	t	p [a]
Cranium Length	3.424	3.62	< 0.05	3.209	4.24	< 0.01	3.597	3.53	< 0.05	2.782	4.63	< 0.001
Cranium Width	2.123	4.00	< 0.01	1.988	4.44	< 0.01	2.525	3.89	< 0.01	1.963	4.63	< 0.001
Sternum	1.503	3.58	< 0.05	1.408	3.64	< 0.01	2.024	3.79	<0.01	1.576	3.55	< 0.05
Synsacrum	1.709	2.74	N.S.	1.602	3.34	< 0.05	2.145	2.39	N.S.	1.650	2.96	< 0.05
Femur	1.528	2.20	N.S.	1.431	2.59	N.S.	2.012	2.40	N.S.	1.557	2.98	< 0.05
Tibiotarsus	1.769	2.75	N.S.	1.657	3.30	< 0.05	2.355	3.03	< 0.05	1.821	4.09	< 0.01

[a] All P-values shown have been corrected by the Bonferroni procedure. The null hypothesis in each case is that b = 1.0

Table V. Allometric exponents (b) of regression of skeletal measures on body mass of 24 species of Bucerotidae in traditional analyses and phlogenetically independent contrasts.

| Dependent variable | Null hypothesis [a] | Traditional analyses (non-phylogenetic) | | | Phylogenetically independent contrasts | | |
		b	t	P b	b	t	p [b]
Maxilla Length	1/3	0.372	1.15	N.S.	0.351	0.45	N.S.
Maxilla Width	1/3	0.317	-0.36	N.S.	0.333	-0.02	N.S.
Cranium Width	1/3	0.280	-3.17	< 0.05	0.302	-1.44	N.S.
Sternum	1/3	0.327	-0.43	N.S.	0.324	-0.56	N.S.
Synsacrum	1/3	0.407	5.39	< 0.001	0.392	3.18	< 0.05
Femur	1/3	0.325	-0.42	N.S	0.349	0.67	N.S.
Tibiotarsus	1/3	0.323	-0.44	N.S.	0.366	1.34	N.S.

[a] The value shown is the isometric expectation of the exponent (b) under the relevant null hypothesis.
[b] All P-values shown have been corrected by the Bonferroni procedure.

linking the Northern Ground Hornbill to the other hornbills (Fig. 2) were excluded, there was a similar relationship between contrasts in log synsacrum and contrasts in log body mass (b = 0.429, test of equality to isometric expectation, p < 0.01).

DISCUSSION

An examination of the relationship among linear measures of major skeletal measures and between those measures and body mass supported an unusual pattern of bill size evolution in the toucan family. Throughout the toucan family, the length and depth of the maxilla increased as a function of body mass at a rate greater than expected under isometry, implying disproportionately large bills per unit body mass in large-bodied toucan species, consistent with the hypothesis that heat dissipation has been an important factor in the evolution of the large bills of toucans (TATTERSALL et al. 2009). Since the capacity for radiation of heat from the bill is a function of surface area, it is further expected that bill surface area will increase with body mass at a rate greater than expected under isometry. The present analyses supported this prediction, since the results showed that product of toucan bill length and depth increases with body mass at a rate greater than the isometric expectation.

In spite of the ecological parallels between toucans and hornbills (KEMP 1995, KINNAIRD & O'BRIEN 2007), the present analyses provided no evidence of a greater than isometric increase in hornbill maxilla length as a function of body mass. These results are consistent with the hypothesis that the bill of hornbills does not play a role in heat dissipation analogous to that of toucans. This hypothesis will require further testing through physiological study of hornbills. It is of interest, however, that hornbills appear to make use of alternative heat-dissipation mechanisms from those seen in toucans; for instance, evaporative water loss from the bare skin under the wings, which is exposed by the hornbills' unique lack of underwing-coverts (KEMP 1995).

In contrast to the maxilla, in hornbills synsacrum length increased with body mass at a greater rate than expected under isometry. This pattern was seen even when the terrestrial Northern Ground Hornbill, in which the synsacrum was unusually large, was excluded from the analysis. The increase in the length of the synsacrum with body mass may reflect an enhanced need for weight support in the larger hornbills. That no similar trend is seen in toucans may reflect their substantially smaller body masses, as well as the fact that even arboreal hornbills spend more time on the ground than toucans

(KEMP 1995), with a consequent requirement to support the body weight on the pelvic girdle.

All phylogenies represent hypotheses, which are subject to revision in the light of additional data (GARLAND *et al.* 2005). In the present case, the fact that traditional and phylogenetic analyses yielded very similar results suggests that the conclusions are likely to be robust to phylogenetic revision. In addition, the phylogenetic perspective provided evidence that bill size increased relative to body size independently in different toucan lineages. In the toucans, 64.3% of variance in size-adjusted skeletal measures was accounted for by a composite variable (PC1) that could be interpreted as reflecting the contrast between bill and body size. PC1 increased markedly the genus *Ramphastos* and the *Pteroglossus/Baillonius* lineage (Fig. 3). Thus, the relationship between bill dimensions and body mass was a recurring feature of evolution across the phylogeny of toucans.

A fuller understanding of the evolution of the bill in toucans and hornbills will require investigation of the thermal biology of these species in a natural setting. At present little is known about the temperature regimes encountered by these birds in nature and the variety of behavioral and physiological strategies which they employ to cope with temperature extremes. Additional studies of morphological evolution, combining data on both within-species and between-species variation, can provide further insights into the selective forces acting on bill morphology. In particular, comparative study of the evolution of bill morphology in males and females will help to elucidate the potential role of sexual selection as a factor in shaping the evolution of the bill in these families.

Typically biological structures are multi-functional; thus, support for the heat-dissipation hypothesis precludes neither the hypothesis that reaching for and grasping fruit played a key role in the origin of the toucan's large bill, nor the hypothesis that the bill has secondarily evolved roles in aposematic and intraspecific signaling, including a role in sexual selection (BÜHLER 1995, SHORT & HORNE 2001). The apparent convergence between the bills of toucans and hornbills lends plausibility to the hypothesis that the original selective pressure favoring large bills in the toucan lineage arose from frugivory (BÜHLER 1995). At the same time, some role in heat dissipation is likely to be a plesiomorphic character of the bills of birds (HAGAN & HEATH 1980, SCOTT *et al.* 2008; GREENBERG et al. al. 2012a, b, GREENBERG & DANNER 2013). Thus, the relatively elaborate mechanisms of heat-dissipation seen in toucans may have arisen as an exaptation (GOULD & VRBA 1982); that is, the co-option of an existing structure for a new function. The present results, because they reveal that the relationship between bill dimensions and body mass has persisted across the toucan phylogeny, suggest that the co-option of the toucan bill for heat dissipation may represent an ancient feature within this family, which has acted in concert with other selective factors favoring large bill size.

ACKNOWLEDGMENTS

I am grateful to the staff of the U.S. National Museum, Bird Division, especially Chris Milensky, for access to specimens. Comments by Glenn J. Tattersall and an anonymous reviewer greatly improved the manuscript.

REFERENCES

ALEXANDER, R.M. 1971. **Size and Shape**. London, Edward Arnold.
BAUMEL, J.J. 1993. **Handbook of Avian Anatomy: Nomina Anatomica Avium**. Cambridge, Nuttall Ornithological Club, 2nd ed.
BONACCORSO, E.; J.M. GUAYASAMIN; A.T. PETERSON & A.G. NAVARRO-SIGÜENZA. 2011. Molecular phylogeny and systematics of Neotropical toucanets in the genus *Aulacorhynchus* (Aves, Ramphastidae). **Zoologica Scripta 40**: 336-349. doi: 10.1111/j.1463-6409.2011.00475.x
BÜHLER, P. 1995. Grösse, Form und Färbung des Tukanschnabels – Grundlage für den evolutiven Erflog der Ramphastiden? **Journal für Ornithologie 136**: 187-193.
DARROCH, J.N.& J.E. MOSIMANN. 1985. Canonical and principal components of shape. **Biometrika 72**: 241-252.
DUNNING, J.C. 2008. CRC handbook of Avian Body Masses. Boca Raton, CRC Press, 2nd ed.
EBERHARD, J.R. & E. BERMONGHAM. 2005. Phylogeny and comparative biogeography of *Pionopsitta* parrots and *Pteroglossus* toucans. Molecular Phylogenetics and Evolution 36: 288-304. doi: 10.1016/j.ympev.2005.01.022
GARLAND JR, T.; P.H. HARVEY & A.R. IVES. 1992. Procedures for the analysis of comparative data using phylogenetically independent contrasts. **Systematic Biology 41**: 18-32. doi: 10.1093/sysbio/42.3.265
GARLAND JR, T.; A.W. DICKERMAN; C.M. JANIS & J.A. JONES. 1993. Phylogenetic analysis of covariance by computer simulation. **Systematic Biology 42**: 265-292.
GARLAND JR, T.; A.F. BENNET & E.L. REZENDE. 2005. Phylogenetic approaches in comparative physiology. **Journal of Experimental Biology 208**: 3015-3035. doi: 10.1242/jeb.01745
GLAZIER, D.S. 2008. Effects of metabolic level on the body size scaling of metabolic rate in birds and mammals. **Proceedings of the Royal Society B 275**: 1405-1410. doi:10.1098/rspb.2008.0118
GONZALEZ, J.-C.; B.C. SHELDON; N.J. COLLAR & J.A. TOBIAS. 2013. A comprehensive molecular phylogeny for the hornbills (Aves: Bucerotidae). **Molecular Phylogenetics and Evolution 67**: 468-483. doi: 10.1016/j.ympev.2013.02.012
GOULD, S.J. & E.S. VRBA. 1982. Exaptation – a missing term in the science of form. **Paleobiology 8**: 4-15.
GREENBERG, R. & R.M. DANNER. 2013. Climate, ecological release and bill dimorphism in an island songbird. **Biology Letters 9**: 20130118. doi:10.1098/rsbl.2013.0118
GREENBERG, R.; V. CADENA; R.M. DANNER & G. TATTERSALL. 2012a. Heat loss may explain bill size differences between birds

occupying different habitats. **Plos One 7** (7): e40933. doi: 10.1371/journal.pone.0040933

GREENBERG, R.; R. DANNER; B. OLSEN & D. LUTER. 2012b. High summer temperature explains bill size variation in salt marsh sparrows. **Ecography 35**: 146-152. doi: 10.1111/j.1600-0587.2011.07002.x

GRUBB, P.J. & T.C. WHITMORE. 1966. A comparison of montane and lowland rain forest in Ecuador: II. The climate and its effects on the distribution and physiognomy of the forests. **Journal of Ecology 54**: 303-333.

HAFFER, J. 1974. **Avian Speciation in Tropical South America.** Cambridge, Nuttal Ornithological Club.

HAFFER, J. 1997. Foreword: species concepts and species limits in ornithology, p. 11-24. *In*: J. DEL HOYO; A. ELLIOTT & J. SARGATAL (Eds). **Handbook of the Birds of the World.** Barcelona, Lynx Ediciones, vol. 4.

HAGAN, A.A. & J.E. HEATH. 1980. Regulation of heat loss in the duck by vasomotion in the bill. **Journal of Thermal Biology 5**: 95-101. doi: 10.1016/0306-4565(80)90006-6

HÖFLING, E. 1991. Étude comparative du crâne chez les Ramphastidae (Aves, Piciformes). **Bonner Zoologische Beiträger 42**: 55-65.

HUGHES, A.L. 2013. Indices of Anseriform body shape based on relative size of major skeletal elements and the relationship to reproductive effort. **Ibis 155**: 835-846. doi: 10.1111/ibi.12087

JUNGERS, W.L.; A.B. FALSETTI & C.E. WALL. 1995. Shape, relative size, and size-adjustments in morphometrics. **American Journal of Physical Anthropology 38**: 137-161. doi: 10.1002/ajpa.1330380608

KEMP, A. 1995. **The Hornbills.** Oxford, Oxford University Press.

KINNAIRD, M.F. & T.G. O'BRIEN. 2007. **The Ecology and Conservation of Asian Hornbills: Farmers of the Forest.** Chicago, University of Chicago Press.

LUTZ, H.L.; J.D. WECKSTEIN; J.S. PATANÉ; J.M. BATES & A. ALEIXO. 2013. Biogeography and spatio-temporal diversification of *Selenidera* and *Andigena* toucans (Aves: Ramphastidae). **Molecular Phylogenetics and Evolution 69**: 873-883. doi: 10.1016/j.ympev.2013.06.017

MADDISON, W.P. & D.R. MADDISON. 2011. Mesquite: a modular system for evolutionary analysis. Version 2.75, available online at: http://mesquiteproject.org [Accessed: 30/IX/2011].

MOSIMANN, J. 1970. Size allometry: size and shape variables with characterizations of the lognormal and generalized gamma distributions. **Journal of the American Statistical Association 65**: 930-945.

NAHUM, L.A.; S.L. PEREIRA; F.M. FERNANDES; S.R. MATIOLI & A.WAJNTAL. 2003. Diversification of Ramphastinae (Aves, Ramphastidae) prior to the Cretaceous/Tertiary boundary as shown by molecular clock of mtDNA sequences. **Genetics and Molecular Biology 26**: 411-418. doi: 10.1590/S1415-47572003000400003.

PATANÉ, J.S.; J.D. WECKSTEIN; A. ALEIXO & J.M. BATES. 2009. Evolutionary history of *Ramphastos* toucans: molecular phylogenetics, temporal diversification, and biogeography. **Molecular Phylogenetics and Evolution 53**: 923-934. doi: 10.1016/j.ympev.2009.08.017

PATEL, S.; J.D. WECKSTEIN; J.S. PATANÉ; J.M. BATES & A. ALEIXO. 2011. Temporal and spatial diversification of *Pteroglossus* araçaris (Aves: Ramphastidae) in the neotropics: constant rate of diversification does not support an increase in radiation during the Pleistocene. **Molecular Phylogenetics and Evolution 58**: 105-115. doi: 10.1016/j.ympev.2010.10.016

SCOTT, G.R.; V. CADENA; G.R. TATTERSALL & W.K. MILSOM. 2008. Body temperature depression and peripheral heat loss accompany the metabolic and ventilatory responses to hypoxia in low and high altitude birds. **Journal of Experimental Biology 211**: 1326-1335. doi: 10.1242/ jeb.015958

SHORT, L.L. & J.F. HORNE. 2001. **Toucans, Barbets and Honeyguides.** Oxford, Oxford University Press.

SICK, H. 1993. **Birds in Brazil.** Princeton, Princeton University Press.

SMITH, R.J. 2009. Use and misuse of the reduced major axis for line-fitting. **American Journal of Physical Anthropology 140**: 476-486. doi: 10.1002/ajpa.21090

SOKAL, R.R. & F.J. ROHLF. 1995. **Biometry.** San Francisco, W.H. Freeman, 3rd ed.

SYMONS, R.E. & G.J. TATTERSALL. 2010. Geographical variation in bill size across bird species provides evidence for Allen's Rule. **The American Naturalist 176**: 188-197. doi: 10.1086/653666

TATTERSALL, G.J.; D.V. ANDRADE & A.S. ABE. 2009. Heat exchange from the toucan bill reveals a controllable vascular thermal radiator. **Science 325**: 468-470. doi: 10.1126/science.1175553

VAN TYNE, J. 1929. The life history of the toucan *Ramphastos brevicarinatus*. **University of Michigan Museum of Zoology, Miscellaneous Publications 19**: 1-43.

WECKSTEIN, J.D. 2005. Molecular phylogenetics of the Ramphastos toucans: implications for the evolution of morphology, vocalizations, and coloration. **Auk 122**: 1191-1209. doi: doi: 10.1642/0004-8038(2005)122[1191:MPOTRT]2.0.CO;2

3

Alpaida (Araneae: Araneidae) from the Amazon Basin and Ecuador: new species, new records and complementary descriptions

Regiane Saturnino[1,*], Bruno V.B. Rodrigues[1] & Alexandre B. Bonaldo[1]

[1]*Laboratório de Aracnologia, Coordenação de Zoologia, Museu Paraense Emílio Goeldi. Avenida Perimetral 1901, Terra Firme, 66077-830 Belém, Pará, Brazil.*
[*]*Corresponding author. E-mail: sf.regiane@gmail.com*

ABSTRACT. Two new species of *Alpaida*, *A. levii* and *A. yanayacu*, the male of *A. iquitos* Levi, 1988 and the female of *A. gurupi* Levi, 1988 are described and illustrated for the first time. *Alpaida levii*, described from the states of Pará and Amazonas, is closely related to *A. delicata* (Keyserling, 1892), but differs in that males have a curved and distally pointed terminal apophysis, and females have the epigynum longer than wide and a drop-shaped median lobe. *Alpaida yanayacu* is only known from Ecuador and is characterized by long and rounded lateral lobes in ventral view and median lobe wide at base. A brief discussion about the morphological similarity among *A. levii*, *A. delicata* and *A. truncata* (Keyserling, 1865) is presented. Based on the information provided, new diagnoses are proposed for *A. delicata* and *A. truncata*. New records of *A. antonio* Levi, 1988, *A. bicornuta* (Taczanowski, 1878), *A. boa* Levi, 1988, *A. deborae* Levi, 1988, *A. delicata*, *A. erythrothorax* (Taczanowski, 1873), *A. guimaraes* Levi, 1988, *A. guto* Abrahim & Bonaldo, 2008, *A. gurupi*, *A. iquitos*, *A. leucogramma* (White, 1841), *A. murtinho* Levi, 1988, *A. negro* Levi, 1988, *A. rossi* Levi, 1988, *A. septemmammata* (O. Pickard-Cambridge, 1889), *A. simla* Levi, 1988, *A. tayos* Levi, 1988, *A. truncata*, *A. urucuca* Levi, 1988, *A. utiariti* Levi, 1988 and *A. veniliae* Levi, 1988 are presented.

KEY WORDS. Arachnida, distribution, Neotropical Region, spiders, taxonomy.

Species of *Alpaida* O. Pickard-Cambridge, 1889 are diurnal orb-weaving spiders occurring only in the Neotropical region. Currently, the genus is composed of 148 species (WORLD SPIDER CATALOG 2014), although it is estimated to contain about 200-300 species (LEVI 1988). Species of this genus are characterized by the glabrous body, orange to red carapace, abdomen and carapace without bristles; male palp with radix, embolus and terminal apophysis fused into one sclerite; mushroom-shaped paramedian apophysis connected to conductor; epigynum usually represented by a transverse sclerotized structure, with posterior lips, a median scape and copulatory openings located on each side between plate and lips (LEVI 1988).

Alpaida was revised by LEVI (1988), who dealt with 134 species, among which 94 were described as new. Many of the 40 previously known species had been erroneously transferred to *Alpaida* from different genera. Since LEVI's (1988) contribution only 11 species have been added to the genus: *A. guto* Abrahim & Bonaldo, 2008 described from the Brazilian Amazon; *A. itacolomi* Santos & Santos, 2010, *A. tonze* Santos & Santos, 2010, *A. caramba* Buckup & Rodrigues, 2011 and *A. arvoredo* Buckup & Rodrigues, 2011 described from the Atlantic Forest; *A. teresinha* Braga-Pereira & Santos, 2013 and *A. toninho* Braga-Pereira & Santos, 2013 from coastal forest areas in Brazil; *A. monzon audiberti* Dierkens, 2014 and *A. oyapockensis* Dierkens, 2014 from French Guiana; *A. losamigos* Deza & Andía, 2014 and *A. penca* Deza & Andía, 2014 from Peru.

As discussed by BRAGA-PEREIRA & SANTOS (2013), even though *Alpaida* is amongst the largest Neotropical spider genera, much of its diversity is still to be described (see LEVI 1988). Since 52 species are known only from females (including the two subspecies of *A. truncata* and *A. monzon audiberti*) and 17 species only from males, there is still much work to be done in the taxonomy of the genus. Since the revision of LEVI (1988), previously unknown opposite sexes were described for only seven species: the male of *A. scriba* (Mello-Leitão, 1940) by BUCKUP & MEYER (1993), the males of *A. citrina* (Keyserling, 1892) and *A. octolobata* Levi, 1988 by RODRIGUES & MENDONÇA (2011), the males of *A. hoffmanni* Levi, 1988, *A. kochalkai* Levi, 1988, *A. lomba* Levi, 1988 and the female of *A. arvoredo* by BUCKUP & RODRIGUES (2011).

Several species of the genus have large distribution ranges. Considering that only 53% of the described species are known from both sexes, it may be questionable to describe a new species in the absence of strong evidence for matching it with the opposite sex of a known species.

As documented in a growing number of faunistic inventories (e.g., HÖFER & BRESCOVIT 2001, BONALDO et al. 2009, CAFOFO et al. 2013), the diversity of *Alpaida* species in the Amazon Basin is high. In some particularly well-collected sites, up to 17 sympatric species have been found (unpublished data). In view of this, we have endeavored to examine all specimens of *Alpaida* deposited at the Museu Paraense Emílio Goeldi, which contain expressive material from the Amazon region. As a result, two new species of *Alpaida* are described, one based on both sexes from Brazil, the other on a female from Ecuador. One of the new species is morphologically related to *A. delicata* (Keyserling, 1892), which was compared to *A. truncata* (Keyserling, 1865) by Levi (1988). A discussion about the morphological similarity among the three species is presented and new diagnoses are proposed for *A. delicata* and *A. truncata*. Additionally, the male of *A. iquitos* Levi, 1988 and the female of *A. gurupi* Levi, 1988 are described for the first time and new records for twenty one species of *Alpaida* are documented: *A. antonio* Levi, 1988, *A. bicornuta* (Taczanowski, 1878), *A. boa* Levi, 1988, *A. deborae* Levi, 1988, *A. delicata*, *A. erythrothorax* (Taczanowski, 1873), *A. guimaraes* Levi, 1988, *A. guto* Abrahim & Bonaldo, 2008, *A. gurupi*, *A. iquitos*, *A. leucogramma* (White, 1841), *A. murtinho* Levi, 1988, *A. negro* Levi, 1988, *A. rossi* Levi, 1988, *A. septemmammata* (O. Pickard-Cambridge, 1889), *A. simla* Levi, 1988, *A. tayos* Levi, 1988, *A. truncata*, *A. urucuca* Levi, 1988, *A. utiariti* Levi, 1988 and *A. veniliae* Levi, 1988.

MATERIAL AND METHODS

The specimens examined were deposited in the Museu Paraense Emílio Goeldi, Belém, Brazil (MPEG, curator: Alexandre Bonaldo), Instituto Nacional de Pesquisas da Amazônia, Manaus, Brazil (INPA, curator: Célio Magalhães) and Museo de Zoología, Sección de Invertebrados, Pontificia Universidad Catolica, Quito, Ecuador (QCAZ, curator: Clifford Keil). The description format and palpal terminology follows Levi (1988). Male palps were expanded by alternated immersion in 10% KOH (potassium hydroxide) solution in distilled water for a few minutes. All measurements are in millimeters. The specimens were photographed and measured using a Leica M205A, with LAS automontage software. Additional information not originally inserted in labels is included between [brackets]. The following abbreviations are used: (A) terminal apophysis, (ALE) anterior lateral eyes, (AME) anterior median eyes, (C) conductor, (E) embolus, (L) distal lobe, (LL) lateral lobes, (MA) median apophysis, (N) notch, (PLE) posterior lateral eyes, (PME) posterior median eyes, (PM) paramedian apophysis, (PMP) posterior median plate, (R) radix, (S) median lobe of the scape. The distribution maps were made using the program QGIS (QGIS Development Team 2012) and include records only from the specimens listed here.

TAXONOMY

Alpaida levii **sp. nov.**
Figs. 1-9, 29

Type material. Male holotype from Brazil, *Pará*: Juruti (Mutum), 02°36'10.6"S 56°12'25.8"W, 14.IX.2002, D. Guimarães leg., deposited in MPEG (24370). Paratypes: Brazil, *Amazonas*: Manaus (Reserva Adolpho Ducke), 02°59'05.97"S

59"55'42.58"W, 1 female, 01.VIII.2008, J. ten Caten leg. (INPA); 1 male, 14.VIII.2008, R. Saturnino leg. (INPA); Canutama, 08°39'14.8"S 64°21'34.3"W, 1 male and 1 female, 30.IV.2007, R. Saturnino leg. (INPA); 1 female (INPA); 08°39'11.6"S 64°21'34.6"W, 1 male and 1 female, 06.V.2007, R. Saturnino leg. (INPA); Pará: Juruti (Capiranga), 02°28'0.6"S 56°12'42.2"W, 1 female, 15.IX.2002, A.B. Bonaldo leg. (MPEG 24371); 1 male (MPEG 24375); 02°29'19"S 56°06'34"W, 1 female, 15.IV.2008, B.V.B. Rodrigues leg. (MPEG 24374); 02°30'25.4"S 56°11'04.8"W, 1 male, 09.II.2007, J.A.P. Barreiros leg. (MPEG 24376); (Mutum), 02°36'10.6"S 56°12'25.8"W, 1 male and 1 female, 10.IX.2002, D. Guimarães leg. (MPEG 24372); Portel (Floresta Nacional de Caxiuanã), 01°57'38.9"S 51°36'45.3"W, 1 female, 12.V.2005, D.F. Candiani leg. (MPEG 24373).

Diagnosis. Males of *A. levii* resemble those of *A. delicata* by the dorsal abdominal coloration pattern (Figs. 1, 11), by the finger-shaped distal lobe (L, Fig. 3) and by the strong spines on tibiae I and II (Fig. 2). They differ by the distally pointed and curved retrolateral apical sector of the terminal apophysis (A, Fig. 3) (spoon-shaped in *A. delicata*; Fig. 12) – see Levi (1988: Figs. 476-478) for comparison – and by the reduced cymbial prolateral projection (Fig. 3) (well-developed in *A. delicata*), see Levi (1988: Fig. 477) for comparison. Females of *A. levii* resemble those of *A. delicata* by the two black humps on posterior end of abdomen (Figs. 7, 10), but differ by the drop-shaped median lobe of the scape (Fig. 8), epigynum longer than wide (Fig. 8) (wider than long in *A. delicata*), posterior median plate medially constricted (Fig. 9) (wide in *A. delicata*); lateral lobes of epigynum diamond-shaped in posterior view (Figs. 8, 9), see Levi (1988: figs. 472-474) for comparison.

Description. Male (MPEG 24370). Total length 6.1. Carapace length 2.6, width 2.0, height 0.6. Clypeus height 0.05. Sternum length 1.3, width 0.9. Abdomen length 3.6, width 1.8, height 1.5. Leg formula I/IV/II/III. Leg lengths: femur, I 3.3, II 2.7, III 1.9, IV 3.0; patella, I 1.1, II 1.0, III 0.8, IV 0.9; tibia, I 2.8, II 2.2, III 1.4, IV 2.5; 1metatarsus, I 3.0, II 2.5, III 1.4, IV 2.8; tarsus, I 1.1, II 1.0, III 0.7, IV 0.9. Eye diameters and interdistances: AME 0.1, ALE 0.1, PME 0.11, PLE 0.1; AME-PME 0.08, AME-ALE 0.34, PME-PLE 0.4, AME-AME 0.14, PME-PME 0.07. Carapace pale yellow, with two gray diagonal stripes on each side of the carapace (Fig. 1). Sternum pale yellow with brown margins, and a central black stripe. Endites and labium pale brown with white apices and brown margins. Chelicerae and legs yellow. Apices of femur, patella, tibia and metatarsus gray. Abdomen longer than wide, rectangular, with rounded borders. Dorsal side pale gray, with many white dots with different sizes (Fig.1). Two black humps and four black spots posteriorly. Two dark spots anteriorly. Venter pale gray with dark gray stripes, epigastric area slightly darker. Palp: palpal patella with two long macrosetae; median apophysis triangular-shaped in mesal view; paramedian apophysis formed by a distally expanded branch (Figs. 3, 4); embolus longer than terminal apophysis basal prong (Figs. 3-5); distal lobe of the terminal

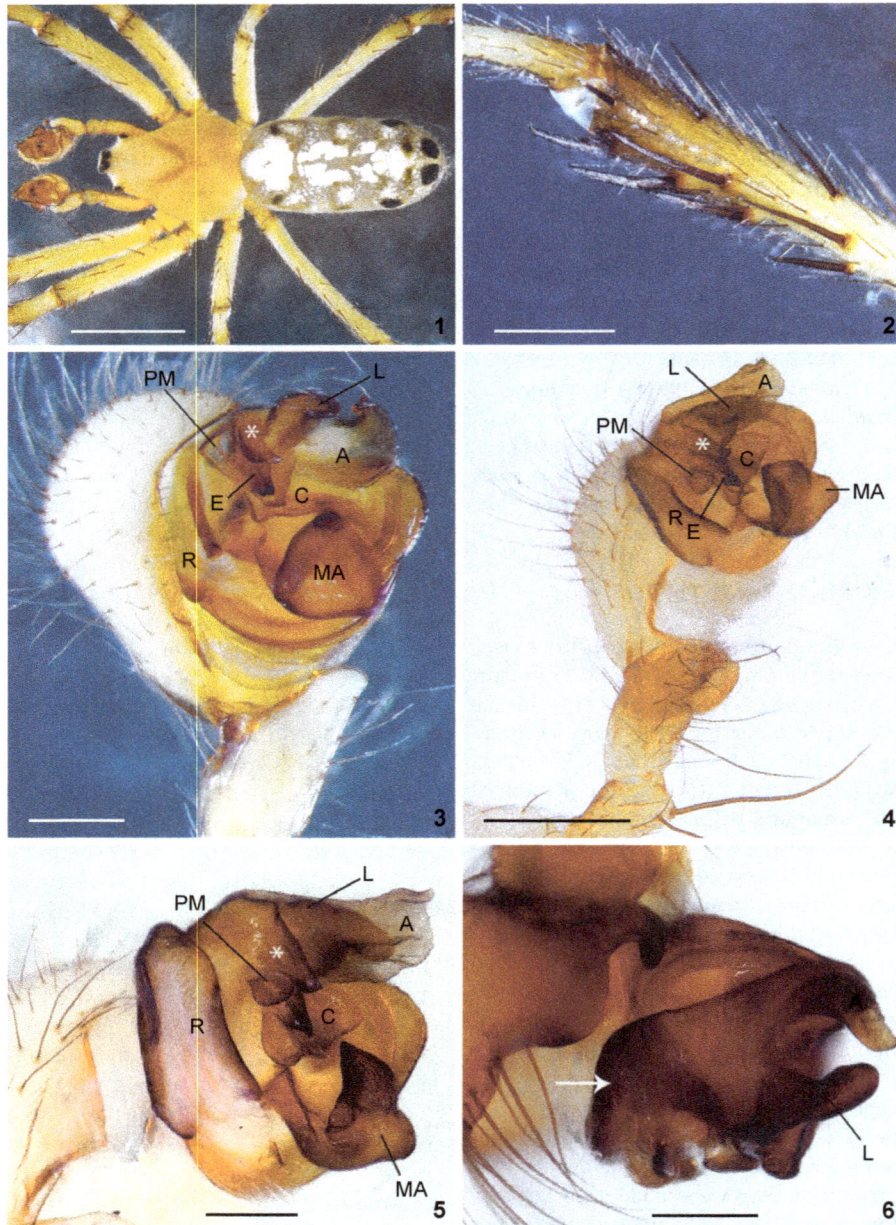

Figures 1-6. *Alpaida levii* **sp. nov.**, male: (1) habitus dorsal; (2) spines of the tibia I; (3) palpus, mesal view; expanded palpus (MPEG 24376): (4) mesal view; (5) detail of the mesal view; (6) detail of the terminal apophysis. (A) Terminal apophysis, (C) conductor, (E) embolus, (L) distal lobe, (MA) median apophysis, (PM) paramedian apophysis, (*) basal prong of the terminal apophysis, (R) radix. Scale bars: (1) = 2 mm, (2, 4) = 0.5 mm, (3, 5, 6) = 0.2 mm.

apophysis finger-shaped in mesal view (Figs. 3, 6); terminal apophysis thin and curved (Fig. 3).

Female (MPEG 24374). Total length 7.8. Carapace length 3.2, width 2.5, height 0.7. Clypeus height 0.13. Sternum length 1.3, width 1.1. Abdomen length 6.0, width 2.6, height 2.2. Leg formula I/IV/II/III. Leg lengths: femur, I 3.5, II 3.0, III 2.0, IV 3.6; patella, I 1.3, II 1.1, III 0.8, IV 1.1; tibia, I 3.1, II 2.5, III 1.4, IV 2.8; metatarsus, I 3.3, II 2.6, III 1.5, IV 2.9; tarsus, I 1.1, II 1.1, III 0.7, IV 1.0. Eye diameters and interdistances: AME 0.15, ALE 0.13, PME 0.15, PLE 0.14; AME-PME 0.1, AME-ALE 0.6,

PME-PLE 0.6, AME-AME 0.16, PME-PME 0.13. Carapace pale yellow, with two lateral gray diagonal stripes (Fig. 4). Sternum pale yellow with brown margins, and a central gray stripe. Endites and labium pale brown with white apices and brown margins. Chelicerae and legs yellow. Apices of femur, patella, tibia and metatarsus gray. Abdomen longer than wide, rectangular, with rounded borders. Dorsum pale gray, almost entirely covered by white and yellow spots and differently sized stripes (Fig.4). Posteriorly with two black humps and four black spots. Venter pale gray, epigastric area slightly darker. Epigynum longer than wide, with inconspicuous notch (Fig. 8); posterior median plate narrow (Fig. 9); lateral lobes diamond-shaped (Figs. 8, 9); scape drop-shaped.

Additional material examined. BRAZIL, *Amazonas*: Manaus (Reserva Adolpho Ducke), 02°59'11.33"S 59°56'14.69"W, 1 female, 30.VII.2008, J. ten Caten leg. (INPA); Coari (Base de Operações Geólogo Pedro de Moura, Porto Urucu), 04°52'07.6"S 65°15'53.6"W, 1 male, 22.VII.2003, A.B. Bonaldo leg. (MPEG 19874); 04°50'32"S 65°04'80"W, 1 female, 05.IX.2006, S. C. Dias leg. (MPEG 13762); Manicoré, 04°54.705'S 61°06.788'W, 1 female, 14.VII.2007, L.T. Miglio leg. (INPA). *Pará*: Melgaço (Floresta Nacional de Caxiuanã, TEAM 2), 01°43'43.2"S 51°29'00.6"W, 1 female, 26.IV.2006, J.A.P. Barreiros leg. (MPEG 24348); 1 female, R.B. Lopes leg. (MPEG 24369); (Floresta Nacional de Caxiuanã, TEAM 3), 01°43'59.2"S 51°30'38.6"W, 1 female, 04.X.2005, N. Abrahim leg. (MPEG 24339); 1 female, 17.IV.2006, C.B. Lopes leg. (MPEG 24353); (Floresta Nacional de Caxiuanã, TEAM 6), 01°44'18.02"S 51°27'48.01"W, 1 female (MPEG 24358); 1 female, 15.IV.2006, R.B. Lopes leg. (MPEG 24360); 1 female, 02.VIII.2011, Equipe MPEG (MPEG 24350); 1 female (MPEG 24366); (Floresta Nacional de Caxiuanã, Estação Científica Ferreira Penna), 01°44'15.5"S 51°26'42.0"W, 1 female, 10.VII.2002, J.P. Sifuerte leg. (MPEG 24356); (Floresta Nacional de Caxiuanã, TEAM 4), 01°45'12.8"S 51°31'14.7"W, 1 female, 12.X.2005, N. Abrahim leg. (MPEG 24351); 1 female, 23.IV.2006, E.J. Sales leg. (MPEG 24346); Juruti (Capiranga), 02°28'0.6"S 56°12'42.2"W, 2 females, 07.IX.2002, D. Guimarães leg. (MPEG 24357); 1 female, 12.IX.2002, D. Guimarães leg. (MPEG 24368); 1 female, A.B. Bonaldo leg. (MPEG 24365); 1 female, 15.IX.2002, A.B. Bonaldo leg. (MPEG 24363); 02°28'22.1"S 56°12'29.4"W, 1 female, 11.VIII.2008, N.C. Bastos leg. (MPEG 24359); 1 female, 10.III.2006, S.C. Dias leg. (MPEG 9133); (Barroso), 02°27'0.11"S 56°00'60"W, 1 female, 09.II.2009, B.V.B. Rodrigues leg. (MPEG 24354); 02°27'45.5"S 56°00'51"W, 1 female, 12.IV.2008, N.C. Bastos leg. (MPEG 24361); 1 male, 12.VIII.2008, N.F. Lo Man Hung leg. (MPEG 24340); 02°27'51.4"S 56°00'08.6"W, 1 female, 22.V.2009, N. Abrahim leg. (MPEG 24333); (Beneficiamento), 02°30'25.4"S 56°11'04.8"W, 1 female, 09.II.2007, N.F. Lo Man Hung leg. (MPEG 24364); 02°30'27.4"S 56°10'39.5"W, 1 female, 10.VIII.2010, B.V.B. Rodrigues leg. (MPEG 24336); 1 female, N.C. Bastos leg. (MPEG 24343); 1 female, 20.II.2011, N.C. Bastos leg. (MPEG 24334); (Mutum), 02°33'06.9"S 56°13'29.0"W, 1 female, 11.VIII.2010, B.V.B. Rodrigues leg. (MPEG 24335); 1 female,

(MPEG 24349); 1 female, N.C. Bastos leg. (MPEG 24347); 1 female, N. Abrahim leg. (MPEG 24352); 02°33'07.2"S 56°13'06.2"W, 1 female, 07.IX.2002, D. Guimarães leg. (MPEG 24337); 02°36'11.2"S 56°12'36.3"W, 1 female, 04.VIII.2004, D.R. Santos-Souzá leg. (MPEG 24341); 1 female (MPEG 24342); 02°36'44.7"S 56°11'39.2"W, 1 female, 27.V.2009, N. Abrahim leg. (MPEG 24338); 1 female, N.F. Lo Man Hung leg. (MPEG 24355); 1 female, 19.VIII.2011, R. Saturnino leg. (MPEG 24345); Novo Progresso (Serra do Cachimbo), 09°22'02.9"S 55°01'11.9"W, 1 male, 06.IV.2004, D. Guimarães leg. (MPEG 6358).

Distribution. Brazil, states of Pará and Amazonas.

Etymology. The specific name is a patronym to honor the late arachnologist Herbert W. Levi, and to recognize his immense contribution to spider taxonomy. His seminal work inspired and will continue to inspire generations of arachnologists.

Variation in paratypes. Six males, total length. 4.8 to 7.4; carapace: 2.3 to 3.0; number of spines on tibiae I and II: 5 to 10. Seven females, total length: 7.4 to 9.0; carapace: 2.0 to 3.2.

Alpaida delicata (Keyserling, 1892)
Figs. 10-13, 31

Epeira delicata Keyserling, 1892: 183, pl. 9, fig. 135, 6 (females and 4 males syntypes from Espírito Santo, Brazil, deposited in Natural History Museum (BMNH), London, not examined).

Araneus taczanowskii Simon, 1897: 473 (female holotype from Tefé, Est. Amazonas, Brazil, deposited in MNHN, not examined); Bonnet, 1955: 609. Synonymized by Levi, 1988.

Alpaida delicata: Levi, 1988: 458, figs. 472-478; Dierkens, 2014: 17, figs. 7, 8, 30, 41.

Diagnosis. Males of *A. delicata* resemble those of *A. levii* by the dorsal abdominal coloration pattern (Figs. 1, 11), by the finger-shaped distal lobe (L, Figs. 3-6) and by the strong spines on tibiae I and II (Fig. 2). They differ by the spoon-shaped retrolateral apical sector of the terminal apophysis (Figs. 12, 13) (distally pointed and curved in *A. levii*; Figs. 3-5) and by the well-developed cymbial prolateral projection (reduced in *A. levii*), see LEVI (1988: figs. 476-478) for comparison. Females of *A. delicata* resemble those of *A. levii* by the two black humps on posterior end of abdomen (Figs. 7, 10) and by the epigynum with the notch not well demarcated, but differs by the sinuous lips in ventral view, epigynum wider than long (longer than wide in *A. levii*; Figs. 8, 9); posterior median plate medially wide (medially constricted in *A. levii*; Fig. 9); lateral lobes rounded in posterior view (diamond-shaped in *A. levii*; Figs. 8, 9), see LEVI (1988: figs. 472-474) for comparison.

Material examined. BRAZIL, *Amazonas*: Presidente Figueiredo (Reserva Biológica de Uatumã), 01°49'05.55"S 59°14'34.23"W, 1 female (INPA); 01°49'37.24"S 59°14'31.81"W, 1 male (INPA); Manaus (Reserva Adolpho Ducke), 02°57'30.23"S 59°55'57.45"W, 1 female (INPA); 02°59'21.44"S 59°57'18.62"W, 1 female (INPA); Autazes, 04°09'26.3"S 60°07'53"W, 1 male and 4 females (INPA); 04°09'55.4"S 60°07'53.9"W, 1 male (INPA);

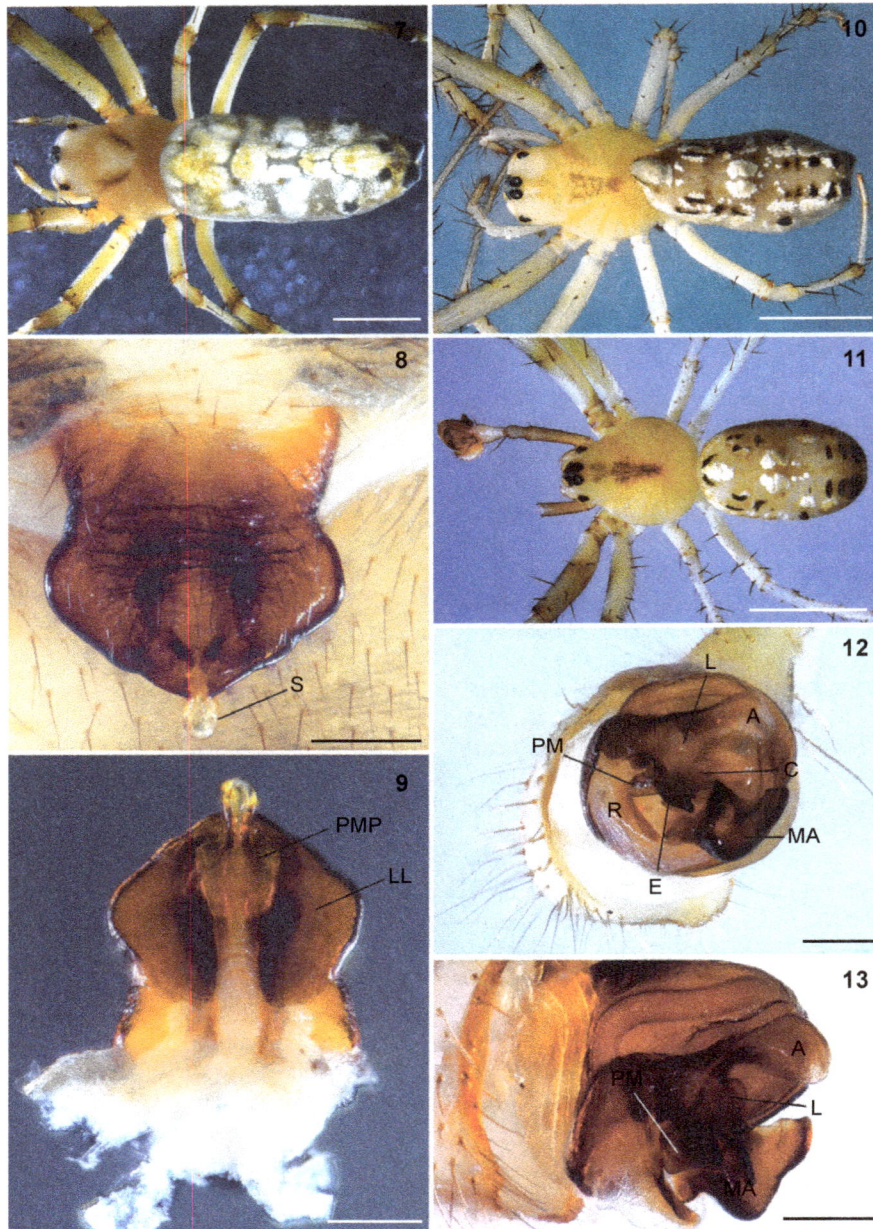

Figures 7-13. (7-9) *Alpaida levii* **sp. nov.**, female: (7) habitus dorsal; (8) epigynum ventral view; (9) epigynum posterior view. (10-13) *Alpaida delicata*: (10) female (MPEG 24408), habitus dorsal; (11) male (MPEG 9136), habitus dorsal; (12-13) expanded palpus: (12) mesal view; (13) detail of the terminal apophysis, median apophysis and paramedian apophysis. (A) Terminal apophysis, (C) conductor, (E) embolus, (L) distal lobe, (LL) lateral lobes, (MA) median apophysis, (PM) paramedian apophysis, (PMP) posterior median plate, (R) radix, (S) median lobe of the scape, (*) basal prong of the terminal apophysis. Scale bars: (7, 10, 11) = 2 mm, (8, 9, 12, 13) = 0.2 mm.

Canutama, 08°38′49″S 64°22′05.5″W, 1 female (INPA); 1 male (INPA); 08°39′05.8″S 64°22′05.6″W, 2 females (INPA); 08°39′06.8″S 64°22′05.8″W, 1 female (INPA); 1 male (INPA); 08°39′11.6″S 64°21′34.6″W, 1 male and 1 female (INPA); 1male (INPA); 1 female (INPA); 2 females (INPA); 08°39′14.8″S 64°21′34.3″W, 1 male (INPA); 1 female (INPA). *Pará*: Melgaço (Floresta Nacional de Caxiuanã, TEAM 1), 01°42′24.00″S 51°27′34.30″W, 1 male (MPEG 24539); (Floresta Nacional de

Caxiuanã, TEAM 2), 01°43'43.20"S 51°29'0.70"W, 1 male (MPEG 24541); 1 male (MPEG 24540); 1 male (MPEG 24522); 1 female (MPEG 24528); (Floresta Nacional de Caxiuanã, TEAM 3), 01°43'59.20"S 51°30'38.60"W, 1 female (MPEG 24542); 1 female (MPEG 24545); 1 male (MPEG 24543); 1 male (MPEG 24544); 1 female (MPEG 24523); 1 female (MPEG 24527); 1 female (MPEG 24526); 1 male (MPEG 24524); 1 male and 1 female (MPEG 24525); (Castanhal do Jacaré), 01°44'13.5"S 51°25'32.8"W, 1 male (MPEG 24560); (Floresta Nacional de Caxiuanã, Estação Científica Ferreira Penna), 01°44'15.5"S 51°26'42.0"W, 1 female (MPEG 24561); (Floresta Nacional de Caxiuanã, TEAM 4), 01°45'12.80"S 51°31'14.70"W, 1 female (MPEG 24529); 2 females (MPEG 24531); 1 female (MPEG 24532); 1 male (MPEG 24530); 1 female (MPEG 24535); 1 female (MPEG 24536); 1 female (MPEG 24546); 1 female (MPEG 24547); 1 male (MPEG 24548); (Floresta Nacional de Caxiuanã, Caiçara), 01°46'41.4"S 51°25'28.7"W, 1 female (MPEG 24558); (Floresta Nacional de Caxiuanã, TEAM 5), 01°47'23.66"S 51°34'52.18"W, 1 female (MPEG 24533); 1 male (MPEG 24534); 1 female (MPEG 24537); 1 female (MPEG 24538); 1 female (MPEG 24550); 1 male (MPEG 24552); 1 female (MPEG 24551); 1 male and 1 female (MPEG 24556); 1 female (MPEG 24557); 1 male and 1 female (MPEG 24555); 1 female (MPEG 24554); 1 female (MPEG 24549); 2 females (MPEG 24553); (Terra Preta), 01°51'19.30"S 51°25' 57.50"W, 1 male (MPEG 24559); (Portel, Plote PPBio), 01°57'38.9"S 51°36'45.3"W, 1 male and 1 female (MPEG 24562); 1 male (MPEG 24563); 1 male (MPEG 24564); (Cametá, Curuçambaba, Área de Floresta), 02°06'27.2"S 49°18'33.1"W, 1 male (MPEG 24577); 1 male (MPEG 24575); 02°07'27.6"S 49°18'52.7"W, 1 male (MPEG 24574); 02°06'39.4"S 49°18' 40.7"W, 1 female (MPEG 24578); 1 female (MPEG 24580); (Curuçambaba, Área de Praia), 02°06'31.4"S 49°18'55.8"W, 1 female (MPEG 24579); 1 female (MPEG 24576); 1 male and 1 female (MPEG 24581); 2 females (MPEG 24573); (Moju, Campo experimental da Embrapa), 02°11'12.44"S 48°47'34.31"W, 1 male (MPEG 24321); (Juruti, Capiranga), 02°28'0.60"S 56°12'42.20"W, 2 females (MPEG 24412); 1 female (MPEG 24380); 02°28'22.1"S 56°12'29.4"W, 1 female (MPEG 24430); 1 female (MPEG 24421); 1 female (MPEG 24414); 1 male (MPEG 9203); 1 male (MPEG 9132); (Barroso), 02°27'41.7"S 56°00'11.6"W, 1 female (MPEG 24420); 02°27'51.4"S 56°00'08.6"W, 1 female (MPEG 24422); 1 female (MPEG 24426); 1 female (MPEG 24403); 1 female (MPEG 24390); 1 female (MPEG 24404); (Ferrovia Km 23), 02°29'19"S 56°06'34"W, 1 male. (MPEG 24401); (Beneficiamento), 02°30'25.4"S 56°11'04.8"W, 1 male (MPEG 24416); 02°30'27.4"S 56°10'39.5"W, 1 male (MPEG 24378); 1 female (MPEG 24387); 1 female (MPEG 24393); (Mutum), 02°33'06.9"S 56°13'29.0"W, 1 female (MPEG 24398); 1 male and 1 female (MPEG 24399); 1 female (MPEG 24417); 1 female (MPEG 24394); 1 male (MPEG 24383); 02°33'18.0"S 56°13'22.4"W, 1 male (MPEG 24384); 1 female (MPEG 24418); 02°36'10.6"S 56°12'25.8"W, 1 female (MPEG 24396); 1 male (MPEG 24410); 02°36'11.2"S 56°12'36.3"W, 2 females (MPEG 24405); 1 male (MPEG 24411); 2 fe-

males (MPEG 24391); 1 female (MPEG 24392); 02°36'44.7"S 56°11'39.2"W, 1 male (MPEG 24386); 1 male (MPEG 9136); 1 female (MPEG 9137); 1 female (MPEG 24379); 1 female (MPEG 24423); 1 male (MPEG 24419); 1 male (MPEG 24427); 1 female (MPEG 24425); 1 female (MPEG 24429); 1 male (MPEG 24382); 1 female (MPEG 24400); 2 males (MPEG 24406); 1 male (MPEG 24409); 1 male (MPEG 24388); 1 female (MPEG 24385); 1 male (MPEG 24377); 1 female (MPEG 24389); 1 female (MPEG 24381); 1 male (MPEG 24397); 1 female (MPEG 24402); 02°36'45.2"S 56°11'27.5"W, 1 male (MPEG 24413); 1 female (MPEG 24407); 1 female (MPEG 24408); 1 male and 1 female (MPEG 24395); 1 female (MPEG 24415); 02°36'45.7"S 56°11'38.2"W, 1 female (MPEG 24424); 1 male (MPEG 24428).

Distribution. Previously known from Colombia, Peru, Bolivia, French Guiana and Brazil (Amazonas: Tefé; Pará: Melgaço; Espírito Santo). Recorded here also from Presidente Figueiredo, Manaus, Autazes and Canutama, state of Amazonas; Moju and Juruti, state of Pará, Brazil.

Alpaida truncata (Keyserling, 1865)
Figs. 14-17, 32

Epeira truncata Keyserling, 1865: 807, pl. 19, figs. 21-22 (female from Uruguay, deposited in BMNH, not examined).

Alpaida truncata Levi, 1988: 472, figs. 570-578; Levi, 2002: 538, figs. 75-77, 260-261; Dierkens, 2014: 22, figs. 19-20, 34, 47.

Diagnosis. Males of *A. truncata* resemble those of *A. levii* and *A. delicata* by having the posteriorly hump-shaped abdomen (Figs. 7, 10, 14) and *A. queremal* by the extremely long and distally pointed median apophysis (Fig. 16), see Levi (1988: fig. 569) for comparison. They differ from other species of *Alpaida* by the c-shaped and digitiform paramedian apophysis (Figs. 16, 17), and a modified second tibia, flattened and wide, bearing two macrosetae; differ from *A. queremal* by a notch in the base of the median apophysis – modified from Levi (1988: see figs. 569, 577-578 for comparison). Females also resemble those of *A. levii*, *A. delicata* and *A. queremal* by the posteriorly hump-shaped abdomen, but differ from *A. levii* by the epigynum wider than longer, from *A. delicata* by the median lobe not well-developed and from *A. queremal* by the lack of the lateral lobes of the epigynum – modified from Levi (1988: see figs. 472-473, 564-565, 570-571 for comparison).

Description. Male and female. See Levi (1988): 472, figs. 570-578.

Material examined. Brazil, *Pará*: Santa Bárbara, 01°13'37.32"S 48°17'45.59"W, 2 females (MPEG 24519); Benevides, 01°21'43.87"S 48°14'37.79"W, 1 male (MPEG 2972); Belém (Jardim Botânico Rodrigues Alves), 01°25'49.0"S 48°27'22.3"W, 1 female (MPEG 24516); 1 female (MPEG 24517); (Museu Paraense Emílio Goeldi, Campus de Pesquisa), 01°27'03.03"S 48°26'40.2"W, 1 male (MPEG 4907); 1 male (MPEG 24512); (Parque Zoobotânico Emílio Goeldi), 01°27'12"S 48°28'35"W, 1 male (MPEG 24513); 1 male (MPEG 24514); 1 female (MPEG

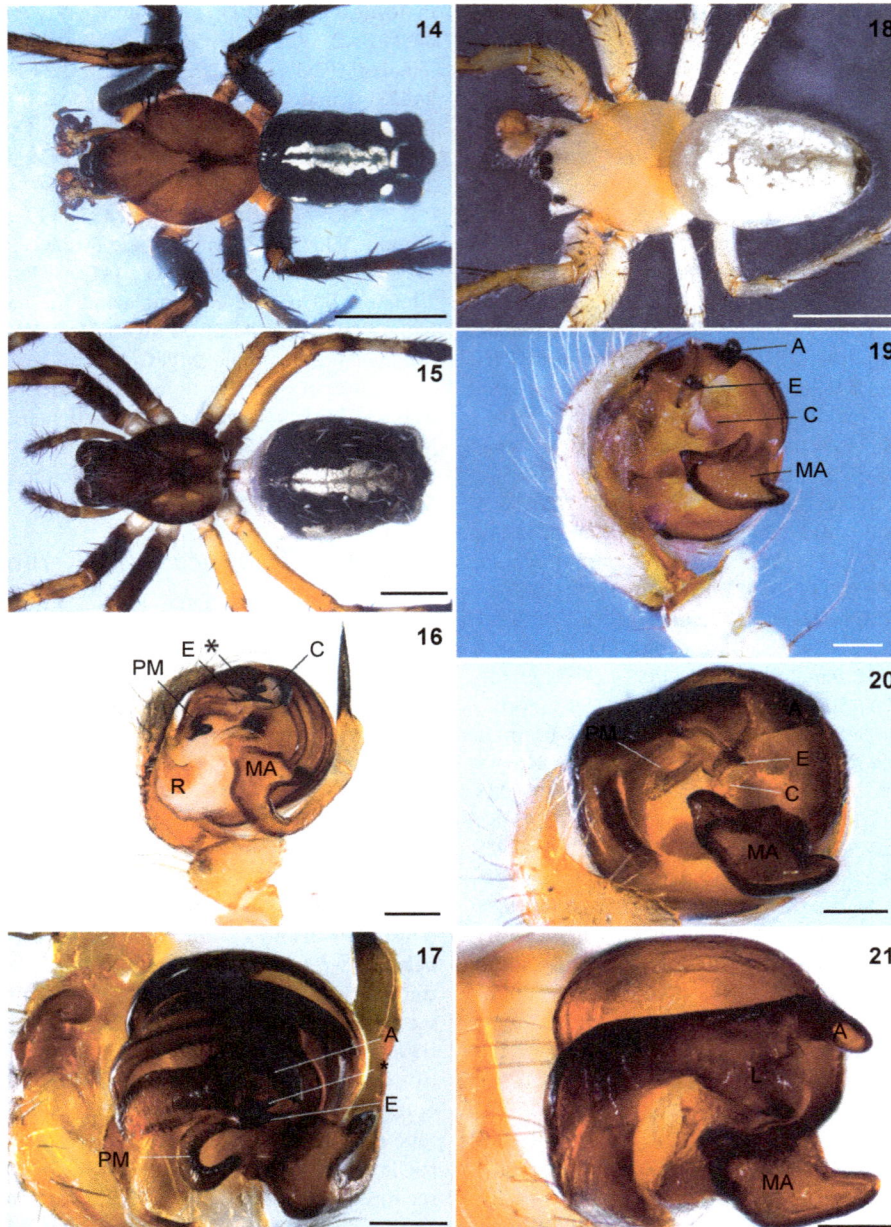

Figures 14-21. (14-17) *Alpaida truncata*: (14, 16, 17) male (MPEG 24508): (14) habitus dorsal; (16) palpus, mesal view; (17) expanded palpus, detail of the terminal apophysis, basal prong, embolus, median and paramedian apophyses; (15) female (MPEG 24503), habitus dorsal. (18-21) *Alpaida iquitos*, male: (18) habitus dorsal; (19) palpus, mesal view; (20-21) expanded palpus: (20) mesal view; (21) detail of the terminal apophysis and median apophysis. (A) Terminal apophysis, (C) conductor, (E) embolus, (L) distal lobe, (MA) median apophysis, (PM) paramedian apophysis, (R) radix, (*) basal prong of the terminal apophysis. Scale bars: (14, 15) = 2 mm, (16, 17) = 0.2 mm, (18) = 1 mm, (19-21) = 0.1 mm.

24515); (Mata do Betina, Universidade Federal do Pará), 01°28'02"S 48°26'33"W, 1 female (MPEG 24518); Melgaço (Floresta Nacional de Caxiuanã, Estação Científica Ferreira Penna), 01°44'15.5"S 51°26'42"W, 1 female (MPEG 8010); 1 female (MPEG 8011); 1 female (MPEG 8009); 01°44'18.02"S 51°27'48.01"W, 1 female (MPEG 22499); 01°47'32.7"S

51°25'59.2"W, 1 male (MPEG 24520); Juruti (Barroso), 02°27'51.4"S 56°00'08.6"W, 1 female (MPEG 24497); 1 male (MPEG 24500); 1 female (MPEG 24504); 1 male (MPEG 24505); (Capiranga), 02°28'22.1"S 56°12'29.4"W, 1 male (MPEG 24499); 1 female (MPEG 24501); 02°29'57.8"S 56°12'60"W, 1 male (MPEG 24502); (Beneficiamento), 02°30'04.9"S 56°09'46.6"W, 1 male (MPEG 24507); 02°30'25.4"S 56°11'04.8"W, 1 male and 1 female (MPEG 24498); 1 female (MPEG 24503); 02°30'27.4"S 56°10'39.5"W (MPEG 24506); 1 male (MPEG 24508); 1 female (MPEG 24511); 2 females (MPEG 24510); 1 female (MPEG 24509); (Mutum), 02°36'45.2"S 56°11'27.5"W, 1 female (MPEG 24596); Marabá (Serra Norte), 06°0'23.1"S 50°17'50.3"W, 1 male (MPEG 4243); 1 male (MPEG 4194); Novo Progresso, 07°08'07"S 55°24'51"W, 1 male (MPEG 4495); Serra do Cachimbo, 09°16'18.6"S 54°56'22.9"W (MPEG 6330).

Distribution. Previously known from Mexico to Argentina. Recorded here also from Benevides, Juruti and Marabá, state of Pará, Brazil.

Alpaida iquitos Levi, 1988
Figs. 18-21, 30

Alpaida iquitos Levi, 1988: 416, figs. 194-197 (female holotype and female paratype from Iquitos, Peru, V.1920, deposited in Museum of Comparative Zoology (MCZ), Harvard University, not examined).

Note. The male, described here, is identified as belonging to *A. iquitos* based on the morphological similarity and also in the sympatric distribution with the female previously described by Levi (1988).

Diagnosis. Males of *A. iquitos* resemble those of *A. bicornuta* by the wider than long median apophysis (MA, Figs. 19-21) and by the flattened distal lobe (L, Fig. 21), but differs from this and all other species of *Alpaida* by the sharply pointed, opposed proximal ends of median apophysis (MA, Figs. 19-21) and by the distally rounded retrolateral apical sector of the terminal apophysis (A, Figs. 19-21) – see Levi (1988: Figs. 17-18) for comparison. As diagnosed by Levi (1988), females differ from *A. variabilis* and *A. kochalkai* by having the posterior plate of the epignum constricted in the middle and from *A. variabilis* by having the epyginum longer (Levi 1988: see figs. 194-197).

Description. Male (MPEG 24164). Total length 3.3. Carapace length 1.7, width 1.3, height 0.6. Clypeus height 0.05. Sternum length 0.7, width 0.6. Abdomen length 2.0, width 1.2, height 1.0. Leg formula I/II/IV/III. Leg lengths: femur, I 1.6, II 1.4, III 1.1, IV 1.3; patella, I 0.6, II 0.6, III 0.4, IV 0.4; tibia, I 1.3, II 1.0, III 0.6, IV 1.1; metatarsus, I 1.2, II 1.1, III 0.6, IV 1.0; tarsus, I 0.6, II 0.5, III 0.4, IV 0.4. Eyes diameters and interdistances: AME 0.12, ALE 0.09, PME 0.11, PLE 0.09; AME-PME 0.09, AME-ALE 0.21, PME-PLE 0.24, AME-AME 0.09, PME-PME 0.07. Carapace pale yellow. Sternum, endites and labium pale yellow with brown margins. Chelicerae pale yellow. Legs pale yellow; tibia, metatarsus and tarsus of legs I-II darker. Abdomen longer than wide, oval. Laterals of dorsal side, pale yellow, center white, with

a dark gray stripe posteriorly (Fig. 18). Venter pale yellow. Palp: palpal patella and palpal tibiae with one long macrosetae; median apophysis wider than long, sharply pointed, opposed proximal ends of median apophysis two-pointed (Figs. 19-21); paramedian apophysis distally expanded (Fig. 20); embolus short, covered by terminal apophysis (Figs. 19-20); basal prong of the terminal apophysis absent; distal lobe of the terminal apophysis well-developed, visible only with the expanded palpus in apical view of the terminal apophysis (Fig. 21); retrolateral portion of the terminal apophysis rounded (Figs. 19-21).

Female. See Levi (1988: 416, figs. 194-197).

Material examined. BRAZIL, *Pará*: Portel (Floresta Nacional de Caxiuanã, Plote PPBio), 01°57'38.9"S 51°36'45.3"W, 1 female, 10.V.2005, C.A. Lopes (MPEG 24184); 1 female, 12.V.2005, D.F. Candiani (MPEG 24179); Melgaço (Floresta Nacional de Caxiuanã, TEAM 1), 01°42'24.0"S 51°27'34.3"W, 1 male, 26.IX.2005, J.A.P. Barreiros leg. (MPEG 24167); (Floresta Nacional de Caxiuanã, TEAM 5), 01°43'21.6"S 51°25'51.2"W, 1 female, 08.X.2005, J.A.P. Barreiros leg. (MPEG 24175); 1 female (MPEG 24177); 1 male, 13.X.2005, Robinho leg. (MPEG 24164); (Floresta Nacional de Caxiuanã), TEAM 2, 01°43'43.2"S 51°29'00.6"W, 1 male, 03.X.2005, J.A.P. Barreiros leg. (MPEG 24168); 1 female, 28.IX.2005, B.B. Santos leg. (MPEG 24170); 1 female, 05.X.2005, J.A.P. Barreiros leg. (MPEG 24171); 3 females, N. Abrahim leg. (MPEG 24169); (Floresta Nacional de Caxiuanã, TEAM 3), 01°43'59.2"S 51°30'38.6"W, 1 female, 29.IX.2005, B.B. Santos leg. (MPEG 24173); 1 female, J.A.P. Barreiros leg. (MPEG 24176); 1 female, 04.X.2005, J.A.P. Barreiros leg. (MPEG 24174); 1 female, 17.IV.2006, C.A. Souza leg. (MPEG 24163); 1 female, R.B. Lopes leg. (MPEG 24162); (Floresta Nacional de Caxiuanã, Estação Científica Ferreira Penna), 01°44'15.5"S 51°26'42.0"W, 1 female, 16.XI.2001, Aires leg. (MPEG 24180); 1 female, 07-13.II.2002, A.B. Bonaldo leg. (MPEG 24181); 1 female, 17.IV.2002, M. Andrade leg. (MPEG 24182); 1 female, 05.VI.2004, A.B. Bonaldo leg. (MPEG 24183); 01°44'18.02"S 51°27'48.01"W, 1 female (MPEG 24185); 1 female (MPEG 24187); 1 female, 25.III.2002, N. Abrahim leg. (MPEG 24178); 1 female, 05.VI.2004, C. Trinca leg. (MPEG 24186); (Floresta Nacional de Caxiuanã, TEAM 4), 01°45'12.8"S 51°31'14.7"W, 1 female, 07.X.2005, Robinho leg. (MPEG 24172); 1 female, 12.X.2005, J.A.P. Barreiros leg. (MPEG 24165); 1 male, N. Abrahim leg. (MPEG 24166).

Distribution. Previously known from Brazil (Pará: Melgaço and Canindé [Paragominas]; Mato Grosso: Chapada dos Guimarães), Ecuador, French Guiana and Peru. Recorded here also from Portel, state of Pará, Brazil.

Alpaida gurupi Levi, 1988
Figs. 22-24, 30

Alpaida gurupi Levi, 1988: 429, figs. 278-279 (male holotype from Canindé, Rio Gurupi, Pará, Brazil, 27-28.II.1966, deposited in American Museum of Natural History (AMNH), not examined).

Note. The females, described here, were identified as belonging to *A. gurupi* based on the morphological similarity and

64

55

78

56
27 S28

LL

PMP

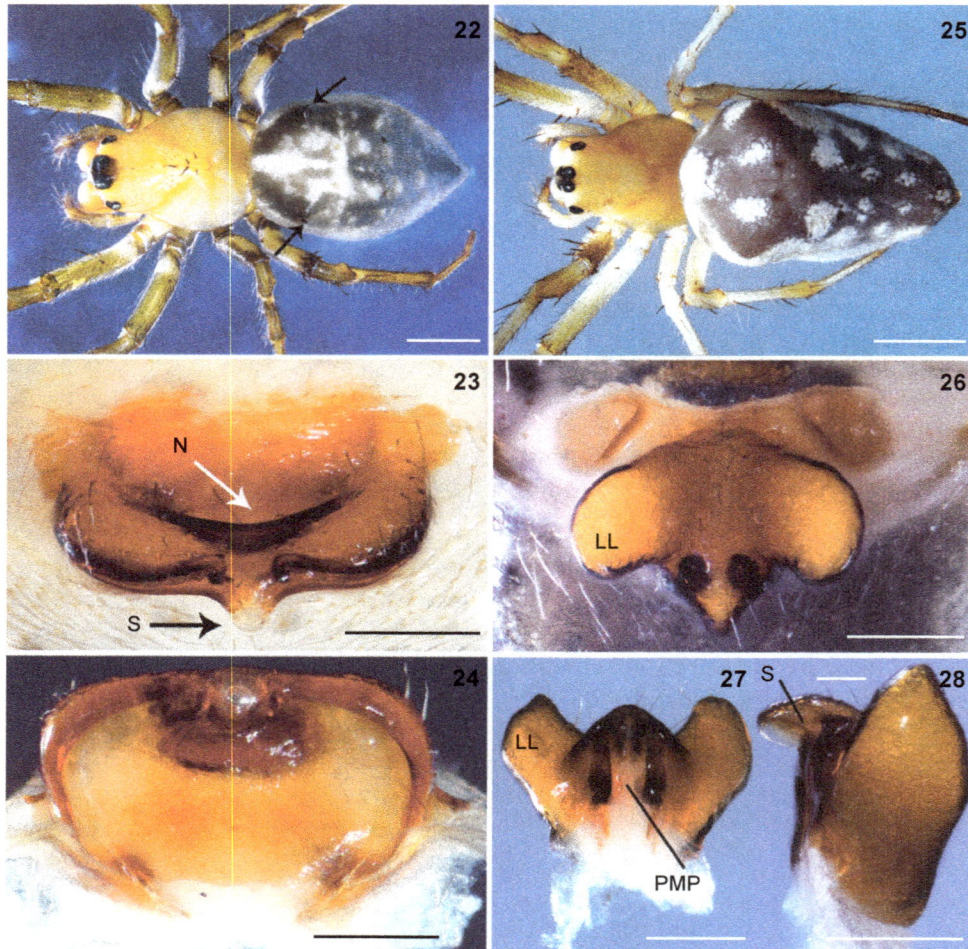

Figures 22-28. (22-24) *Alpaida gurupi*, female: (22) habitus dorsal, arrow = spines; (23-24) epigynum: (23) ventral view; (24) posterior view. (25-28) *Alpaida yanayacu* new species, female: (25) habitus dorsal; (26-28) epigynum: (26) ventral view; (27) lateral view; (28) posterior view. (LL) Lateral lobes, (N) notch, (PMP) posterior median plate, (S) median lobe of the scape. Scale bars: (22, 25) = 1 mm, (23-28) = 0.2 mm.

also in the sympatric distribution with the male previously described by Levi (1988).

Diagnosis. Females of *A. gurupi* resemble those of *A. amambay* by the epigynum wider than long, by the well demarcated notch and by short and rounded scape median lobe (Fig. 23), but differs by the notch occupying more than half of the epigynum width (Fig. 23), posterior median plate wide, without sinuous borders (Fig. 24), see Levi (1988: Figs. 274-275) for comparison. As diagnosed by Levi (1988), males differ from other species of *Alpaida* by the gently curved median apophysis and the hooded appearance of the terminal apophysis see Levi (1988: figs. 278-279).

Description. Female (MPEG 24190). Total length 5.4. Carapace length 2.2, width 1.8, height 0.8. Clypeus height 0.08. Sternum length 0.9, width 0.9. Abdomen length 3.3, width 2.0,

height 1.8. Leg formula IV/I/II/III. Leg lengths: femur, I 2.0, II 1.8, III 1.4, IV 2.1; patella, I 0.8, II 0.8, III 0.6, IV 0.8; tibia, I 1.6, II 1.3, III 1.0, IV 1.8; metatarsus, I 1.5, II 1.4, III 1.0, IV 1.6; tarsus, I 0.8, II 0.7, III 0.6, IV 0.8. Eyes diameters and interdistances: AME 0.1, ALE 0.12, PME 0.17, PLE 0.14; AME-ALE 0.14, PME-PLE 0.3, AME-AME 0.12, PME-PME 0.2. Carapace pale yellow, with darker yellow stripes on cephalic and thoracic region of the carapace (Fig. 9). Sternum pale yellow with orange margins. Endites and labium pale orange with white apices and brown margins. Chelicerae yellow. Legs orange-brownish. Femur, patella, tibia and metatarsus gray. Abdomen longer than wide, oval, posteriorly pointed with rounded borders; anteriorly with two spines, one on each side (see arrows Fig. 22). Dorsal side dark gray, many white and pale gray dots with different sizes (Fig. 22). Ventral side dark gray with light

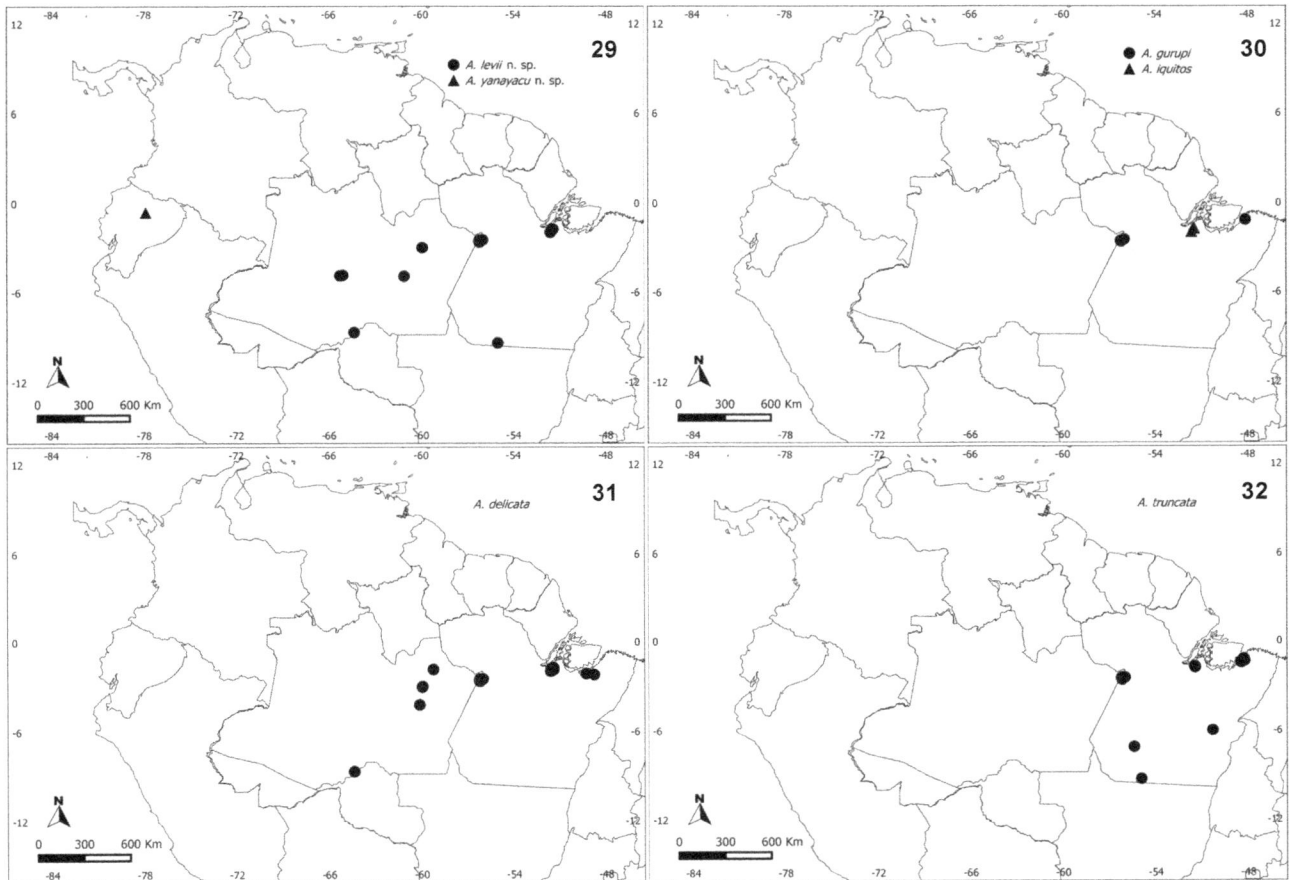

Figures 29-32. Distribution of *Alpaida* species in the Amazon region, North Brazil, and Ecuador: (29) *Alpaida levii* **sp. nov.** and *A. yanayacu* **sp. nov.**; (30) *A. gurupi* and *A. iquitos*; (31) *A. delicata*; (32) *A. truncata*.

gray stripes and many white dots. Epigynum wider than long, with evident notch; (Fig. 23); posterior median plate wide (Fig. 24); median lobe of scape short, drop-shaped (Fig. 23).

Male. See Levi (1988): 429, figs. 278-279.

Material examined. Brazil, *Pará*: Santo Antônio do Pará 01°09'09"S 48°07'45"W, 1 male, 07.IV.1975, R.F. da Silva leg. (MPEG 2973); Juruti (Barroso), 02°27'51.4"S 56°00'08.6"W, 1 female, 08.II.2007, N.F. Lo Man Hung leg. (MPEG 24199); 1 female, 16.XI.2007, N.F. Lo Man Hung leg. (MPEG 24197); 1 female (MPEG 24201); 1 male and 1 female, 08.VIII.2008, N.F. Lo Man Hung leg. (MPEG 24189); (Mutum), 02°33'06.9"S 56°13'29.0"W, 2 females, 12.VIII.2010, N.C. Bastos leg. (MPEG 24195); 02°33'18"S 56°13'22.4"W, 1 female, 05.V.2010, N.C. Bastos leg. (MPEG 24191); 02°36'11.2"S 56°12'36.3"W, 1 female, 04.VIII.2004, D.F. Candiani and D.R. Santos-Souza leg. (MPEG 24193); 1 female, 09.VIII.2004, D.F. Candiani leg. (MPEG 24196); 1 male, 09.VIII.2004, D.R. Santos-Souza leg. (MPEG 24194); 02°36'44.7"S 56°11'39.2"W, 1 female, 12.II.2007, N.F. Lo Man Hung leg. (MPEG 24198); 1 female (MPEG 24200); 1

female, 08.VIII.2008, N.C. Bastos leg. (MPEG 24188); 1 female, N.F. Lo Man Hung leg. (MPEG 24190); 1 female, L.T. Miglio leg. (MPEG 24192).

Distribution. Previously known from Colombia and Brazil (Pará: Canindé [Paragominas]). Recorded here also from Santo Antônio do Pará and Juruti, state of Pará, Brazil.

Variation. Six females, total length. 5.4 to 7.2; carapace: 2.4 to 2.9.

Alpaida yanayacu **sp. nov.**
Figs. 25-28, 29

Type material. Female holotype from Ecuador: Napo (Yanayacu Biological Station), 0°36'29.76"S 77°52'56.82"W, 25.XI.2009, A.B. Bonaldo leg., deposited in QCAZ.

Diagnosis. Females of *A. yanayacu* resemble those of *A. machala* Levi, 1988 and *A. eberhardi* Levi, 1988 by having the lateral lobes long and median lobe wide at its base, which is proportionally closer to *A. eberhardi*, since the median lobe is larger in both species than in *A. machala*; they differ by the lateral lobes rounded in ventral view (Fig. 26) (almost straight in *A. machala* and not visible in this view in *A. eberhardi*); by the borders of the posterior median plate parallel (Fig. 27); and by the less curved median lobe scape (Fig. 28) than in *A. machala*, see Levi (1988: 555-556, 559-560) for comparison.

Description. Female (QCAZ). Total length 4.4. Carapace length 1.7, width 1.35, height 0.5. Clypeus height 0.05. Sternum length 0.75, width 0.67. Abdomen length 3.25, width 1.9, height 1.8. Leg formula I/II/IV/III. Leg lengths: femur, I 1.55, II 1.5, III 1.1, IV 1.6; patella, I 0.7, II 0.57, III 0.45, IV 0.47; tibia, I 1.47, II 1.17, III 0.75, IV 1.17; metatarsus, I 1.3, II 1.15, III 0.62, IV 1.15; tarsus, I 0.62, II 6.0, III 0.45, IV 0.52. Eyes diameters and interdistances: AME 0.1, ALE 0.07, PME 0.08, PLE 0.07; AME-PME 0.075, AME-ALE 0.26, PME-PLE 0.3, AME-AME 0.08, PME-PME 0.11. Carapace yellow, with two gray stripes on the lateral edges. Sternum black. Endites and labium dark brown with pale apices. Chelicerae yellow with distal third brown. Legs yellow, except the tarsus brown. Abdomen longer than wide, rounded anteriorly and pointed posteriorly. Dorsal side gray with white pigments of different sizes. Two pairs of white lateral stripes. Venter dusky gray from epigastric area to behind spinnerets. Epigynum wider than long, with notch not demarcated (Fig. 26); posterior median plate narrow with parallel borders (Fig. 27); scape relatively long and curved (Fig. 28).

Distribution. Known only from the type locality.

Etymology. The specific name is a noun in apposition taken from the type locality.

Alpaida antonio Levi, 1988
Fig. 33

Alpaida antonio Levi, 1988: 446, figs. 392-397 (female holotype from Fazenda Santo Antônio, Uruçuca, Bahia, Brazil, 27.XI.1977, deposited in Museu de Ciências Naturais (MCN), Fundação Zoobotânica do Rio Grande do Sul, not examined); Dierkens 2014: 15, figs. 1, 38.

Material examined. Brazil, *Amazonas*: Manicoré, 04°54'57"S 61°06'45.4"W, 1 male (INPA); 1 male (INPA); 1 male (INPA).

Distribution. Previously known from Guyana, French Guiana and Brazil (Pará: Melgaço, Canindé [Paragominas]; Bahia: Uruçuca, Camacã; Espírito Santo: Rio São José). Recorded here also from Manicoré, state of Amazonas, Brazil.

Alpaida bicornuta (Taczanowski, 1878)
Fig. 34

Epeira bicornuta Taczanowski, 1878: 168, pl. 2, fig. 18 (female lectotype and paralectotypes designated by Levi, 1988 from Pumamarca and Amable María, Junín, Peru, deposited in Polska Akademia Nauk (PAN), not examined).
Alpaida bicornuta: Levi, 1988: 387, figs. 11-18; Dierkens, 2014: 16, figs. 3, 4, 28, 39.

Material examined. Brazil, *Pará*: Bragança (Ilha das Canelas), 0°47'8.08"S 46°43'20.88"W, 1 female (MPEG 4980); 5 males and

2 females (MPEG 11187); 01°3'S 46°46'W, 1 female (MPEG 2971); Juruti (Área de várzea), 02°12'36.1"S 56°07'20.7"W, 1 female (MPEG 24493); 1 male (MPEG 24490); 1 male (MPEG 24491); 1 female (MPEG 24492); 1 female (MPEG 24494); 1 female (MPEG 24495); 02°24'33.2"S 56°26'10.6"W, 1 male (MPEG 24486); 1 male (MPEG 24487); (Barroso), 02°28'28.9"S 55°59'58.8"W, 1 female (MPEG 24489); 1 female (MPEG 24488); Marabá (Serra Norte, Fofoca), 05°58'13.81"S 50°21'28.16"W, 1 female (MPEG 4262).

Distribution. Previously known from Costa Rica to Argentina. Recorded here also from Bragança, Juruti and Marabá, state of Pará, Brazil.

Alpaida boa Levi, 1988
Fig. 35

Alpaida boa Levi, 1988: 447, figs. 408-409 (male holotype from Fonte Boa, Amazonas, Brazil, IX.1975, deposited in AMNH, not examined); Dierkens 2014: 16, figs. 5, 29.

Material examined. Brazil, *Pará*: Almerim (Jari), 0°53'16.53"S 52°50'41.59"W, 1 male (MPEG 7604).

Distribution. Previously known from French Guyana and Brazil (Amazonas: Fonte Boa). Recorded here also from Almerim, state of Pará, Brazil.

Alpaida deborae Levi, 1988
Fig. 36

Alpaida deborae Levi, 1988: 442, figs. 364-366 (female holotype from Browns Berg, 05°N, 55°27'W, Brokopondo Prov., Surinam, 20.II.1982, deposited in MCZ, not examined); Dierkens 2014: 16, figs. 6, 40.

Material examined. Brazil, *Pará*: Juruti (Área de Várzea), 02°24'33.2"S 56°26'10.6"W, 1 female (MPEG 24313).

Distribution. Previously known from Surinam, French Guiana and Brazil (Pará: Belém). Recorded here also from Juruti, state of Pará, Brazil.

Alpaida erythrothorax (Taczanowski, 1873)
Fig. 35

Singa erythrothorax Taczanowski, 1873: 126 (female lectotype, 2 males and 1 juvenile paralectotypes from Cayenne, French Guiana, deposited in PAN, not examined.
Alpaida erythrothorax: Levi, 1988: 444, figs. 376-378; Dierkens, 2014: 22.

Material examined. Brazil, *Pará*: Melgaço (Floresta Nacional de Caxiuanã), 01°44'18.02"S 51°27'48.01"W, 1 female (MPEG 24330); 1 female (MPEG 24332); 1 female (MPEG 24331); 01°44'15.5"S 51°26'42.0"W, 1 female (MPEG 24329); 1 female (MPEG 24328); Novo Progresso (Serra do Cachimbo), 09°16'18.6"S 54°56'22.9"W, 1 female (MPEG 6182); 1 female (MPEG 6363).

Distribution. Previously known from French Guiana and Brazil (Pará: Melgaço). Recorded here also from Novo Progresso, state of Pará, Brazil.

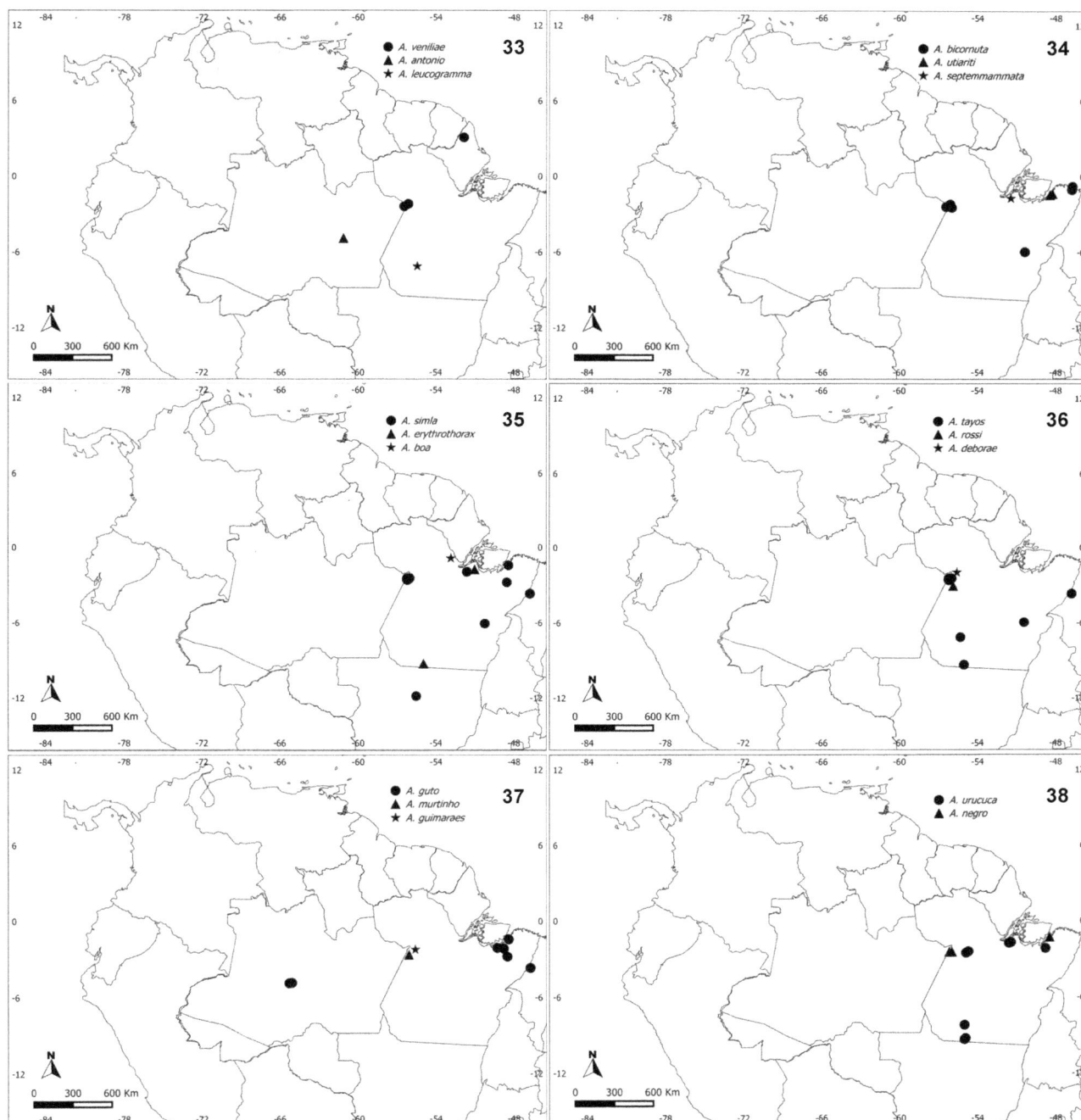

Figures 33-38. Distribution of new records of *Alpaida* species in the Amazon region, North Brazil: (33) *Alpaida antonio, A. leucogramma* and *A. veniliae*; (34) *A. bicornuta, A. septemmammata* and *A. utiariti*; (35) *A. boa, A. erythrothorax* and *A. simla*; (36) *A. deborae, A. rossi* and *A. tayos*; (37) *A. guimaraes, A. guto* and *A. murtinho*; (38) *A. negro* and *A. urucuca*.

Alpaida guimaraes Levi, 1988
Fig. 37

Alpaida guimaraes Levi, 1988: 390, figs. 19-24 (female holotype from Chapada dos Guimarães, Mato Grosso, Brazil, 01.XII.1983, deposited in MCN, not examined); Dierkens, 2014: 23.

Material examined. BRAZIL, *Pará*: Juruti (Área de Várzea), 02°12'36.1"S 56°07'20.7"W, 1 female (MPEG 24317); 1 female

(MPEG 24319); 1 female (MPEG 24318); 1 female (MPEG 24320).

Distribution. Previously known from Guyana and Brazil (Pará: Jacareacanga; Bahia: Uruçuca; Mato Grosso: Xavantina). Recorded here also from Juruti, state of Pará, Brazil.

Alpaida guto Abrahim & Bonaldo, 2008
Fig. 37

Alpaida guto Abrahim & Bonaldo, 2008: 398, figs. 1-4 (male holotype from Floresta Nacional de Caxiuanã, Melgaço, Pará, Brazil, 09.V.2005, deposited in MPEG 5241, examined); Dierkens, 2014: 14.

Material examined. BRAZIL, *Amazonas*: Coari (Base de Operações Geólogo Pedro de Moura, Porto Urucu), 04°50'01"S 65°03'53"W, 1 female (MPEG 13776); 04°52'06"S 65°15'52"W, 1 female (MPEG 13736). *Pará*: Belém (Parque Estadual do Utinga), 01°25'18.8"S 48°25'48.3"W, 1 male and 3 females (MPEG 24203); 1 male (MPEG 24217); 3 females (MPEG 24218); 1 female (MPEG 24210); 1 female (MPEG 24222); 1 male (MPEG 24212); 1 female (MPEG 24202); 1 male and 4 females (MPEG 24204); 2 females (MPEG 24205); 1 male (MPEG 24206); 1 female (MPEG 24207); 1 female (MPEG 24208); 1 male (MPEG 24209); 1 female (MPEG 24211); 1 male (MPEG 24213); 1 female (MPEG 24214); 2 females (MPEG 24215); 1 male (MPEG 24216); 2 males and 5 females (MPEG 24219); 1 female (MPEG 24220); 1 male (MPEG 24221); 1 male and 6 females (MPEG 24223); (Reserva Mocambo), 01°26'48"S 48°25'1"W, 1 female (MPEG 24246); 2 males and 1 female (MPEG 24254); 1 female (MPEG 24245); 1 female (MPEG 24253); 1 male and 1 female (MPEG 24259); 1 female (MPEG 24255); 3 females (MPEG 24264); 1 female (MPEG 24252); 1 male (MPEG 24239); 1 female (MPEG 24251); 2 females (MPEG 24238); 1 male and 3 females (MPEG 24260); 1 male and 1 female (MPEG 24249); 1 male and 1 female (MPEG 24256); 3 males (MPEG 24240); 1 male (MPEG 24247); 1 male (MPEG 24248); 1 female (MPEG 24261); 1 male and 1 female (MPEG 24241); 2 males (MPEG 24242); 1 male and 1 female (MPEG 24243); 1 female (MPEG 24244); 1 male (MPEG 24250); 1 female (MPEG 24258); 1 female (MPEG 24262); 1 male and 1 female (MPEG 24257); 1 male (MPEG 24263); Cametá (Curuçambaba, Área de Floresta), 02°06'27.2"S 49°18'33.1"W, 4 females (MPEG 24586); 2 males (MPEG 24590); 1 male (MPEG 24588); 2 females (MPEG 24582); 1 male and 1 female (MPEG 24587); 1 male (MPEG 24589); 02°07'27.6"S 49°18'52.7"W, 2 females (MPEG 24585); 1 female (MPEG 24584); 1 female (MPEG 24583); Moju (Campo experimental da Embrapa), 02°09'38.9"S 48°47'50.64"W, 1 male (MPEG 24234); 02°10'41.52"S 48°47'37.13"W, 1 male (MPEG 24232); 02°11'44.37"S 48°47'38.79"W, 1 male (MPEG 24235); Tailândia (Fazenda Marupiara), 02°47'44.9"S 48°32'39.2"W, 1 male (MPEG 24230); 1 male (MPEG 24233); 02°48'43.7"S 48°30'44"W, 1 male (MPEG 24237); 1 male (MPEG 24231); 1 male (MPEG 24236). *Maranhão*: Centro Novo do Maranhão (Reserva Biológica do Gurupi), 03°41'21"S 46°45'16.5"W, 1 female (MPEG 24227); 2 males (MPEG 24228); 1 male (MPEG 24229); 03°41'33.84"S 46°44'46.62"W, 1 female

(MPEG 24226); 03°41'47.22"S 46°44'17.4"W, 1 male (MPEG 24224); 1 female (MPEG 24225).

Distribution. Previously known from Brazil (Pará: Melgaço and Santa Bárbara). Recorded here also from Coari, state of Amazonas; Belém, Moju and Tailândia, state of Pará; and Centro Novo do Maranhão, state of Maranhão, Brazil.

Alpaida leucogramma (White, 1841)
Fig. 33

Epeira (Singa) leucogramma White, 1841: 474 (female holotype from Rio de Janeiro, Brazil, deposited in BMNH, not examined).
Alpaida leucogramma: Levi, 1988: 391, figs. 32-38; Dierkens, 2014: 17, figs. 10, 31.

Material examined. BRAZIL, *Pará*: Novo Progresso, 07°08'07"S 55°24'51"W, 1 male (MPEG 4478).

Distribution. Previously known from Panama to Argentina. Recorded here also from Novo Progresso, state of Pará, Brazil.

Alpaida murtinho Levi, 1988
Fig. 37

Alpaida murtinho Levi, 1988: 399, figs. 84-85 (male holotype from Vila Murtinho, Rondônia, Brazil, 03.IV.1922, ex MCZ, deposited in Museu de Zoologia da Universidade de São Paulo (MZSP), not examined).

Material examined. BRAZIL, *Pará*: Juruti (Beneficiamento), 02°30'08.8"S 56°09'48.87"W, 12 males (MPEG 24314).

Distribution. Previously known from Brazil (Rondônia). Recorded here also from Juruti, state of Pará, Brazil.

Alpaida negro Levi, 1988
Fig. 38

Alpaida negro Levi, 1988: 448, figs. 410-414 (female holotype from Rio Negro, Paraná, Brazil, deposited in MZSP, not examined).

Material Examined. BRAZIL, *Pará*: Belém (Icoaraci), 01°17'59.5" S 48°28'42.1"W, 1 male (MPEG 3348); Juruti (Barroso), 02°27'51.4"S 56°00'08.6"W, 1 female (MPEG 24485); (Capiranga), 02°28'22.1"S 56°12'29.4"W, 1 female (MPEG 24484).

Distribution. Previously known from Brazil (Mato Grosso and Paraná). Recorded here also from Belém and Juruti, state of Pará, Brazil.

Alpaida rossi Levi, 1988
Fig. 36

Alpaida rossi Levi, 1988: 447, figs. 404-407 (female holotype from Monzón Valley, Tingo María, Dpto. Huánuco, Peru, 10.XI.1954, deposited in California Academy of Sciences (CAS), not examined).

Material examined. BRAZIL, *Pará*: Juruti (Mutum), 02°33'18.0"S 56°13'22.4"W, 1 female (MPEG 24315); 02°33'13.8"S 56°13'22.1"W, 1 female (MPEG 24316).

Distribution. Previously known from Peru. Recorded here also from Juruti, state of Pará, Brazil.

Alpaida septemmammata (O. Pickard-Cambridge, 1889)
Fig. 34

Epeira septemmammata O. Pickard-Cambridge, 1889: 42, pl. 7, fig. 6 (fifteen females specimens from Teapa, Mexico, deposited in BMNH, type not located, both material not examined); Keyserling, 1892: 89. Pl. 4, fig. 67.

Alpaida septemmammata: Levi, 1988: 452, figs. 427-434; Dierkens, 2014: 21, figs. 15, 32.

Material examined. BRAZIL, *Pará*: Melgaço (Floresta Nacional de Caxiuanã, Estação Científica Ferreira Penna), 01°43'21.6"S 51°25'51.2"W, 1 male (MPEG 24482); 1 male (MPEG 24483).

Distribution. Previously known from Mexico to Argentina. Recorded here also from Melgaço, state of Pará, Brazil.

Alpaida simla Levi, 1988
Fig. 35

Alpaida simla Levi, 1988: 430, figs. 289-293 (female holotype, male and 6 immature paratypes from Simla, Trinidad, Lesser Antilles, IV.1964, deposited in MCZ, not examined); Bonaldo et al., 2009; Cafofo et al., 2013.

Material examined. BRAZIL, *Pará*: Belém (Reserva Mocambo), 01°26'48"S 48°25'1"W, 1 male (MPEG 24289); 1 male (MPEG 24290); Portel (Floresta Nacional de Caxiuanã, Plote PPBio), 01°57'38.9"S 51°36'45.3"W, 1 female (MPEG 13399); Tailândia (Fazenda Marupiara), 02°47'44.9"S 48°32'39.2"W, 1 female (MPEG 24288); Juruti (Capiranga), 02°28'0.6"S 56°12'42.2"W, 1 female (MPEG 24282); 02°28'22.1"S 56°12'29.4"W, 2 males (MPEG 8141); 1 female (MPEG 8188); 2 males (MPEG 8162); 1 male (MPEG 24265); 1 male (MPEG 24283); 1 female (MPEG 24286); (Barroso), 02°27'41.7"S 56°00'11.6"W, 1 male (MPEG 8148); 1 male (MPEG 8153); 1 male (MPEG 8180); 1 male (MPEG 8192); (Beneficiamento), 02°30'08.8"S 56°09'48.87"W, 2 females (MPEG 24270); 02°30'27.4"S 56°10'39.5"W, 1 female (MPEG 24275); (Mutum), 02°33'04.8"S 56°13'32.5"W, 1 female (MPEG 24267); 1 female (MPEG 24271); 1 female (MPEG 24276); 1 female (MPEG 24284); 1 male (MPEG 24280); 02°33'06.9"S 56°13'29.0"W, 1 male (MPEG 24277); 1 female (MPEG 24266); 1 male (MPEG 24269); 1 male (MPEG 24268); 1 male (MPEG 24272); 02°33'13.8"S 56°13'22.1"W, 1 male (MPEG 24285); 1 male (MPEG 24273); 02°33'18.0"S 56°13'22.4"W, 1 female (MPEG 24278); 1 female (MPEG 24281); 1 female (MPEG 24274); 02°36'44.7"S 56°11'39.2"W, 1 female (MPEG 24279); Marabá (Serra Norte), 06°4'22.10"S 50°14' 47.27"W, 2 males (MPEG 4196). *Maranhão*: Centro Novo do Maranhão (Reserva Biológica do Gurupi), 03°41'07.9"S 46°45' 46"W, 1 female (MPEG 24287). *Mato Grosso*: Sinop, 11°51'38.73"S 55°30'34.85"W, 1 male (MPEG 3350).

Distribution. Previously known from Trinidad & Tobago and Brazil (Pará: Melgaço and Portel). Recorded here also from Belém, Tailândia, Juruti and Marabá, Pará; Centro Novo do Maranhão, state of Maranhão; and Sinop, state of Mato Grosso, Brazil.

Alpaida tayos Levi, 1988
Fig. 36

Alpaida tayos Levi, 1988: 456, figs. 458-467 (female holotype from Los Tayos-Santiago, banana plantation, 03°04'S, 78°02'W, Prov. Morona-Santiago, Ecuador, 03.VIII.1976, deposited in MCZ, not examined).

Material examined. BRAZIL, *Pará*: Juruti (Capiranga), 02°28'0.6"S 56°12'42.2"W, 1 female (MPEG 24158); 02°28'22.1"S 56°12'29.4"W, 1 male (MPEG 24142); (Barroso), 02°27'51.4"S 56°00'08.6"W, 1 female (MPEG 24145); 1 male (MPEG 24161); 1 male (MPEG 24139); (Beneficiamento), 02°30'25.4"S 56°11'04.8"W, 1 male (MPEG 24140); 02°30'27.4"S 56°10'39.5"W 1 male (MPEG 24136); 1 female (MPEG 24132); (Mutum), 02°33'06.9"S 56°13'29.0"W, 1 female (MPEG 24133); 1 female (MPEG 24147); 1 male (MPEG 24134); 1 female (MPEG 24137); 1 female, (MPEG 24148); 1 female (MPEG 24135); 1 male (MPEG 24150); 1 female (MPEG 24146); 02°33'13.8"S56°13'22.1"W, 1 female (MPEG 24159); 1 male and 2 females (MPEG 24141); 02°33'18.0"S 56°13'22.4"W, 1 female (MPEG 24156); 02°36'11.2"S 56°12'36.3"W, 1 male (MPEG 24155); 02°36'44.7"S 56°11'39.2"W, 1 female (MPEG 9204); 1 female (MPEG 9134); 1 male (MPEG 163); 1 male (MPEG 24138); 1 female (MPEG 24143); 1 female (MPEG 24149); 1 female (MPEG 24151); 1 male (MPEG 24153); 2 females (MPEG 24152); 1 male (MPEG 24160); 02°36'45.2"S 56°11'27.5"W, 1 male (MPEG 24154); 1 female. (MPEG 24157); 02°36'45.7"S 56°11'38.2"W, 1 female (MPEG 24144); Marabá (Serra Norte), 05°57'48.56"S 50°24'1.7"W, 1 male (MPEG 4234); Novo Progresso, 07°09'53"S 55°18'53"W, 1 female (MPEG 4499); (Serra do Cachimbo), 09°21'59"S 55°02'01"W, 1 male (MPEG 6169). *Maranhão*: Centro Novo do Maranhão (Reserva Biológica do Gurupi), 03°41'07.92"S 46°45'46.08"W, 1 male (MPEG 24131).

Distribution. Guyana, Ecuador, Peru, French Guiana and Brazil (Pará: Ananindeua and Canindé [Paragominas]). Recorded here also from Juruti, Marabá, Novo Progresso, state of Pará; and Centro Novo do Maranhão, state of Maranhão, Brazil.

Alpaida urucuca Levi, 1988
Fig. 38

Alpaida urucuca Levi, 1988: 454, figs. 440-445 (female holotype from Fazenda Antonio, Uruçuca, Bahia, Brazil, 24.X.1979, deposited in MCN, not examined).

Material examined. BRAZIL, *Pará*: Melgaço (Floresta Nacional de Caxiuanã), 01°42'24"S 51°27'34.3"W, 1 female (MPEG 24325); 01°43'21.6"S 51°25'51.2"W, 1 female (MPEG 24326); 01°45'12.8"S 51°31'14.7"W, 1 female (MPEG 24323); 1 female (MPEG 24324); 01°46'36.00"S 51°35'12.21"W, 1 female (MPEG 24322); Moju (Campo experimental da Embrapa), 02°11'44.37"S 48°47'38.79"W, 1 female (MPEG 24327); Santarém (Alter-do-Chão), 02°26'33.18"S 54°43'8.70"W, 1 female (MPEG 16153); 1 female (MPEG 16154); 1 female (MPEG 16151); 02°32'59.02"S 54°54'05.20"W, 1 female (MPEG 16152); 1 female (MPEG 16408); 1 female (MPEG 16150); Altamira (Castelo dos Sonhos),

08°13′03″S 55°00′57″W, 1 female (MPEG 4488); Novo Progresso (Serra do Cachimbo), 09°16′18.6″S 54°56′22.9″W, 1 female (MPEG 6125); 09°22′02.9″S 55°01′ 11.9″W, 1 female (MPEG 6086).

Distribution. Previously known from Brazil (Pará: Melgaço; Bahia: Uruçuca). Recorded here also from Moju, Santarém, Altamira and Novo Progresso, state of Pará, Brazil.

Alpaida utiariti Levi, 1988
Fig. 34

Alpaida utiariti Levi, 1988: 466, figs. 523-524 (male holotype from Utiariti, Mato Grosso, Brazil, 30.VII.1961, deposited in MZSP, not examined).

Material examined. BRAZIL, *Pará*: Belém (MPEG, Campus de Pesquisa), 01°27′03.0″S 48°26′40.2″W, 1 male (MPEG 22467); 1 male (MPEG 24521); Benevides, [01°21′43.87″S 48°14′37.79″W], 1 male (MPEG 4665).

Distribution. Previously known from Brazil (Mato Grosso: Utiariti). Recorded here also from Benevides, state of Pará, Brazil.

Alpaida veniliae Levi, 1988
Fig. 33

Epeira veniliae Keyserling, 1865: 817, pl. 19, fig. 23 (seven females and one male syntypes from New Granada, deposited in BMNH, not examined); Keyserling, 1893: 256, pl. 13, fig. 191.

Epeira pantherina Taczanowski, 1872: 132. Male lectoype designated by LEVI (1988) from Uaça, Amapá, Brazil (PAN), not examined. Synonymyzed by LEVI (1988).

Alpaida veniliae: Levi, 1988: 402, figs. 103-109; Dierkens, 2014: 22, figs. 21, 22, 35, 48.

Material examined. BRAZIL, *Amapá*: Oiapoque, 03°4′25.63″S 51°51′8.18″W, 2 males and 5 females (MPEG 5000). *Pará*: Juruti (Área de Várzea), 02°12′36.1″S 56°07′20.7″W, 1 male and 1 female (MPEG 24308); 1 female (MPEG 24297); 1 male (MPEG 24305) 1 female (MPEG 24309); 1 female (MPEG 24301); 1 female (MPEG 24311); 2 females (MPEG 24299); 2 males (MPEG 24304); 1 male (MPEG 24303); 1 female (MPEG 24312); 02°24′33.2″S 56°26′10.6″W, 3 females (MPEG 24291); 2 males and 1 female (MPEG 24293); 4 females (MPEG 24294); 2 females (MPEG 24295); 2 females (MPEG 24298); 1 male and 1 female (MPEG 24302); 1 female (MPEG 24306); 1 female (MPEG 24307); 2 females (MPEG 24310); 6 males and 11 females (MPEG 24292); 3 males and 4 females (MPEG 24296).

Distribution. Previously known from Panama to Argentina. Recorded here also from Oiapoque, state of Amapá; and Juruti, state of Pará, Brazil.

DISCUSSION

Alpaida levii new species shares somatic and genitalic characters with *A. delicata*: abdomen hump-shaped, cymbium prolaterally expanded and strong spines on tibiae I and II, sug-gesting that *A. levii* and *A. delicata* may be sister species. Due to these similarities, records of *A. delicata* in the faunistic literature may not be accurate, as observed for at least some of the specimens of *A. levii* examined by us, which were previously determined as *A. delicata*.

LEVI (1988) compared *A. delicata* with *A. truncata* for diagnostic purposes, based on the hump-shaped abdomen, which is shared by both species. However, considering the new information provided here, we propose a new diagnosis for *A. delicata*, which we compare with *A. levii*. The median apophysis of *A. truncata* is extremely long, with a distally pointed tip (Fig. 16), more similar to that of by *A. queremal* Levi, 1988, while the median apophysis of *A. delicata* is short, medially excavated and quadrangular, similar to *A. levii*. *Alpaida delicata* and *A. levii* also share strong spines on tibiae I and II. The terminal apophysis of *Alpaida* species can present a distal lobe, a basal prong and a retrolateral apical sector, recognized here for the first time. The retrolateral apical sector can be very developed in some species, such as *A. delicata* (A, Fig. 12) and *A. levii* (A, Figs. 3-5) or reduced, as in *A. truncata* (A, Fig. 17). The distal lobe and the basal prong of the terminal apophysis are absent in some *Alpaida* species and the distribution of those characters may be important in a phylogenetic context.

Due to the high complexity of the palps of *Alpaida*, especially with regards to the terminal apophysis, and giving LEVI's (1988) choice to document the palps only in mesal view, it is difficult to identify all sclerites of all species and the identification of some species may be uncertain, especially when there is intraespecific variation. For this reason we document the expanded palp and details of some sclerites for the new species and for the males of *A. iquitos* and *A. truncata*. This refined information will facilitate the recognition of these taxa and an eventual phylogenetic analysis to clarify the relationship among species of *Alpaida*.

ACKNOWLEDGEMENTS

The authors would like to thank the reviewers and editor for their comments that help improve the manuscript. The authors are supported by grants: ABB – PQ grant #304965/2012-0, BVBR# 302358/2013-7 – CNPq, RS #3362 – CELPA/FADESP Monitoramento (REF.: 061/2013). Project 3362 is supported by Centrais Elétricas do Pará S.A. – CELPA and involves monitoring studies in the Marajó Archipelago, entitled "Monitoramento dos possíveis impactos da linha de transmissão do Marajó sobre a fauna".

REFERENCES

ABRAHIM N, BONALDO AB (2008) A new species of *Alpaida* (Araneae, Araneidae) from Caxiuanã National Forest, Oriental Amazonia, Brazil. **Iheringia, Série Zoologia, 98**(3): 397-399. doi: 10.1590/S0073-47212008000300015

BONALDO AB, CARVALHO LS, PINTO-DA-ROCHA R, TOURINHO A, MIGLIO LT, CANDIANNI DF, LO-MAN-HUNG NF, ABRAHIM N, RODRIGUES BVB, BRESCOVIT AD, SATURNINO R, BASTOS NC, DIAS SC, SILVA BJF, PEREIRA-FILHO JMB, RHEIMS CA, LUCAS SM, POLOTOW D, RUIZ G, INDICATTI R (2009) Inventário e história natural dos aracnídeos da Floresta Nacional de Caxiuanã, p. 577-621. In: LISBOA PLB (Org.). **Caxiuanã: desafios para a conservação de uma Floresta Nacional na Amazônia.** Belém, Museu Paraense Emílio Goeldi.

BONNET P (1955) Bibliographia araneorum. **Toulouse** 2(1): 1-918.

BRAGA-PEREIRA GF, SANTOS AJ (2013) Two new species of the spider genus *Alpaida* (Araneae: Araneidae) from restinga areas in Brazil. **Revista Brasileira de Zoologia** 30(3): 324-328. doi: 10.1590/S1984-46702013000300010

BUCKUP EH, MEYER AC (1993) Sobre o macho de *Alpaida scriba* (Araneae, Araneidae). **Revista Brasileira de Entomologia** 37(2): 353-354.

BUCKUP EH, RODRIGUES ENL (2011) Espécies novas de *Alpaida* (Araneae: Araneidae), descrições complementares e nota taxonômica. **Iheringia, Série Zoologia,** 101(3): 262-267. doi: 10.1590/S0073-47212011000200013

CAFOFO EG, SATURNINO R, SANTOS AJ, BONALDO AB (2013) Riqueza e composição em espécies de aranhas da Floresta Nacional de Caxiuanã, p. 539-562. In: PLB LISBOA (Ed.). **Caxiuanã: Paraíso ainda preservado.** Belém, Museu Paraense Emílio Goeldi.

DIERKENS M (2014) Contribution à l'étude des Araneidae de Guyane française. V – Les genres *Alpaida* et *Ocrepeira*. **Bulletin Mensuel de la Societe Linneenne de Lyon** 83(1-2): 14-30.

HÖFER H, BRESCOVIT AD (2001) Species and guild structure of a Neotropical spider assemblage (Araneae; Reserva Ducke, Amazonas, Brazil). **Andrias** 15: 99-120.

KEYSERLING E (1865) Die Beiträge zur Kenntniss der Orbitelae Latr. **Verhandlungen der Kaiserlich-Königlichen Zoologisch-Botanischen Gesellschaft in Wien** 15: 799-856.

KEYSERLING E (1892) Die Spinnen Amerikas, Epeiridae. **Nürnberg** 4: 1-208.

KEYSERLING E (1893) Die Spinnen Amerikas. Epeiridae. **Nürnberg** 4: 209-377.

LEVI HW (1988) The neotropical orb-weaving spiders of the genus *Alpaida* (Araneae: Araneidae). **Bulletin of the Museum of Comparative Zoology** 151: 365-487.

LEVI HW (2002) Keys to the genera of araneid orbweavers (Araneae, Araneidae) of the Americas. **Journal of Arachnology** 30(3): 527-562.

PICKARD-CAMBRIDGE O (1889) Arachnida. Araneida. **Biologia Centrali-Americana, Zoology** 1: 1-56.

QGIS DEVELOPMENT TEAM (2012) **Quantum GIS Geographic Information System.** Open Source Geospatial Foundation Project, v. 1.8.0. Available online at: http://qgis.osgeo.org [Accessed: 28/11/2015]

RODRIGUES ENL, MENDONÇA JR MS (2011) Araneid orb-weavers (Araneae, Araneidae) associated with riparian forests in southern Brazil: a new species, complementary descriptions and new records. **Zootaxa** 2759: 60-68.

SIMON E (1897) Etudes arachnologiques. 27e Mémoire. XLII. Descriptions d'espèces nouvelles de l'ordre des Araneae. **Annales de la Société Entomologique de France** 65: 465-510.

TACZANOWSKI L (1872) Les aranéides de la Guyane française. **Horae Societatis Entomologicae Rossicae** 8: 32-132.

TACZANOWSKI L (1873) Les aranéides de la Guyane française. **Horae Societatis Entomologicae Rossicae** 9: 64-150.

TACZANOWSKI L (1878) Les Aranéides du Pérou central. **Horae Societatis Entomologicae Rossicae** 14: 140-175.

WHITE A (1841) Description of new or little known Arachnida. **Annals and Magazine of Natural History** 7: 471-477.

WORLD SPIDER CATALOG (2014) **World Spider Catalog.** Bern, Natural History Museum Bern, version 15.5. Available online at: http://wsc.nmbe.ch [Accessed: 7/10/2014].

4

Effect of humic acid on survival, ionoregulation and hematology of the silver catfish, *Rhamdia quelen* (Siluriformes: Heptapteridae), exposed to different pHs

Silvio T. da Costa[1,*], Luciane T. Gressler[2], Fernando J. Sutili[2], Luíza Loebens[1], Rafael Lazzari[1] & Bernardo Baldisserotto[3]

[1]*Departamento de Zootecnia e Ciências Biológicas, Centro de Educação Norte do Rio Grande do Sul, Universidade Federal de Santa Maria. 98300-000 Palmeira das Missões, RS, Brazil.*
[2]*Programa de Pós-graduação em Farmacologia, Universidade Federal de Santa Maria. 97105-900 Santa Maria, RS, Brazil.*
[3]*Departamento de Fisiologia e Farmacologia, Universidade Federal de Santa Maria. 97105-900 Santa Maria, RS, Brazil.*
[*]*Corresponding author. E-mail: silvio.teixeira.da.costa@gmail.com*

ABSTRACT. This study evaluates whether humic acid (HA; Aldrich) protects the silver catfish, *Rhamdia quelen* (Quoy & Gaimard, 1824), against exposure to acidic pH. Survival, levels of Na^+, Cl^- and K^+ plasma, hematocrit, hemoglobin and erythrocyte morphometry were measured. Fish were exposed to 0, 10, 25 and 50 mg L^{-1} HA at four pH levels: 3.8, 4.0, 4.2 and 7.0 up to 96 hours. None of the fish exposed to pH 3.8 survived for 96 hours into the experiment, and survival of fish subjected to pH 4.0 decreased when HA concentration increased. Plasma Na^+ levels decreased when pH was acidic, with no influence of HA, while Cl^- levels declined at low pH with increased HA concentration. The levels of K^+ at pH 4.0 and 4.2 increased without HA. Hematocrit and hemoglobin augmented under the effect of HA. At pH 4.0 and 4.2, erythrocytes of fish not exposed to HA were smaller, an effect that was partially offset by the presence of HA, since the values at pH 7.0 were higher. Although HA showed some positive effects changes in hematological and plasma K^{+a} in silver catfish caused by exposure to acidic pH, the overall findings suggest that HA does not protect this species against acidic pH because it increased mortality and Cl^- loss at pH 4.0.

KEY WORDS. Blood parameters; humic acid; plasma ion levels; survival.

Dissolved organic matter, an integral part of all ecosystems, results from the decay of plant and animal debris (Thurman 1985). It comprises humic, fulvic, and other organic acids, and is usually quantified as dissolved organic carbon (DOC) (Wood et al. 2011). DOC is known to positively regulate several biotic/abiotic processes (Steinberg et al. 2007, Wood et al. 2011). In blackwaters, such as those found in forest streams in the Amazon, coastal lagoons in southeastern Brazil, Finnish and Swedish lakes, and Canadian wetlands, DOC may range from 10 to 300 mg CL^{-1}, while its average content in freshwater systems elsewhere is 0.5-4.0 mg CL^{-1} (Thurman 1985, Küchler et al. 2000, Farjalla et al. 2009). The high levels of DOC account for the acidity of the aquatic environment. In order to thrive in acidic environments, organisms need a certain degree of specialization in their osmoregulatory organs (Matsuo & Val 2007).

Low pH (pH 4-5) induces ion loss (Zaions & Baldisserotto 2000, Wood et al. 1998, 2002, 2003, Gonzalez et al. 1998, 2002, Bolner & Baldisserotto 2007, Matsuo & Val 2007, Duarte et al. 2013) and the interference with gill ionoregulatory mechanisms may also trigger hematological disturbances. Ionic dilution,

potentiated by the plasma acidosis prompted by H^+ entry, affects body fluid distribution. This could promote reduction in plasma volume, swelling of erythrocytes or splenic contraction, resulting in elevation of the hematocrit (Milligan & Wood 1982). Despite being highly responsible for the acidic nature of blackwaters, there is evidence that DOC protects native fish from the deleterious effects of low pH, reducing ion loss (Wood et al. 1998, 2002, 2003, 2011, Gonzalez et al. 1998, 2002, Matsuo & Val 2007). According to some studies, the occurrence of various charged functional groups in the heterogeneous compounds of DOC may change fundamental properties of the gill epithelium, such as the transepithelial potential, thus altering membrane permeability and stimulating ion uptake (Wood et al. 2011). Humic acid also reduces respiratory stress in fish exposed to slightly acidic water, but increases it at more acidic waters (Holland et al. 2014) and decreases lipid peroxidation and modulates the antioxidant system (Riffel et al. 2014). In opposition to the several findings regarding the effects of DOC on ionoregulatory disturbances, respiratory stress and antioxidant system, no evidence has been documented about its influence on the hematology of fish subjected to acidic pH.

This study evaluated whether humic acid (HA), one of the major components of DOC, would offer the silver catfish, *Rhamdia quelen* (Quoy & Gaimard, 1824), protection against the physiological disturbances induced by low pH. This species does not naturally inhabit DOC-enriched, acidic waters, so it is not adapted to such conditions. However, different water quality parameters, including pH and DOC, are present in southern Brazil, where this species is widely cultivated. The outcome of the interaction between such variables should be investigated to improve the rearing conditions of this fish. In laboratory settings, silver catfish juveniles survive for at least 96 hours in the pH 4-9 range (Zaions & Baldisserotto 2000), but exposure to pH 5.0 is enough to reduce growth in this species (Copatti et al. 2005). Therefore, if humic acid has a protective effect on silver catfish exposed to acidic waters, it could reduce the deleterious effect of low pH and improve growth in this species.

MATERIAL AND METHODS

Juvenile silver catfish (n = 240, 73.43 ± 3.5 g, 20.32 ± 1.22 cm, voucher number 19612, Ichthyology Laboratory, Universidade Federal do Rio Grande do Sul) were acquired from a commercial fishery in Santa Maria, southern Brazil, and acclimated in the Laboratório de Fisiologia de Peixes, Universidade Federal de Santa Maria (UFSM) for three weeks. The fish were equally distributed in 8 tanks of 250 L and kept in dechlorinated tap water under constant aeration (22.14 ± 1.5°C, 6.05 ± 0.45 mg L^{-1} dissolved oxygen (DO), pH 7.45 ± 0.13 and hardness 24.7 ± 3.9 mg CaCO$_3$ L^{-1}). The water was totally renewed every second day and siphoning was performed daily two hours after feeding. The fish were fed commercial food for juveniles with 42% crude protein once a day.

Lyophilized HA (CAT: 0.675-2 Aldrich® H1 – HA sodium salt) was the source of DOC used in the tests. It was dissolved in water (the same water used in the acclimation tanks) and agitated for 12 hours in a magnetic stirrer to prepare the stock solution. It was not possible to measure the concentration of DOC in the experimental solutions, but estimation of DOC concentration was made based on the fact that the commercial HA corresponded to ~40% DOC (McGeer et al. 2002). HA was tested at 0 (control), 10, 25 and 50 mg L^{-1} HA, the latter corresponding to the nominal DOC concentration of 20 mg C L^{-1}. These concentrations were chosen because they are within the range observed in the water of the rio Negro Basin (Küchler et al. 2000). At each concentration of HA, four pH ranges were tested, with the following minimum values: 3.8, 4.0, 4.2 and 7.0 (Table 1). The acidic pH tested in the present study were near the most acidic pH (pH 4.0) that allows 100% survival in silver catfish (Zaions & Baldisserotto 2000). A pH meter DMPH-2 (Digimed, São Paulo, Brazil) was used to measure the variable four times a day and adjustments to the minimum values within each range were made with sulfuric acid 1 M when necessary. The water in the experimental aquaria was not renewed during exposure time.

Table 1. Ion levels in water at different pH and humic acid (HA) levels for *Rhamdia quelen*.

HA (mg L^{-1})	pH	Na$^+$ (mg L^{-1})	Cl$^-$ (mg L^{-1})	K$^+$ (mg L^{-1})
0	3.84 ± 0.5	3.1 ± 0.3	5.9 ± 1.0	0.04 ± 0.01
	4.08 ± 0.4	3.2 ± 0.7	6.1 ± 0.9	0.04 ± 0.01
	4.25 ± 0.4	3.4 ± 0.8	6.0 ± 0.9	0.03 ± 0.02
	7.02 ± 0.3	3.3 ± 0.4	6.0 ± 0.8	0.04 ± 0.02
10	3.87 ± 0.5	3.7 ± 0.3	5.6 ± 1.3	0.03 ± 0.01
	4.09 ± 0.6	3.7 ± 0.3	5.7 ± 1.4	0.03 ± 0.01
	4.22 ± 0.3	3.6 ± 0.2	6.8 ± 1.6	0.04 ± 0.02
	7.03 ± 0.4	4.3 ± 0.4	6.0 ± 0.7	0.05 ± 0.02
25	3.83 ± 0.4	3.8 ± 0.6	6.1 ± 0.6	0.04 ± 0.02
	4.05 ± 0.5	4.3 ± 1.1	5.8 ± 0.8	0.04 ± 0.01
	4.27 ± 0.6	3.9 ± 0.5	6.3 ± 1.0	0.05 ± 0.02
	7.05 ± 0.5	3.9 ± 0.2	6.2 ± 1.2	0.03 ± 0.02
50	3.81 ± 0.2	4.9 ± 0.4	6.1 ± 1.6	0.04 ± 0.02
	4.02 ± 0.2	4.7 ± 0.5	6.4 ± 0.8	0.03 ± 0.01
	4.21 ± 0.1	4.8 ± 0.9	6.5 ± 1.3	0.05 ± 0.02
	7.02 ± 0.3	5.5 ± 1.3	6.0 ± 1.6	0.05 ± 0.02

Mean values ± SE (n = 4/group for ions). There was no significant difference between treatments.

Juveniles were fasted for 24 hours prior to being transfered to 40 L aquaria (16 treatments, three replicates of each treatment, five fish per replicate) for the 96-h experiment. Survival was observed four times a day and the dead fish were removed from the aquaria. Fish that survived up to the end of the experimental period were anesthetized with eugenol 50 mg L^{-1} (Cunha et al. 2010) and their blood was rapidly collected from the caudal vein with heparinized syringes. After sampling, fish were killed by sectioning the spinal cord. All procedures were conducted with the approval of the Ethics Committee on Animal Experimentation of the UFSM (registration #128/2010).

DO levels and temperature were measured daily with Orion 810 oxygen meter (Thermo Electron Corporation, Waltham, Al, USA). Water samples were collected every second day to verify total ammonia (Verdouw et al. 1978), un-ionized ammonia, hardness (Eaton et al. 2005), nitrite (Boyd & Tucker 1992), Cl$^-$ (Zall et al. 1956), and Na$^+$ and K$^+$ levels, which were measured in a flame photometer (Micronal B262, São Paulo, Brazil). Details on the composition of the water are provided in Tables 1 and 2. There were no significant differences in water quality parameters between treatments.

To obtain the hematocrit, microcapillary tubes were filled with blood immediately after euthanasia and centrifuged at 10000 Xg for 5 minutes, and the results were obtained using a hematocrit card reader. The concentration of hemoglobin was determined by the cyanmethemoglobin method using a spectrophotometer (Brown 1976). For the morphometric analyses, blood smears were prepared immediately from the whole blood, air-dried, fixed in methanol and stained with May-Grünwald (Tavares-Dias et al. 2004). The surface area and the major and minor axes of the erythrocyte as well as of its nucleus were determined (Dorafshan et al. 2008). Briefly, ten high-power fields were randomly selected on each blood smear, and morphometry of ten erythrocytes were determined in each of these fields. All analyses were performed using the Zeiss Axio Vision System with Remote Capture 4.7 Rel DC – Cannon Power shot G9.

Blood samples were spun at 3000 Xg for 10 minutes and plasma was stored at -25°C until analyses of Na^+, Cl^- and K^+. The ion levels in the plasma were determined as previously described for the water ion levels.

Homogeneity of variances was assessed via Levene test and the comparison between treatments was carried out by two-way ANOVA and Tukey test. The Kruskal-Wallis test, followed by multiple comparisons of mean ranks, was used for analyses of plasma ion levels (Statistica 7.0 software). Minimum level of significance was 95% ($p < 0.05$). Data are presented as mean ± standard error (SE).

RESULTS

Survival

None of the fish exposed to pH 3.8 survived the 96 hours of experiment. At pH 4.0 there was a progressive decrease in survival (100, 86, 60 and 40%) with increased HA level (0, 10, 25 and 50 mg L^{-1} HA, respectively). The survival rates at pH 4.2 (93.33%) and 7.0 (100%) were not affected by HA concentration. Survival at pH 4.0 was lower than at pH 4.2 and 7.0 at all treatments with the presence of HA (i.e. 10, 25 and 50 mg L^{-1} of HA) (Fig. 1).

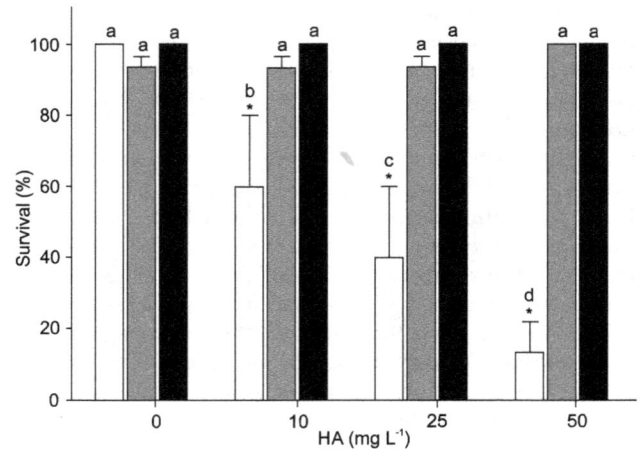

Figure 1. Effect of humic acid (HA) and pH on survival of silver catfish (*Rhamdia quelen*). Different letters indicate significant difference between HA concentrations at the same pH. * indicate significant difference from pH 7.0 at the same HA concentration ($p < 0.05$). Mean values ± SE (n = 6-15/group). (□) pH 4.0, (■) pH 4.2, (■) pH 7.0.

Hematocrit and hemoglobin

Overall, HA triggered an increase in the percentages of hematocrit and hemoglobin. The presence of HA promoted an increase in the percentage hematocrit at pH 4.0 and 4.2. Upon exposure to pH 7.0, fish experienced a gradual increase in hematocrit from 0 to 25 mg L^{-1} HA. Hematocrit declined with the

Table 2. Water quality parameters at different pH and humic acid (HA) levels for *Rhamdia quelen*.

HA (mg L^{-1})	pH	Total ammonia (mg L^{-1})	Un-ionized ammonia (mg L^{-1})	Nitrite (mg L^{-1})	Hardness (mg $CaCO_3$ L^{-1})	Temperature (°C)	Dissolved oxygen (mg L^{-1})
0	3.8	0.85 ± 0.0015	0.0342 ± 0.0009	0.3181 ± 0.012	25.8 ± 4.2	21.4 ± 2.1	6.25 ± 0.61
	4.0	0.44 ± 0.0010	0.0251 ± 0.0003	0.3492 ± 0.025	25.7 ± 4.1	21.2 ± 2.2	6.21 ± 0.64
	4.2	0.16 ± 0.0004	0.0270 ± 0.0004	0.3758 ± 0.032	27.3 ± 6.1	21.2 ± 2.6	6.10 ± 0.45
	7.0	0.32 ± 0.0010	0.0279 ± 0.0005	0.3625 ± 0.041	26.1 ± 5.8	21.3 ± 2.6	6.08 ± 0.54
10	3.8	0.76 ± 0.01	0.0027 ± 0.0003	0.4011 ± 0.013	26.8 ± 5.6	20.1 ± 2.0	6.12 ± 0.61
	4.0	0.92 ± 0.021	0.0028 ± 0.0002	0.3442 ± 0.022	25.1 ± 4.8	20.2 ± 2.1	6.14 ± 0.48
	4.2	0.83 ± 0.014	0.0027 ± 0.0004	0.2034 ± 0.024	27.6 ± 5.1	20 ± 2.1	6.18 ± 0.45
	7.0	1.09 ± 0.032	0.1820 ± 0.0019	0.3285 ± 0.045	26.3 ± 4.9	20.4 ± 2.0	6.17 ± 0.48
25	3.8	0.74 ± 0.012	0.015 ± 0.0003	0.4101 ± 0.023	26.1 ± 5.1	21.4 ± 1.9	6.26 ± 0.62
	4.0	0.87 ± 0.017	0.027 ± 0.0003	0.4119 ± 0.025	25.9 ± 5.2	22 ± 1.3	6.11 ± 0.53
	4.2	1.22 ± 0.024*	0.215 ± 0.0150	0.3289 ± 0.034	26.4 ± 5.4	21.5 ± 2.4	6.15 ± 0.60
	7.0	0.81 ± 0.014	0.021 ± 0.0003	0.3032 ± 0.054	26.8 ± 5.1	21.9 ± 2.6	6.14 ± 0.70
50	3.8	0.64 ± 0.011	0.019 ± 0.0006	0.4516 ± 0.024	26.4 ± 5.0	21.1 ± 2.3	6.21 ± 0.65
	4.0	0.66 ± 0.012	0.019 ± 0.0005	0.3442 ± 0.039	27.1 ± 4.9	21.1 ± 2.1	6.24 ± 0.39
	4.2	0.75 ± 0.015	0.021 ± 0.0007	0.2034 ± 0.054	26.1 ± 5.1	21.0 ± 2.0	6.18 ± 0.53
	7.0	0.81 ± 0.014	0.022 ± 0.0009	0.3285 ± 0.061	26.2 ± 5.2	20.9 ± 2.0	6.19 ± 0.55

*Significantly different from pH 4.2 and HA 0 mg L^{-1} ($p < 0.05$). Mean values ± SE (n = 4/group).

increase in pH at 0 mg L⁻¹ HA. Exposure to 10 mg L⁻¹ HA induced significantly higher hematocrit percentage at pH 4.0 and 4.2, while at 25 mg L⁻¹ HA the pH had negligible influence on hematocrit. Treatment with 50 mg L⁻¹ HA caused a significantly reduction in hematocrit concentration at pH 7.0 (Fig. 2).

There was no difference in the percentage of hemoglobin between HA treatments at pH 4.2. Exposure to pH 4.0 induced a higher hemoglobin level at 25 mg L⁻¹ HA than at 0 mg L⁻¹ HA. When fish were exposed to pH 7.0 the hemoglobin was lower at 50 than at 10 mg L⁻¹ HA, and at 0 mg L⁻¹ HA was also lower than at 25 mg L⁻¹ HA. In the absence of HA the levels of hemoglobin decreased as the pH increased. Exposure to 10 mg L⁻¹ HA did not induce differences in hemoglobin values between the different pH. Hemoglobin levels at 25 mg L⁻¹ HA were higher at pH 4.0 than at pH 4.2 and 7.0. On exposure to 50 mg L⁻¹ HA the hemoglobin levels were higher at pH 4.0 and 4.2 than at pH 7.0 (Fig. 3).

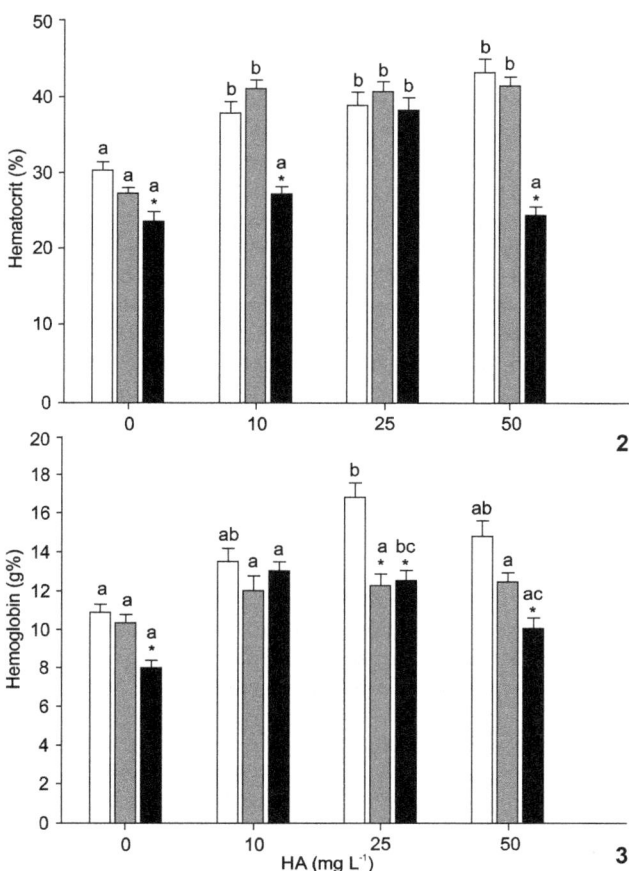

Figures 2-3. Effect of humic acid (HA) and pH on hematocrit (2) and hemoglobin (3) of silver catfish (*Rhamdia quelen*). Different letters indicate significant difference between HA concentrations at the same pH. * indicate significant difference from pH 7.0 at the same HA concentration (p < 0.05). Mean values ± SE (n = 6-15/group). (□) pH 4.0, (■) pH 4.2, (■) pH 7.0.

Erythrocyte morphometry

Fish subjected to pH 4.0 showed greater cell area and cell minor and major axes in the presence of HA than in the absence of it. At pH 4.2, cell area was larger at 10 and 50 mg L⁻¹ HA than at 25 mg L⁻¹ HA, and it decreased further when HA was not present. At the same pH (4.2), cell minor axis was bigger at 50 than at 0 and 25 mg L⁻¹ HA; it was also bigger at 10 mg L⁻¹ HA compared to 0 mg L⁻¹ HA. At 0 mg L⁻¹ HA, cell area and its minor and major axes were bigger at pH 7.0 than at pH 4.0. The group exposed to 25 mg L⁻¹ HA presented greater cell area at pH 7.0 than at all the acidic pH and greater cell minor axis at pH 7.0 than at pH 4.2. Fish treated with 50 mg L⁻¹ HA had bigger cell minor axis at pH 4.2 than at pH 4.0, and bigger cell major axis at pH 7.0 comparing with pH 4.0 (Table 3).

Plasma Na⁺, Cl⁻ and K⁺

In Na⁺ levels, no significant differences were observed in fish exposed to the different HA treatments at pH 4.0 and 7.0, but at pH 4.2, exposure to 25 mg L⁻¹ HA increased Na⁺ levels compared to the group non exposed to HA. Silver catfish exposed to pH 4.0 and 4.2 without HA presented significantly lower Na⁺ levels than those at pH 7.0 without HA, but at 10 mg L⁻¹ HA plasma Na⁺ in fish exposed to pH 4.2 were not significantly different from pH 7.0. Fish at 25 mg L⁻¹ HA and pH 4.2 presented significantly higher Na⁺ levels than at pH 7.0 (Fig. 4).

The levels of Cl⁻ at pH 4.0 were significantly greater in fish exposed to 10 mg L⁻¹ HA, while at pH 4.2 and 7.0 HA did not affect significantly plasma Cl⁻. Fish subjected to 10 mg L⁻¹ had higher Cl⁻ levels at pH 7.0 and 4.0 than at pH 4.2. Plasma Cl⁻ levels were significantly lower at 25 mg L⁻¹ HA and pH 4.0 and at 50 mg L⁻¹ and pH 4.0 and 4.2 than at pH 7.0 and the same HA levels (Fig. 5).

K⁺ levels were not affect by HA treatments at pH 7.0. However, significantly higher K⁺ levels were observed at 0 mg L⁻¹ HA than at 10 mg L⁻¹ HA and pH 4.0, and at 25 and 50 mg L⁻¹ HA and pH 4.2. The levels of K⁺ were significantly higher at pH 4.0 and 4.2 than at pH 7.0 in fish kept in water without HA (Fig. 6).

DISCUSSION

All water parameters analyzed were within the limits that permit normal growth and survival of silver catfish (e.g. nitrite and un-ionized ammonia levels below 1.2 mg L⁻¹ and 0.1 mg L⁻¹ respectively) (LIMA et al. 2011, MIRON et al. 2011).

According to ZAIONS & BALDISSEROTTO (2000), even though silver catfish presents a marked loss of Na⁺ at pH 4.0, this is the acidic pH threshold for the species survival, at least for 96 h. In the present assessment this assertion was confirmed by the 0% survival of fish exposed to pH 3.8 regardless the HA concentrations. As stated by WOOD & MCDONALD (1982), nonacidophilic species suffocate at pH levels below 4.0 due to gill structural damage, edema and mucification. Moreover, fish mortality in acid waters is largely associated with a failure to

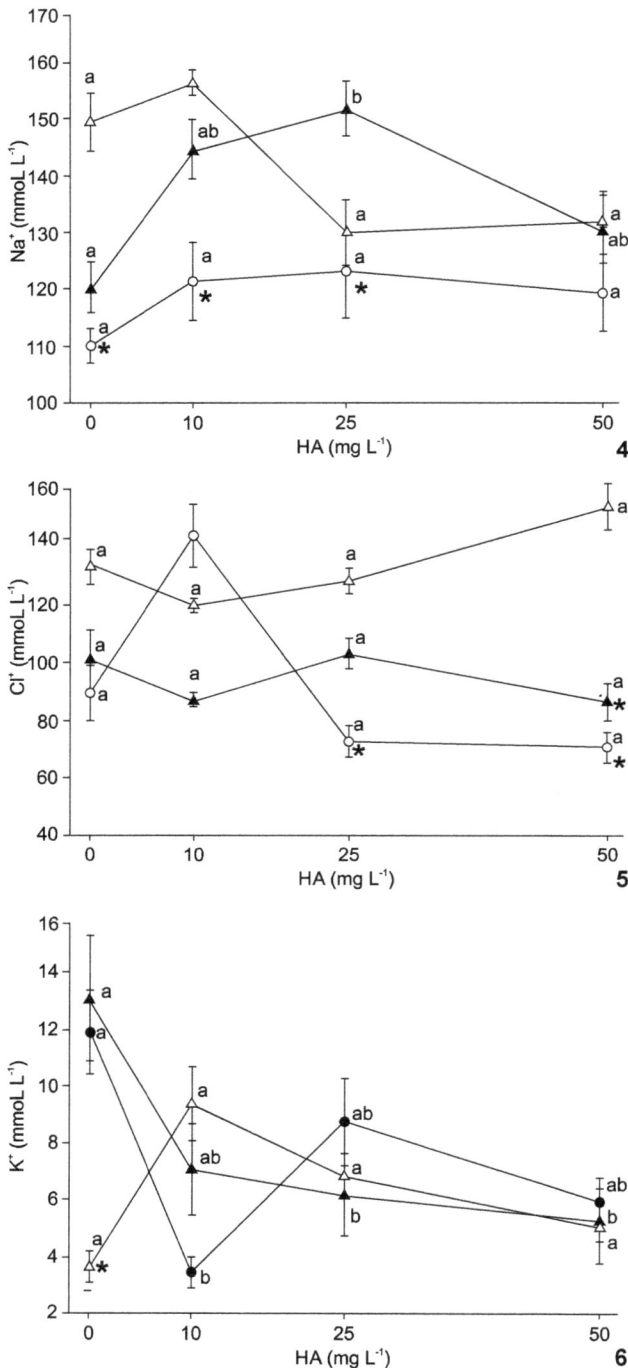

Figures 4-6. Effect of humic acid (HA) and pH on plasma Na+ (4), Cl- (5) and K+ (6) of silver catfish (*Rhamdia quelen*). Different letters indicate significant difference between HA concentrations at the same pH. * indicate significant difference from pH 7.0 at the same HA concentration (p < 0.05). Mean values ± SE (n = 6-15/group). (○) pH 4.0, (▲) pH 4.2, (△) pH 7.0.

ionoregulate, especially due to stimulation of Na+ efflux (MILLIGAN & WOOD 1982). A study on shiners *Notropis cornutus* (Mitchill, 1817), rainbow trout *Oncorhynchus mykiss* (Walbaum, 1792) and perch *Perca flavescens* (Mitchill, 1814) clearly proved that principle by showing that a great amount of Na+ and Cl- (50-60%) had been lost at death after exposure to pH 4.0 (FREDA & MCDONALD 1988). Similarly, HOLLAND et al. (2014) reported increased morbidity of the eastern rainbow fish *Melanotaenia splendida splendida* (Peters, 1866) as the pH dropped to 3.5-4.0 in the presence of commercial HA, despite having observed a protective effect of the substance at higher acidic levels. The authors suggested that HA may have enhanced the toxicity of low pH by increasing ion loss in the fish.

At pH 4.0, HA displayed a deleterious effect on the physiology of silver catfish; increased concentration of HA was associated with a decline in fish survival. COSTA et al. (pers. comm.) observed that the presence of HA induces proliferation of chloride cells in the lamellae of the gill of silver catfish. Gas transfer in pavement cells might be impaired when chloride cells inundate the lamellae, since thickening of the lamellar epithelium increases blood-to-water diffusion distance (GRECO et al. 1996). BINDON et al. (1994) previously reported a reduction in the lamellar epithelium as a consequence of chloride cell proliferation. Thus, the silver catfish may have been unable to cope with the combination of limited gas exchange, due to increased concentrations of HA, and ionic loss, the result of extreme pH. At higher pH levels, however, the detrimental effect of HA was not observed, most likely because at low pH the excess positive charge titrates away the negatively charged groups associated with the extracellular surface of epithelial membranes (CAMPBELL et al. 1997). This reduces the electrochemical repulsion between the membrane and the negatively charged HA, allowing the two parameters to associate and induce an effect. However, as pH increases the decrease in positive charge means that the HA is now less electrochemically favored to associate with the membrane and thus the effect is diminished.

As stated by ARIDE et al. (2007), acid stress triggers various changes in hematological parameters of freshwater fish. When it takes place, the generated osmotic and ionic gradient favors the entry of water into intracellular space and electrolyte flux in the opposite direction. With that, blood volume decreases, erythrocyte physiology changes, hematocrit, hemoglobin and plasma protein levels rise, and ion loss through the gills is further enhanced (WOOD et al. 1998, ARIDE et al. 2007). Water acidification is also associated with blood acidosis. It affects the oxygenation capacity of the hemoglobin and PO_2 is thus reduced, which in turn triggers an increase in hematocrit and hemoglobin in order to restore proper homeostatic control (MCDONALD & WOOD 1981, DHEER et al. 1987). MILLIGAN & WOOD (1982) found increases in both hematocrit and hemoglobin in rainbow trout during acid exposure. The authors stated that hematocrit elevation probably resulted from a re-

Table 3. Effect of humic acid and pH on erythrocyte morphology of *Rhamdia quelen*.

	Humic acid (mg L⁻¹)			
	0	10	25	50
pH 4.0				
Cell area (µm²)	98.73 ± 15.9[a*]	146.82 ± 20.3[b]	135.92 ± 13.3[b*]	136.30 ± 6.48[b*]
Cell minor axis (µm)	9.55 ± 1.00[a*]	11.75 ± 0.75[b]	11.26 ± 0.70[b]	11.32 ± 0.31[b]
Cell major axis (µm)	13.32 ± 0.92[a*]	16.13 ± 1.11[b]	15.67 ± 0.73[b]	15.60 ± 0.51[b*]
Nucleus area (µm²)	18.34 ± 3.62[a]	20.62 ± 4.37[a]	21.57 ± 2.09[a]	20.41 ± 3.84[a]
Nucleus minor axis (µm)	4.16 ± 0.47[a]	4.42 ± 0.29[a]	4.58 ± 0.19[a]	5.77 ± 2.81[a]
Nucleus major axis (µm)	5.79 ± 0.51[a]	6.12 ± 0.80[a]	6.21 ± 0.37[a]	7.70 ± 3.66[a]
pH 4.2				
Cell area (µm²)	122.70 ± 6.15[a]	139.88 ± 7.38[b]	132.82 ± 4.57[c*]	144.73 ± 6.47[b]
Cell minor axis (µm)	10.52 ± 0.36[a]	11.45 ± 0.27[bc]	11.09 ± 0.53[ac*]	11.94 ± 0.39[b]
Cell major axis (µm)	15.11 ± 0.50[a]	15.85 ± 0.84[a]	15.53 ± 0.52[a]	15.74 ± 0.27[a]
Nucleus area (µm²)	20.07 ± 1.60[a]	20.23 ± 2.61[a]	20.87 ± 1.31[a]	18.45 ± 1.47[a]
Nucleus minor axis (µm)	4.34 ± 0.18[a]	4.43 ± 0.32[a]	4.48 ± 0.15[a]	4.27 ± 0.26[a]
Nucleus major axis (µm)	6.03 ± 0.28[a]	5.99 ± 0.33[a]	6.30 ± 0.39[a]	5.73 ± 0.27[a]
pH 7.0				
Cell area (µm²)	135.11 ± 19.63[a]	151.22 ± 20.83[a]	155.43 ± 12.71[a]	153.34 ± 14.78[a]
Cell minor axis (µm)	11.21 ± 0.95[a]	11.88 ± 0.79	12.21 ± 0.57[a]	11.81 ± 0.23[a]
Cell major axis (µm)	15.52 ± 1.08[a]	16.45 ± 1.19[a]	16.51 ± 0.66[a]	16.85 ± 1.30[a]
Nucleus area (µm²)	20.63 ± 3.53[a]	23.73 ± 3.63[a]	22.43 ± 2.40[a]	22.02 ± 1.78[a]
Nucleus minor axis (µm)	4.48 ± 0.35[a]	4.67 ± 0.38[a]	4.67 ± 0.29[a]	4.55 ± 0.09[a]
Nucleus major axis (µm)	6.04 ± 0.55[a]	6.66 ± 0.51[a]	6.31 ± 0.32[ab]	6.38 ± 0.40[a]

Different letters indicate significant difference between HA concentrations at the same pH. * Indicate significant difference from pH 7.0 at the same HA concentration ($p < 0.05$). Mean values ± SE (n = 8-15/group).

duction in plasma volume, erythrocyte swelling and release of erythrocytes from the spleen due to increased circulating catecholamines.

In this investigation, both hematocrit and hemoglobin were highly affected by the experimental variables, considering the basal range previously reported for silver catfish, 17.00-34.00 and 4.95-9.09 respectively (TAVARES-DIAS et al. 2002). Some of the groups exposed to higher pH levels increased hematocrit and hemoglobin values in the presence of HA, which may be a result of the before-mentioned limited gas exchange induced by HA. Inefficient gill ventilation triggers mechanisms such as splenic contraction in an attempted to absorb more oxygen, therefore elevating hematocrit and hemoglobin (SAMPAIO et al. 2008). RIFFEL et al. (2014) have similarly reported that the addition of HA to the water, though at low concentrations, induced hematocrit and hemoglobin rises in silver catfish at neutral pH.

Somewhat different results were found at pH 7.0 for the hematocrit in the group subjected to 10 mg L⁻¹ HA and for both the hematocrit and hemoglobin in the group exposed to 50 mg L⁻¹ HA. It seems that the fish in those groups, especially in the latter one, were able to compensate for the decreased ventilatory drive caused by HA at the neutral pH, which was not observed at 25 mg L⁻¹ HA.

Exposure of tambaqui to an extreme pH of 3.0 had no influence on blood oxygenation or hemoglobin concentration, demonstrating that this fish, which migrates from circumneutral to acidic waters in its natural habitat, does not encounter challenges in oxygen delivery at such pH level (WOOD et al. 1998). Likewise, ARIDE et al. (2007) observed similar hemoglobin levels between tambaqui subjected to either circumneutral or acid pH, though there was elevation in hematocrit during acid exposure.

As already mentioned, MILLIGAN & WOOD (1982) found that acid exposure triggered disturbances in hematological homeostasis and fluid volume distribution in rainbow trout. Elevation in erythrocyte volume in that species was most likely a result of fluid redistribution from extra- to intracellular compartments due to the ionic dilution of the plasma. In contrast, ARIDE et al. (2007) observed no changes in erythrocyte volume in tambaqui subjected to acid exposure. In the present study it was demonstrated that: a) regardless the HA concentration, the size of the erythrocytes and their nuclei remained stable throughout the groups at pH 7.0; b) at pH 4.0 and 4.2, the significant differences indicate smaller values in the absence of HA; and c) within a given concentration of HA, most differences pointed to higher values at pH 7.0 than at acidic pH. The overall response suggests that, unlike the studies cited above, low pH caused a shrinking effect on the erythrocytes of silver catfish. The presence of HA did not fully counteract such outcome, since the differences were significant comparing to pH 7.0 This effect could be due to output of water and hydromineral disturbance, which typically arise from stress in fish (WENDELAAR BONGA 1997).

Plasma levels of Na^+ of silver catfish at pH 4.0 and 4.2 were lower than those observed at pH 7.0. The disruptive process in this extreme aquatic environment primarily involves active inhibition of ion uptake and increased ion loss in the gills (MILLIGAN & WOOD 1982). FREDA & MCDONALD (1988) observed the complete inhibition of Na^+ influx in shiners and trout exposed to pH 4.0, in addition to an increase in the ion outward flux. LIN & RANDALL (1993) claimed that inhibition of Na^+ uptake at low pH results from the reduced activity of an apical electrogenic H^+ATPase that energizes an apical Na^+ channel in chloride cells, an effect attributed to the H^+ gradient. Stimulation of Na^+ efflux, which is the primary determinant of low pH tolerance, is usually a consequence of the H^+-induced Ca^{2+} leaching from the paracellular channels in the gills (GONZALEZ et al. 1997).

MCDONALD & WOOD (1981) and WOOD et al. (1998) stated that disturbance of ionoregulation by high external H^+ is likely to occur in nonacidophilic species when they are subjected to a sudden acid stress. On the other hand, fish that inhabit naturally acidified, diluted waters, such as those found in the Amazon basin or along the eastern coast of the United States, show a greater tolerance to high concentration of water H^+ and have a lower pH threshold at which marked ion losses occur (GONZALEZ & DUNSON 1989, GONZALEZ et al. 1998, WOOD et al. 1998, MATSUO & VAL 2007). In some species the adaptation to thrive in these waters involves increased branchial affinity for Ca^{2+} at the paracellular junction, thus counteracting low pH-induced displacement (FREDA & MCDONALD 1988, GONZALEZ & DUNSON 1989). Further, some fish are able to take up ions at high rates when there is high diffusive ion leakage (GONZALEZ et al. 1997, 1998), and at least two Amazon species have pH-insensitive Na^+ transporter (GONZALEZ & WILSON 2001). For FREDA & MCDONALD (1988), two important abilities may respond for the interspecific differences in acid tolerance: limitation of the ionic leakiness prompted by low pH, and ion transporter recovery from the low pH inhibition.

Besides their own endogenous mechanisms, fish native to DOC-enriched habitats may relay on the great amount of organic substances found there to improve ion homeostasis (GONZALEZ et al. 1998, 2002, WOOD et al. 2002, 2003). GONZALEZ et al. (2002) and MATSUO & VAL (2007) observed that the presence of DOC in acidic water reduced both Na^+ influx inhibition and diffusive efflux stimulation in teleosts native to

Amazonia. The role of DOC to bind fish gills at low pH and promote physiological benefits (CAMPBELL et al. 1997) may be comparable with the above-mentioned action of elevated waterborne levels of Ca^{2+}, that is, stabilization of tight junctions and prevention of ion losses. That would override any protective effect otherwise achieved by the divalent ion (WOOD et al. 2003, 2011). Besides, it could result from the ability of the organic molecules to bind to ion apical transporters and help concentrate Na^+ and Cl^- ions by complexation, or to help deliver the ions to the uptake sites, which is normally credited to mucus (GONZALEZ et al. 2002, STEINBERG et al. 2007).

Except when pH was 4.2, at which the presence of HA was associated with a slightly higher Na^+ plasma level in silver catfish, there were no differences in the ion levels between the different HA concentrations at any given pH. This could be explained by the observation that this fish species is not native to waters with high DOC content, so its gill physiology may not be sensitive to the DOC's protective mechanism (MATSUO et al. 2004). Another possibility is that commercial HA is not as useful to silver catfish as natural black water, as observed by WOOD et al. (2003) in stingrays (Potamotrygon sp.): HA stimulated Na^+ and Cl^- leakage, probably because its high affinity for cations ends up stripping Ca^{2+} from the gills. Consequently, the authors concluded that this source of DOC may have different binding characteristics than does natural black water DOC.

The inhibitory mechanism of Cl^- uptake under low pH is possibly associated with the already described mechanism of Na^+ uptake inhibition. Besides, it could be due to a reduction in intracellular HCO_3^- at the chloride cells, thus exhausting the apical Cl^-/HCO_3^- exchanger (WOOD 2001). Cl^- loss in silver catfish was exacerbated at 25 and 50 mg L^{-1} HA in the fish exposed to pH 4.0, thus demonstrating a greater involvement of HA in Cl^- than in Na^+ flux. Such difference may be linked to their distinct ionoregulatory mechanisms across the gill epithelium, since Cl^- and Na^+ are exchanged for base and acid equivalents, respectively (GOSS & WOOD 1990).

Comparing with the responses on Na^+ and Cl^- balance, a different effect of HA was observed with regard to the dynamics of K^+ regulation. When there was no HA in the test water, the levels of K^+ in silver catfish exposed to pH 4.0 and 4.2 were higher in comparison to ion levels in fish subjected to pH 7.0. A similar outcome was observed in the pirapitinga, Piaractus brachypomus, in a recent investigation (GARCIA et al. 2014). Further, ZAIONS & BALDISSEROTTO (2000) found lower body levels of K^+ in silver catfish subjected to pH 7.0 than in those subjected to either acidic or alkaline pH. MATHAN et al. (2010) also observed higher plasma K^+ levels in the common carp, Cyprinus carpio, after exposure to acidic pH, and suggested that it could be due to a release of K^+ from the muscle cells as H^+ enters them. Another study assessing plasma ion levels in silver catfish exposed to Aldrich HA (0, 2.5 and 5 mg L^{-1}) at pH ~ 7.0 observed a progressive increase in K^+ levels with increased HA concentrations, suggesting that HA could limit gill permeability (Riffel et al. 2014). In this study HA did not influence K^+ levels at pH 7.0, while at the intermediate pH levels (4.0 and 4.2) its presence caused a marked decrease in ion levels. Thus, all concentrations of HA were able to counteract increased K^+ levels caused by pH 4.0 and 4.2, bringing K^+ levels back to normal values for the species (Bolner & Baldisserotto 2007), at pH 7.0.

Low pH exposure induced continuous net branchial losses of Na^+, Cl^- and K^+, and a progressive decline in plasma Na^+ and Cl^- levels in rainbow trout (McDonald & Wood 1981). In spite of an improved tolerance to acidity reported for fish that are exposed to gradual water acidification in the wild, it is possible that the same fish will undergo ion loss when faced with sudden environmental acidification. For instance, Aride et al. (2007) found that plasma levels of Na^+ and K^+ in the tambaqui were reduced in acidic water compared to a circumneutral water. Wilson et al. (1999) observed that acid exposure produced different patterns of Na^+, Cl^- and K^+ fluxes in three Amazon fish, which implies that acid tolerance is not necessarily a typical feature of the fish that inhabit this region. Instead, it is largely related to the occurrence of these fish in the blackwater areas of that ecosystem, which are known to impose higher levels of acidity on the species.

Although HA showed some positive effects on hematological and plasma K^+ changes provoked in silver catfish by acidic pH exposure, the overall findings suggest that HA does not protect this species against acidic pH burden, since it increased mortality and Cl^- loss at pH 4.0.

ACKNOWLEDGMENTS

The authors thank Conselho Nacional de Desenvolvimento Científico e Tecnológico (CNPq) for the research fellowship to B. Baldisserotto and Fundação de Amparo à Pesquisa do Estado do Rio Grande do Sul (FAPERGS) and Coordenação de Aperfeiçoamento de Pessoal de Nível Superior, Brazil (Capes) for the graduate fellowships to L.T. Gressler and F.J. Sutili respectively. This work was funded by CNPq and Fundação de Amparo à Pesquisa do Estado do Amazonas (FAPEAM – INCT ADAPTA).

REFERENCES

Aride PHR, Roubach R, Val AL (2007) Tolerance response of tambaqui *Colossoma macropomum* (Cuvier) to water pH. **Aquaculture Research** 38: 588-594. doi: 10.1111/j.1365-2109.2007.01693.x

Bindon SD, Gilmour KM, Fenwick JC, Perry SF (1994) The effects of branquial chloride cell proliferation on respiratory function in the rainbow trout *Oncorhyncus mykiss*. **Journal of Experimental Biology** 197: 47-63.

Bolner KCS, Baldisserotto B (2007) Water pH and urinary excretion in silver catfish *Rhamdia quelen*. **Journal of Fish Biology** 70: 50-64. doi: 10.1111/j.1095-8649.2006.01253.x

Boyd CE, Tucker CS (1992) **Water quality and pond soil analyses for aquaculture.** Auburn, Alabama Agricultural Experiment Station, Auburn University, 183p.

Brown BA (1976) **Hematology: Principles and procedures.** Philadelphia, Lea & Febiger, 336p.

Campbell PGC, Twiss MR, Wilkinson KJ (1997) Accumulation of natural organic matter on the surfaces of living cells: implications for the interaction of toxic solutes with aquatic biota. **Canadian Journal of Fisheries and Aquatic Sciences** 54: 2543-2554.

Copatti CE, Codebella IJ, Radünz Neto J, Garcia LO, Rocha MC, Baldisserotto B (2005) Effect of dietary calcium on growth and survival of silver catfish fingerlings, *Rhamdia quelen* (Heptapteridae), exposed to different water pH. **Aquaculture Nutrition** 11: 345-350. doi: 10.1111/j.1365-2095.2005.00355.x

Cunha MA, Zeppenfeld CC, Garcia LO, Loro VL, Fonseca MB, Emanuelli T, Veeck APD, Copatti CE, Baldisserotto B (2010) Anesthesia of silver catfish with eugenol: time of induction, cortisol response and sensory analysis of fillet. **Ciência Rural** 40: 2107-2114. doi: 10.1590/S0103-84782010005000154

Dheer JMS, Dheer TR, Mahajan CL (1987) Haematological and haematopoetic responses to acid stress in an air-breathing freshwater fish, *Channa punctatus*. **Journal of Fish Biology** 30: 577-588.

Dorafshan S, KalbAssi MR, Pourkazemi M, Amiri BM, Karimi SS (2008) Effects of triploidy on the Caspian salmon *Salmo trutta caspius* haematology. **Fish Physiology and Biochemistry** 34: 195-200. doi: 10.1007/s10695-007-9176-z

Duarte RM, Ferreira MS, Wood CM, Val AL (2013) Effect of low pH exposure on Na^+ regulation in two cichlid fish species of the amazon. **Comparative Biochemistry and Physiology a-Molecular & Integrative Physiology** 166: 441-448. doi: 10.1016/j.cbpa.2013.07.022

Eaton AD, Clesceri LS, Rice EW, Grennberg AE (2005) **Standard methods for the examination of water and wastewater.** Springfield, American Public Health Association, 21st ed., 1600p.

Farjalla VF, Amado AM, Suhett AL, Meirelles-Pereira F (2009) DOC removal paradigms in highly humic aquatic ecosystems. **Environmental Science and Pollution Research** 16: 531-538. doi: 10.1007/s11356-009-0165-x

Freda J, McDonald DG (1988) Physiological correlates of interspecific variation in acid tolerance in fish. **The Journal of Experimental Biology** 136: 243-258.

Garcia LO, Gutiérres-Espinosa MC, Vásques-Torres W, Baldisserotto B (2014) Dietary protein levels in *Piaractus brachypomus* submitted to extremely acidic or alkaline pH. **Ciência Rural** 44: 301-306. doi: 10.1590/S0103-84782014000200017

Gonzalez RJ, Dunson WA (1989) Acclimation of sodium regulation to low pH and the role of calcium in the acid-tolerant sunfish *Enneacanthus obesus*. **Physiologycal Zoology** 62: 977-992.

Gonzalez RJ, Wilson RW (2001) Patterns of ion regulation in acidophilic fish native to the ion-poor acidic Rio Negro.

Journal of Fish Biology 58: 1680-1690. doi: 10.1111/j.1095-8649.2001.tb02322.x

GONZALEZ RJ, DALTON VM, PATRICK ML (1997) Ion regulation in ion-poor, acidic water by the blackskirt tetra (*Gymnocorymbus ternetzi*), a fish native to the Amazon River. **Physiological Zoology 70:** 428-435.

GONZALEZ RJ, WOOD CM, WILSON RW, PATRICK ML, BERGMAN HL, NARAHARA A, VAL AL (1998) Effects of water pH and calcium concentration on ion balance in fish of the Rio Negro, Amazon. **Physiological Zoology 71:** 15-22.

GONZALEZ RJ, WILSON RW, WOOD CM, PATRICK ML, VAL AL (2002) Diverse strategies for ion regulation in fish collected from the ion-poor, acidic Rio Negro. **Physiological and Biochemical Zoology 75:** 37-47. doi: 10.1086/339216

GOSS GG, WOOD CM (1990) Kinetic analysis of the relationships between ion exchange and acid-base regulation at the gills of freshwater fish, p. 119-136. In: TRUCHOT JP, LAHLOU B (Eds.). **Animal Nutrition and Transport Processes. 2. Transport, Respiration and Excretion: Comparative and Environmental Aspects.** Basel, Karger Publishers.

GRECO AM, FENWICK JC, PERRY SF (1996) The effects of soft-water acclimation on gill structure in the rainbow trout *Oncorhynchus mykiss*. **Cell and Tissue Research 285:** 75-82.

HOLLAND A, DUIVENVOORDEN LJ, KINNEAR SHW (2014) The double-edged sword of humic substances: contrasting their effect on respiratory stress in eastern rainbow fish exposed to low pH. **Environmental Science and Pollution Research 21:** 1701-1707. doi: 10.1007/s11356-013-2031-0

KÜCHLER IL, MIEKELEY N, FORSBERG BR (2000) A contribution to the chemical characterization of rivers in the rio Negro basin, Brazil. **Journal of the Brazilian Chemical Society 11:** 286-292. doi: 10.1590/S0103-50532000000300015

LIMA RL, BRAUN N, KOCHHANN D, LAZZARI R, RADÜNZ-NETO J, MORAES BS, LORO V, BALDISSEROTTO B (2011) Survival, growth and metabolic parameters of silver catfish, *Rhamdia quelen*, juveniles exposed to different waterborne nitrite levels. **Neotropical Ichthyology 9:** 147-152. doi: 10.1590/S1679-62252011005000004

LIN H, DJ RANDALL (1993) Proton ATPase activity in crude homogenates of fish gill tissue: inhibitor sensitivity and environmental and hormonal regulation. **Journal of Experimental Biology 180:** 163-174.

MATHAN R, KURUNTHACHALAM SK, PRIYA M (2010) Alterations in plasma electrolyte levels of a freshwater fish *Cyprinus carpio* exposed to acidic pH. **Toxicological and Environmental Chemistry 92:** 149-157. doi: 10.1080/02772240902810419

MATSUO AYO, VAL AL (2007) Acclimation to humic substances prevents whole body sodium loss and stimulates branchial calcium uptake capacity in cardinal tetras *Paracheirodon axelrodi* (Schultz) subjected to extremely low pH. **Journal of Fish Biology 70:** 989-1000. doi: 10.1111/j.1095-8649.2007.01358.x

MATSUO AYO, PLAYLE RC, VAL AL, WOOD CM (2004) Physiological action of dissolved organic matter in rainbow trout in the presence and absence of copper: sodium uptake kinetics and unidirectional flux rates in hard and softwater. **Aquatic Toxicology 70:** 63-81. doi: 10.1016/j.aquatox.2004.07.005

MCDONALD D, WOOD CM (1981) Branchial and renal acid and ion fluxes in the rainbow trout, *Salmo gairdneri*, at low environmental pH. **Journal of Experimental Biology 93:** 101-118.

MCGEER JC, SZEBEDINSZKY C, MCDONALD DG, WOOD CM (2002) The role of dissolved organic carbon in moderating the bioavailability and toxicity of Cu to rainbow trout during chronic waterborne exposure. **Comparative Biochemistry and Physiology C 133:** 147-160. doi: 10.1016/S1532-0456(02)00084-4

MILLIGAN CL, WOOD CM (1982) Disturbances in haematology, fluid volume distribution and circulatory function associated with low environmental pH in the rainbow trout, *Salmo Gairdneri*. **Journal of Experimental Biology 99:** 397-415.

MIRON DS, BECKER AG, LORO VL, BALDISSEROTTO B (2011) Waterborne ammonia and silver catfish, *Rhamdia quelen*: survival and growth. **Ciência Rural 41:** 349-353. doi: 10.1590/S0103-84782011000200028

RIFFEL APK, SACCOL EMH, FINAMOR IA, OURIQUE GM, GRESSLER LT, PARODI T, GOULART LOR, LLESUY S, BALDISSEROTTO B, PAVANATO MA (2014) Humic acid and moderate hypoxia alter oxidative and physiological parameters in different tissues of silver catfish (*Rhamdia quelen*). **Journal of Comparative Physiology B 184:** 469-482. doi: 10.1007/s00360-014-0808-1

SAMPAIO FG, BOIJINK CL, OBA ET, SANTOS LRB, KALININ AL, RANTIN FT (2008) Antioxidant defenses and biochemical changes in pacu (*Piaractus mesopotamicus*) in response to single and combined copper and hypoxia exposure. **Comparative Biochemistry and Physiology C 147:** 43-51. doi: 10.1016/j.cbpc.2012.07.002

STEINBERG CEW, SAUL N, PIETSCH K, MEINELT T, RIENAU S, MENZEL R (2007) Dissolved humic substances facilitate fish life in extreme aquatic environments and have the potential to extend the lifespan of *Caenorhabditis elegans*. **Annals of Environmental Science 1:** 81-90.

TAVARES-DIAS M, MELO JFB, MORAES G, MORAES FR (2002) Características hematológicas de teleósteos brasileiros: VI. Variáveis do jundiá *Rhamdia quelen* (Pimelodidae). **Ciência Rural 32:** 693-698. doi: 10.1590/S0103-84782002000400024

TAVARES-DIAS M, BOZZO FR, SANDRIN EFS, CAMPOS-FILHO E, MOARES FR (2004) Células sanguíneas, eletrólitos séricos, relação hepato e esplenossomática de carpa comum, *Cyprinus carpio* (Cyprinidae) na primeira maturação gonadal. **Acta Scientiarum Biological Sciences 26:** 73-80. doi: 10.4025/actascibiolsci.v26i1.1661

THURMAN EM (1985) Organic geochemistry of natural waters. Dordrecht, Martinus Nijhof, Dr. W. Junk Publishers, 507p.

VERDOUW H, VAN ECHTELD CJA, DEKKERS EMJ (1978) Ammonia determination based on indophenols formation with sodium salicylate. **Water Research 12**: 399-402. doi: 10.1016/0043-1354(78)90107-0

WENDELAAR BONGA SE (1997) The stress response in fish. **Physiology Reviews 77**: 591-625.

WILSON RW, WOOD CM, GONZALEZ RJ, PATRICK ML, BERGMAN HL, NARAHARA A, VAL AL (1999) Ion acid-base balance in three species of Amazonian fish during gradual acidification of extremely soft water. **Physiological and Biochemical Zoology 72**: 277-285.

WOOD CM, MCDONALD DG (1982) Physiological mechanisms of acid toxicity to fish, p. 197-226. In: JOHNSON RE (Ed.). **Acid Rain/Fisheries: Proceedings of an International Symposium on Acid Precipitation and Fishery Impacts in North-Eastern North America.** Ithaca, American Fisheries Society.

WOOD CM, WILSON RW, GONZALEZ RJ, PATRICK ML, BERGAMAN HL, NARAHARA A, VAL AL (1998) Responses of an Amazonian teleost, the tambaqui (*Colossoma macropomum*) to low pH in extremely soft water. **Physiologycal and Biochemical Zoology 71**: 658-670.

WOOD CM (2001) Toxic response of the gill, p. 1-89. In: SCHLENK D, BENSON WH (Eds.). **Target organ toxicity in marine and freshwater teleosts.** London, Taylor & Francis.

WOOD CM, MATSUO AYO, GONZALEZ RJ, WILSON RW, PATRICK ML, VAL AL (2002) Mechanisms of ion transport in *Potamotrygon*, a stenohaline freshwater elasmobranch native to the ion-poor blackwaters of the Rio Negro. **Journal of Experimental Biology 205**: 3039-3054.

WOOD CM, MATSUO AYO, WILSON RW, GONZALEZ RJ, PATRICK ML, PLAYLE RC, VAL AL (2003) Protection by natural blackwater against disturbances in ion fluxes caused by low pH exposure in freshwater stingrays endemic to the Rio Negro. **Physiological and Biochemical Zoology 76**: 12-27. doi: 10.1086/367946

WOOD CM, AL-REASI HA, SCOTT DS (2011) The two faces of DOC. **Aquatic Toxicology 105S**: 3-8. doi: 10.1016/j.aquatox.2011.03.007

ZAIONS MI, BALDISSEROTTO B (2000) Na+ and K+ body levels and survival of fingerlings of *Rhamdia quelen* (Siluriformes, Pimelodidae) exposed to acute changes of water pH. **Ciência Rural 30**: 1041-1045. doi: 10.1590/S0103-84782000000600020

ZALL DM, FISHER M, GARNER MQ (1956) Photometric determination of chlorides in water. **Analytical Chemistry 28**: 1665-1678. doi: 10.1021/ac60119a009

5

Intra- and inter-annual variations in Chironomidae (Insecta: Diptera) communities in subtropical streams

Diane Nava[1], Rozane M. Restello[1] & Luiz U. Hepp[1,*]

[1]Programa de Pós-graduação em Ecologia, Universidade Regional Integrada do Alto Uruguai e das Missões. Avenida Sete de Setembro 1621, 99709-910 Erechim, RS, Brazil.
*Corresponding author. E-mail: luizuhepp@gmail.com

ABSTRACT. The structure and composition of stream benthic communities are strongly influenced by spatial and temporal factors. This study evaluated the intra and inter-annual variations in Chironomidae communities in subtropical streams. The organisms were sampled from 10 small-order streams during the summer and winter of 2010-2012. The number of chironomid specimens sampled was 7,568, distributed in 49 genera. Chironomid abundance and richness varied intra and inter-annually and community composition varied intra-annually (2010 and 2011). Water temperature, total organic carbon, nitrogen, and rainfall were correlated with chironomid community composition. The intra-annual variation of the community was dependent on climatic variations (temperature and rainfall) and changes caused by intensive agricultural use. We conclude that the temporal variation observed in the Chironomidae community correlates with climatic variations (rainfall) and changes in the total organic carbon and total nitrogen, caused by intensive agricultural land use.

KEY WORDS. Agriculture impacts; bioindicators; macroinvertebrates; rainfall.

The structure and composition of aquatic communities are influenced by spatial and temporal factors (SUAREZ 2008). Knowing how these factors act on biological communities facilitates the understanding of how local and regional factors influence species occurrence (POFF et al. 2006, SUAREZ 2008). The distribution of benthic macroinvertebrates is affected by factors such as type of substrate (HEPP et al. 2012), habitat characteristics (GALDEAN et al. 2000, BUSS et al. 2004), land use (HEPP et al. 2010), and climatic variations over a timescale (SCHEFFER & VAN NES 2007). Climatic variations have decisive effects on the distribution of benthic organisms (SMITH et al. 2003) and can occur at different timescales, both intra-annual (seasons) and inter-annual (between years).

Chironomidae (Insecta: Diptera) occur in great abundance and high diversity in most aquatic ecosystems in all continents (EPLER 2001, FERRINGTON 2008). They play an important role in the food web of aquatic communities, establishing links between producers and consumers, as well as participating in nutrient cycles (HENRIQUES-OLIVEIRA et al. 2003). Chironomids are tolerant to various changes in the environment (ROSIN & TAKEDA 2007, RESTELLO et al. 2012), and depending on the species, they may display negative or positive responses to human impacts (FERRINGTON 2008).

Chironomid communities can be effected by the integrity of the riparian zone. The state and the extent of the riparian vegetation correlates with differences in the abundance, richnness, and composition of chironomid communities in streams (SENSOLO et al. 2012). For instance, suppression of the riparian vegetation results in decreased overall diversity and increased numbers of tolerant taxa (AL-SHAMI et al. 2010a). Different chironomids inhabit different habitats and substrates (SANSEVERINO & NESSIMIAN 2001) although they are most frequent in heterogeneous and stable environments, where they attain high diversity (ROSA et al. 2011, 2013).Temporal variations in biological communities are mainly linked to climate-related changes (e.g., temperature and rainfall). Climate affects ecological processes such as competition, predation and recruitment (GRESENS et al. 2007). In addition to climatic factors, temporal variations in the structure and composition of chironomid communities may reflect the biological characteristics of the species that compose these communities (HEINIS & DAVIDS 1993, SIQUEIRA et al. 2008) or temporal changes in physical and chemical characteristics (AL-SHAMI et al. 2010b). Natural disturbances, such as spates caused by increased rainfall in human-impacted areas may carry chemicals from adjacent areas to the streams, thus affecting chironomid communities (GRESENS et al. 2007).

In this study, the intra- and inter-annual variation of Chironomidae communities in subtropical streams was assessed over three years. We tested the hypothesis that environmental factors related to human activities may be important in structuring communities in streams. Thus, the objectives of this study were (1) to evaluate the intra and inter-annual variations in Chironomidae communities and (2) to determine whether these temporal variations are associated with environmental factors.

MATERIAL AND METHODS

This study was conducted in the upper portion of the Uruguay River Basin in southern Brazil (27°12′59″ and 28°00′47″S, 52°48′12″ and 51°49′34″W, Fig. 1). The region is characterized by a subtropical climate (Koppen Cfb) with average annual rainfall of 1912.3 mm and average annual temperature of 17.6°C. The vegetation is a subtropical forest mix. It is mostly composed of species with tropical-subtropical distribution in the Upper Uruguay, and Araucaria Forest with a predominance of Araucaria (OLIVEIRA-FILHO et al. 2015). The predominant land use is intensive agricultural practice (~77% of the total area), with soybeans, corn, wheat crops, and large forested areas (DECIAN et al. 2009). Thus, all 10 selected streams are embedded in a complex agricultural matrix. All streams studied were small-order streams (<3rd order) and had similar limnological characteristics. The average percentage of vegetation in the riparian zone of the streams was 23% (range 11-49%).

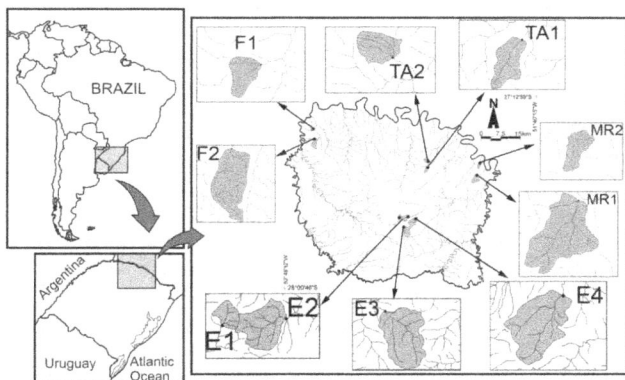

Figure 1. Geographical location of sampling sites at Alto Uruguay region, RS. F: Faxinalzinho, TA: Três Arroios, ERE: Erechim, MR: Marcelino Ramos.

We obtained the following variables from each stream: water temperature, turbidity, conductivity, total dissolved solids, dissolved oxygen and pH, with the aid of a multiparameter analyser HORIBA® U50. A Shimadzu® TOC-VCSH analyzer was used to measure total organic carbon (TOC) and total nitrogen.

Chironomidae larvae were collected in August (winter) and December (summer) of the years 2010, 2011 and 2012. At each stream, three sampling units were obtained with a Surber sampler (mesh 250 µm, area 0.09 m²) on a rock substrate. The material was fixed in the field using 80% alcohol. In the laboratory, Chironomidae larvae were dipped in a of 10% bleach solution of potassium hydroxide for 24 hours. Individuals were then mounted on semi-permanent slides with Hoyer solution and were identified under optical microscope with a magnifi-

cation of 1,000 times. Specimens were identified at the genus level using the identification keys of TRIVINHO-STRIXINO & STRIXINO (1995) and TRIVINHO-STRIXINO (2011).

To assess the intra- and inter-annual variations in abiotic variables Multivariate Analysis of Variance (MANOVA) was used. Variations in chironomid abundance and richness between the seasons (intra-annual) and between the years (inter-annual) were evaluated using a repeated measure Analysis of Variance (RM-ANOVA). Non-Metric Multidimensional Scaling (NMDS) (KRUSKAL 1964) was used to order the chironomid communities. The NMDS was performed with a biological matrix based on the presence or absence of genera in each stream using the Jaccard index. The relationship between environmental and biological data was tested by fitting vectors of environmental variables to the NMDS ordination (function 'envfit' of the vegan package). Analysis of Similarity (ANOSIM) was used to evaluate the level of segregation in community composition between years and within years,. All analyses were performed using R software (R CORE TEAM 2013) with the 'vegan' package (OKSANEN et al. 2013).

RESULTS

The studied streams have well-oxygenated (10.85 ± 2.51 mg L⁻¹), slightly acidic water (pH 6.62 ± 0.22) with electrical conductivity of 0.030 ± 0.059 mS cm⁻¹ (mean of three years). The highest average turbidity was recorded in summer (11.75 ± 4.90 NTU). However, the highest average total organic carbon was recorded in winter (218.34 ± 216.16 mg L⁻¹) (Table 1). The total organic carbon was very high in the winter of 2011 (Table 1). The highest average monthly rainfall occurred in 2011 (172.20 ± 85.16 mm) followed by 2010 (122.6 ± 101.7 mm) and 2012 (36.7 ± 60.0 mm; Fig. 2). However, in 2010 and 2012there was as much rainfall in the winter and the summer. In 2011, the difference in rainfall between the winter and summer seasons was ca. 150 mm. Overall, the abiotic variables differed among the years and between the seasons ($F_{(2,56)}$ = 18.22, p = 0.001 and $F_{(1,56)}$ = 25.10, p = 0.001, respectively, Table 2).

We obtained a total of 7,568 chironomid larvae distributed in 49 genera. The highest abundance was recorded in 2012 (3,304 larvae, 43.7% of the total), followed by 2011 (2,430 larvae, 32.1%) and 2010 (1,834 larvae, 24.2%). In two of the three years studied (2010 and 2012, Fig. 3, Table 3) chironomids were more abundant in the winter. The greatest number of chironomid genera (43 genera) was identified in 2011, followed by 2012 (33 genera) and 2010 (25 genera). Thus, abundance varied intra-annually while richness varied intra- and inter-annually (Table 3, Fig. 4).

Among the genera identified, *Pentaneura* Philippi, 1865, *Polypedilum* Kieffer, 1912, and *Rheotanytarsus* Thienemann & Bause in Bause, 1913 were the most frequent in the samples. *Aedokritus* Roback, 1958, *Antillocladius* Saether, 1981, *Denopelopia* Roback & Rutter, 1988, *Djalmabatista* Fittkau, 1968,

Table 1. Mean and standard deviation of limnological variables quantified the drainage areas of the 10 studied streams in the region Alto Uruguay Rio Grande Sul, in the period 2010-2012.

Variables	2010		2011		2012	
	Summer	Winter	Summer	Winter	Summer	Winter
Water temperature (°C)	20.72 ± 0.87	15.05 ± 1.84	21.45 ± 3.03	14.40 ± 1.57	22.78 ± 1.98	16.31±1.79
pH	6.87 ± 0.84	6.45 ± 0.56	6.45 ± 0.59	6.89 ± 0.79	6.71 ± 0.43	6.37±0.48
Electrical Conductivity (mS cm^{-1})	0.05 ± 0.03	0.05 ± 0.02	0.08 ± 0.04	0.051 ± 0.02	1.52 ± 4.59	0.06±0.03
Turbidity (UNT)	13.37 ± 17.26	3.07 ± 8.99	6.25 ± 2.82	8.46 ± 5.05	15.65 ± 19.87	8.85±5.40
DO (mg L^{-1})	10.38 ± 1.06	8.36 ± 0.90	9.54 ± 2.57	9.49 ± 0.73	11.92 ± 3.15	15.39±1.67
TDS (mg L^{-1})	0.04 ± 0.02	0.03 ± 0.01	0.05 ± 0.03	0.03 ± 0.01	0.04 ± 0.02	0.03±0.02
Nitrogen (mg L^{-1})	15.66 ± 3.37	6.36 ± 2.80	0.44 ± 0.47	1.32 ± 0.68	1.13 ± 0.89	1.05±0.69
TOC (mg L^{-1})	17.14 ± 5.35	25.05 ± 21.80	83.90 ± 39.06	451.76 ± 100.39	61.28 ± 58.23	178.21±50.15

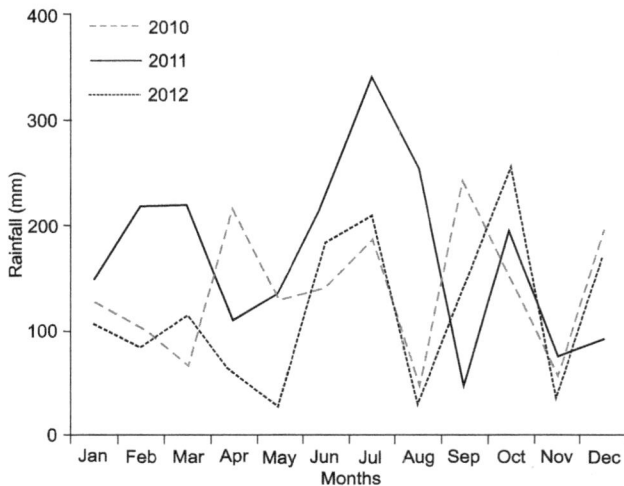

Figure 2. Monthly rainfall in the years 2010, 2011 and 2012 at Alto Uruguay region, RS. The horizontal lines indicate the annual average for the respective years (INMET).

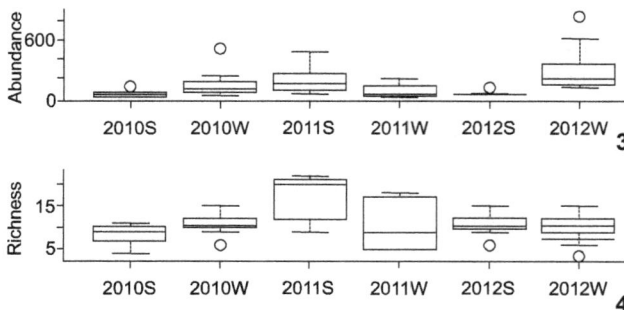

Figures 3-4. Box-plot (median and quartiles) showing the variation of (3) and abundance (4) wealth of inter-annual chironomid larvae (2010, 2011 and 2012) and intra-annual (summer: S, winter: W) in subtropical streams.

Table 2. MANOVA results for limnological intra and inter-annual among the studied streams were considered as factors the years studied (inter-annual) and the seasons (intra-annual).

	DF	SS	MS	F	p
Season	1	1.013	1.013	25.101	0.001
Year	2	1.471	0.735	18.225	0.001
Residuals	56	2.260	0.040	0.476	
Total	59	4.744			

Table 3. Repeated measures ANOVA results for the variation in the abundance (log) and richness of Chironomidae community between seasons and among years at 10 stream sites in Alto Uruguay region, Rio Grande Sul.

	DF	SS	F	p
Abundance (log)				
Year	2	1.01	0.75	0.477
Seasons	1	4.73	7.04	0.010
Year:Seasons	2	32.22	23.97	<0.001
Residuals	54	36.30		
Richness				
Year	2	235.90	8.44	<0.001
Seasons	1	43.35	3.10	0.008
Year:Seasons	2	237.90	8.51	<0.001
Residuals	54	754.50		

Manoa Fittkau, 1963, *Paracladius* Hirvenoja, 1973, *Paramerina* Fittkau & Stur, 1997, and *Pseudochironomus* Riethia Malloch, 1915 occurred only in winter, while *Endotribelos* Grodhaus, 1987, *Microchironomus* Kieffer, 1918, *Parapentaneura* Stur, Fittkau & Serrano, 2006, and *Ubatubaneura* Wiedenbrug & Trivinho-Strixino, 2009 occurred only in summer (Table 4). Community composition was similar between seasons when the three years were analysed together (ANOSIM, R = -0.01, p = 0.596,

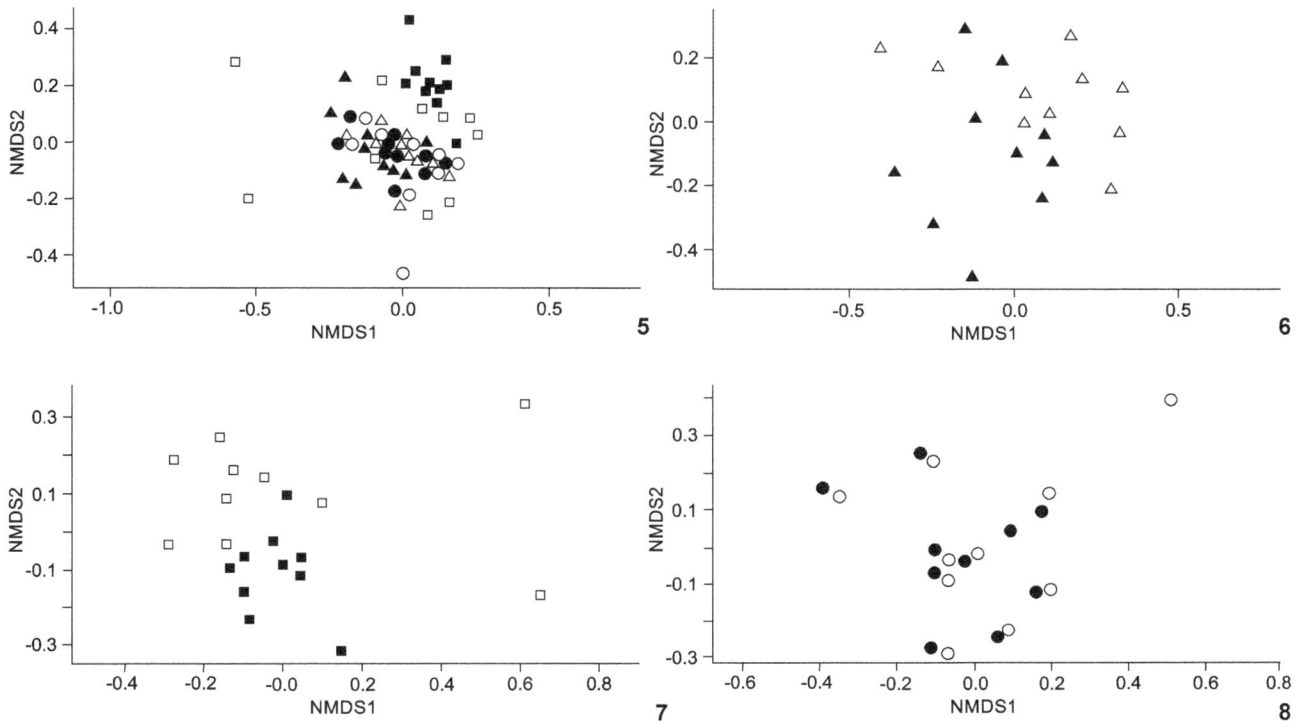

Figures 5-8. Non-Metric Multidimensional Scaling Analysis (NMDS) of the temporal distribution of chironomid community in subtropical streams. (5) inter-annual distribution (2010, 2011, 2012), (6) intra-annual distribution for 2010, (7) intra-annual distribution for 2011, (8) intra-annual distribution for 2012. Open symbols: winner, closed symbols: summer.

Figs. 5-8). However, the generi composition showed segregation between summer and winter in 2011 (ANOSIM, R = 0.32, p = 0.001) and 2010 (ANOSIM, R = 0.12, p = 0.034), but not in 2012 (ANOSIM, R = -0.09, p = 0.976) (Figs. 5-8).

The environmental variables that were correlated with the scores of the first two NMDS dimensions were those that variated seasonally (temperature), human activities (total organic carbon and nitrogen) and rainfall (Table 5). When data from all years were pooled, the community composition correlated with C:N ratio (r = 0.09, p = 0.05) and rainfall (r = 0.11, p = 0.04). Moreover, in 2010, the NMDS scores showed significant correlation with temperature (r = 0.31, p = 0.04) and nitrogen (r = 0.31, p = 0.03). In 2011 there was a relationship between water temperature (r = 0.47, p = 0.004), nitrogen (r = 0.42, p = 0.009), TOC (r = 0.35, p = 0.02) and rainfall (r = 0.30, p = 0.03). In 2012 there was no correlation with any environmental variable (Table 5).

DISCUSSION

Chironomidae abundance and generic richness varied either intra-annually (between seasons) and inter-annually (between years). In our study, the abundance of individuals and number of chironomid genera were associated with limnological characteristics. The effect of these characteristics on the chironomid community may have occurred in association with a significant rainfall event in 2011. Increased precipitation causes changes in aquatic ecosystems, such as changes in the physical and chemical characteristics of the water, as well as changes in the distribution of communities (SMITH et al. 2003). When rainfall increases, chemical compounds present in the soil are dragged into the streams and cause changes in the abundance, richness, and composition of chironomid communities (GRESENS et al. 2007). The effect of runoff in this study was especially observed in TOC concentrations in the winter of 2011.

In 2010 and 2012 rainfall was normal with respect to historical records. Streams and their chironomid communities were stable as a result of this. Rainfall increases the amount of water and unstable and homogeneous substrates into the streams (ROSA et al. 2013, SALLES & FERREIRA-JUNIOR 2014). Therefore, the increased rainfall (2011) during the study period was an important contributor to the observed variations in inter-annual richness. Fluctuations in water regimen allow great habitat diversification, i.e. more tolerant species can occupy distinct regions of the drainage area, in addition to changing the limnological characteristics of streams (SHUVART et al. 2005, ABURAYA & CALLIL 2007, ROQUE et al. 2007). SILVA et al. (2014) also observed that Chironomidae diversity was higher during the years when the abiotic charac-

Table 4. Chironomidae identified intra-annual (2010-2012) and inter-annual (summer and winter) in subtropical streams.

	2010		2011		2012	
	S	W	S	W	S	W
Chironominae						
Aedokritus Roback, 1958				*		
Caladomyia Säwedal, 1981	*	*	*		*	*
Chironomus Meigen, 1803			*	*		
Dicrotendipes Kieffer, Epler, 1988			*	*		
Endotribelos Grodhaus, 1987			*			
Goeldichironomus Fittkau, 1965	*		*			
Manoa Fittkau, 1963				*		
Microchironomus Kieffer, 1918			*			
Parachironomus Lenz, 1921	*		*			
Paratanytarsus Thienemman & Bause, 1951			*	*		
Paratendipes Kieffer, 1911	*	*	*	*	*	*
Phaenopsectra Kieffer, 1921	*		*	*		
Polypedilum Kieffer, 1912	*	*	*	*	*	*
Pseudochironomus, Riethia Malloch, 1915		*	*	*	*	*
Rheotanytarsus Thienemann & Bause, 1913	*	*	*	*	*	*
Saetheria Saether, 1983			*			
Stenochironomus Kieffer, 1919			*			
Tanytarsus Van der Wulp, 1874	*	*	*	*	*	*
Xestochironomus Sublette & Wirth, 1972			*			
Zavreliella Kieffer, 1920			*	*		
Orthocladiinae						
Antillocladius Saether, 1981				*		
Cardiocladius Kieffer, 1912			*	*		
Corynoneura Winnertz, 1846	*	*	*	*	*	*
Cricotopus Van der Wulp, 1874	*	*	*		*	*
Cricotopus, Orthocladius Lopescladius Oliveira, 1967	*	*	*	*	*	*
Gymnometriocnemus Goetghebuer, 1932	*	*	*			
Lopescladius Oliveira, 1967	*	*	*		*	*
Metriocnemus Kieffer 1921			*	*		
Nanocladius Kieffer, 1912	*	*	*	*	*	*
Onconeura Andersen & Saether, 2005	*	*	*	*	*	*
Orthocladiinae A Kieffer, 1911			*	*		
Orthocladiinae B Kieffer, 1911			*	*		
Paracladius Hirvenoja, 1973			*			
Parakiefferiella Thienneman 1926		*	*	*	*	*
Parametriocnemus Goetghebuer, 1932	*	*	*	*	*	*
Paraphaenocladius Thienemann, 1924			*	*		
Rheocricotopus Thienemann & Harnisch, 2004	*	*	*	*	*	*
Thienemannia Kieffer, 1909			*	*		
Thienemanniella Kieffer, 1911		*	*	*	*	*
Ubatubaneura Wiedenbrug & Trivinho-Strixino, 2009			*			
Tanypodinae						
Denopelopia Roback e Rutter (1988)				*		
Djalmabatista Fittkau (1968)			*			
Hudsonimyia Roback, 1979			*	*		
Labrudinia Fittkau, 1962			*			
Larsia Roback & Coffman (1989)			*	*		
Nilotanypus Kieffer, 1923	*	*	*	*	*	*
Paramerina Stur and Fittkau, 1997			*			
Parapentaneura Stur, Fittkau & Serrano, 2006			*			
Pentaneura Philippi, 1865	*	*	*	*	*	*

Table 5. Analysis of the structure between the abiotic data and the biological matrix (NMDS), inter- and intra-annual variation tested from the non-parametric multivariate analysis in subtropical streams.

	NMDS1	NMDS2	R2	p
2010 to 2012				
Water temperature	0.331	0.943	0.078	0.12
pH	-0.967	-0.251	0.094	0.058
Electrical Conductivity	0.559	-0.828	0.011	0.669
Turbidity	-0.812	-0.583	0.016	0.575
DO	-0.168	-0.985	0.088	0.088
TDS	-0.125	0.992	0.042	0.285
Nitrogen	-0.999	0.043	0.041	0.287
TOC	-0.968	-0.247	0.004	0.907
C:N ratio	-0.288	0.957	0.099	0.050
Water velocity	0.891	-0.452	0.016	0.551
Rainfall	-0.847	-0.531	0.110	0.041
2010				
Water temperature	-0.834	-0.551	0.310	0.040
pH	-0.898	0.439	0.118	0.364
Electrical Conductivity	-0.990	-0.134	0.092	0.457
Turbidity	-0.163	-0.986	0.071	0.535
DO	-0.426	-0.904	0.285	0.064
TDS	-0.880	-0.473	0.111	0.378
Nitrogen	-0.711	-0.702	0.316	0.039
TOC	-0.609	0.792	0.284	0.057
C:N ratio	-0.408	0.912	0.260	0.090
Water velocity	0.749	-0.661	0.062	0.588
Rainfall	-0.695	-0.718	0.051	0.626
2011				
Water temperature	-0.039	-0.999	0.474	0.004
pH	0.590	0.806	0.091	0.441
Electrical Conductivity	0.335	-0.941	0.107	0.378
Turbidity	-0.523	0.851	0.238	0.105
DO	-0.271	0.962	0.065	0.527
TDS	0.342	-0.939	0.110	0.371
Nitrogen	-0.328	0.944	0.426	0.009
TOC	0.113	0.993	0.352	0.021
C:N ratio	0.980	0.194	0.196	0.156
Water velocity	-0.579	0.814	0.065	0.462
Rainfall	0.134	0.990	0.309	0.036
2012				
Water temperature	0.986	-0.165	0.007	0.938
pH	0.753	0.658	0.017	0.854
Electrical Conductivity	0.291	-0.956	0.097	0.338
Turbidity	-0.785	0.618	0.226	0.114
DO	0.947	0.320	0.006	0.949
TDS	0.917	0.397	0.023	0.832
Nitrogen	0.948	0.315	0.012	0.903
TOC	0.142	0.989	0.075	0.543
C:N ratio	-0.671	0.741	0.249	0.104
Water velocity	0.332	0.943	0.039	0.726
Rainfall	0.705	0.708	0.019	0.855

teristics of their studied lake changed. Rainfall is one abiotic variable that can create favorable conditions for certain species, not only as a function of the new habitat conditions (BISPO et al. 2006). Water temperature affects the metabolism of organisms and the availability of food, causing changes in community composition (HAHN & FIGI 2007, GRAY & ELLIOTT 2009). The highest concentrations of carbon and nitrogen were found in areas with intense human activity (e.g. agricultural practices) (NEILL et al. 2001, SILVA et al. 2007). Currently, many streams display similar signs of anthropogenic change (mainly as the result of agricultural practices), a phenomenon known as eutrophication of aquatic ecosystems (GALLOWAY et al. 2003, SILVEIRA et al. 2006). The drainage areas of the streams studied were populated with crops and exposed soil (in preparation for cultivation). In these areas, the use of pesticides and fertilizers is high, causing soil contamination particularly when rainfall is intense. These pesticides, plus organic matter and nutrients, are carried into the streams by the rain water.

Organic matter is primarily composed of carbon, but it can be associated with other chemical compounds, such as metals (ALI et al. 2002, AL-SHAMI et al. 2010a, SENSOLO et al. 2012). On the other hand, nitrogen is among the most limiting nutrients to primary productivity and the availability of this nutrient affects the abundance of some aquatic organisms (GALLOWAY et al. 2003). *Cricotopus* species feed on algae, which in turn are dependent on certain concentrations of dissolved nutrients (SENSOLO et al. 2012). *Polypedilum* and *Rheotanytarsus* were the most common organisms in all samples. Studies report that these genera are easily sampled in streams and are reported as cosmopolitan/tolerant (AMORIM et al. 2004, MARCHESE et al. 2005, ABURAYA & CALLIL 2007). Furthermore, the high density of individuals of *Rheotanytarsus* is due to their eating habits: they are filter-feeding organisms, consuming exclusively organic matter present in the water (COFFMAN & FERRINGTON 1996). *Polypedilum* species stand out for being tolerant to a wide range of environmental conditions, as they may occur both in sites impacted with organic compounds and in non-impacted sites (HEINIS & DAVIDS 1993, ROSIN & TAKEDA 2007).

In conclusion, in small temporal scales, local environmental factors have great relative influence on community composition. On the other hand, in larger timescales, climatic factors generate variation (SILVA et al. 2014). In this study, we observed that in those three years, with semi-annual collecting, intra-annual variations are most evident in the chironomid communities. In this study, the temporal variation of the community was dependent on climatic variations (rainfall) as well as the changes in the TOC and TN caused by intensive agricultural land use.

ACKNOWLEDGMENTS

DN received financial support from the Program PROSUP/CAPES. RMR receives financial support from CNPq (Proc. 477274/2011-0 and proc. 475251/2009-1). LUH receives financial support from CNPq (Edital Universal, process 471572/2012-8) and FAPERGS (process 12/1354-0). The authors thank Rodrigo Fornel for their help editing images, and Adriano Melo and two anonymous reviewers for their suggestions and criticisms.

REFERENCES

ABURAYA FH, CALLIL CT (2007) Variação temporal de larvas de Chironomidae (Diptera) no Alto Rio Paraguai (Cáceres Mato Grosso, Brasil). **Revista Brasileira de Zoologia 24**(3): 565-572. doi: 10.1590/S0101-81752007000300007

ALI A, FROUZ J, LOBINSKE RJ (2002) Spatio-temporal effects of selected physico-chemical variables of water, algae and sediment chemistry on the larval community of nuisance Chironomidae (Diptera) in a natural and a man-made lake in central Florida. **Hydrobiologia 470**: 181-193.

AL-SHAMI SA, RAWI CSM, HASSANAHMAD A, NOR SAM (2010a) Distribution of Chironomidae (Insecta: Diptera) in polluted rivers of the Juru River Basin, Penang, Malaysia. **Journal of Environmental Sciences 22**(11) 1718-1727.

AL-SHAMI SA, SALMAH MRC, HASSAN AA, AZIZAH MNS (2010b) Temporal distribution of larval Chironomidae (Diptera) in experimental rice fields in Penang, Malaysia. **Journal of Asia-Pacific Entomology 13**: 17-22. doi: 10.1016/j.aspen.2009.11.006

AMORIM RM, HENRIQUES-OLIVEIRA AL, NESSIMIAN JL (2004) Distribuição espacial e temporal das larvas de Chironomidae (Insecta: Diptera) na seção ritral do rio Cascatinha, Nova Friburgo, Rio de Janeiro, Brasil. **Lundiana 5**(2): 119-127.

BISPO PC, OLIVEIRA LG, BINI LM, SOUSA KG (2006) Ephemeroptera, Plecoptera and Trichoptera assemblages from riffles in mountain streams of central Brazil: environmental factors influencing the distribution and abundance of immature. **Brazilian Journal of Biology 66**(2): 611-622. doi: 10.1590/S1519-69842006000400005

BUSS DF, BAPTISTA DF, NESSIMIAN JL, EGLER M (2004) Substrate specificity, environmental degradation and disturbance structuring macroinvertebrate assemblages inneotropical streams. **Hydrobiologia 518**(1-3): 179-188. doi: 10.1023/B:HYDR.0000025067.66126.1c

COFFMAN WP, FERRINGTON JR LC (1996) Chironomidae, p. 635-754. In: MERRITT KW, CUMMINS RW (Eds.). **An introduction to the aquatic insects of North America.** Dubuque, Kendall, Hunt Publishing.

DECIAN V, ZANIN EM, HENKE C, QUADROS FR, FERRARI CA (2009) Uso da terra na região Alto Uruguai do Rio Grande do Sul e obtenção de banco de dados relacionado a fragmentos de vegetação arbórea. **Perspectiva 33**(121): 165-176.

EPLER J (2001) **Identification manual for the larval Chironomidae (Diptera) of North and South Carolina.** Orlando, Departament of Enviromental and Natural Resources, 526p.

FERRINGTON LC (2008) Global diversity of non-biting midges (Chironomidae; Insecta-Diptera) in freshwater. **Hydrobiologia 595**: 447-455. doi: 10.1007/s10750-007-9130-1

GALDEAN N, CALLISTO M, BARBOSA FAR, ROCHA LA (2000) Lotic ecosystems of Serra do Cipó, southeast Brazil: water quality and a tentative classification based on the benthic macroinvertebrate community. Journal Aquatic Ecosystem Health and Management 3: 545-552. doi: 10.1016/S1463-4988(00)00044-0

GALLOWAY JN, ABER JD, ERSIMAN JW, SEITZINGER SP, HOWARTH RW, COWLING EB, J COSBY (2003) The Nitrogen Cascade. BioScience 53(4): 341-356. doi: 10.1641/0006-3568(2003)053[0341:TNC]2.0.CO;2

GRAY JS, ELLIOTT M (2009) Ecology of Marine Sediments. From Science to Management. Oxford, Oxford University Press, 2nd ed., 225p.

GRESENS SE, BELT KT, TANG JA, GWINN DC, BANKS PA (2007) Temporal and spatial responses of Chironomidae (Diptera) and other benthic invertebrates to urban stormwater runoff. Hydrobiologia 575: 173-190. doi 10.1007/s10750-006-0366-y

HANH NS, FIGI R (2007) Alimentação de peixes em reservatórios brasileiros: alterações e consequências nos estágios iniciais de represamento. Oecologia Brasiliensis 4(11): 469-480. doi: 10.4257/oeco.2007.1104.01

HEINIS F, DAVIDS C (1993) Factors Governing the spatial and temporal distribution of chironomid larvae in the Maarseveen Lakes with special emphasis on the tole of oxygen conditions. Netherlands Journal of Aquatic Ecology 27(1): 21-34.

HENRIQUES-OLIVEIRA AL, NESSIMIAN JL, DORVILLÉ LFM (2003) Feeding habits of chironomid larvae (Insecta: Diptera) from a stream in the floresta da Tijuca, Rio de Janeiro, Brazil. Brazilian Journal Biology 63(2): 269-281. doi: 10.1590/S1519-69842003000200012

HEPP LU, MILESI SV, BIASI C, RESTELLO RM (2010) Effects of agricultural and urban impacts on macroinvertebrates assemblages in streams (Rio Grande do Sul, Brazil). Revista Brasileira de Zoologia 27(1): 106-113. doi: 10.1590/S1984-46702010000100016

HEPP LU, LANDEIRO VL, MELO AS (2012) Experimental Assessment of the Effects of Environmental Factors and Longitudinal Position on Alpha and Beta Diversities of Aquatic Insects in a Neotropical Stream. International Review of Hydrobiology 97(2): 157-167. doi: 10.1002/iroh.201111405

KRUSKAL JB (1964) Multidimensional caling by optimizing goodness of fit to a nonmetric hypothesis. Psychometrika 9(1): 1-27.

MARCHESE MR, WATZEN KM, EZCURRA-DE-DRAGO I (2005) Benthic invertebrate assemblages and species diversity patterns of the upper Paraguay River. River Research and Applications 21(5): 485-499. doi: 10.1002/rra.814

NEILL C, DEEGAN LD, CERRI CC, THOMAZ S (2001) Deforestation for pasture alters nitrogen and phosphorus in small Amazonian streams. Ecological Applications 11(6): 1817-1828. doi: 10.1890/1051-0761(2001)011[1817:DFPANA]2.0.CO;2

OLIVEIRA-FILHO AT, BUDKE JC, JARENKOW JA, EISENLOHR PV, NEVES DRM (2015) Delving into the variations in tree species composition e richness across South American subtropical Atlantic and Pampean forests. Journal of Plant Ecology 8(3): 242-260. doi: 10.1093/jpe/rtt058

OKSANEN J, BLANCHET F, KINDT R, LEGENDRE P, O'HARA RG, SIMPSON GL, SOLYMOS P, STEVENS MHH, WAGNER H (2013) Vegan: Community Ecology Package. R package, v. 1.17-0. Available online at: http://CRAN.R-project.org/package = vegan [Accessed: 12/11/2013]

POFF NL, OLDEN JD, VIEIRA NKM, FINN DS, SIMMONS MP, KONDRATIEFF BC (2006) Functional trait niches of North American lotic insects: traits based ecological applications in light of phylogenetic relationships. Journal of the North American Benthological Society 25(4): 730-755. doi: 10.1899/0887-3593(2006)025[0730:FTNONA]2.0.CO;2

R CORE TEAM (2013) A Language and Environment for Statistical Computing. R Foundation for Statistical Computing. Vienna, ISBN 3-900051-07-0. Available online at: http://www.R-project.org [Accessed: 12/11/2013]

RESTELLO RM, HEPP LU, MENEGATTI C, DECIAN V, HENKE-OLIVEIRA C (2012) Efeito das características da área de drenagem sobre a distribuição de Chironomidae (Diptera) em riachos do Sul do Brasil, p. 324-340. In: SANTOS JE, ZANIN EM, MOSCHINI LE (Orgs.). Faces da Polissemia da Paisagem – Ecologia, Planejamento e Percepção. São Carlos, Rima, vol. 4.

ROQUE FO, TRIVINHO-STRIXINO S, MILAN L, LEITE JG (2007) Chironomid species richness in low-order streams in the Brazilian Atlantic Forest: a first approximation through a Bayesian approach. Journal of North American Benthological Society 26(2): 221-231.

ROSA BFJV, VASQUES M, ALVES RG (2011) Structure and spatial distribution of the Chironomidae community in mesohabitats in a first order stream at the Poc'o D'Anta Municipal Biological Reserve in Brazil. Journal of Insect Science 11: 1-13.

ROSA BFJV, VASQUES M, ALVES RG (2013) Chironomidae (Insecta, Diptera) associated with stones in a first-order Atlantic Forest stream. Revista Chilena de Historia Natural 86: 291-300.

ROSIN GC, TAKEDA AM (2007) Larvas de Chironomidae (Diptera) da planície de inundação do alto rio Paraná: distribuição e composição em diferentes ambientes e períodos hidrológicos. Acta Scientiarum Biological Sciences 29(1): 57-63. doi: 10.4025/actascibiolsci.v29i1.127

SALLES FF, FERREIRA-JÚNIOR N (2014) Hábitat e Hábitos, p. 39-49. In: HAMADA N, JL NESSIMIAN, QUERINO RB (Eds.). Insetos aquáticos na Amazônia brasileira: taxonomia, biologia e ecologia. Manaus, INPA, 724p.

SANSEVERINO AM, NESSIMIAN JL (2001) Haìbitats de larvas de Chironomidae (Insecta, Diptera) em riachos de Mata Atlantica no Estado do Rio de Janeiro. Acta Limnologica Brasiliensia 13: 29-38.

SCHEFFER M, VAN NES EH (2007) Shallow lakes theory revisited: various alternative regimes driven by climate, nutrients, depth and lake size. Hydrobiologia 584: 455-466. doi: 10.1007/978-1-4020-6399-2_41

Sensolo D, Hepp LU, Decian V, Restello RM (2012) Influence of landscape on assemblages of Chironomidae in Neotropical streams. **Annales de Limnologie – International Journal of Limnology 48**: 391-400. doi: 10.1051/limn/2012031

Shuvartz M, Oliveira LG, Diniz-Filho JAF, Bini LM (2005) Relações entre distribuição e abundância de larvas de Trichoptera (Insecta), em córregos de Cerrado no entorno do Parque Estadual da Serra de Caldas (Caldas Novas, Estado de Goiás). **Acta Scientiarum Biological Sciences 27**(1): 51-55. doi: 10.4025/actascibiolsci.v27i1.1360

Silva DML, Ometto JPHB, Lobo GA, Lima WP, Scaranello MA, Mazi E, Roch HR (2007) Can Land Use Changes Alter Carbon, Nitrogen and Major Ion Transport in Subtropical Brazilian Streams? **Scientia Agricola 64**(4): 317-324. doi: 10.1590/S0103-90162007000400002

Silva JS, Albertoni EF, Silva CP (2014) Temporal variation of phytophilous Chironomidae over a 11-year period in a shallow Neotropical lake in southern Brazil. **Hydrobiologia 737**(1): 1-14. doi: 10.1007/s10750-014-1972-8

Silveira MP, Buss DF, Nessimian JL, Baptista DF (2006) Spatial and temporal distribution of benthic macroinvertebrates in southeastern Brazilian river. **Brazilian Journal of Biology 66**(2): 623-632. doi: 10.1590/S1519-69842006000400006

Siqueira T, Roque FO, Trivinho-Strixino S (2008) Phenological patterns of neotropical lotic chironomids: Is emergence constrained by environmental factors? **Austral Ecology 33**: 902-910. doi: 10.1111/j.1442-9993.2008.01885.x

Smith H, Wood PJ, Gunn J (2003) The influence of habitat structure e flow permanence on invertebrate communities in karst spring systems. **Hydrobiologia 510**(1): 53-66. doi: 10.1023/B:HYDR.0000008501.55798.20

Súarez YR (2008) Variação espacial e temporal na diversidade e composição de espécies de peixes em riachos da bacia do Rio Ivinhema, Alto Rio Paraná. **Biota Neotropica 8**(3): 197-204. doi: 10.1590/S1676-06032009000100012

Trivinho-Strixino S (2011) **Larvas de Chironomidae. Guia de identificação.** São Carlos, Departamento de Hidrobiologia, Laboratório de Entomologia Aquática, UFSCar, 371p.

Trivinho-Strixino S, Strixino G (1995) **Larvas de Chironomidae (Diptera) do estado de São Paulo: guia de identificação e diagnose dos gêneros.** São Carlos, PPG-RRN/UFSCar, 229p.

6

The activity time of the lesser bamboo bat, *Tylonycteris pachypus* (Chiroptera: Vespertilionidae)

Li-Biao Zhang[1,*], Fu-Min Wang[2], Qi Liu[1] & Li Wei[3]

[1]*Guangdong Public Laboratory of Wild Animal Conservation and Utilization & Guangdong Key Laboratory of Integrated Pest Management in Agriculture, Guangdong Entomological Institute, Guangzhou 510260, China.*
[2]*Guangdong Provincial Wildlife Rescue Center, Guangzhou 510520, China.*
[3]*College of Ecology, Lishui University, Lishui 323000, China.*
Corresponding author. E-mail: zhanglb@gdei.gd.cn

ABSTRACT. The activity time of the lesser bamboo bat, *Tylonycteris pachypus* (Temminck, 1840), was investigated at two observation locations in southern China: Longzhou and Guiping. Two bouts of activity (post dusk and predawn), with an intervening period of night roosting at diurnal roosts, were identified. The period of activity within each bout was usually less than 30 minutes. The activity periods of individuals belonging to the Longzhou population right after dusk and just before dawn lasted longer than those of the the Guiping population. We also found that the nocturnal emergence time of *T. pachypus* from the Longzhou population happened earlier than in the Guiping population. These findings indicate that the activity time of *T. pachypus* was quite short at night, and that different locations may affect the nocturnal activity rhythm of this species.

KEY WORDS. Activity period; emergence; return; *Tylonycteris pachypus*.

Most bats are nocturnal, foraging at night and resting in roosts during the day. Between activity bouts, they also spend time in night roosts (ANTHONY & KUNZ 1997). The patterns of nocturnal activity vary dramatically among different species. O'SHEA & VAUGHAN (1977) have reported that the pallid bat *Antrozous pallidus* (Le Conte, 1856) utilizes two foraging periods with an intervention period of night roosting. Some other species, such as *Euderma maculatum* (J.A. Allen, 1891), spend the entire night flying and foraging (WAI-PING & FENTON 1988). Likewise, the duration of bouts has also been found to be different. For instance, *Eptesicus fuscus* (Beauvois, 1796) spends only 2 hours flying each night (BRIGHAM 1991), while *Nacteris grandis* spends even less time in this activity (FENTON et al. 1990).

The timing and pattern of bat nocturnal activity may be influenced by environmental factors such as light levels (LEE & MCCRACKEN 2001), prey abundance (ERKERT 1982), temperature (CATTO et al. 1995), cloud (KUNZ & ANTHONY 1996) and rain (MCANEY & FAIRLEY 1988). Moreover, intrinsic biological factors such as predation risk (MCWILLIAM 1989, SPEAKMAN 1991), colony size, age, sex, the reproductive status of individuals (AVERY 1986, RYDELL 1989, KORINE et al. 1994, CLARK et al. 2002, O'DONNELL 2002), and interspecific competition (SWIFT & RACEY 1983, BONACCORSO et al. 2006) may also impact the nocturnal activity of bats. The intensity of competition between or among sympatric related species is expected to be greater because they are morphologically similar, which is assumed to reflect niche similarity (FINDLEY & BLACK 1983, ALDRIDGE & RAUTENBACH 1987, ARITA

1997). When common resources are limited, the niche theory predicts that resource partitioning (such as spatial and temporal niche) is necessary for species to coexist within a guild. They may forage in different habitats (ARLETTAZ 1999) and then feed on different diet items (ARLETTAZ et al. 1997), or forage during different times (BONACCORSO et al. 2006). In contrast, when their shared resources are not limited, species may forage concomitantly and for longer periods.

The lesser bamboo bat, *Tylonycteris pachypus* (Temminck, 1840) (Chiroptera: Vespertilionidae), and its sibling species, *Tylonycteris robustula* Thomas, 1915, are genetically closely related. The two species have similar morphological features (MEDWAY & MARSHALL 1978). According to our field observations, which have been published in another contribution, the two species rarely roost together in the same internode, although they overlap in distribution. Also, they can alternate their use of the same internodes at different times or seasons (ZHANG et al. 2004) and forage on similar categories of insects that occur sympatrically (ZHANG et al. 2005a).

The nocturnal activity rhythm of *T. pachypus* and *T. robustula* is still poorly known. In the present study we investigated the activity time (emergence time and foraging duration) of the lesser bamboo bat, *T. pachypus*, in two populations inhabiting Longzhou and Guiping Counties, Guangxi, south China, approximately 250 km apart from each other. *T. pachypus* is sympatric with *T. robustula* in Longzhou County, but occurs alone in Guiping County.

MATERIAL AND METHODS

The study was carried out from March to November, in 2008 in Guangxi Province, south China. Two locations (Longzhou County, 21°10'N, 106°50'E, 116 m in elevation, and Guiping County, 23°09'N, 110°10'E, 68 m in elevation) were selected. ZHANG et al. (2005a) had previously described the climate and vegetation of Longzhou County as having average annual temperature of 22.8°C and average annual precipitation of 1,180 mm (FANG 1995). Guiping County has a similar climate but the habitat is hilly rather than the typical karst of Longzhou County, with average annual temperature of 21.4°C and average annual precipitation of 1,727 mm (FANG 1995). Within both study areas, the bamboo, *Bambusa spinosa* Roxb, is abundant. This plant has enough internodes to provide enough suitable roost sites for both bat species. During data collection, *T. pachypus* and *T. robustula* were never found in the same roost at the same time, except on one occasion, when a single *T. pachypus* male and a single *T. robustula* male roosted in the same internode. In Longzhou County, bamboo is found in and around villages, while in Guiping County it is distributed along streams in areas that are somewhat far from villages.

Nocturnal observations were conducted from dusk to the next morning. Observers were split into two groups (two persons per group) and each group observed the activity time of *T. pachypus* in the two locations, at the same time. Normally, bamboo bats are faithful to their bamboo internodes for a short period (ZHANG et al. 2004), which allowed us to continuously observe groups in the same internode from dusk to dawn. We selected a fixed bamboo forest in each location to conduct our observations. The bamboo bat colony in each forest had more than two hundred individuals. We swapped among different bamboo internodes on different days during the same month, and the size of the group in each of these internodes was normally over eight bats. In each location, the time of emergence and returning were recorded, respectively. Emergence time was defined as the time when the first bat individual flied out of its bamboo internode; the returning time was defined as the time when the first bat individual flied back into the bamboo roost, or attempted to do so. If no bat emerged from a bamboo internode after 20 minutes (confirmed by using a wire to check bat lairs), we terminated the observation on emergence time. Likewise, if no bat returned back to its bamboo internode we terminated the observation on returning time. The duration of activity was defined as the time period between emergence and returning, since it is difficult to observe the entire behavior of the bat after it flies out. Additionally, *T. pachypus* forages around or over their roost bamboo forest; for example, their foraging sites are nearby their roosts (ZHANG et al. 2007). We also recorded air temperature at sunset and sunrise, and position of the roost site (via GPS, eTrex, Garmin Corp., Taiwan). The time of sunset and sunrise were read from a GPS. During the observation period, nocturnal observations were conducted for at least one week per

month, normally in the middle of the month. Since *T. pachypus* is sympatric with *T. robustula* in Longzhou, we identified and confirmed the identify of *T. pachypus* using characteristics of its echolocation calls using a sound detector (D-980, Petterson Electronic AB, Uppsala) when individuals flied out or returned. This study was conducted according to the protocols approved by the Guangdong Entomological Institute Administrative Panel on Laboratory Animal Care.

A total of 75 night observations were conducted both in Longzhou and Guiping. When it was raining during the bat's normal activity period, the data of the corresponding night were omitted from both sites in the analysis. As a result, data on twelve night observations were discharged, and 63 night observations, one week per month, were analyzed. All data were tested for normality and homogeneity of variances using Kolmogorov-Smirnov test and Bartlett test. We conducted Independent-Samples t-test for comparisons of emergence time, returning time and duration of activity between the Longzhou and Guiping populations, respectively. Then for comparisons of monthly variations in activity duration, we used One-Way ANOVA. Regression was used to analyze activity time and local air temperature. All statistical analyses were performed in SPSS 17.0 for Windows. Descriptive data were expressed as Mean ± SD, and the significant difference at 95% confidence level was calculated by Central Limit Theorem.

RESULTS

Based on our observations, the lesser bamboo bat *T. pachypus* normally feeds in and around bamboos in the forests in which they roost, and fly to foraging sites directly from their roost. The activity time of *T. pachypus* individuals in the two populations was characterized by two distinct bouts, one immediately after dusk and the second just before dawn. The duration of the activity within each bout was usually less than 30 minutes. The dusk (23.7 ± 8.39 min) and predawn (27.9 ± 14.79 min) activity periods of *T. pachypus* individuals were significantly longer in Longzhou than in Guiping (dusk: 21.2 ± 8.09 min; predawn: 25.4 ± 14.43 min) (dusk: t = 2.374, p < 0.05; predawn: t = 2.857, p < 0.05) (Fig. 1). A strong positive correlation was also observed between activity duration and air temperature at dusk and predawn in the two populations, respectively (Longzhou: r_{dusk} = 0.891, $r_{predawn}$ = 0.904, respectively, both p < 0.001; Guiping: r_{dusk} = 0.881, $r_{predawn}$ = 0.837, both p < 0.001) (Figs. 2-3). In addition, we found that the durations of activity of *T. pachypus* varied significantly in both populations from one month to another (Longzhou: F = 4.069, p < 0.001; Guiping: F = 3.619, p < 0.001) (Figs. 2-3).

The time of emergence obtained for the Longzhou population at post dusk (12.7 ± 0.62 min after sunset) was significantly earlier than for the Guiping population (14.7 ± 0.48 min) (t = 2.37, p < 0.05), but the returning time at predawn did not significantly differ between them (Longzhou: 17.3 ± 0.94

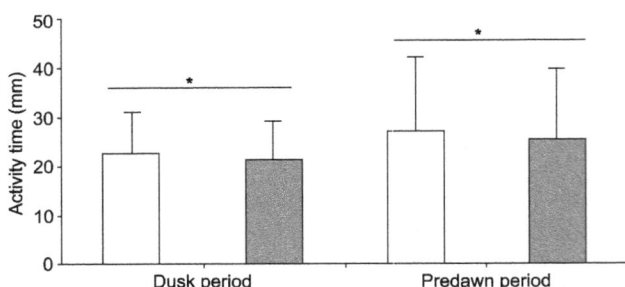

Figure 1. The foraging time of dusk (n = 63) and predawn (n = 63) periods of *Tylonycteris pachypus* in Longzhou County (empty column) and Guiping County (grey column), Guangxi, south China. All data are expressed as Mean ± SD. *p < 0.05.

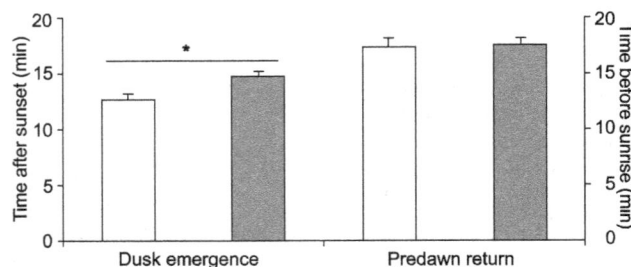

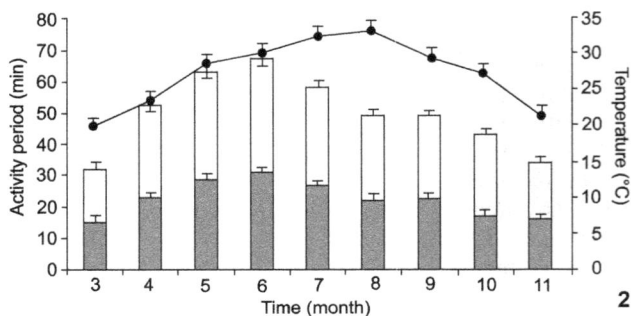

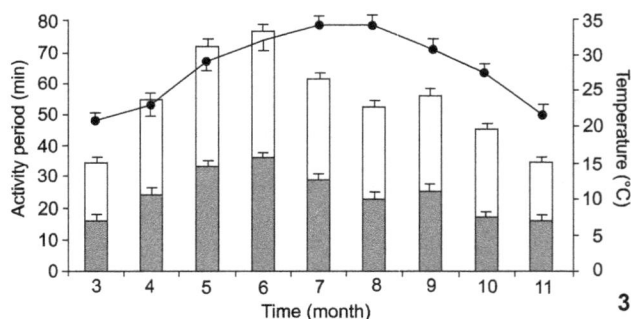

Figure 4. Time of dusk emergence (n = 63) and predawn returning (n = 63) of *Tylonycteris pachypus* in Longzhou County (empty column) and Guiping County (grey column), Guangxi, south China. All data are expressed as Mean ± SD. *p < 0.05.

Figure 2-3. The total foraging time over one night (average value of seven nights for each month) of *Tylonycteris pachypus* in Longzhou County (2) and Guiping County (3). Solid portions indicate dusk foraging periods while hollow portions indicate dawn foraging periods. The dots indicate the average temperature.

min before sunrise, Guiping: 17.5 ± 0.62 min) (t = 1.38, p > 0.05) (Fig. 4). We also found that there is a significant correlation between the time of evening emergence and the average air temperature at sunset, and between time of predawn return and average air temperature at sunrise for both populations, respectively (Longzhou: $r_{emergence}$ = 0.882, r_{return} = -0.915, respectively, both p < 0.001; Guiping: $r_{emergence}$ = 0.735, r_{return} = -0.826, both p < 0.001).

DISCUSSION

Mammals that are characterized by energetically expensive modes of locomotion and which encounter limited (temporally) food supply tend to regulate their foraging behavior. Bat species usually emerge to feed after sunset and return before sunrise (Bateman & Vaughan 1974, O'Shea & Vaughan 1977, Duvergé et al. 2000). The timing of these activity bouts correspond to the time at dusk and predawn when insects are most abundant (Kunz 1974, Racey 1982, Racey & Swift 1985, Rydell 1993, Swift 1997). Based on previous studies, most bat species spend more than one hour each night in predation activities, and their nocturnal activity rhythm can be highly variable among different species. For instance, O'Shea & Vaughan (1977) have reported that the activity of pallid bats was characterized by two foraging periods with an intervening period of night roosting. Anthony et al. (1981) have also reported that the little brown bat typically has two foraging periods at night, which are divided by a short break in night roost. In the present study, we found that the night foraging time patterns of *T. pachypus* were characterized by two bouts: dusk and predawn, respectively. Within each bout, activity duration (time period from emergence to returning) was remarkably short, less than 30 minutes. This may be correlated with the fact that the lesser bamboo bat is one of the smallest bat species, and therefore individuals can spend relatively short periods of activity time to balance energy intake and the costs to maintain their high metabolic rate during flight (Speakman 2005). On the other hand, the roosting behavior of *T. pachypus* may also contribute to their short activity time. *T. pachypus* roosts within bamboo internodes, and these restricted spaces may limit the activity of bats and subsequently reduce their energy consumption. Moreover, the short foraging distance away from their internode roosts decrease flight time as well as energy consumption (Zhang et al. 2007).

Our results indicate that in Longzhou County *T. pachypus* emerges earlier and forages for longer periods of time than that in Guiping County. On average, the activity time of individuals in Longzhou each night is approximately five minutes, which is longer than that in Guiping. Although this increase is small in magnitude, it represents a 10% increase in average activity time. The different activity behavior in different locations may have resulted from many factors such as the effect of interspecific competition (SWIFT & RACEY 1983, BONACCORSO et al. 2006), variations in prey abundance (ERKERT 1982), predation risk (MCWILLIAM 1989, SPEAKMAN 1991), and colony size (AVERY 1986, RYDELL 1989, KORINE et al. 1994, CLARK et al. 2002, O'DONNELL 2002). Both *T. pachypus* and *T. robustula* eat similar categories of insects when they occur sympatrically, although the prey of the later is somewhat larger than that of the former (ZHANG et al. 2005a). In conclusion, in Longzhou where both sibling species are sympatric, the activity time of *T. pachypus* may be affected by *T. robustula*.

Many authors have documented seasonal fluctuations in emergence and returning time of bats (BATEMAN & VAUGHAN 1974, O'SHEA & VAUGHAN 1977, DUVERGÉ et al. 2000). Pallid bats emerge earlier in the summer sunset when compared with spring and autumn sunsets (O'SHEA & VAUGHAN 1977). LEE & MCCRACKEN (2001) have also reported that the timing of evening emergence and returning of the Mexican free-tailed bats, *Tadarida brasiliensis* (I. Geoffroy, 1824), are correlated with the time of sunset and sunrise, and that bats are more likely to emerge earlier in relation to earlier sunset time during late summer, when compared with spring to early summer. They also return progressively later at dawn, which is associated with sunrise in the entire season. We found that the dawn returning times of *T. pachypus* were correlated with sunrise from March to November. In this study, *T. pachypus* emerged only 5-10 minutes after sunset during summer, but 10-20 minutes during spring and autumn. Pallid bats emerged 20-40 minutes after sunset (O'SHEA & VAUGHAN 1977) and greater horseshoe bats emerged 45-53 minutes after sunset (DUVERGÉ et al. 2000). KUNZ et al. (1995) have pointed out that increased energetic demands during pregnancy and lactation could cause females, especially those lactating, to spend twice or even three times as much time foraging than bats that are neither pregnant nor lactating. As a result, emerging earlier in the evening and returning later at dawn will give these bats more time to feed, but at the cost of a greater danger of predation (FENTON 1995, SPEAKMAN et al. 1995, RYDELL et al. 1996). Our results indicate that the activity time of flat-headed bats is longer in summer than in spring and in autumn. This may be influenced by prey activity, reproductive status and energy demands (RICHARDS 1989). Female flat-headed bats usually become pregnant in May, and give birth and lactate in June (ZHANG et al. 2005b). During this time, reproductive females would require more energy to maintain the increased physiological requirements of pregnancy and lactation, which is reflected in their increased activity time.

Seasonal fluctuations in activity patterns can be influenced by many factors, such as ambient temperature and/or prey availability (ANTHONY et al. 1981, RYDELL 1989, MAIER 1992). In the present study, the bamboo bats emerged earlier and returned latter when the ambient temperature was higher, which resulted in a positive association between high ambient temperature and increased activity. In March and November, when temperatures were relatively low, the activity periods of bamboo bats were obviously short than from April to October. Reduced activity in lower temperatures has also been demonstrated for many other insectivorous bat species (e.g., ANTHONY et al. 1981, KRONWITTER 1988, RYDELL 1989, MAIER 1992, CATTO et al. 1995). This negative influence may be a reflection of decreased food availability, because insect activity throughout the night is positively correlated with temperatures (TAYLOR 1963, LEWIS & TAYLOR 1964).

In conclusion, our findings suggest that the lesser bamboo bat spends relatively short time being active, even less than half hour each bout. This behavior may be correlated with the high-energy demand for this tiny mammal to be able to fly. Secondly, our findings suggest that the activity behavior of the lesser bamboo bat, including emergence time and activity duration, varies in different locations. The variation in activity may result from many factors such as the effect of sibling sympactric species, and variations in prey abundance, predation risk, and colony size in different locations. Further studies should be conducted to confirm the factors that influence the variations in bat activity.

ACKNOWLEDGEMENTS

We thank Hui Qin, Shen-Ming Huang (Guangxi Normal University) for their assistance in the field, and Stuart Parsons (University of Auckland) and Yi Chen for assistance in writing this manuscript. This study was financed by the Special Planning of Major Scientific and Technological Production (Cultivation Project), Guangdong Academy of Sciences (ZDCCYD201307), Special Foundation for Innovative Scientists of Guangdong Entomological Institute (GDEI-cxrc201303), and Science & Technology Planning Project of Guangdong (2013B050800024). Li-Biao Zhang and Fu-Min Wang contributed equally to this paper.

REFERENCES

ALDRIDGE HDJN & RAUTENBACH IL (1987) Morphology, echolocation and resource partitioning in insectivorous bats. **Journal of Animal Ecology** 56: 763-778. doi: 10.2307/4947.

ANTHONY ELP, KUNZ TH (1997) Feeding strategies of the little brown bat, *Myotis lucifugus*, in southern New Hampshire. **Ecology** 58: 775-786. doi: 10.2307/1936213

ANTHONY ELP, STACK MH, TH KUNZ (1981) Night roosting and the nocturnal time budget of the little brown bat, *Myotis lucifugus*: effects of reproductive status, prey density, and environmental conditions. **Oecologia** 51: 151-156. doi: 10.1007/BF00540593

ARITA E (1997) Species composition and morphological structure of the bat fauna of Yucatan, Mexico. **Journal of Animal Ecology** 66: 83-97. doi.org/10.2307/5967

ARLETTAZ R (1999) Habitat selection as a major resource partitioning mechanism between the two sympatric sibling bat species *Myotis myotis* and *Myotis blythii*. **Journal of Animal Ecology** 68: 460-471. doi: 10.1046/j.1365-2656.1999.00293.x

ARLETTAZ R, PERRIN N, HAUSSER J (1997) Trophic resource partitioning and competition between the two sibling bat species *Myotis myotis* and *Myotis blythii*. **Journal of Animal Ecology** 66: 897-991. doi: 10.2307/6005

AVERY MI (1986) Factors affecting the emergence times of *Pipistrelle* bats. **Journal of Zoology (London)** 209: 293-296. doi: 10.1111/j.1469-7998.1986.tb03589.x

BATEMAN GC, VAUGHAN TA (1974) Nightly activities of mormoopid bats. **Journal of Mammalogy** 55: 45-65. doi: 10.2307/1379256

BONACCORSO FJ, WINKELMANN JR, SHIN D, AGRAWAL CI, ASLAMI N, BONNEY C, HSU A, JEKIELEK PE, KNOX AK, KOPACH SJ, JENNINGS TD, LASKY JR, MENESALE SA, RICHARDS JH, RUTLAND JA, SESSA AK, ZHAUROVA L, KUNZ TH (2006) Evidence for exploitative competition: comparative foraging behavior and roosting ecology of short-tailed fruit bats (Phyllostomidae). **Biotropica** 39: 249-256. doi: 10.1111/j.1744-7429.2006.00251.x

BRIGHAM RM (1991) Flexibility in foraging and roosting behavior by the big brown bat (*Eptesicus fuscus*). **Canada Journal of Zoology** 69: 117-121. doi: 10.1139/z91-017

CATTO CMC, RACEY PA, STEPHENSON PJ (1995) Activity patterns of the serotine bat (*Eptesicus serotinus*) at a roost in southern England. **Journal of Zoology (London)** 235: 635-644. doi: 10.1111/j.1469-7998.1995.tb01774.x

CLARK BS, CLARK BK, LESIE JR DM (2002) Seasonal variation in activity patterns of the endangered Ozark big-eared bat (*Corynorhinus townsendii ingens*). **Journal of Mammalogy** 83: 590-598. doi: 10.1644/1545-1542

DUVERGÉ PL, JONES G, RYDELL J, RANSOME RD (2000) Functional significance of emergence timing in bats. **Ecography** 23: 32-40. doi: 10.1034/j.1600-0587.2000.230104.x

ERKERT HG (1982) Ecological aspects of bat activity rhythms, p. 201-242. In: KUNZ TH (Ed.). **Ecology of bats.** New York, Plenum Press. doi: 10.1007/978-1-4613-3421-7-5

FANG WZ (1995) **Natural resource of Guangxi, China.** Beijing, China Environment Science Press, 369p.

FENTON MB (1995) Constraint and flexibility-bats as predators, bats as prey. **Symposium of the Zoological Society of London** 67: 277-290.

FENTON MB, SWANEPOEL CM, BRIGHAM RM, CEBEK E, HICKEY MB (1990) Foraging behavior and prey selection by large slit-faced bats (*Nycteris grandis*; Chiroptera: Nycteridae). **Biotropica** 22: 2-8. doi: 10.2307/2388713

FINDLEY JS, BLACK H (1983) Morphological and dietary structuring of a Zambian insectivorous bat community. **Ecology** 64: 625-630. doi: 10.2307/1937180

KORINE C, IZHAKI I, MAKIIN D (1994) Population structure and emergence order in the fruit-bat (*Rousettus aegyptiacus*: Mammalia, Chiroptera). **Journal of Zoology (London)** 232: 163-174. doi: 10.1111/j.1469-7998.1994.tb01566.x

KRONWITTER F (1988) Population structure, habitat use, and activity patterns of the noctule bat, *Nyctalus noctula* Schreib. 1774 (Chiroptera: Vespertilionidae) revealed by radio-tracking. **Myotis** 26: 23-85.

KUNZ TH (1974) Feeding ecology of a temperate insectivorous bat (*Myotis velifer*). **Ecology** 55: 693-711.

KUNZ TH, ANTHONY ELP (1996) Variation in the timing of nightly emergence behavior in the little brown bat, *Myotis lucifugus* (Chiroptera: Vespertilionidae), p. 225-235. In: GENOWAYS HH, BAKER RJ (Eds.). **Contribution in Mammalogy: a memorial volume honoring Dr. J. Knox.** Lubbock, The Museum of Texas Tech University, 315p.

KUNZ TH, WHITAKER JR JO, WADANOLI MD (1995) Dietary energetics of the insectivorous Mexican free-tailed bats (*Tadarida brasiliensis*) during pregnancy and lactation. **Oecologica** 101: 407-415. doi: 10.1007/BF00329419

LEE Y-F, MCCRACKEN GF (2001) Timing and variation in the emergence and return of Mecican free-tailed bat, *Tadarida brasiliensis mexicana*. **Zoological Study** 40: 309-316.

LEWIS T, TAYLOR LR (1964) Diurnal periodicity of flight by insects. **Transactions of the Entomological Society of London** 116: 396-435. doi: 10.1111/j.1365-2311.1965.tb02304.x

MAIER C (1992) Activity patterns of pipistrelle bats (*Pipistrellus pipistrellus*) in Oxfordshire. **Journal of Zoology (London)** 116: 396-435. doi: 10.1111/j.1469-7998.1992.tb04433.x

MCANEY CM, FAIRLEY JS (1988) Activity patterns of the lesser horseshoe bat *Rhinolophus hipposideros* at summer roosts. **Journal of Zoology (London)** 216: 325-338. doi: 10.1111/j.1469-7998.1988.tb02433.x

MCWILLIAM AM (1989) Emergence behavior of the bat *Tadarida (Chaerephon) pumila* (Chiroptera: Molossidae) in Ghana, West Africa. **Journal of Zoology (London)** 219: 698-701. doi: 10.1111/j.1469-7998.1989.tb02615.x

MEDWAY L, MARSHALL AG (1978) Roost-site selection among flat-headed bats (*Tylonycteris* spp.). **Journal of Zoology (London)** 161: 237-245. doi: 10.1111/j.1469-7998.1970.tb02038.x

O'DONNELL CFJ (2002) Influence of sex and reproductive status on nocturnal activity of long-tailed bats (*Chalinolobus tuberculatus*). **Journal of Mammalogy** 83: 794-803. doi: 10.1644/1545-1542

O'SHEA TJ, VAUGHAN TA (1977) Nocturnal and seasonal activities of the pallid bat, *Antrozous pallidus*. **Journal of Mammalogy** 58: 269-284. doi: 10.2307/1379326

RACEY PA (1982) Ecology of bat reproduction, p. 335-427. In: T.H. KUNZ (Ed.). **Ecology of bats.** New York, Plenum Publishing. doi: 10.1007/978-1-4613-3421-7_2

RACEY PA, SWIFT SM (1985) Feeding ecology of *Pipistrellus pipistrellus* (Chiroptera: Vespertilionidae) during pregnancy and lactation. I. Foraging behavior. **Journal of Animal Ecology** 54: 205-215. doi: 10.2307/4631

RICHARDS GC (1989) Nocturnal activity of insectivorous bats relative to temperature and prey availability in tropical Queensland. **Australian Wildlife Research** 16: 151-158. doi: 10.1071/WR9890151

Rydell J (1989) Feeding activity of the northern bat *Eptesicus nilssoni* during pregnancy and lactation. **Oecologica 80:** 562-565. doi: 10.1007/BF00380082

Rydell (J) 1993. Variation in foraging activity of an aerial insectivorous bat during reproduction. **Journal of Mammalogy 74:** 503-509. doi: 10.2307/1382411

Rydell J, A Entwistle, PA Racey (1996) Timing of foraging flights of three species of bats in relation to insect activity and predation risk. **Oikos 76:** 243-252. doi: 10.2307/3546196

Speakman JR (1991) Why do insectivorous bats in Britain not fly in daylight more frequently? **Functional Ecology 5:** 518-524. doi: 10.2307/2389634

Speakman JR (2005) Body size, energy metabolism and lifespan. **Journal of Experimental Biology 208:** 1717-1730. doi: 10.1242/jeb.01556

Speakman JR, Stone RE, Kerslake JE (1995) Temporal patterns in the emergence behavior of pipistrelle bats, *Pipistrellus pipistrellus*, from maternity colonies are consistent with an anti-predator response. **Animal Behavior 50:** 1147-1156. doi: 10.1016/0003-3472(95)80030-1

Swift SM (1997) Roosting and foraging behavior of Natterer's bats (*Myotis nattereri*) close to the northern border of their distribution. **Journal of Zoology (London) 242:** 375-384. doi: 10.1111/j.1469-7998.1997.tb05809.x

Swift SM, Racey PA (1983) Resource partitioning in two species of vespertilionid bats (Chiroptera) occupying the same roost.

Journal of Zoology (London) 200: 249-2593. doi: 10.1111/j.1469-7998.1983.tb05787.x

Taylor LR (1963) Analysis of the effect of temperature on insects in flight. **Journal of Animal Ecology 32:** 99-117.

Wai-Ping V, Fenton MB (1988) Ecology of spotted bat (*Euderma maculatum*): roosting and foraging behavior. **Journal of Mammalogy 70:** 617-622. doi: 10.2307/1381434

Zhang L-B, Liang B, Jones G, Parsons S, Wei L, Zhang S-Y (2007) Morphology, echolocation and foraging behavior in two sympatric sibling species of bat (*Tylonycteris pachypus* and *T. robustula*) (Chiroptera: Vespertilionidae). **Journal of Zoology (London) 271:** 344-351. doi: 10.1111/j.1469-7998.2006.00210.x

Zhang L-B, Liang B, Zhou S-Y, Lu L-R, Zhang S-Y (2004) Group structure of lesser flat-headed bat *Tylonycteris pachypus* and greater flat-headed bat *T. robustula*. **Acta Zoology Sinica 50:** 326-333.

Zhang L-B, Jones G, Rossiter S, Ades G, Liang B, Zhang S-Y (2005a) Diet of flat-headed bats, *Tylonycteris pachypus* and *T. robustula*, in Guangxi, South China. **Journal of Mammalogy 86:** 61-66. doi: 10.1644/1545-1542

Zhang L-B, Jones G, Parsons S, Liang B, Zhang S-Y (2005b) Development of vocalizations in the flat-headed bats, *Tylonycteris pachypus* and *T. robustula* (Chiroptera: Vespertilionidae). **Acta Chiropterologica 7:** 91-99. doi: 10.3161/1733-5329

7

A new species of *Kingsleya* (Crustacea: Decapoda: Pseudothelphusidae) from the Xingu River and range extension for *Kingsleya junki*, freshwater crabs from the Southern Amazon basin

Manuel Pedraza[1], José Eduardo Martinelli-Filho[2] & Célio Magalhães[3,4]

[1]*Programa de Pós-Graduação, Museu de Zoologia, Universidade de São Paulo. Avenida. Nazaré 481, Ipiranga, 04263-000 São Paulo, SP, Brazil. E-mail: manupedrazam@gmail.com*
[2]*Faculdade de Oceanografia, Instituto de Geociências da Universidade Federal do Pará. Campus Universitário do Guamá, 66075-110 Belém, PA. Brazil.*
[3]*Instituto Nacional de Pesquisas da Amazônia. Caixa Postal 2223, 69080-971 Manaus, AM, Brazil.*
[4]*Corresponding author. E-mail: celiomag@inpa.gov.br*

ABSTRACT. *Kingsleya castrensis* **sp. nov.**, a pseudothelphusid crab is described and illustrated from the Xingu River, state of Pará, southern Amazon region, Brazil. The new species is characterized by the male first gonopod bearing a large, well-developed apical plate, with a broadly rounded, thick distal lobe. New records of *Kingsleya junki* Magalhães, 2003 extend the distribution of this species eastward to the Tocantins River basin, in the state of Pará, Brazil.

KEY WORDS. Amazon; Brachyura; Kingsleyini; Neotropical region; taxonomy.

Kingsleya Ortmann, 1897 currently comprises seven species that are all distributed in the highlands of the Guyanan and Central Brazilian Shields. This area encompasses a large portion of northern South America from southern Venezuela, Guyana, Suriname, and French Guiana to the northern Brazilian states of Amazonas, Pará and Roraima (MAGALHÃES 2003a, MAGALHÃES & TÜRKAY 2008). In Brazil, species of this genus occurs in tributaries of the Amazon River draining the Guyana Shield where it is represented by *K. latifrons* (Randall, 1840) (in Rio Branco, Rio Negro, and Rio Trombetas), and by *K. siolii* Bott, 1967 (in Rio Trombetas and Rio Paru do Oeste); and the Central Brazilian Shield: *Kingsleya gustavoi* Magalhães, 2005 (Rio Tocantins), and *K. junki* Magalhães, 2003 (Xingu River). *Kingsleya ytupora* Magalhães, 1986 (found in the Rio Uatumã, Rio Trombetas, Rio Curuá-Una, Rio Xingu) is the only species known to occur on both sides of the Amazon valley (MAGALHÃES 1986, 2003a, b, MAGALHÃES & TÜRKAY 2008).

Although pseudothelphusids living in high altitude localities typically have restricted distributions, this may not be the case for species living in the Amazon basin which have wide distributions (although much of the southern Amazon River tributaries are still poorly surveyed for decapods). Crab samples sporadically collected during ichthyological and entomological expeditions to the middle and lower course of the Xingu River were studied by MAGALHÃES (2003b) and indicated that five species occur in this stretch of the river basin: two pseudothelphusids (*K. junki* and *K. ytupora*) and three trichodactylids – *Sylviocarcinus devillei* H. Milne Edwards, 1853,

S. pictus (H. Milne-Edwards, 1953), and *Trichodactylus ehrhardti* Bott, 1969. Recent collections from southern tributaries of the Amazon river revealed the presence of *Kingsleya*, including an undescribed species of this genus from the surroundings of the city of Altamira, on the left bank of the middle course of the Xingu River. The new species is herein described and illustrated, and a range extension for *K. junki* is reported.

MATERIAL AND METHODS

Specimens are deposited at the Instituto Nacional de Pesquisas da Amazônia, Manaus, Brazil (INPA), Museu Nacional, Universidade Federal do Rio de Janeiro, Rio de Janeiro (MNRJ), Museu Paraense Emilio Goeldi, Belém, Brazil (MPEG), Museu de Zoologia, Universidade de São Paulo, São Paulo, Brazil (MZUSP), and Senckenberg Research Institute and Natural History Museum (SMF). The following abbreviations are used: carapace width (cw), measured across the carapace at its widest point; carapace length (cl), measured along the midline, from the frontal to the posterior margin; carapace height (ch), the maximum height of the cephalothorax, measured as the distance between the dorsal and ventral edges of the shell; frontal width (fw), the width of the front measured along its upper border; male first (G1) and second (G2) gonopods; third maxilliped (Mxp3); cheliped (P1); pereiopods 2 to 5 (P2-P5); and sternal sulcus (s). Geographic coordinates inserted between brackets were taken from Google Earth. Illustrations were made using a Leica M8 stereomicroscope with a camera lucida; the computerized photographs were taken using a stereomicroscope Zeiss Discovery V12 (Automontage® system). Measurements of carapace width and carapace length, in millimeters, were made with a calipers and are given in parentheses after the number of specimens examined. Terminology for describing the morphology of the G1 was adapted from Smalley (1964) and Magalhães & Türkay (2008).

TAXONOMY

Kingsleya castrensis **sp. nov.**
Figs. 1-4, 7-14

Diagnosis. G1 with large, roughly rounded, thick apical plate, widest medially; proximal lobe of apical plate subtriangular, well developed, situated on mesio-caudal side; distal margin straight, stretching diagonally over the distal lobe, fusing to mesiodistal portion of apical plate; distal lobe of apical plate broad, with lateral margin angulate in mesial view, caudal margin straight, distal margin slightly concave, mesial margin rounded, thick.

Description. Carapace outline ellipsoid, widest medially (cb/cl 1.68); dorsal surface smooth, slightly convex, regions partially defined (Fig. 7). Two distinct gastric pits, close to each other, on metagastric region. Cervical grooves deep, narrow, nearly straight, faint proximally, distal end failing to reach anterolateral margin. Postfrontal lobules small, quite distinct; median groove indistinct. Surface of carapace between front and postfrontal lobules smooth and slightly inclined anteriorly and medially. Upper border of front smooth, angulate, slightly convex in dorsal view, median notch absent; lower border carinate, slightly sinuous in both frontal and dorsal view, more projected anteriorly than upper one, except medially. Upper orbital margin smooth, lower orbital margin slightly crenualte; exorbital angle low, obtuse (Fig. 13). Anterolateral margin of carapace nearly smooth, with very shallow depression just behind exorbital angle, followed by a set of faint, minute teeth increasing in size from the anterior to posterior portion; posterolateral margin smooth, barely defined. Epistome narrow; epistomial tooth triangular, deflexed, with carinate, smooth borders. Suborbital and subhepatic regions of carapace sidewall smooth; pterygostomial regions with narrow pilose patches along outer borders of bucal cavity (Figs. 8 and 13).

Endopod of Mxp3 with outer margin of ischium slightly convex, inner margin straight; outer margin of merus rounded, inner surface of palp covered with large setae; exopod of Mxp3 short, narrow, 0.18 times length of outer margin of ischium (Fig. 14). Aperture of efferent branchial channel wide, upper margin subquadrate, lacking setae (Fig. 13).

First pereiopods heterochelous in both males and females, similarly armed, right P1 usually largest (holotype left P1 major). Major cheliped merus subtriangular in cross section; superior margin rounded with irregular row of tubercles, fainter distally; medial margin lined by longitudinal row of rounded,

low teeth, slightly increasing in size distally; inferior lateral margin marked by row of faint tubercles, smooth distally; distal margin arched, smooth laterally, with straight row of faint tubercles mesially. Carpus with inner margin granular proximally, with prominent median spine, smooth distally; outer margin rounded, smooth. Palm narrow (length/breadth 1.61 in holotype), smooth on both sides. Fingers moderately gaping, tips not crossing; both fingers with large triangular teeth sometimes interspaced with small ones, smaller distally. Dactylus distinctly arched, longer then palm (dactylus/palm 1.36 in holotype, measured dorsally), upper, outer surface of dactylus smooth, distomedian portion darker than proximal. Propodal finger with smooth surfaces. P2-5 slender, ratios dactylus/propodus, dactylus/merus (left side measurements in holotype), respectively, as follows: P2 = 1.80 and 0.90, P3 = 1.73 and 0.88, P4 = 1.66 and 0.88, P5 = 1.64 and 0.92. P2-5 with dactyli shorter than propodi, bearing five longitudinal rows of sharp, corneous spines, increasing in size distally, 2 faint grooves on the proximal external surface.

Thoracic sternum slightly longer than broad. Thoracic sternites of Mxp3 and P1 completely fused, except for small notches at lateral edges of sternum; s4/s5, s5/s6, s6/s7 distinct, interrupted medially, just failing to reach midline of thoracic sternum; s7/s8 complete, reaching midline. Midline of thoracic sternum marked by deep groove between sternites VII, VIII, deeper at interception with sternal suture 7/8. Episternites 4-6 triangular posteriorly, episternite 7 posteriorly truncate. Sterno-abdominal cavity strongly concave, with few, scattered pubescence. Penis noticeably long, emerging from nearby coxosternal condyle articulation, located in shallow depression on sternite 8, proximally thick, abruptly tapering distally.

All abdominal segments free. Lateral margins of male telson slightly concave, slightly cranulate, tip rounded (Fig. 8).

G1 (Figs. 1-4 and 9-12) sinuous, broadened distally, with strong median curvature on caudal surface in mesial view, bearing well-developed mesial process. Marginal suture sinuous, displaced to mesial side in distally, bearing several setae proximally. Lateral suture deep, extending 2/3 of gonopod length from proximal portion. Marginal process short, broad, subrectangular in mesial view, not projecting distally beyond field of apical spines area, distal notch in latero-caudal surface. Mesial process well developed, roughly subretangular, approximately 1.8 times longer than apical plate in mesial view, proximal portion rounded, distal portion produced into sharp conical spine pointing in mesial direction; mesial process juxtaposed to the apical plate, both structures clearly separated by a deep incision. Apical plate well developed, large, thick, expanded along caudo-cephalic axis, with 2 juxtaposed lobes; proximal lobe subtriangular, well developed, situated subdistally on mesio-caudal side, narrower than distal lobe, distal margin straight, stretching diagonally over the distal lobe, gradually merging to mesiodistal portion of the apical plate; distal lobe enlarged, caudal margin rather angulate in mesial

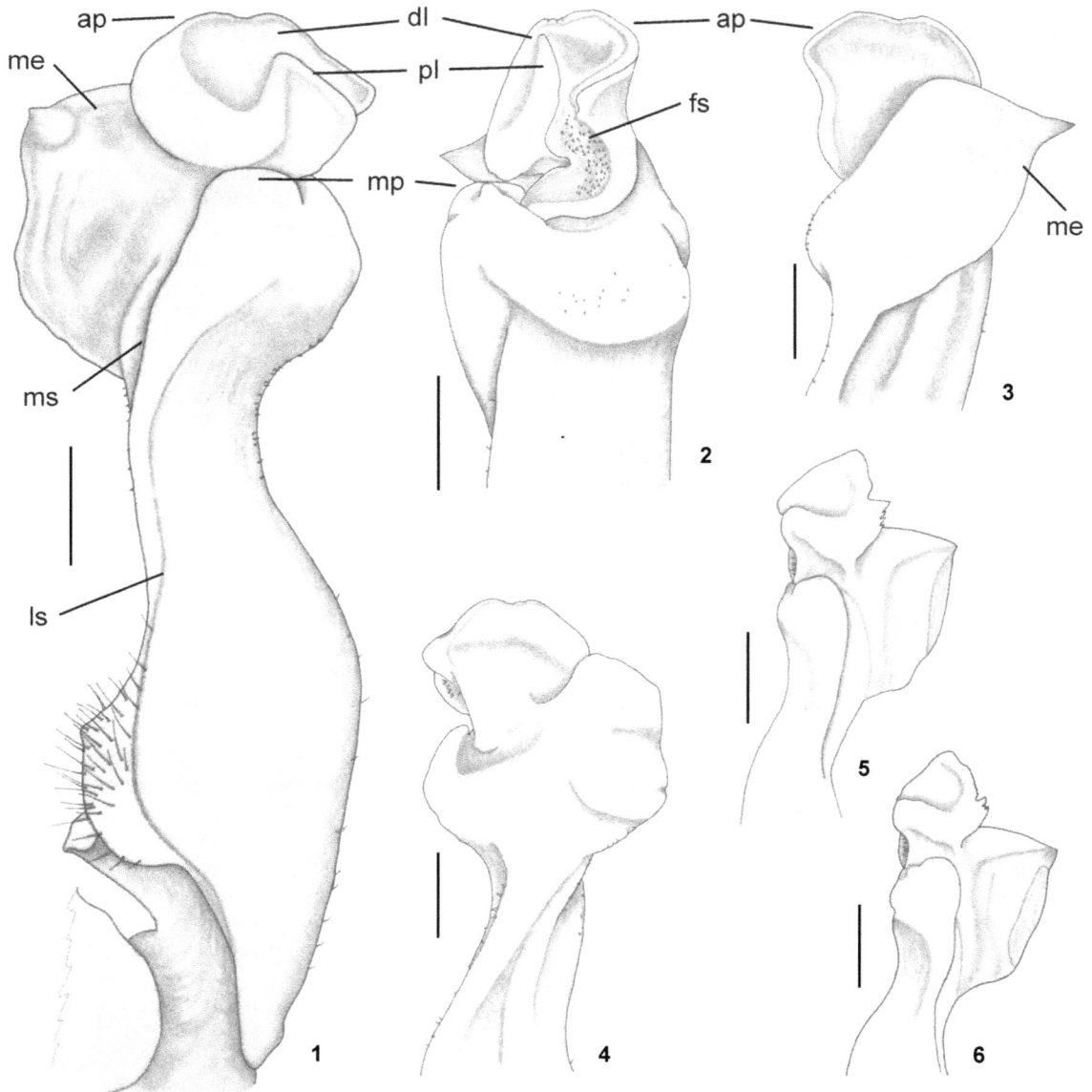

Figures 1-6. (1-4) *Kingsleya castrensis* **sp. nov.**, male, holotype (cw 46.4mm, cl 27.7mm), left first gonopod, INPA 2010: (1) whole limb, mesial view; (2) distal part, caudal view; (3) distal part, cephalic view; (4) distal part, lateral view. (5-6) *Kingsleya junki*, male, right first gonopod: (5), distal part, caudal-mesial view, INPA 1708; (6) distal part, caudal-mesial view, INPA 2012. (ap) Apical plate, (dl) distal lobe of apical plate, (fs) field of apical spines, (ls) lateral suture, (me) mesial process, (mp) marginal process, (ms) marginal suture. Scale bars: 1 mm.

view, distal margin slightly concave, mesial margin rounded. Apical spine field well developed, curved, narrow patch of minute spines, longitudinally directed along caudal side of apical plate, delimited by mesial, lateral borders of apical plate, distally opened by distinct notch at apex of apical plates proximal lobe. Sperm channel opening proximally at base of apical spine field.

G2 straight, almost as long as G1 (ca. 0.8 times length of G1), flagellum slender, strongly tapering after distal quarter, tip flattened, with short spinules on sternal surface.

Type material. BRAZIL, *Pará*: Altamira (51° B.I.S. – Batalhão de Infantaria de Selva camp area, 3°1147"S, 52°0958"W), male (cw 46.4, cl 27.7, ch 18.3, fw 13.7), holotype, 16.VIII.2011, José E. Martinelli Filho and Cléber S. de Sousa *leg.*, INPA 2010; same

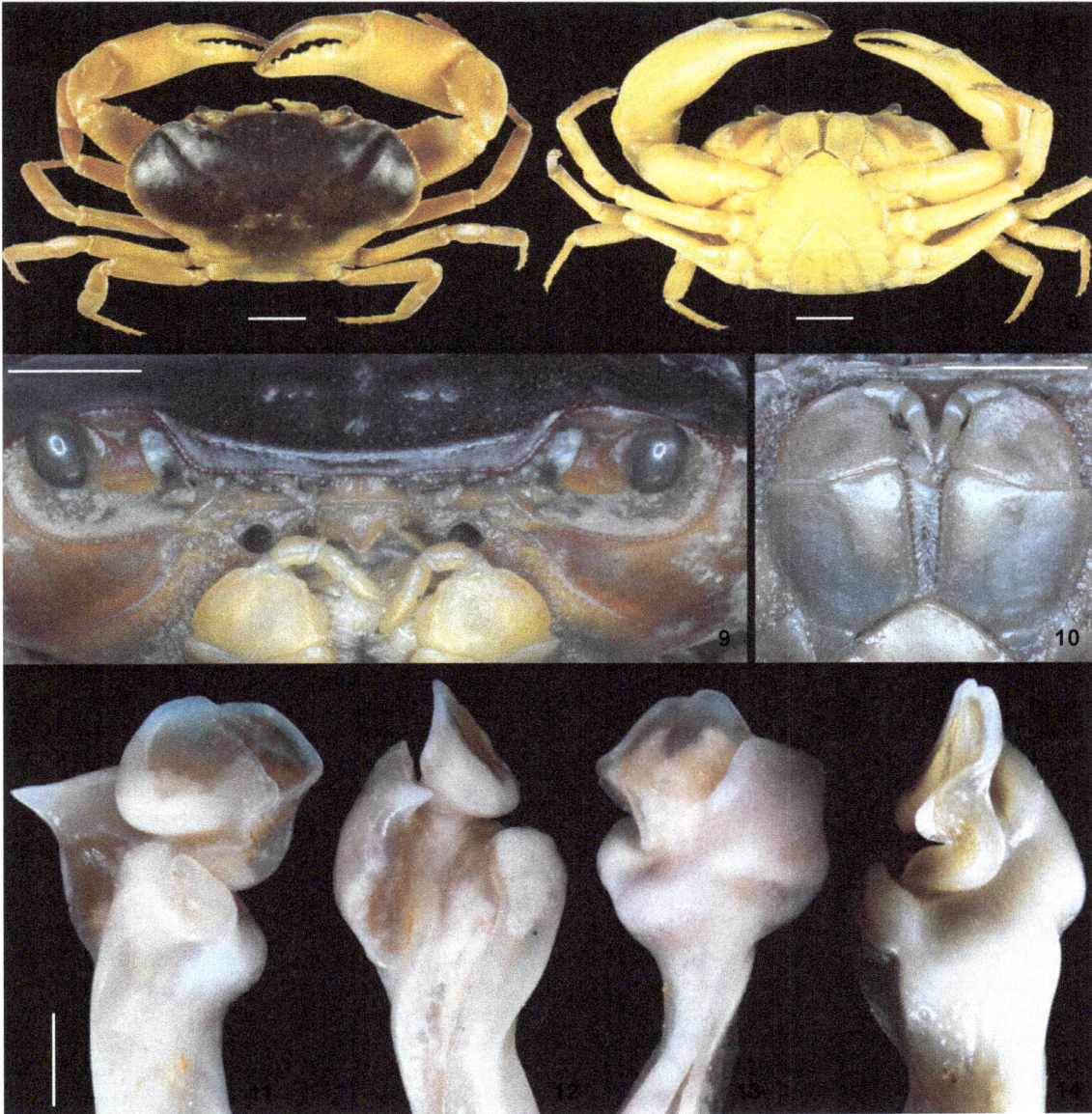

Figures 7-14. *Kingsleya castrensis* **sp. nov.**, male, paratype, MZUSP 26394: (7) habitus, dorsal view; (8) habitus, ventral view; (9) frontal view; (10) pair of third maxillipeds, frontal view. Male, paratype, left first gonopod, MZUSP 23393: (11) caudal-mesial view; (12) mesial-cephalic view; (13) lateral view; (14) idem, caudal view. Scale bars: 7, 8 = 10 mm, 9, 10 = 5 mm; 11-14 = 1 mm.

data as holotype, 1 male (cw 40.8, cl 25.1), 3 females (cw 24.7, cl 16.6; cw 46.3, cl 28.4; cw 21.5, cl 14.5), paratypes, 16.VIII.2011, José E. Martinelli Filho and Cléber S. de Sousa *leg.*, INPA 2011; same data as holotype, 4 males (cw 37.2, cl 22.7; cw 37.2, cl 23.3; cw 41.9, cl 25.2; cw 44.1, cl 26.5), paratypes, 26.VIII.2011, Cléber S. de Sousa *leg.*, MPEG 1013; same data as holotype, 2 males (cw 39.7, cl 24.8; cw 42.1, cl 25.7), 1 female (cw 39.0, cl 25.0), paratypes, 26.VIII.2011, Cléber

S. de Sousa *leg.*, SMF 47661; same data as holotype, 1 male (cw 32.2, 20.4), paratype, 28.VIII.2011, José E. Martinelli Filho and Cléber S. de Sousa *leg.*, MNRJ 25117; Altamira (Recanto Cardoso, 3°0920"S, 52°1533"W), 1 male (cw 42.9, cl 27.1), paratype, 9.IV.2012, Ronan Santos *leg.*, INPA 2058; Altamira (Princesa do Xingu road, 3°0935"S 52°1439"W), 1 male, (cw 32.9, cl 21.2), paratype, 6.I.2012, Cléber S. de Sousa *leg.*, INPA 2056; Altamira (Princesa do Xingu road, 3°0942"S, 52°1433"W), 1 male (cw

34.3, 22.2), paratype, 06.I.2012, Cléber S. de Sousa *leg.*, INPA 2057; Altamira (Abrigo Pedra do Navio, 3°1706.4"S, 52°13 42.2"W), 1 male (cw 32.57, cl 21.08), 11.IV.2012, R. Pinto-da-Rocha *leg.*, MZUSP 26943; Brasil Novo (Travessão 16, 3°1820"S, 52°3540"W), 1 male (cw 31.2, cl 19.8), 1 female (40.2, 25.2), paratypes, 13.I.2012, Cléber S. de Sousa *leg.*, MZUSP 32737; Altamira (Pedra da Cachoeira cave, 3°1914.8"S, 52°1953.1"W), 1 male (cw: 49.8, cl: 29.7), paratypes, 20.VI.2012, R. Pinto-da-Rocha *leg.*, MZUSP 26394.

Additional material examined: Brazil, *Pará*: Altamira (Princesa do Xingu road, 3°0935"S, 52°1439"W), 1 female (cw 39.8, cl 25.6), 20.I.2012, Cléber S. de Sousa *leg.*, INPA 2061; Altamira (Princesa do Xingu road, 3°0935"S, 52°1439"W), 1 male (cw 28.8, cl 18.4), 21.VIII.2011, Cléber S. de Sousa *leg.*, MPEG 1014; Altamira (Princesa do Xingu road, 3°1004"S, 52°2156"W), 1 male (cw 38.7, cl 24.2), 19.VI.2011, Cléber S. de Sousa *leg.*, MZUSP 32738; Altamira (51° B.I.S. – Batalhão de Infantaria de Selva camp area, 3°11478"S, 52°0956"W), 1 male (cw 41.5, cl 24.7), 10.XII.2010, José E. Martinelli Filho and Cléber S. de Sousa *leg.*, INPA 2060; Altamira (Arapujá island, 3°1330"S, 52°1213"W), 1 female (cw 37.0, cl 24.0), 10.V.2010, Anderson Prates *leg.*, INPA 2059; Brasil Novo (Travessão 8, 3°2157"S, 52°3218"W), 1 male (cw 38.1, cl 24.2), 2 females (cw 28.7, cl 18.9; cw 35.8, cl 28.2), 15.VIII.2012, Cléber S. de Sousa *leg.*, INPA 2062; Altamira (Planaltina cave, 3°2239"S, 52°3431"W), 1 female (cw 27.1, cl 17.4), 17.VIII.2011, Cléber S. de Sousa *leg.*, MPEG 1141.

Type locality and distribution. Brazil, state of Pará, city of Altamira. Most specimens were collected within the camp area of the "51° Batalhão de Infantaria de Selva" (B.I.S.), a unit of the Brazilian Armys Battalion of Jungle Infantry, headquartered in the city of Altamira. Additional specimens were also collected in in Altamira and Brasil Novo cities.

Ecological notes. Most of the crabs were collected on the margins of a 1-3m wide, third order tributary stream of the Xingu river. The stream is located inside a secondary forest fragment dominated by palm trees: *Euterpe oleracea* Mart. and *Attalea phalerata* Mart. ex Spreng in the flooded area, and *Astrocaryum gynacanthum* Mart. and *A. aculeatum* G. Mey in the well-drained soil area (*terra firme*). A floodplain along the borders of the stream varies from a few meters wide to almost a hundred meters according to topography and season (Salm et al. 2015). Adult crabs dig holes in the mud or hide between the aerial roots of the palm trees. Occasionally adult male and female crabs were found together in the same hole. Juvenile crabs were found on palm leaves and trunks or beneath leaf litter. A few specimens were collected outside the flooded margins of the stream, in the *terra firme* area.

Female crabs carried young crabs under their abdominal brood pouch. Morning field investigations revealed that 68.5% of the 108 observed crabs were males. The species is probably aggressive and territorial, since one of the chelipeds was lacking in 20% of the males and 13% of the females. The loss of

chelipeds may also be attributed to autotomy as a response to predators, since skeletal remains and pereiopods of *Kingsleya* were frequently found in the studied area.

Etymology. The specific epithet refers to *castra*, the Latin word for military camp, in reference to the Brazilian Army battalion camp where this species was found.

Remarks. The new species is attributed to *Kingsleya* since its G1 shows the diagnostic characters of the genus, namely the marginal process distally enlarged, not overreaching the apical field of spines, the bi-lobed apical plate, the mesial process clearly separated from the apical plate and standing out from the cephalic surface of the stem; the apical plate with two partially superimposed lobes; and the field of apical spines distally divided by a terminal notch (Magalhães & Türkay 2008).

Kingsleya castrensis **sp. nov.** can be easily distinguished from *K. junki* Magalhães, 2003 and *K. ytupora* Magalhães, 1986, the other two species of the genus that occur in the Xingu River (Magalhães 2003b) by characters of the G1s apical plate. In both *K. junki* and *K. ytupora* (see Magalhães 2003b: 384, figs. 1B-D, and 385, fig. 2B, respectively) the apical plate is narrow and produced distally in relation to the mesial process, whereas the apical plate of *K. castrensis* **sp. nov.** is distinctly enlarged and short in relation to the mesial process (Figs. 1, 3, 5 and 6). Moreover, in *K. castrensis* **sp. nov.** the mesial margin of the apical plates distal lobe is smooth (Fig. 1), whereas in *K. junki* the distal plate is clearly indented (Figs. 5 and 6). The distal margin of the apical plate, in mesial view, is rounded and rather narrow in *K. ytupora* (see Magalhães 2003: 385, fig. 2B), whereas in *K. castrensis* **sp. nov.** this margin is much broader rounded and enlarged (Figs. 1, 3 and 4). Another character that readily separates these two species is the presence (in *K. ytupora*) or absence (in *K. castrensis* **sp. nov.**) of a set of six to seven large, sharp teeth on the anterolateral margin of the carapace; in the latter species, this margin is fringed with a set of faint, minute teeth that lends to this margin an almost smooth appearance.

Kingsleya castrensis **sp. nov.** is unique among the species of the genus because of the distinctly enlarged, broadly rounded apical plate of its G1. All other species of *Kingsleya* have a G1 with an apical plate that is, in spite of their specific differences, much narrower, tapering and roughly subtriangular in shape (Magalhães 1986, 1990, 2003b, 2005, Magalhães & Türkay 2008).

Kingsleya junki Magalhães, 2003
Figs. 5-6

Kingsleya junki Magalhães, 2003b: 378, fig. 1.

Material examined. 1 male (cw 26.7, cl 16.7), INPA 1708, Brazil, Pará, 2 km south of Jacundá [4°27S 49°07W], right bank of Tocantins River, 7.V.1984, W. Overal *leg.*; 1 male (cw 30.5, cl 19.5), 1 female (cw 53.2, cl 33.7), INPA 2012, Brazil, Pará, Altamira, Leonardo da Vinci stream, 3°0908"S, 52°0432"W, 26.III.2012, C.S. Souza *leg.*; 1 male (cw: 46.1, cb: 28.7), MZUSP 32818, Brazil, Pará, Altamira, Abrigo do Chuveiro cave, IV.2009, leg. unknow.

Distribution. The species was known only from its type locality, Vitória do Xingu, downstream from Altamira, on the left bank of Xingu River (MAGALHÃES 2003b). The present record from Jacundá extends its distribution to the eastern Amazon region, in the middle course of the Tocantins River basin.

Ecological notes. The specimens of *K. junki* were found in the vegetated margins of the Leonardo da Vinci stream, a small, clear-water tributary on the left bank of the Xingu River. All of the crabs were collected inside holes or beneath rocks. Two couples were found, and males and females apparently shared the same hole.

Remarks. The G1 of the specimens reported herein (Figs. 5 and 6) is similar to that of the holotype of *K. junki* (see MAGALHÃES 2003b: 384, fig. 1B), although some variability can be noticed in the morphology of the apical plate, particularly in the mesial margin of the distal lobe. In the holotype, this margin is indented both proximally and distally, whereas it is more distinctly indented only in the proximal portion of this margin in the present specimens. Since such a situation can be verified in the specimens from both the Xingu River and Tocantins River basins, this might be due to intraspecific variability and, therefore, the specimen from Tocantins River was considered to be conspecific with those from the Xingu River basin.

ACKNOWLEDGMENTS

MP thanks FAPESP (Fundação de Amparo à Pesquisa do Estado de São Paulo) for providing financial support through a doctoral fellowship (2012/01334-7). JEMF is grateful to Rodolfo Salm and to biologists Anderson Prates and Cleber Sousa for their assistance during field observations. JEMF was supported by Universidade Federal do Pará (research grants PROPESP #04/2014 and PROPESP/FADESP #09/2014). CM thanks the Conselho Nacional de Desenvolvimento Científico e Tecnológico for an ongoing Research Grant (Proc. 303837/ 2012-6). We also thank Barbara Robertson, Michael Türkay, and two anonymous reviewers for corrections and suggestions that greatly improved the manuscript.

REFERENCES

MAGALHÃES C (1986) Revisão taxonômica dos caranguejos de água doce brasileiros da família Pseudothelphusidae (Crustacea, Decapoda). **Amazoniana** 9(4): 609-636.

MAGALHÃES C (1990) A new species of the genus *Kingsleya* from Amazonia, with a modified key for the Brazilian Pseudothelphusidae (Crustacea: Decapoda: Brachyura). **Zoologische Mededelingen** 63(21): 275-281.

MAGALHÃES C (2003a) Brachyura: Pseudothelphusidae e Trichodactylidae, p. 143-297. In: MELO GAS (Ed.). **Manual de Identificação dos Crustacea Decapoda de Água Doce do Brasil.** São Paulo, Edições Loyola.

MAGALHÃES C (2003b) The occurrence of freshwater crabs (Crustacea: Decapoda: Pseudothelphusidae, Trichodactylidae) in the Rio Xingu, Amazon Region, Brazil, with description of a new species of Pseudothelphusidae. **Amazoniana** 17(3/4): 377-386.

MAGALHÃES C (2005) A new species of freshwater crab (Crustacea: Decapoda: Pseudothelphusidae) from the southeastern Amazon Basin. **Nauplius** 12(2): 99-107.

MAGALHÃES C, TÜRKAY M (2008) A new species of *Kingsleya* from the Yanomami Indians area in the Upper Rio Orinoco, Venezuela. (Crustacea: Decapoda: Brachyura: Pseudothelphusidae). **Senckenbergiana biologica** 88(2): 1-7.

SALM R, PRATES A, SIMÕES NR, FEDER L (2015) Palm community transitions along a topographic gradient from floodplain to terra firme in the eastern Amazon. **Acta Amazonica** 45(1): 65-74. doi: 10.1590/1809-4392201401533

SMALLEY A (1964) A terminology for the gonopods of the American river crabs. **Systematic Zoology** 13: 28-31.

8

Trogolaphysa formosensis sp. nov. (Collembola: Paronellidae) from Atlantic Forest, Northeast Region of Brazil

Diego Dias da Silva[1] & Bruno Cavalcante Bellini[1,2,3]

[1]*Programa de Pós-graduação em Sistemática e Evolução, Centro de Biociências, Universidade Federal do Rio Grande do Norte. Campus Universitário Lagoa Nova, 59072-970 Natal, RN, Brazil. E-mail: diegocollembola@gmail.com*
[2]*Departamento de Botânica, Ecologia e Zoologia, Centro de Biociências, Universidade Federal do Rio Grande do Norte. 59072-970 Natal, RN, Brazil.*
[3]*Corresponding author: E-mail: entobellini@gmail.com*

ABSTRACT. *Trogolaphysa formosensis* **sp. nov.** (holotype male deposited in DBEZ from Brazil, state of Rio Grande do Norte State, municipality of Bani Formosa), a new springtail from the Atlantic Forest domain, Rio Grande do Norte, Brazil, is described and illustrated. This species is diagnosed by unique coloration pattern, presence of 8+8 eyes, reduced number of setae on metatrochanteral organ, unguiculi truncated and dorsal chaetotaxy. *Trogolaphysa formosensis* **sp. nov.** is the first species of the genus from Brazil with all eye lenses. All other Brazilian species present 0+0 or 2+2 eyes. It is also the first species of *Trogolaphysa* described from the Northeast Region of Brazil.

KEY WORDS. Chaetotaxy; Paronellinae; soil fauna; springtail; taxonomy.

The species of *Trogolaphysa* Mills, 1869 (Paronellinae), one of the most common genera of Paronellidae in the Neotropical Region, are characterized by the following: scales covering dorsal head, dorsal body and ventral face of furcula; fourth antennal segment smooth or annulated, but never subdivided; eyes with 0-8 lenses; labial seta L2 normal; abdominal segments II-IV with 2+2, 3+3, 3+3 trichobothria respectively; manubrium lacking spines; dens with 1-2 rows of spines; and mucro short with 3-5 teeth (Soto-Adames & Taylor 2013). *Trogolaphysa* currently comprises 40 described species worldwide, 37 of which are from the New World (Soto-Adames & Taylor 2013). It is the largest known genus of Paronellidae in Brazil, holding one third of the known species of the family (five species) (Arlé 1939, Arlé & Guimarães 1979, Yoshii 1988, Abrantes et al. 2010, 2012).

Herein we describe a new species of *Trogolaphysa* from the Atlantic Forest of Rio Grande do Norte State, Northeast Brazil. All other Brazilian species are from the Southeast Region. Furthermore, it is the first Brazilian species analyzed based on detailed description of the dorsal chaetotaxy, the most reliable set of characters used to compare species of Entomobryoidea (Szeptycki 1979, Soto-Adames 2008, Soto-Adames & Taylor 2013).

MATERIAL AND METHODS

The specimens were collected during the rainy season. The climate of the area is "As" following Koeppen's system, with two distinct seasons: a dry summer and a wet winter (Kottek et al. 2006). Specimens were collected in leaf litter; fixed and preserved in 70% ethanol; cleared with hydrochloric acid and potassium dichromate; and mounted on glass slides in Hoyer's Medium for study under the optical microscope. Measurements were taken from the holotype; the overall morphology was described based on the entire type series.

The detailed chaetotaxy schemes follow Szeptycki (1979), Soto-Adames (2008) and Soto-Adames & Taylor (2013) for dorsal head and body, and Gisin (1963, 1964) for labial triangle. Chaetotaxy symbols are presented in Fig. 10. List of abbreviations used in the text: (Abd) abdominal segment; (Ant) antennal segment; (Th) thoracic segment; DBEZ/UFRN: Departamento de Botânica, Ecologia e Zoologia da Universidade Federal do Rio Grande do Norte; CRFS/UEPB: Coleção de Referência para a Fauna de Solo da Universidade Estadual da Paraíba.

TAXONOMY

Trogolaphysa formosensis **sp. nov.**
Figs. 1-20

Description. Habitus entomobryid (Figs. 1 and 2). Color of specimens after fixation in 70% ethanol, white with dark-blue pigment covering Ant I-IV; frontal head, eyepatches, lateral borders of Th II-III and Abd I-III, and one third of Abd IV (Fig. 1). Scales present over Ant I-II, base of Ant III, head, trunk, ventral side of ventral tube, legs, manubrium and ventral dentes.

Head. Ant IV not annulated or subdivided, lacking apical bulb (Fig. 3); Ant III sense organ with two rods and three guard sensillae (Fig. 4); Ant II with two subapical small blunt setae (Fig. 5). Eyes with 8+8 lenses, posterior lenses (G and H)

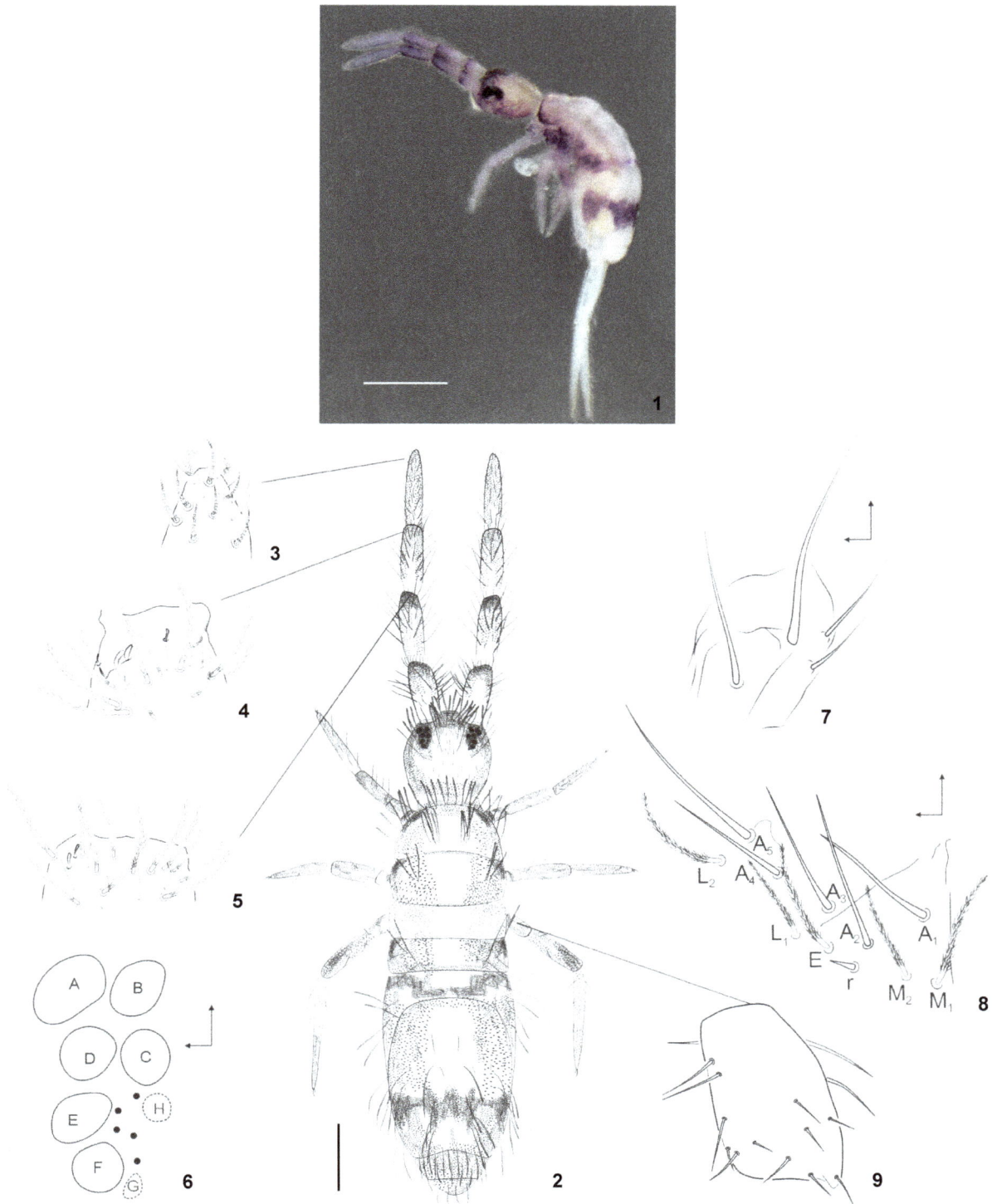

Figure 1-9. *Trogolaphysa formosensis* **sp. nov.**: (1) habitus of a fixed specimen in ethanol, lateral view; (2) habitus (dorsal view); (3) apical region of Ant IV (right); (4) apical region of Ant III (right); (5) apical region of Ant II (right); (6) eye patch (left); (7) maxillary palp and sublobal plate (left); (8) labial triangle chaetotaxy (left); (9) metatrochanteral organ. Scale bars: 0.2 mm.

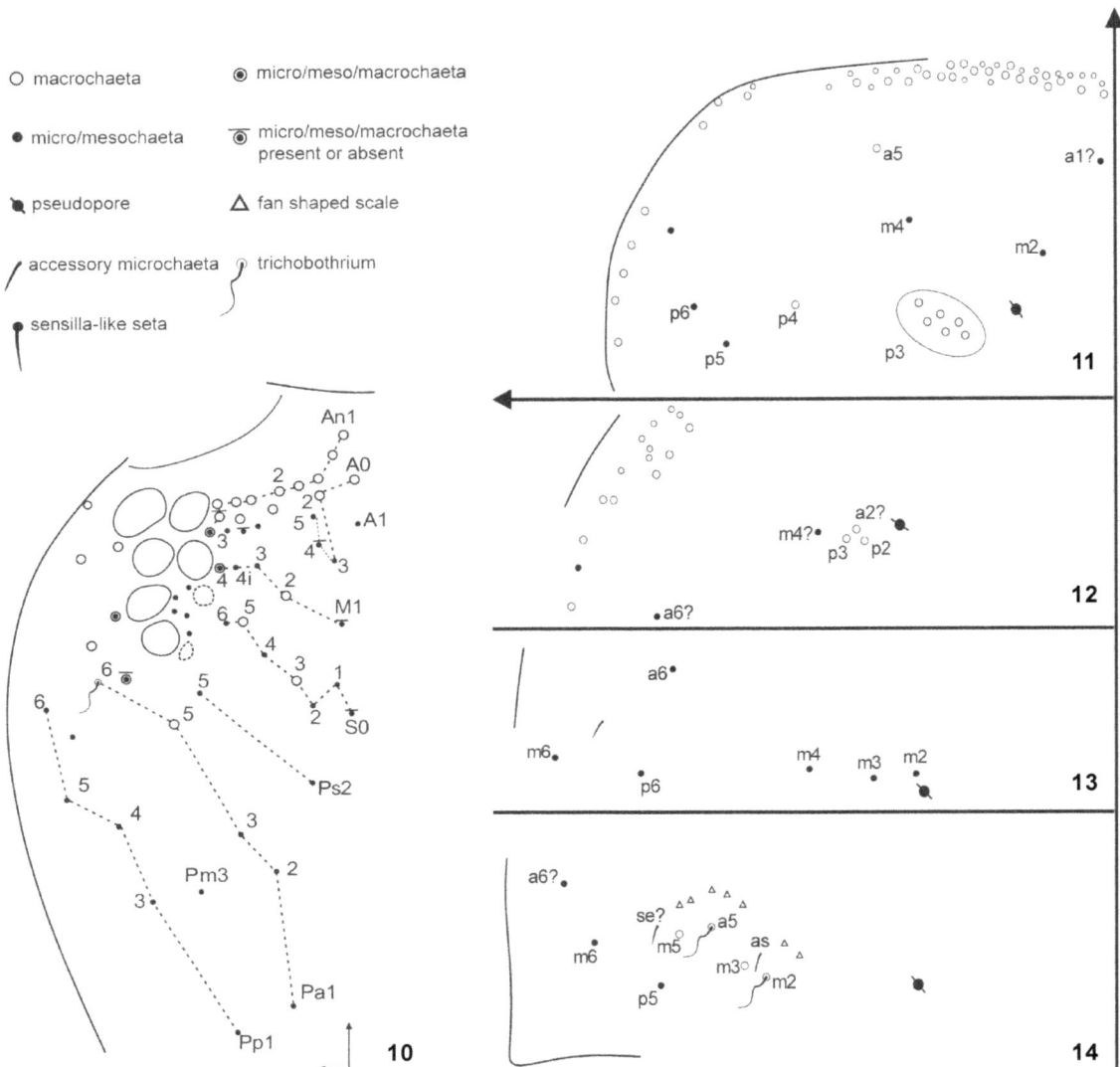

Figures 10-14. *Trogolaphysa formosensis* **sp. nov.**: dorsal chaetotaxy (left): (10) head and setae symbols; (11) Th II; (12) Th III; (13) Abd I; (14) Abd II.

vestigial (Fig. 6). Labral setae smooth, prelabral setae feathered; distal margin of labrum lacking spines. Maxillary palp with apical and basal setae smooth; apical seta larger; sublobal plate with 2 smooth appendages (Fig. 7). Labial triangle formula: M1M2rEL1L2/A1-5; r smooth and reduced (Fig. 8). Dorsal head chaetotaxy: 'An' series with 10+10 macrochaetae; 'A' series with A0 and A2 as macrochaetae and A1, A3-5 as microchaetae; 'M' series with 5+5 setae, M1, M3, M4i as microchaetae, M2 as macrochaeta, M4 as macro or microchaeta; 'S' series with S0 present as microchaeta, S1, S2, S4, and S6 as microchaetae, S3 and S5 as macrochaetae; interocular series with 5+5 setae; 'Ps' series with 2+2 microchaetae (Ps2 and Ps5);

'Pa' series with 5+5 setae, Pa1-3 as microchaetae, Pa5 as macrochaeta and Pa6 as trichobothrium; 'Pm' series with 1+1 microchaetae (Pm3); 'Pp' series with 5+5 microchaetae (Pp6, Pp5, Pp4, Pp3 and Pp1), Pp2 absent (Fig. 10).

Thorax. Dorsal Th II chaetotaxy: 'a' series with 1 internal microchaeta (a1?) and a5 as macrochaeta; 'm' series with m2 and m4 as microchaetae; 'p' series with p3 complex arranged as typical for the genus (see Soto-Adames & Taylor 2013), p4 as macrochaeta, p5 and p6 as microchaetae (Fig. 11). Dorsal Th III cheatotaxy: 3 macrochaetae (a2?, p2 and p3), 1 medial microchaeta (m4?) and 1 external microchaeta (a6?) (Fig. 12). Metatrochanteral organ reduced, with 12 short spines (Fig. 9).

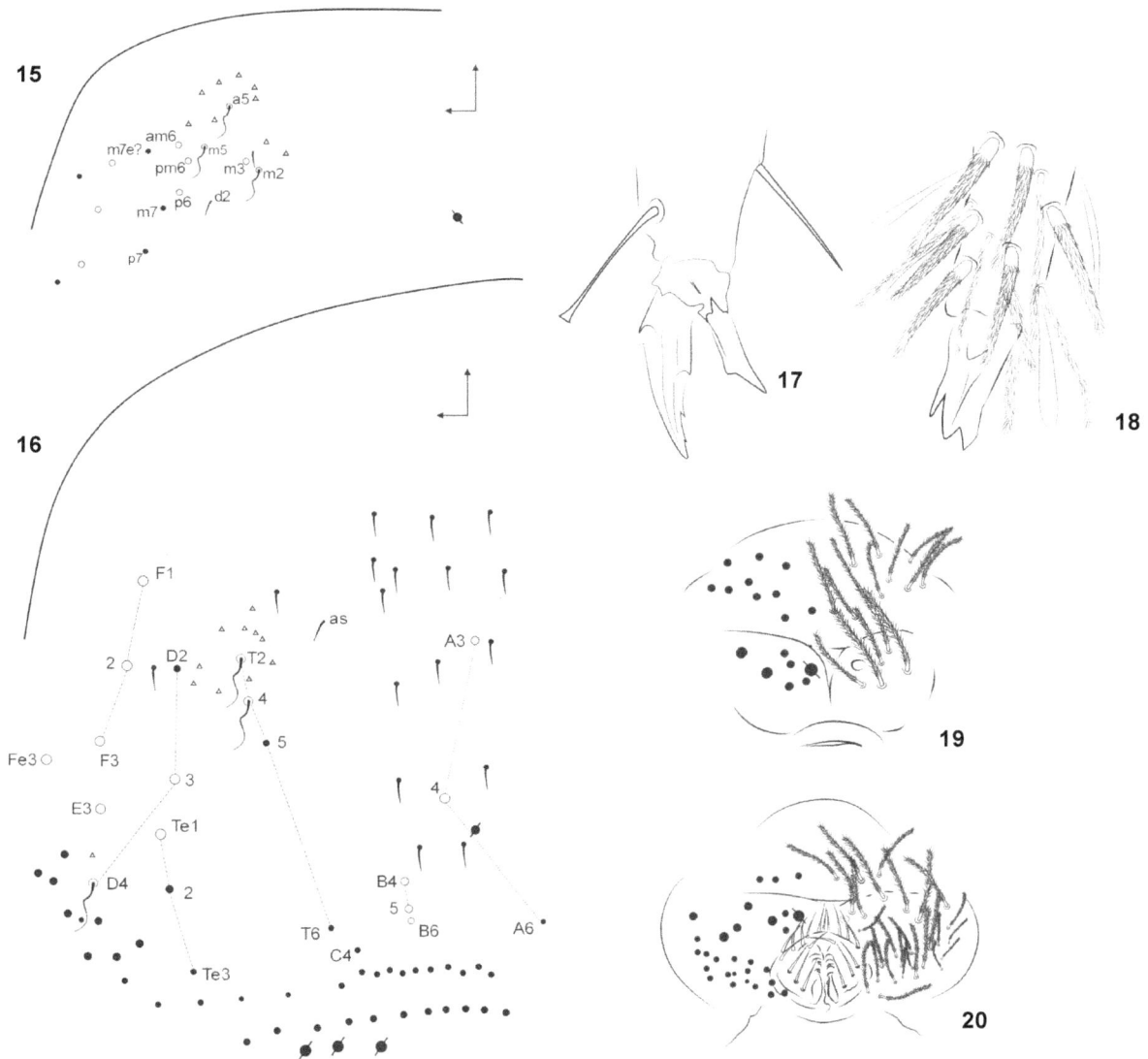

Figures 15-20. *Trogolaphysa formosensis* **sp. nov.**: (15) dorsal chaetotaxy of Abd III (left); (16) dorsal chaetotaxy of Abd IV (left); (17) hind foot complex; (18) distal dens and mucro; (19) genital plate (female); (20) genital plate (male).

Terminal tibiotarsi with capitate tenent hair. External and internal lamellae of ungues with one and three teeth, respectively; outer tooth present (not represented in the figure); unguiculi truncated (Fig. 17).

Abdomen. Dorsal Abd I chaetotaxy: 1 anterior microchaeta (a6) and 5 median microchaetae (m2, m3, m4, p6 and m6) (Fig. 13). Dorsal Abd II chaetotaxy: two trichobothria (m2 and a5), surrounded by 2 and 5 fan shaped scales respectively, two macrochaetae (m3 and m5), two accessory sensilla (as and se), and three external microchaetae (p5, m6 and a6?) (Fig. 14). Dorsal Abd III chaetotaxy: three trichobothria (m2, a5 and m5)

surrounded by 2, 5 and 2 fan shaped scales respectively, m3, p6, pm6, am6 and m7e? as macrochaetae, am6, m7 and p7 as microchaetae, d2 sensilla present (Fig. 15). Dorsal Abd IV chaetotaxy: 'A' series with 3+3 setae, A3 and A4 as macrochaetae, A6 as micro or mesochaeta; 'B' series with 3+3 macrochaetae (B4-6); 'C' series with 1+1 micro or mesochaetae (C4); 'T' series with 4+4 setae, T2 and T4 as trichobothria, surrounded by 10 fan shaped scales, T5 as microchaeta and T6 as micro or mesochaeta; 'Te' series with 3+3 setae, Te1 as macrochaeta, Te2 and Te3 as micro or mesochaetae; 'D' series with 3+3 setae, D2 as micro or mesochaeta, D3 as macrochaeta and D4 as trichobothrium; 'E' series with 1+1 macrochaetae (E3); 'F' series with 3+3 macrochaetae (F1-3); 'Fe' series with 1+1 macrochaetae (Fe3); several sensillae present among the main setae, 15+15 posterior setae present (Fig. 16). Dentes with 2 rows of ciliated spines; mucro with 4 teeth, with a crest-like structure similar to a fifth tooth (Fig. 18). Genital papillae of male with approximately 20 setae; chaetotaxy of anal valves of female and male as in Figs. 19 and 20, respectively.

Measurements. Total length (head + body) 1.44 mm; body 1.17 mm; longitudinal and transversal head length 0.27 mm and 0.24 mm respectively; Ant I-IV respectively: 0.16 mm, 0.22 mm, 0.22 mm, 0.25 mm; manubrium, dentes and mucro respectively: 0.32 mm, 0.24 mm, 0.02 mm.

Material examined. Holotype male on slide (DBEZ/UFRN), Brazil, *Rio Grande do Norte State*: Baía Formosa Municipality, "Mata Estrela" (06°22'10"S, 35°00'28"W), 7.IX.2012, pitfall-trap, D.D. Silva *leg*. Paratypes: 3 females on different slides (DBEZ/UFRN), 2 females and 2 males on same slide (CRFS-UEPB), same date as holotype. Type material deposited at Collembola Collection of DBEZ/UFRN, number 1001; and at CRFS-UEPB, number 5173.

Etymology. The species was named after its type locality, Baía Formosa, Rio Grande do Norte State, Brazil.

Remarks. The new species can be separated from its congeners by the following unique set of characters: color pattern; presence of 8+8 eyes; reduced number of setae on metatrochanteral organ; unguiculi truncated; and configuration of dorsal chaetotaxy. *Trogolaphysa formosensis* **sp. nov.** present setae M1 and M2 in head, Pa series shows five setae, Pa1 is present, Pp series bears five setae and there are seven macrochaetae on anterior cephalic region (A0, A2, M2, S3, S5 and two unnamed near An series); on thorax, Th II p4 seta is a macrochaeta and there is one extra microseta on Th III (possibly a misplaced a6); on abdomen, Abd II present another extra microseta (also named a6?); on Abd III the sensillum d2 is posterior to pm6 macrochaeta; and Abd IV present several extra sensillae and B4, E3 and trichobothrium D4 are apparently misplaced. None of these features were seen in other species of the genus (which were analyzed concerning the dorsal chaetotaxy), as shown by the revision of *Trogolaphysa* by Soto-Adames & Taylor (2013). This particular morphology, combined with Th III wider than Th II; and a distinct mucro shape, with a crest-like structure similar to a fifth tooth, suggest that *T. formosensis* **sp. nov.** represents an unusual lineage within the genus. However, given that the chaetotaxy of most Brazilian *Trogolaphysa* have not been described in detail, we do not know if this particular set of characters is unique to this species or to a group of species within *Trogolaphysa*.

Trogolaphysa formosensis **sp. nov.** is the only Brazilian species of the genus bearing all eye lenses. All other species present 2+2 or 0+0 eyes. It differs from *T. aelleni* Yoshii, 1988, *T. hauseri* Yoshii, 1988 and *T. hirtipes* Handschin 1924 also in the color pattern (colored in *Trogolaphysa formosensis* **sp. nov.**,

unpigmented or weakly pigmented on thorax in the other three species) and shape of the unguiculi (truncated in the new species against lanceolated). *Trogolaphysa millsi* Arlé, 1939 has truncated unguiculi, which is also unpigmented, and its mucro is longer and lacks the crest-like structure seen in the new species. Lastly, the most similar species to *Trogolaphysa formosensis* **sp. nov.** among the Brazilian taxa, is *T. tijucana* Arlé & Guimarães, 1979, which is also well pigmented and bears truncated unguiculi. However, the color pattern of *T. tijucana* is restricted to blue pigment covering the antennal segments, Th II-III, Abd I-II, proximal legs and ventral tube. This contrasts with the situation found in *Trogolaphysa formosensis* **sp. nov.** which presents blue pigments also over the head, Abd III and IV; furthermore while the new species has dentes with 2 rows of ciliated spines, *T. tijucana* present dentes with only 1 row of smooth spines (Arlé 1939, Arlé & Guimarães 1979, Cassagnau 1963, Yoshii 1988).

Trogolaphysa formosensis **sp. nov.** was collected in the Atlantic Forest, a forestry phytogeographic domain originally ranging from the northern Atlantic coast of Brazil, through Paraguay and northeastern Argentina; within Good's biogeographic zone 27 (Good 1974). All other species of *Trogolaphysa* seen in Brazil were also collected from this domain, but from the southeastern Region of the country (Arlé 1939, Arlé & Guimarães 1979, Cassagnau 1963, Yoshii 1988).

ACKNOWLEGMENTS

Felipe N. Soto-Adames made important suggestions on the drawings. The second author has a grant from the Conselho Nacional de Desenvolvimento Científico e Tecnológico (CNPq)/ Programa de Pesquisa em Biodiversidade – Invertebrados (PPBio).

REFERENCES

Abrantes EA, Bellini BC, Bernardo AN, Fernandes LH, Mendonça MC, Oliveira EP, Queiroz GC, Sautter KD, Silveira TC, Zeppelini D (2010) Synthesis on Brazilian Collembola: na update to the species list. **Zootaxa 2388**: 1-22.

Abrantes EA, Bellini BC, Bernardo AN, Fernandes LH, Mendonça MC, Oliveira EP, Queiroz GC, Sautter KD, Silveira TC, Zeppelini D (2012) Errata Corrigenda and update for the "Synthesis of Brazilian Collembola: an update to the species list." Abrantes et al. (2010), Zootaxa 2388: 1-22. **Zootaxa 3168**: 1-21.

Arlé R (1939) Collemboles nouveaux de Rio de Janeiro. **Annais da Academia Brasileira de Ciências 11**: 25-32.

Arlé R, Guimarães AE (1979) Nova espécie do gênero *Paronella* Schött, 1893 do Rio de Janeiro (Collembola). **Revista Brasileira de Entomologia 23** (4): 213-217.

Cassagnau P (1963) Collemboles d'Amérique du Sud, II. Orchesellini, Paronellinae, Cyphoderinae. **Biologie de la Amérique Australe 2**: 127-148.

GISIN H (1963) Collemboles d'Europe. V. **Revue Suisse de Zoologie 70**(5): 77-101.

GISIN H (1964) Collemboles d'Europe. VI. **Revue Suisse de Zoologie 71**(20): 383-400.

GOOD R (1974) **The geography of flowering plants.** London, Longman Group, 574p.

KOTTEK M, GRIESER J, BECK C, RUDOLF B, RUBEL F (2006) World Map of the Köppen-Geiger climate classification updated. **Meteorologische Zeitschrift 15:** 259-263.

SOTO-ADAMES FN (2008) Postembryonic development of the dorsal chaetotaxy in *Seira downlingi* (Collembola, Entomobryidae), with an analysis of the diagnostic and phylogenetic significance of the primary chaetotaxy in *Seira*. **Zootaxa 1683:** 1-31.

SOTO-ADAMES FN, TAYLOR SJ (2013) The dorsal chaetotaxy of *Trogolaphysa* (Collembola, Paronellidae), with descriptions of two new species from caves in Belize. **ZooKeys 323:** 35-74.

SZEPTYCKI A (1979) **Morpho-systematic studies on Collembola. IV. Chaetotaxy of the Entomobryidae and its phylogenetical significance.** Kraków, Polska Akademia Nauk, 216p.

YOSHII R (1988) Paronellid Collembola from caves of Central and South America collected by P. Strinati. **Revue suisse Zoologie 95** (2): 449-459.

Advertisement call of *Dendropsophus microps* (Anura: Hylidae) from two populations from Southeastern Brazil

Lucas Rodriguez Forti[1,*], Rafael Márquez[2] & Jaime Bertoluci[3]

[1]*Programa de Pós-Graduação Interunidades em Ecologia Aplicada, Escola Superior de Agricultura Luiz de Queiroz, Universidade de São Paulo. Avenida Centenário 303, 13400-970 Piracicaba, SP, Brazil.*
[2]*Fonoteca Zoológica, Departamento de Biodiversidad y Biología Evolutiva, Museo Nacional de Ciencias Naturales. José Gutiérrez Abascal 2, Madrid, Spain. E-mail: rmarquez@mncn.csic.es*
[3]*Departamento de Ciências Biológicas, Escola Superior de Agricultura Luiz de Queiroz, Universidade de São Paulo. Avenida Pádua Dias 11, 13418-900 Piracicaba, SP, Brazil. E-mail: jaime.bertoluci@usp.br*
Corresponding author. E-mail: lucas_forti@yahoo.com.br

ABSTRACT. In anurans, acoustic communication is a major mechanism of pre-zygotic isolation, since it carries information about species recognition. Detailed descriptions of the acoustic properties of anuran advertisement calls provide important data to taxonomist and to the understanding of the evolution of the group. Herein we re-describe the advertisement call of the hylid frog *Dendropsophus microps* (Peters, 1872) after analyzing a larger sample than that of previous descriptions. We also compare the acoustic properties of the call in two populations and discuss the effect of the presence of the sister species, *Dendropsophus giesleri* (Mertens, 1950), a potential competitor, in one of the populations. Additionally, we provide information on calling sites and size of males. Males of *D. microps* emit two types of calls, which differ mainly in pulse repetition rate. Type "A" call has a mean frequency band varying from 4574 to 5452 Hz, (mean dominant frequency = 4972 Hz). Type "B" call has a mean frequency band varying from 4488 to 5417 Hz (mean dominant frequency = 4913 Hz). The calls of *D. microps* and *D. giesleri* are the only in the *D. parviceps* species group that have harmonic structure. The spectral properties of the call showed low intra-individual variation, being considered static, while the temporal properties were highly variable. Compared with males from the Boracéia population, males from the Ribeirão Grande population called from lower perches, and their calls had slightly lower frequency bands and significantly higher pulse rates in their type "B" calls. Inter-populational differences in acoustic properties, body size and use of calling sites could be related to selective forces associated with the presence of the sister species, a potential competitor for the population from Ribeirão Grande.

KEY WORDS. Acoustic traits; Amphibia; bioacoustics; call evolution; competition.

Anuran vocalizations play a key role in mate recognition and may correspond to their main pre-zygotic isolation mechanism, besides being an essential component of sexual selection (GERHARDT 1994, WELLS 1977). Given the complex acoustic repertoires (TOLEDO & HADDAD 2005, FORTI et al. 2010, MORAIS et al. 2012) of frog species, it is likely that accurate quantitative descriptions of anuran advertisement calls will help to solve taxonomic problems and to recover the relationships among taxa (RYAN & RAND 1993, WELLS 2007). The advertisement calls of many anuran species have been described. However, sample sizes for these descriptions are often small, which limits their usefulness, since intraspecific variation may not be sufficiently accounted for. For this reason, re-descriptions of advertisement calls based on larger sample sizes are needed. Larger samples sizes are also important to quantify among population variations, and to understand how the advertisement call has evolved in a given species (GERHARDT 2012). Many factors can cause evolutionary changes in the properties of the call, and one of the fundamental sources of among population variation is acoustic interaction between sister species (LITTLEJOHN 1976, HÖBEL & GERHARDT 2003, GERHARDT 2012). The presence of similar species calling on the same reproductive habitat may represent a selective pressure on call properties. Differences in them are expected to improve the quality of co-specific communication and to prevent the formation of heterospecific couples (HADDAD et al. 1994, HÖBEL & GERHARDT 2003).

Herein we re-describe the advertisement call of the hylid species *Dendropsophus microps* (Peters, 1872), using data from two populations, and provide measurements of intra-individual and intra-population variation from a large sample. We also compare the acoustic properties of the call and their variations between two populations and discuss the effect of the presence of the sister species and potential competitor, *Dendropsophus giesleri* (Mertens, 1950), in one of them. Additionally we provide information on male body size and calling sites.

MATERIAL AND METHODS

The hylid frog *Dendropsophus microps* belongs to the *D. parviceps* group, whose species occur mainly in the Amazon region, with two exceptions, *D. giesleri*, and *D. microps*, which occur in the Atlantic forest (FAIVOVICH et al. 2005, FROST 2014). The natural habits of the species of the group are poorly known.

Dendropsophus microps (Figs. 1-2) is distributed from the southern portion of the of state Bahia to the northern portion of the state of Rio Grande do Sul, southeastern Brazil, in areas of Atlantic forest and adjacent Cerrado (FROST 2014). Its advertisement call was described from a single male recorded from Teresópolis, state of Rio de Janeiro (HEYER 1980), and from a male from Boracéia, state of São Paulo (HEYER et al. 1990).

Data were collected from two sites in the Atlantic forest, state of São Paulo, southeastern Brazil. Both sites are located within the mountain complex of Serra do Mar, which is characterized by high levels of rainfall and exuberant evergreen vegetation. Males were recorded in two reproductive sites within the Boracéia Biological Station, at 847 and 872 masl, in the municipality of Biritiba Mirim (23°37′S, 45°52′W) (for more complete descriptions of the site see: HEYER et al. 1990), and also in two breeding sites in the Parque Estadual Intervales (791 and 864 masl), municipality of Ribeirão Grande (24°15′S, 48°24′W) – for more complete descriptions of the site see BERTOLUCI & RODRIGUES (2002).

Recordings of a total of 20 males of *D. microps* were obtained from the two localities mentioned above.

The abiotic conditions during call recordings are listed in Table 1. Male calling sites were characterized by perch nature and height from the water surface. Digital recordings were made at a sampling rate of 48 kHz and 16 bit resolution with a Marantz PMD660 recorder and a Yoga EM 9600 microphone positioned about 80-60 cm from the calling males. The analysis of acoustic properties of 124 calls were completed with Raven pro 64 1.4 software for Windows (Cornell Lab of Ornithology), using FFT (Fast Fourier Transformation) = 1024 and Overlap = 50. We analyzed the following call properties: highest frequency of call A (HFCA), highest frequency of call B (HFCB), lowest frequency of call A (LFCA), lowest frequency of call B (LFCB),

Figures 1-2. Males of *Dendropsophus microps* from Boracéia (1) and Ribeirão Grande (2), São Paulo state, southeastern Brazil.

dominant frequency of call A (DFCA), dominant frequency of call B (DFCB), duration of the first note of call A (DFNCA), duration of call B (DCB), pulses rate of the first note of call A (PRFNCA) and pulses rate of call B (PRCB). The spectral measurements were obtained by selecting three variables in the menu "choose measurements" in Raven 1.4: (1) Frequency 5% (Hz), (2) Frequency 95% (Hz) – these two measurements include highest frequency and lowest frequency, ignoring 5% downward and upward over the frequency band formed by the distribution of energy; and (3) Max Frequency (Hz) – it shows the dominant frequency (the frequency in which the power is maximum within the call). The snout-vent length (SVL) of the calling males was measured with digital calipers (to the nearest 0.1 mm). Eleven individuals were collected and were deposited as voucher specimens in the Herpetological Collection of Escola Superior de Agricultura Luiz de Queiroz,

Table 1. Abiotic conditions during call recordings of *Dendropsophus microps* in the two localities. Values presented as mean ± standard deviation (range).

Abiotic variables	Boracéia (N = 9)	Ribeirão Grande (N = 11)
Air temperature (°C)	15.5 ± 2.2 (13-20.2)	15.5 ± 2.1 (12.5-18.6)
Air humidity (%)	89.3 ± 2.9 (85-93)	90.3 ± 2.9 (84-93)
Recording period	7:30-9:25 p.m.	7:10-10:20 p.m.
Recording date	October 19-22th 2010 and January 13th 2011	October 27-31th 2010 and February 7-9th 2011

Universidade de São Paulo, Brazil (ICMBio license number 23799-1), with the accessing codes VESALQ 572, 698, 802, 811, 838, 848, 856, 922, 938, 943, and 985. Acoustic recordings of 13 males were deposited in the Fonoteca Neotropical Jacques Vielliard with collection numbers 30876 to 30888.

Two-sample T-tests (with 0.06 significance level) were used to detect significant differences between the types of call considering pulse repetition rate, and between population average considering body size (mm), calling perch (cm) and all call properties mentioned above.

Quantitative acoustic variables were calculated through descriptive statistics including coefficient of variation (SD/mean) for intra-individual and intra-population levels. All the statistical analysis was carried out in the software SYSTAT 13.

RESULTS

Body size

Male SVL values were significantly greater in Ribeirão Grande (24.3 ± 1.7 mm) than in Boracéia (22.5 ± 1.1 mm) (N = 18, t = -2.578, p < 0.06).

Calling perches

Males of both populations called in or near lentic water bodies. In the population of Boracéia, males used mainly the adjacent vegetation (75%, N = 11) as calling perches, while in Ribeirão Grande, males also called from the emergent vegetation (45%, N = 8). Males from Boracéia called from higher perches (average of 71 ± 36.9 cm) than males from Ribeirão Grande (average of 43 ± 20.3 cm) (N = 19, t = 2.126, p < 0.06).

Acoustic analysis

Dendropsophus microps has two call types that are distinct in their pulse repetition rate (N = 19, t = 6.625, p < 0.01). Type "A" calls have a pulse rate of 60-203 pulses/s (N = 34), and type "B" calls have a pulse rate of 23-60 pulses/s (N = 45). Type "A" calls are formed by one to five notes, with a considerable range in duration (0.08-1.93 s, N = 29) and two kinds of pulse structures: one longer with fused pulses and the other short with evident pulses (Figs. 3-4). We considered "note" an uninterrupted sound element that compose the call. Type "B" call is formed by a single pulse train, with average duration of 0.67 s, (range 0.17-2.29 s, N = 46) (Fig. 5-6).

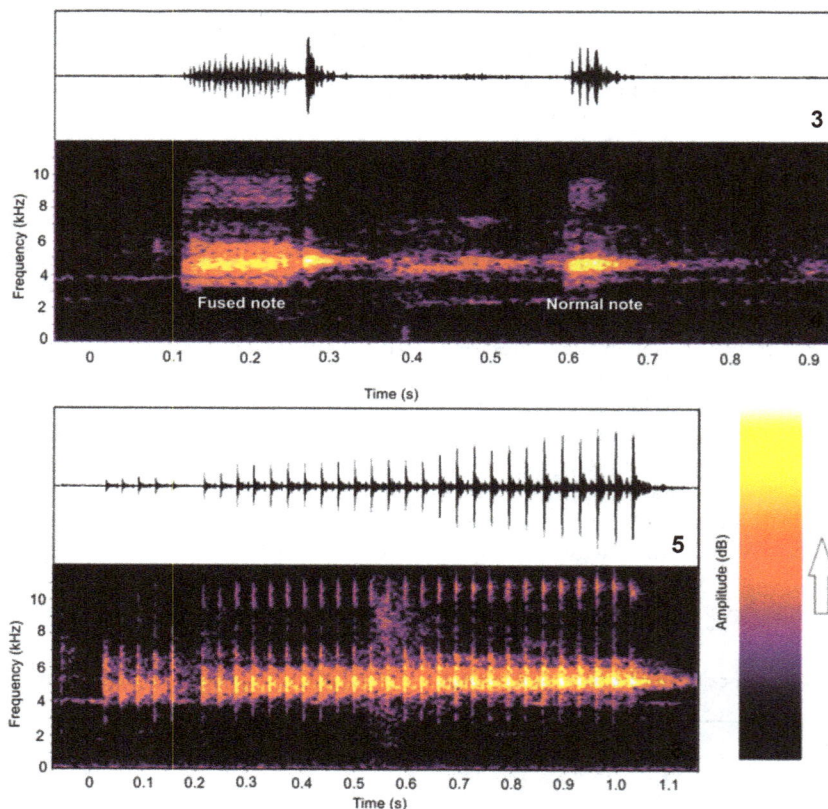

Figures 3-6. *Dendropsophus microps* calls: (3) waveform and (4) spectrogram with notes of type "A" call of a male from Ribeirão Grande (SVL 22.1 mm), recorded at 7:45 p.m. on February 9th 2011; air temperature = 19.6°C; air humidity = 88%; (5) waveform and (6) spectrogram of type "B" call of a male from Boracéia (SVL 21.2 mm), recorded at 7:47 p.m. on October 19th 2010; air temperature = 15°C, air humidity = 85%.

Both call types have slight upward frequency modulation and increasing intensity. The average frequency band of both call types was similar: call type "A" 4574-5452 Hz (N = 10), average dominant frequency 4972 ± 195 Hz (N = 10); call type "B" 4488-5417 Hz (N = 13), average dominant frequency 4913 ± 242 Hz (N = 13). Both call types showed a less intense harmonic band between 8000 and 11500 Hz. More than one half of the recorded males emitted both call types (A and B) in the same call sequence (64%). The emission proportion of both call types inside the same sequence was similar in both populations (57% in males from Boracéia and 71% in males from Ribeirão Grande).

Many quantitative traits of calls were similar between the populations from Boracéia and Ribeirão Grande, but in call B we found significant differences between populations in highest frequency, duration, and pulse rate. Quantitative acoustical data and the results of statistical tests are presented in Table 2.

Variation of acoustic properties

In general, spectral acoustic properties are less variable than temporal properties in *Dendropsophus microps* (Figs. 7-9). The majority of call traits, both in the intra-individual and intra-population levels related to frequency showed coefficients of variation below 12% and may be considered "static" (those below 5%) or "intermediary" (more than 5% and less than 12% of variation) sensu GERHARDT (1991) (Figs. 7 and 9). On the other hand, all temporal properties measured, at the intra-individual level, showed CVs higher than that value and could be considered "dynamic" GERHARDT (1991) (Fig. 8).

The variation in spectral acoustic properties followed a similar pattern between populations. However, variations in some temporal properties were different between populations, as pulse rate of both types of call (Fig. 9). Pulse rate of call B (PRCB) in the population of Ribeirão Grande was the only temporal variable considered "static" at the population level.

DISCUSSION

Body size

The mean SVL of *Dendropsophus microps* males from Boracéia was 22.5 mm, consistent with the results of HEYER et al. (1990) for the same locality. The differences in SVL between the two populations of this study may be associated with the selective pressure on morphological or acoustical traits caused by the sister species, *D. giesleri*, living in sympatry with the Ribeirão Grande population. Reproductive isolation in anuran species is expected to occur at the acoustic level rather than body size. However, selection may act on pleitropic genes that affect both size and call frequency. PFENNIG & PFENNIG (2005) observed that *Spea multiplicata* (Scaphiopodidae) individuals were smaller in areas where they were sympatric with *Spea bombifrons*. They concluded that this difference was the result

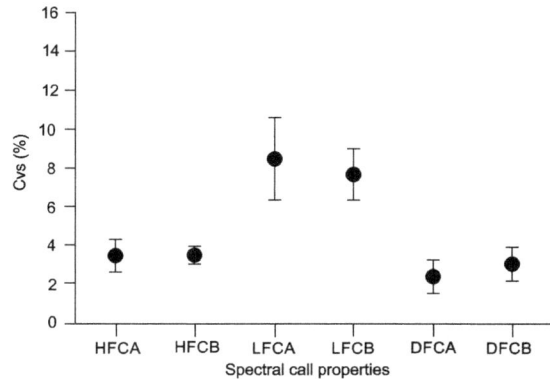

7

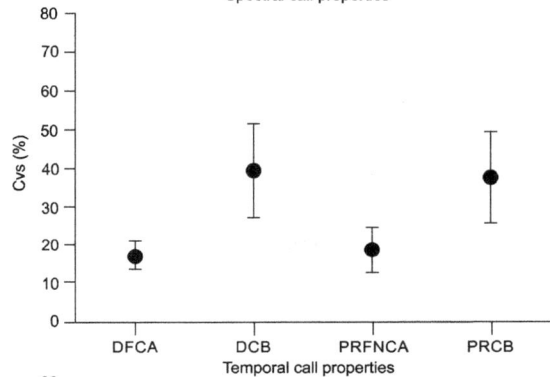

8

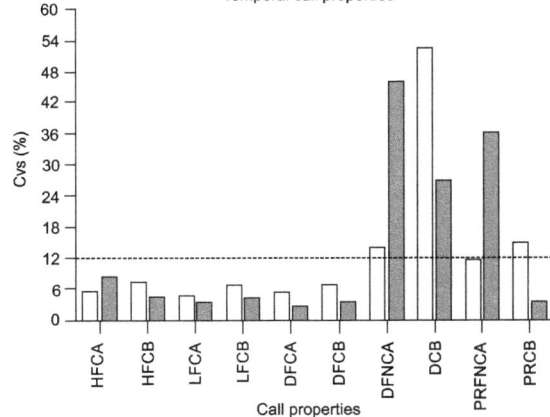

9

Figures 7-9. Intraindividual coefficient of variation (%) of *Dendropsophus microps* spectral (7) and temporal (8) call properties. Error bars represent the confidence interval of 95% around the average (black circles). Intrapopulational coefficient of variation (%) of *D. microps* call properties of the two populations studied (9). The red dashed line separates dynamic (above 12% of variation) and static acoustic properties sensu Gerhardt (2001). Abbreviations: highest frequency of call A (HFCA), highest frequency of call B (HFCB), lowest frequency of call A (LFCA), lowest frequency of call B (LFCB), dominant frequency of call A (DFCA), dominant frequency of call B (DFCB), duration of the first note of call A (DFNCA), duration of call B (DCB), pulses rate of the first note of call A (PRFNCA) and pulses rate of call B (PRCB). (□) Boracéia, (■) Ribeirão Grande

Table 2. Call properties of the two populations of *Dendropsophus microps* studied. The values represent mean ± SD, and N = number of males. Results of the two-sample t-tests are included.

Acoustic Properties	Boracéia	Ribeirão Grande	t-value	p-value
Highest frequency of call A (Hz)	5532 ± 277 (N = 4)	5398 ± 422 (N = 6)	0.604	0.562
Highest frequency of call B (Hz)	5580 ± 377 (N = 7)	5226 ± 203 (N = 6)	2.145	0.059*
Lowest frequency of call A (Hz)	4664 ± 191 (N = 4)	4513 ± m129 (N = 6)	1.385	0.227
Lowest frequency of call B (Hz)	4513 ± 283 (N = 7)	4458 ± 168 (N = 6)	0.430	0.676
Dominant frequency of call A (Hz)	5109 ± 237 (N = 4)	4880 ± 96 (N = 6)	1.834	0.147
Dominant frequency of call B (Hz)	4944 ± 311 (N = 7)	4876 ± 146 (N = 6)	0.514	0.620
Duration of the first note of call A (s)	0.405 ± 0.054 (N = 4)	0.288 ± 0.130 (N = 6)	1.950	0.091
Duration of call B (s)	1.013 ± 0.529 (N = 7)	0.533 ± 0.140 (N = 6)	2.305	0.055*
Pulse rate of the first note of call A (p/s)	101 ± 11 (N = 3)	107 ± 38 (N = 6)	-0.357	0.732
Pulse rate of call B (p/s)	32.0 ± 4.5 (N = 6)	44.0 ± 1.5 (N = 6)	-6.274	0.001**

of competition, which caused divergence in mating behavior and larval development. The smaller size of individuals in the sympatric populations resulted in decreased offspring survival, female fecundity, and the sexual selection pressure on males (PFENNIG & PFENNIG 2005). A result of this adaptive equilibrium is the coexistence of the two species in sympatry.

Calling perches

Calling sites are typical attributes of species and are correlated primarily with their morphology and size (DIXON & HEYER 1968, CRUMP 1971, HÖDL 1977). The fact that *Dendropsophus microps* males from Ribeirão Grande call from emergent vegetation may be related to structural differences between the habitats or a shift in micro-habitat selection, possibly due to competition for calling sites with other species. This hypothesis may also account for the observed differences in perch height between the two populations and could be verified with further observations and comparisons.

Due to plasticity in the use of calling sites, it is possible that a variation in the choice of calling perches may occur in closely related species in syntopy or in situations of high intra- and interspecific densities. BERTOLUCI & RODRIGUES (2002), for example, observed that *Dendropshophus minutus* and *Physalaemus cuvieri* used alternative calling sites when chorus sizes exceeded available perches.

BERTOLUCI & RODRIGUES (2002) recorded males of *D. microps* in the population of Ribeirão Grande calling from perches 30-70 cm high. This is consistent with the results of the present study obtained for the same population. This lower perch height may be related to the presence of *D. giesleri*, whose males were observed calling from perches 70 ± 31 cm high (N = 6) (LRF pers. obs.). This hypothesis remains to be tested, and if confirmed, it may reflect a phenomenon similar to that observed in *Anolis* lizards, which alter the elevation of their perches when there is another species of similar size living in the same area (LISTER 1976).

Acoustic analysis

The advertisement call of *Dendropsophus microps* described by HEYER (1980) and by HEYER et al. (1990) is very similar to that described here, but we have added information about variation in the acoustical properties of both call types, including pulse repetition rate and call duration. POMBAL JR (2010) also provided a succinct description of the call of *D. microps* from a population from Ribeirão Branco, state of São Paulo, but he did not mention that there were two call types. Nevertheless, it remains unclear how *D. microps* males use the two call types.

In the *D. parviceps* group, the calls of *Dendropsophus microps* and *D. giesleri* are the only calls that have harmonic structure (HEYER 1980). A complete comparison of the advertisement call of *D. microps* with its sister species is given in Table 3. According to the phylogeny reconstructed by MOTTA et al. (2012), using rRNA, genes *D. giesleri* is close to *D. allenorum*, but the relationship among the species inside the *D. parviceps* group is poorly resolved.

The main differences in acoustic traits between the two populations of the present study were in call B, which can be a target of evolutionary pressure. It is possible that these differences are associated with the presence of the sister species, *D. giesleri*, in Ribeirão Grande. The advertisement call of *D. giesleri* is very similar in structure to call B of *D. microps* (Figs. 10-11), making it possible to consider a hypothesis of acoustic character displacement, as already confirmed for other species (LITTLEJOHN 1976). Under this hypothesis the coexistence of sister species might cause adjustments to the calling properties.

Considering all explanations above, call B was most likely under pressure to change because it could represent the main acoustic trait for species recognition and mate selection. Although this idea has yet to be tested, the increased pulse rate showed by the population from Ribeirão Grande could be only a compensation to enhance the visibility of males in the presence of the sister species, *D. giesleri*. As a dynamic trait, the pulse rate of call B of males from Boracéia can be measured

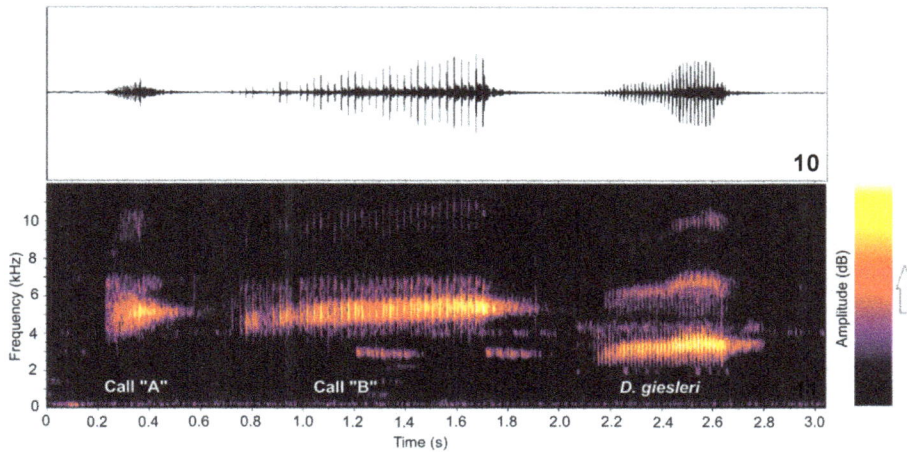

Figures 10-11. (10) Waveform and (11) spectrogram with both *Dendropsophus microps* call types (from Boracéia) and *D. giesleri* advertisement call from Ribeirão Grande recorded at 08:52 p.m. on October 28th 2010, SVL = 29.4 mm, air temperature = 13.5°C, humidity = 93%.

Table 3. Acoustic properties of the advertisement call of some species of *Dendropsophus parviceps* group. The complementary data come from Duellman & Crump (1974), Heyer (1980), Martins & Cardoso (1987), Orrico et al. (2013).

Species	Dominant frequency (Hz)	Call duration (s)	Pulse rate (p/s)	Number of notes
D. allenorum	3430	0.40 to 0.45	22 to 25	10 to 11
D. bokermanni	4000 to 4652	0.23 to 0.28	100 to 190	5 to 19
D. brevifrons	4152 to 5115	0.43 to 0.49	30 to 40	26-46
D. giesleri	3000 to 3600	0.3	100	1
D. microps*	4972 (Call A) and 4913 (Call B)	0.34 (first note Call A) and 0.67 (Call B)	105 (Call A) and 38 (Call B)	1 to 5 (call A) and 1 (call B)
D. parviceps	6072 to 6341	0.12 to 0.14	140	54.5
D. subocularis	2200	0.53	43	3 to 20
D. timbeba	3000 to 4200	0.6	18	10

*Data from the present study.

during a playback experiment simulating a chorus of *D. giesleri* males. If this hypothesis (of compensation) is correct, an increased pulse rate of call B would be expected.

It is necessary to understand that acoustic signals are multidimensional and involve both spectral and temporal traits, whose relative importance for mate recognition may vary from species to species (Erdtmann & Amézquita 2009) and their variation could increase reproductive isolation (Lemmon 2009). In this scenario, static acoustic properties (such as the spectral properties in our study) generally evolve under light directional or stabilizing selection, since they are commonly associated with conspecific recognition (Gerhardt 1991, Márquez et al. 2008).

Variation of acoustic properties

Temporal acoustic properties generally vary more than spectral properties because they may respond to temperature-related changes, and to social conditions, especially in calls with long pulse trains or long notes (Gerhardt & Huber 2002,

Wong et al. 2004). On the other hand, spectral properties are strongly related to patterns associated with the occupation of frequency bands used for specific recognition (Gerhardt & Davis 1988) and therefore tend to have low variation. A high variation in acoustic properties may not be adaptive when associated with pre-zygotic isolation mechanisms.

The fact that both populations show similar patterns of variation in spectral properties suggest that they may not be subjected to different selective pressures affecting these call traits. However, the pattern of variation of pulse rate for both types of calls were remarkably different between populations, with a special attention to a very small variation in the pulse rate of call B in the population of Ribeirão Grande. This result reinforces the hypothesis of compensation, where males subjected to interspecific competition increased pulse rate as an escape for acoustic visibility and recognition.

Our results show substantial differences between two populations of *D. microps* in body size, calling perch, and some call traits. These differences may be associated with the presence of the sister species, *D. giesleri*, in one of the localities. In order to verify this association, a larger number of populations need to be sampled, with the possible addition of experiments measuring female preference in sympatry and allopatry, as performed by MÁRQUEZ & BOSCH (1997) for *Alytes* (Alytidae) and by PFENNIG & RYAN (2006) for Scaphiopodidae. An assessment of the degree of genetic differentiation between populations, considering the wide geographical distribution of *D. microps* in the Brazilian Atlantic Forest, could also help to explain the observed differences. Such an assessment would involve sampling of genetic, morphological and acoustical data from many populations. In addition, carrying out playback experiments to simulate the effect of a sister species as an acoustic competitor could clarify whether the difference in the pulse rate of call B between populations is a punctual consequence of the sympatry, since pulse rate is considered a plastic acoustic trait.

ACKNOWLEDGEMENTS

We are grateful to Fábio A. Martins for help during fieldwork. Special thanks to FAPESP by the doctoral grant and postdoctoral fellowship to LRF (process 2009/13987-2 and 2013/21519-4). JB is a researcher of CNPq (process 304938/2013-0). Partial funding for acoustical analyses was provided by Ministerio de Ciencia e Innovación, Spain, project TATANKA CGL2011-25062 (P.I.R. Márquez).

REFERENCES

BERTOLUCI J, RODRIGUES MT (2002) Utilização de hábitats reprodutivos e micro-hábitats de vocalização em uma taxocenose de anuros (Amphibia) da Mata Atlântica do sudeste do Brasil. **Papéis Avulsos de Zoologia 42**(11): 287-297. doi: 10.1590/S0031-10492002001100001

CRUMP ML (1971) Quantitative analysis of the ecological distribution of a tropical herpetofauna. **Occasional Papers of the Museum of Natural History University of Kansas 3**: 1-62.

DIXON JR, HEYER WR (1968) Anuran succession in a temporary pond in Colima, Mexico. **Bulletin of the Southern California Academy of Sciences 67**: 129-137.

DUELLMAN WE, CRUMP ML (1974) Speciation in frogs of the *Hyla parviceps* group in the upper Amazon Basin. **Occasional Papers of the Museum of Natural History, University of Kansas 23**: 1-40.

ERDTMANN L, AMÉZQUITA A (2009) Differential evolution of advertisement call traits in dart-poison frogs (Anura: Dendrobatidae). **Ethology 115**: 801-811. doi: 10.1111/j.1439-0310.2009.01673.x

FAIVOVICH J, HADDAD CFB, GARCIA PCA, FROST DR, CAMPBELL JA, WHEELER WC (2005) Systematic review of the frog family Hylidae, with special reference to Hylinae: phylogenetic analysis and taxonomic revision. **Bulletin of the American Museum of Natural History 294**: 1-240. doi: 10.1206/0003-0090(2005)294[0001:SROTFF]2.0.CO;2

FORTI LR, STRÜSSMANN C, MOTT T (2010) Acoustic communication and vocalization microhabitat in *Ameerega braccata* (Steindachner, 1864) (Anura, Dendrobatidae) from Midwestern Brazil. **Brazilian Journal of Biology 70**(1): 211-216. doi: 10.1590/S1519-69842010000100029

FROST DR (2014) **Amphibian species of the world: an online reference.** New York, American Museum of Natural History, v. 5.6. Avalaible online at: http://research.amnh.org/herpetology/amphibia/index.html [Accessed: 23/02/2013]

GERHARDT HC (1991) Female mate choice in treefrogs: static and dynamic acoustic criteria. **Animal Behaviour 42**: 615-635. doi: 10.1016/S0003-3472(05)80245-3

GERHARDT HC (1994) The evolution of vocalization in frogs and toads. **Annual Review of Ecology and Systematics 25**: 293-324. doi: 10.1146/annurev.es.25.110194.001453

GERHARDT HC (2012) Evolution of Acoustic Communication: a Multi-Level Analysis of Signal Variation. **Bioacoustics Journal 21**(1): 9-11. doi: 10.1080/09524622.2011.647469

GERHARDT HC, DAVIS MS (1988) Variation in the coding of species identity in the advertisement calls of *Litoria verreauxi* (Anura: Hylidae). **Evolution 42**: 556-565.

GERHARDT HC, HUBER F (2002) **Acoustic communication in insects and anurans: common problems and diverse solutions.** Chicago, The University of Chicago Press, 531p.

HADDAD CFB, POMBAL-JR JP, BATISTIC RF (1994) Natural hybridization between diploid and tetraploid species of Leaf-Frogs, genus *Phyllomedusa* (Amphibia). **Journal of Herpetology 28**(4): 425-430. doi: 10.2307/1564953

HEYER WR (1980) The calls and taxonomic positions of *Hyla giesleri* and *Ololygon opalina* (Amphibia: Anura: Hylidae). **Proceedings of the Biological Society of Washington 93**: 655-661.

HEYER WR, RAND AS, CRUZ CAG, PEIXOTO O, NELSON CE (1990) Frogs of Boracéia. **Arquivos de Zoologia 31**: 231-410.

HÖDL W (1977) Call differences and calling site segregation in anuran species from Central Amazonian floating meadows. **Oecologia 28**: 351-363.

HÖBEL G, GERHARDT HC (2003) Reproductive character displacement in the acoustic communication system of green tree frogs (*Hyla cinerea*). **Evolution 57**: 894-904. doi: 10.1111/j.0014-3820.2003.tb00300.x

LEMMON EM (2009) Diversification of conspecific signals in sympatry: geo graphic overlap drives multidimensional reproductive character displacement in frogs. **Evolution 63**(5): 1155-1170. doi: 10.1111/j.1558-5646.2009.00650.x

LISTER BC (1976) The nature of niche expansion in West Indian *Anolis* lizards I: ecological consequences of reduced competition. **Evolution 30**: 659-676.

LITTLEJOHN MJ (1976) The *Litoria ewingi* complex (Anura: Hylidae) in south-eastern Australia IV. Variation in mating-call structure across a narrow hybrid zone between L. ewingi and L. paraewingi. **Australian Journal of Zoology 24**: 283-293.

MÁRQUEZ R, BOSCH J (1997) Male advertisement call and female preference in sympatric and allopatric midwife toads (*Alytes obstetricans* and *Alytes cisternasii*). **Animal Behaviour 54**: 1333-1345. doi: 10.1006/anbe.1997.0529

MÁRQUEZ R, BOSCH J, EEKHOUT X (2008) Intensity of female preference quantified through playback setpoints: call frequency versus call rate in midwife toads. **Animal Behaviour 75**: 159-166.

MARTINS M, CARDOSO AJ (1987) Novas espécies de hilídeos do Estado do Acre (Amphibia: Anura). **Revista Brasileira de Biologia 47**: 549-558.

MORAIS AR, BATISTA VG, GAMBALE PG, SIGNORELLI L, BASTOS RP (2012) Acoustic communication in a Neotropical frog (*Dendropsophus minutus*): vocal repertoire, variability and individual discrimination. **Herpetological Journal 22**: 249-257.

MOTTA AP, CASTROVIEJO-FISHER S VENEGAS PJ, ORRICO VGD, PADIAL MJ (2012) A new species of the *Dendropsophus parviceps* group from the western Amazon basin (Amphibia: Anura: Hylidae). **Zootaxa 3249**: 18-30.

ORRICO VGD, DUELLMAN WE, SOUZA MB, HADDAD CFB (2013) The Taxonomic Status of *Dendropsophus allenorum* and *Dendropsophus timbeba* (Anura: Hylidae). **Journal of Herpetology 47**(4): 615-618. doi: 10.1670/12-208

PFENNIG KS, PFENNIG DW (2005) Character displacement as the 'best of a bad situation': fitness trade-offs resulting from selection to minimize resource and mate competition. **Evolution 59**: 2200-2208. doi: 10.1111/j.0014-3820.2005.tb00928.x

PFENNIG KS, RYAN MJ (2006) Reproductive character displacement generates reproductive isolation among conspeciûc populations: an artiûcial neural network study. **Proceedings of the Royal Society B 273**: 1361-1368. doi: 10.1098/rspb.2005.3446

POMBAL JR JP (2010) O espaço acústico em uma taxocenose de anuros (Amphibia) do sudeste do Brasil. **Arquivos do Museu Nacional do Rio de Janeiro 68**(1-2): 135-144.

RYAN MJ, RAND AS (1993) Species recognition and sexual selection as a unitary problem in animal communication. **Evolution 47**(2): 647-657.

TOLEDO LF, HADDAD CFB (2005) Acoustic Repertoire and Calling Behavior of *Scinax fuscomarginatus* (Anura, Hylidae). **Journal of Herpetology 39**(3): 455-464. doi: 10.1670/139-04A.1

WELLS KD (1977) The social behavior of anuran amphibians. **Animal Behaviour 25**: 666-693.

WELLS KD (2007) **The ecology and behavior of amphibians**. Chicago, The University of Chicago Press, 1148p.

WONG BBM, COWLING ANN, CUNNINGHAM RB, DONNELLY CF, COOPER PD (2004) Do temperature and social environment interact to affect call rate in frogs (*Crinia signifera*)? **Austral Ecology 29**(2): 209-214. doi: 10.1111/j.1442-9993.2004.01338.x

10

A new species of *Masteria* (Araneae: Dipluridae: Masteriinae) from Southeastern Brazil

Denis Rafael Pedroso[1], Renner Luiz Cerqueira Baptista[2] & Rogério Bertani[3,4]

[1]*Laboratório de Aracnologia, Museu Nacional, Universidade Federal do Rio de Janeiro, Brazil. E-mail: drbpedroso@gmail.com*
[2]*Laboratório de Diversidade de Aracnídeos, Instituto de Biologia, Universidade Federal do Rio de Janeiro, Brazil. E-mail: baptistr@gmail.com*
[3]*Laboratório Especial de Ecologia e Evolução, Instituto Butantan. Avenida Vital Brazil 1500, 05503-900 São Paulo, SP, Brazil.*
[4]*Corresponding author. E-mail: rogerio.bertani@butantan.gov.br, rogerio.bertani@uol.com.br*

ABSTRACT. A new species of *Masteria* L. Koch, 1873 from iron ore caves at Caeté and Santa Bárbara, state of Minas Gerais, Brazil, *Masteria emboaba* **sp. nov.**, is described. It was collected inside caves and in the litter of nearby dry forests. It is the first masteriine species described from southeastern Brazil and the second masteriine species for the country. The new species is the only known *Masteria* with only two eyes. Additionally, the male of *M. emboaba* **sp. nov.** has only two regular, thin spines at the apex of tibia I, lacking the tibial apophysis found in most other *Masteria* species. The only other described *Masteria* species that has spines in the place of tibial apophysis is *M. aimeae* (Alayón, 1995) from Cuba; however, the last species has a longer and sinuous embolus, contrasting the embolus of *M. emboaba* **sp. nov.**, which is much smaller, less sinuous and transversally placed. The only other described Brazilian species, *M. manauara* Bertani, Cruz & Oliveira, 2013, has a double tibial apophysis, with both ends tipped by a strong, short spine, and a very long embolus, parallel to the bulb.

KEYWORDS. Iron ore; lateritic cave; Minas Gerais; Mygalomorphae; Serra da Gandarela.

Masteria L. Koch, 1873 belongs to Dipluridae, Masteriinae (RAVEN 1985), which contains many of the smallest mygalomorph species known (RAVEN 1979, 1981, RAVEN & PLATNICK 1981, PLATNICK & FOSTER 1982). The 23 described species are widely distributed in the Australasian Region and in the New World (PLATNICK 2014). The Australasian region has eight species, including the type species, *M. hirsuta* L. Koch, 1873, from the Fiji Islands and Micronesia. In the New World, there are 15 described species: *M. aimeae* (Alayón, 1995) and *M. golovatchi* Alayón, 1995 – Cuba; *M. lewisi* (Chickering, 1964) and *M. pecki* Gertsch, 1982 – Jamaica; *M. petrunkevitchi* (Chickering, 1964) – Puerto Rico; *M. modesta* (Simon, 1891) – Saint Vincent; *M. barona* (Chickering, 1966) and *M. simla* (Chickering, 1966) – Trinidad; *M. downeyi* (Chickering, 1966) and *M. spinosa* (Petrunkevitch, 1925) – Costa Rica and Panama; *M. colombiensis* Raven, 1981 – Colombia; *M. cyclops* (Simon, 1889), *M. lucifuga* (Simon, 1889) and *M. tovarensis* (Simon, 1889) – Venezuela. Finally, a single species is known from Brazil (state of Amazonas), *M. manauara* Bertani, Cruz & Oliveira, 2013. It inhabits the ground litter (BERTANI et al. 2013), particularly amongst the leaves of small palm trees.

Besides *M. manaura*, there are records of an unidentified *Masteria* species from the state of Piauí (L.S. Carvalho unpubl. data) and several areas of the state of Minas Gerais: Nova Lima (ANGLOGOLD ASHANTI 2009), Conceição de Mato Dentro (LEÃO & AULER 2012), Caeté and Santa Bárbara (COELHO et al. 2010). The records from Minas Gerais are based on material collected in and around caves in iron ore deposits.

There are more than 3,000 caves in iron ore deposits in Brazil (AULER et al. 2014). Most of these caves are located at the two major iron ore provinces: Carajás ridge, in the Amazon, and Iron Quadrangle, in Minas Gerais, southeastern Brazil. The Iron Quadrangle caves are found on iron-rich deposits topped by "canga", an iron-rich breccia surface cemented by ferruginous matrix (AULER et al. 2014). The vegetation of the "canga" areas is open grassland with scattered trees and is dominated by herbs and bushes. The alfa and beta diversity of plant species is high, including dozens of rare and endemic species (CARMO & JACOBI 2013). The plant species are a mix of elements from the Atlantic Forest, Cerrado and Serra do Espinhaço (CARMO & JACOBI 2013).

Despite the small average size of iron ore caves, they have "a high potential as habitat of troglobitic invertebrates in Brazil" (TRAJANO & BICHUETTE 2010). Several troglobitic species have been found in Brazilian iron ore caves, both in the Carajás area (PELLEGRINI & FERREIRA 2011, PEDROSO & BAPTISTA 2014) and the Iron Quadrangle (SOUZA & FERREIRA 2005, COELHO et al. 2010). Caves from iron ore areas are usually near the surface and have an extensive array of microchannels that house a highly diverse associated fauna (FERREIRA 2005). The fauna of the "canga" areas of Minas Gerais also include many endemic taxa (e.g., Bernardi et al. 2013, Ázara & Ferreira 2013).

Herein we describe the second masteriine species from Brazil and the first from Southeast Brazil. In addition, it is the first Masteriinae from iron ore caves collected at Caeté and Santa Barbára, Minas Gerais.

MATERIAL AND METHODS

Specimens were collected inside and near iron ore caves located at the hilltops of "Serra da Gandarela" (Gandarela range), at Caeté and Santa Bárbara, in the Iron Quadrangle area, state of Minas Gerais, southeastern Brazil. The caves are located on "canga" plates; however, the collecting area also included dry patches of Atlantic forest, especially on the small river valleys near the tips of the hill range. Most specimens were collected during a trip of the "Projeto Mina Apolo", a project to study the environmental impact from the installation of a large iron ore mine. The first author participated in the project as a member of the environmental consulting "Amplo Consultoria" team. Additional specimens from the caves in the collecting area had been mentioned by Coelho et al. (2010).

The general description format follows Raven (1981) with some modifications; e.g., eye diameters are given in their real measurements, not in ratios. All measurements are in millimeters and were obtained with a Leica LAS Interactive Measurements module. Leg and palp measurements were taken from the dorsal aspect of the left side (unless appendages were lost or obviously regenerated). A Leica LAS Montage and LAS 3D module, mounted on a Leica M205C dissecting microscope, were used to capture images of the structures of spiders. Illustrations of spermathecae were drawn over images obtained with a Leica DM 2500 compound microscope. The spermathecae soft tissues were digested with trypsin for several days and subsequently cleared with clove oil before they were photographed.

Abbreviations. (ITC) inferior tarsal claw, (PLS) posterior lateral spinnerets, (PME) posterior median eyes, (PMS) posterior median spinnerets, and (STC) superior tarsal claws.

Specimens are deposited in the arachnological collection of the Museu Nacional, Universidade Federal do Rio de Janeiro (MNRJ).

TAXONOMY

Masteria emboaba **sp. nov.**
Figs. 1-16

Diagnosis. Males and females of *Masteria emboaba* **sp. nov.** have only two eyes (Figs. 5 and 14). Most other species have six to eight eyes and two, *M. pecki* and *M. caeca* (Simon, 1892), have no eyes at all. This new species is much smaller than the other known South American species, except for *M. manauara*, a tiny species from the state of Amazonas and the only other species described from Brazil. The male of *M.*

emboaba **sp. nov.** has only two regular, thin spines at the apex of tibia I, lacking the tibial apophysis found in most other *Masteria* species. The only other described *Masteria* species with spines in the place of tibial apophysis is *M. aimeae*, from Cuba; however, the last species has a longer and sinuous embolus (Alayón 1995: figs. 1c-d), contrasting with the embolus of *M. emboaba* **sp. nov.**, which is much smaller, less sinuous and transversally placed (Figs. 6-8). The only other described Brazilian species, *M. manauara*, has a double tibial apophysis, with both ends tipped by a strong, short spine, and a very long embolus, parallel to the bulb (Bertani et al. 2013: figs. 6-9). Table I summarizes the available information on the geographical range and characters of all species of *Masteria*.

Description. Male holotype (MNRJ 4540) (Fig. 1). Entirely pale yellow. Carapace 1.22 long, 0.96 wide, clothed with long (ca. 0.15) prostate gray bristles on interstrial ridges (Figs. 1 and 2). Two eyes on tubercle occupying 0.28 of head width (Fig. 5). Eye group 0.12 wide. Sizes and interdistances: PME 0.04, PME-PME 0.04. Chelicerae 0.29 long, 0.21 wide, with nine spaced teeth on promarginal furrow and six spinules mesobasally. Labium 0.10 long, 0.20 wide. Maxillae 0.31 long, 0.26 wide. Sternum 0.66 long, 0.63 wide; sigilla not evident (Fig. 3).

Palp with elongated cymbium, bearing seven spines on apical edge; bulb pear-shaped, tegulum 0.22 long, 0.12 wide, tapering and giving origin to a short (0.13) and relatively thin (0.012) embolus, almost transversally placed, due to a strong basal curve to retrolateral side, keeping its diameter and with a gentle curvature on apical half, not tapering to apex (Figs. 6-8). Leg lengths and midwidths in Table II.

Leg formula 4123. Tibia I lacking spur, with two large, thin spines ventrally on distal edge. Metatarsus I lacking spur (Fig. 9). Spines elongate: leg I tibia v3, metatarsus v2; leg 2, patella d1, tibia p1, v4, metatarsus v4; leg 3, patella d3, tibia d3, p2, r1, v5, metatarsus d6, p1, v3; leg 4, patella d2, tibia d4, p3, r1, v4, metatarsus d7, p2, r1, v3; palp, tarsus 7. Five to seven teeth on STC; zero to three teeth on ITC. Abdomen 1.42 long, 0.98 wide. Spinnerets (Fig. 4): PMS 0.20 long, 0.81 wide, 0.24 apart; basal, middle, and apical segments of PLS 0.34 long, 0.14 wide; 0.29 long, 0.13 wide; 0.29 long, 0.11 wide, respectively.

Female paratype (MNRJ 4540). As in male, except as noted. Abdominal tegument translucent, allowing recognition of internal structures (Figs. 10 and 13). Carapace 1.20 long, 0.97 wide, clothed with long (ca. 0.11) prostate gray bristles on interstrial ridges (Fig. 11). Two eyes on tubercle occupying 0.36 of head width (Fig. 14). Eye group 0.07 wide. Sizes and interdistances: PME 0.05, PME-PME 0.05. Chelicerae 0.31 long, 0.26 wide, with eleven widely spaced teeth on promarginal furrow and six spinules mesobasally. Labium 0.10 long, 0.22 wide. Maxillae 0.32 long, 0.28 wide. Sternum 0.72 long, 0.65 wide; sigilla not evident (Fig. 12). Leg lengths and midwidths in Table II. Leg formula 4123. Spines elongate: leg 1,metatarsus v2; leg 2, metatarsus v3; leg 3, tibia d4, v2, metatarsus d5, v4; leg 4, tibia d4, v4, metatarsus d6,v3; palpus, tarsus v4. Four

Figures 1-9. *Masteria emboaba* **sp. nov.** Holotype male (MNRJ 4540) (1) habitus; (2) carapace and chelicerae; (3) sternum, maxillae, chelicerae, coxae and labium; (4) abdomen, ventral; (5) eye tubercle; (6-8) left male palp, (6) retrolateral, (7) prolateral, (8) ventral; (9) left leg I, ventral. Scale bars: 1-2, 4 = 1 mm, 5-9 = 0.1 mm, 3 = 0.5 mm.

Table I. Comparison of geographical distribution and main characters of species of *Masteria*. (M) Male, (F) female, (I) immature, "(–)" non-applicable, "(?)" data not available.

Species	Sex	Type-Locality	Eyes	Carapace length	Spermathecae	Apex of male tibia I	Metatarsal apophysis	Paraembolic apophysis	Embolus
M. aimeae (Alayón, 1995)	M	Cuba: Holguín	6	1.3	–	four spines	no	no	long, curved
M. barona (Chickering, 1966)	MF	Trinidad: Arima Valley, Simla	6	M 1.69 F 1.63	?	holotype legs I missing	holotype legs I missing	no	long, thickened, sinuous
M. caeca (Simon, 1892)	F	Philippines: Morong, d'Antipolo Cave	0	?	–	–	–	–	–
M. cavicola (Simon, 1892)	F	Philippines: Manila, Montalvan, San-Mateo Cave	6	?	?	–	–	–	–
M. colombiensis Raven, 1981	MF	Colombia: Magdalena	8	M 1.8 F 2.1	two spermathecae, basally divided, wide stalk, rounded tip	three distal processes on tibia articulation.	yes	no	short, slightly curved
M. cyclops (Simon, 1889)	I	Venezuela: Caracas, Catuche Forest	6	?	?	–	–	–	–
M. downeyi (Chickering, 1966)	MF	Costa Rica: Turrialba	6	M 1.67 F 2.05	?	three distal processes on tibia articulation	no	no	very short, straight
M. emboaba **sp. nov.**	MF	Brazil: Minas Gerais	2	M 1.2 F 1.2	Two spermathecae, double thin stalk, wide flattened tip	two unmodified spines	no	no	short, slightly curved
M. franzi Raven, 1991	M	New Caledonia: Hienghène, Tiouandé	6	1.25	–	three distal processes on tibia articulation	?	yes	long, filiforme, almost straight
M. golovatchi Alayón, 1995	M	Cuba: Guantánamo, Paso Cuba	6	1.5	–	six spines	no	no	very short, curved
M. hirsuta L. Koch, 1873	F	Fiji, Micronesia: Ovalau	6	F 2.75	four spermathecae, the outer ones large and rounded, the inner short and coniform; all lobes lack circular ribbing (Raven, 1991)	–	–	–	–
M. kaltenbachi Raven, 1991	F	New Caledonia: Nékliai	6	1.0	four spermathecae, the outer ones with a spiral ribbing	–	–	–	–
M. lewsi (Chickering, 1964)	MF	Jamaica: St. Catherine Parish	6	1.34	?	three distal processes on tibia articulation	no	no	short, thickened, tapering strongly on its distal portion
M. lucifuga (Simon, 1889)	I	Venezuela: Aragua, Colonia Tovar	6	?	?	–	–	–	–
M. macgregori (Rainbow, 1898)	F	New Guinea: Neneba	6	1.7	four spermathecae	–	–	–	–
M. manauara Bertani, Cruz & Oliveria, 2013	MF	Brasil: Amazonas	6	M 0.7 F 0.8	two spermathecae, convoluted stalk, large rounded tip	one upper and one lower spur, both with an enlarged spine at tip	yes	no	long, very thin
M. modesta (Simon, 1891)	F	St. Vincent	6	?	?	–	–	–	–
M. pallida (Kulczyn'ski, 1908)	F	New Guinea	6	?	?	–	–	–	–
M. pecki Gertsch, 1982	F	Jamaica: Falling Cave	0	2.75	two low suboval spermathecae	–	–	–	–
M. petrunkevitchi (Chickering, 1964)	MF	Puerto Rico: Mayaguez	8	1.69	?	three distal processes on tibia articulation	no	no	short, thickened, tapering to its tip
M. simla (Chickering, 1966)	MF	Trinidad: Arima Valley, Simla	8	M 1.52 F 1.78	?	three distal processes on tibia articulation	no	yes	short, curved
M. spinosa (Petrunkevitch, 1925)	MF	Panama: San Lorenzo River	8	M 1.86 F 2.05	?	three distal processes on tibia articulation	no	yes	short, thin, curved
M. toddae Raven, 1979	MF	Australia: Queensland, Home Rule	6	M 1.56 F 2.26	four spermathecae	three distal processes on tibia articulation	no	yes	short, thin, straigh
M. tovarensis (Simon, 1889)	F	Venezuela: Aragua, Colonia Tovar	6	?	?	–	–	–	–

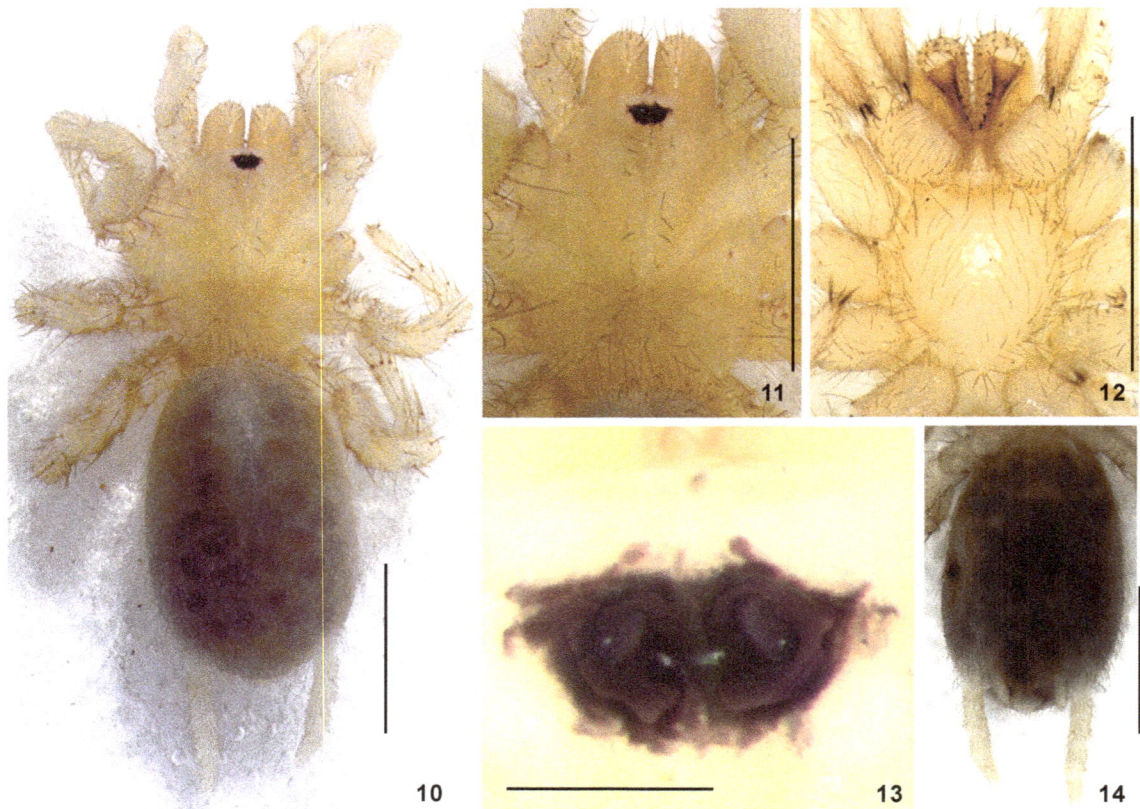

Figures 10-14. *Masteria emboaba* **sp. nov.** Paratype female (MNRJ 4540): (10) habitus; (11) carapace and chelicerae; (12) sternum, maxillae, chelicerae, coxae and labium; (13) eye tubercle; (14) abdomen and spinnerets, ventral. Scale bars: 10-12, 14 = 1 mm, 13 = 0.1 mm.

to eleven teeth on STC; two to four on ITC. Abdomen 1.86 long, 1.22 wide. Spinnerets (Fig. 13): PMS 0.23 long, 0.09 wide, 0.34 apart; basal, middle, and apical segments of PLS 0.33 long, 0.14 wide; 0.19 long, 0.14 wide; 0.22 long, 0.11 wide, respectively. Epigastric plate not posteriorly produced; two spermathecae, each one with one pair of long, thin stalks bearing a distal rounded and somewhat flattened receptacle. The outer stalk-receptacle set is smaller than the inner one (Fig. 15).

Type material. Male Holotype: BRAZIL, *Minas Gerais*: Caeté (inside a natural cavity, 20°01'40"S, 43°40'52"W, 1,484 m a.s.l., highlands of Serra da Gandarela), May 2011, Bichuettte, M.E. *leg.* (MNRJ 4540). Paratypes: BRAZIL, *Minas Gerais*: Caeté, same data as holotype (2 females, 1 immature, MNRJ 4540); near cave AP. 09 (20°01'33"S, 43°40'54"W), 1,439 m a.s.l., Projeto Mina Apolo, sifting forest litter, July 09 2011, Equipe Aracno *leg.* (2 females, 10 immatures, MNRJ 4388); Santa Bárbara: near cave AP. 31 (20°02'14"S, 43°40'38"W), 1,443 m a.s.l., Projeto Mina Apolo, sifting forest litter, July 08 2011, Equipe Aracno *leg.* (1 immature, MNRJ 4380).

Additional material. BRAZIL, *Minas Gerais*: Caeté: (near cave AP. 54, 20°01'40"S, 43°40'52"W, 1,484 m a.s.l.), Projeto Mina

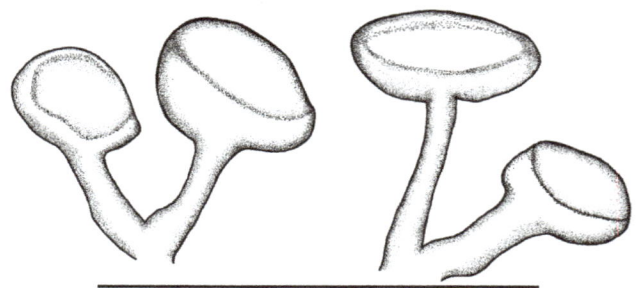

Figure 15. *Masteria emboaba* **sp. nov.** paratype (MNRJ 4540), spermathecae dorsal view. Scale bar: 100 µm.

Apolo, sifting forest litter, July 06 2011, Equipe Aracno *leg.* (1 immature, MNRJ 4378); same locality, under stones and rotten wood, forest, September 24 2011, Equipe Aracno *leg.* (1 female, 1 immature, MNRJ 4436; 1 immature, MNRJ 4437).

Distribution. Only known from small tracts of Atlantic Forest and caves on "canga" areas, on the hilltops of the Gandarela range, Caeté and Santa Bárbara, state of Minas Gerais, Brazil.

Table II. *Masteria emboaba* **sp. nov.** Male holotype and femlale paratype (MNRJ 4540). Length and midwidths of right legs and palpal segments.

Length/Midwidths	Male holotype						Female paratype					
	Femur	Patella	Tibia	Metatarsus	Tarsus	Total (length)	Femur	Patella	Tibia	Metatarsus	Tarsus	Total (length)
Pp	0.54/0.19	0.31/0.16	0.41/0.16	–	0.33/0.14	1.59	0.53/0.17	0.38/0.18	0.40/0.17	–	0.46/0.14	1.77
I	0.84/0.26	0.46/0.18	0.54/0.17	0.49/0.13	0.30/0.11	2.63	0.85/0.29	0.45/0.26	0.59/0.23	0.46/0.15	0.39/0.12	2.74
II	0.77/0.25	0.47/0.19	0.47/0.17	0.43/0.14	0.38/0.11	2.52	0.72/0.23	0.48/0.20	0.44/0.20	0.44/0.15	0.36/0.12	2.44
III	0.70/0.23	0.42/0.19	0.45/0.17	0.50/0.12	0.43/0.09	2.50	0.66/0.23	0.38/0.21	0.44/0.18	0.48/0.13	0.41/0.11	2.37
IV	0.85/0.23	0.50/0.21	0.64/0.18	0.70/0.11	0.48/0.08	3.17	0.88/0.21	0.55/0.23	0.65/0.17	0.64/0.14	0.52/0.10	3.24

Etymology. The specific name, "emboaba" is a noun in apposition and refers to the historical episode "Guerra dos Emboabas" (loosely translated as "War of the Emboabas"). This episode was a series of fights between gold miners from different regions of Brazil in the early 18th century throughout Minas Gerais, especially at the Caeté region (MELLO 1979).

Remarks. The spiders were whitish and almost translucent when alive (Fig. 16), but became yellowish and opaque in alcohol (Figs. 1 and 10). Additional specimens of *M. emboaba* **sp. nov.** were collected during the initial phase of the "Mina Apolo" project (COELHO et al. 2010), inside several of the "canga" caves, located between 20°01'33"S and 21°02'31"S to 43°40'25"W and 43°41'18"W. The *Masteria* specimens from a locality near Caeté, in Nova Lima, Minas Gerais, mentioned by ANGLOGOLD ASHANTI (2009), may also belong to *M. emboaba* **sp. nov.**

Natural history. *Masteria emboaba* **sp. nov.** was collected both inside "canga" caves and in the dry forested tracts near the "canga" area. However, they were not found in open grassland areas covering the "canga" around the caves (Fig. 17). In the dry forest (Fig. 18), they were found sieving through the litter, or were spotted under rotten wood and stones. Therefore, we may assume that this species is associated with forest litter, eventually invading nearby caves. Cave colonization would be an easy step for an animal that is already adapted to small cavities in litter. The small patches of dry forests covering the slopes of drainage valleys do not present a continuous and deep litter layer, but there are litter pockets amassed in suitable areas. The "canga" caves are placed just below the surface and are penetrated by many plant roots. They seem to offer a suitable environment for the species, with high humidity and plenty of spider food. In addition, there are many access points to the cave through the microchannels in the porous iron matrix (FERREIRA 2005). These access points may serve as a refuge for animals during dry periods, allowing the species to survive the long, dry winter of the area.

The *Masteria* species for which habitat information is available live in underground habitats, for instance in the depths of litter or in caves. Their pale color and some degree of eye reduction seem to be correlated with life in dark habitats. Most species of *Masteria* have only six eyes (e.g., SIMON 1889, 1892, RAVEN 1991), and those that have eight eyes have a pronounced reduction of the anterior median pair, e.g. *M. petrunkevitchi* and *M. simla* (CHICKERING 1964, 1966). The complete loss of eyes is found in two troglobite species: *M. caeca*, from Phillipines (SIMON 1892), and *M. pecki*, from Jamaica (GERTSCH 1982). The reduction to only two eyes in *Masteria emboaba* **sp. nov.** may point to a high degree of specialization to underground habitats.

Figures 16-18. (16) *Masteria emboaba* **sp. nov.**, living female, near AP-54; (17) view of typical open grassland vegetation found in "canga" areas; (18) view of dry forest patches covering drainage valleys near "canga" areas. Photos: D. Pedroso.

ACKNOWLEDGMENTS

We thank Norman Platnick for help with the literature. Support: FAPERJ PhD grant for D. Pedroso, FAPESP 2012/01093-0 and CNPq Research Fellow-Brazil for R. Bertani.

REFERENCES

Alayón GG (1995) La subfamilia Masteriinae (Araneae: Dipluridae) en Cuba. **Poeyana 453**: 1-8.

AngloGold Ashanti (2009) **Diversidade da Mata Samuel de Paula.** Nova Lima, AngloGold Ashanti, 296p.

Auler AS, Piló LB, Parker CW, Senko JM, Sasowsky ID, Barton HA (2014) Hypogene cave patterns in iron ore caves: convergence of forms or processes? **Karst Waters Institute Special Publication 18**: 15-19.

Ázara LN, Ferreira RL (2013) The first troglobitic *Cryptops* (*Trigonocryptops*) (Chilopoda: Scolopendromorpha) from South America and the description of a non-troglobitic species from Brazil. **Zootaxa 3709**(5): 432-44. doi: 10.11646/zootaxa.3826.1.10

Bernardi LFO, Klompen H, Zacarias MS, Ferreira RL (2013) A new species of *Neocarus* Chamberlin & Mulaik, 1942 (Opilioacarida, Opilioacaridae) from Brazil, with remarks on its postlarval development. **Zookeys 358**: 69-89. doi: 10.3897/zookeys.358.6384

Bertani R, Cruz WR, Oliveira MEES (2013) *Masteria manauara* **sp. nov.**, the first masteriine species from Brazil (Araneae: Dipluridae: Masteriinae). **Zoologia 30**(4): 437-440. doi: 10.1590/S1984-46702013000400010

Carmo FF, Jacobi CM (2013) A vegetação de canga no Quadrilátero Ferrífero, Minas Gerais: caracterização & contexto fitogeográfico. **Rodriguesia 64**(3): 527-541. doi: 10.1590/S2175-78602013000300005

Chickering AM (1964) Two new species of the genus *Accola* (Araneae, Dipluridae). **Psyche 71**: 174-180.

Chickering AM (1966) Three new species of *Accola* (Araneae, Dipluridae) from Costa Rica and Trinidad, W. I. **Psyche 73**: 157-164.

Coelho A, Piló LB, Auler AS, Bessi R (2010) **Espeleologia da área do Projeto Apolo, Quadrilátero Ferrífero, MG.** Belo Horizonte, Carste Consultores Associados, Relatório Técnico, 179p.

Ferreira RL (2005) A vida subterrânea nos campos ferruginosos. **O Carste 3**(17): 106-115.

Gertsch WJ (1982) The troglobitic mygalomorphs of the Americas (Arachnida, Araneae). **Association for Mexican Cave Studies Bulletin 8**: 79-94.

Leão MR, Auler AS (2012) **Pedido de Supressão da Cavidade ASS-01, Serra do Sapo – Conceição do Mato Dentro.** Belo Horizonte, Carste Consultores Associados, Relatório Técnico, 22p.

Mello JS (1979) **Os Emboabas**. São Paulo, Governo do Estado de São Paulo, 295p.

Pedroso DR, Baptista RLC (2014) A new troglomorphic species of *Harmonicon* (Araneae, Mygalomorphae, Dipluridae) from Pará, Brazil, with notes on the genus. **Zookeys 389**: 77-88. doi: 10.3897/zookeys.389.6693

Pellegrini TG, Ferreira RL (2011) *Coarazuphium tapiaguassu* (Coleoptera: Carabidae: Zuphiini), a new Brazilian troglobitic beetle, with ultrastructural analysis and ecological considerations. **Zootaxa 3116**: 47-58.

Platnick NI (2014) The World Spider Catalog, version 15. American Museum of Natural History, online at http://research.amnh.org/entomology/spiders/catalog/index.html. doi: 10.5531/db.iz.0001

Platnick NI, Foster RR (1982) On the Micromygalinae, A New Subfamily of Mygalomorph Spiders (Araneae, Microstigmatidae). **American Museum Novitates 2734**: 1-13.

Raven RJ (1979) Systematics of the mygalomorph spider genus *Masteria* (Masteriinae: Dipluridae: Arachnida). **Australian Journal of Zoology 27**: 623-636.

Raven RJ (1981) Three new mygalomorph spiders (Dipluridae, Masteriinae) from Colombia. **Bulletin of the American Museum of Natural History 170**: 57-63.

Raven RJ (1985) The spider infraorder Mygalomorphae (Araneae): Cladistics and systematics. **Bulletin of the American Museum of Natural History 182**: 1-180.

Raven RJ (1991) A revision of the mygalomorph spider family Dipluridae in New Caledonia (Araneae). In: Chazeau J, Tillier S (Eds). Zoologia Neocaledonica. **Mémoires du Museum National d'Histoire Naturelle (A) 149**: 87-117.

Raven RJ, Platnick NI (1981) A revision of the American spiders of the family Microstigmatidae (Araneae, Mygalomorphae). **American Museum Novitates 2707**: 1-20.

Simon E (1889) Arachnides. In: Voyage de M.E. Simon au Venezuela (décembre 1887-avril 1888), 4e Mémoire. **Annales de la Societe Entomologique de France 9**: 169-220.

Simon E (1892) Arachnides. In: Raffrey A, Bolivar I, Simon E (Eds) Etudes cavernicoles de l'île Luzon. Voyage de M.E. Simon aux l'îles Phillipines (mars et avril 1890), 4e Mémoire. **Annales de la Société Entomologique de France 61**: 35-52.

Souza MFVR, Ferreira RL (2005) *Eukoenenia* (Palpigradi: Eukoeneniidae) in Brazilian caves with the first troglobiotic palpigrade from South America. **Journal of Arachnology 38**: 415-424.

Trajano E, Bichuette ME (2010) Diversity of Brazilian subterranean invertebrates, with a list of troglomorphic taxa. **Subterranean Biology 7**: 1-16.

11

Description of the males of *Lincus singularis* and *Lincus incisus* (Hemiptera: Pentatomidae: Discocephalinae)

Aline S. Maciel[1], Thereza de A. Garbelotto[1], Ingrid C. Winter[1], Talita Roell[1] & Luiz A. Campos[1,2]

[1]*Departamento de Zoologia, Universidade Federal do Rio Grande do Sul. Avenida Bento Gonçalves 9500, Agronomia, 91501-970 Porto Alegre, RS, Brazil.*
[2]*Corresponding author. E-mail: luiz.campos@ufrgs.br*

ABSTRACT. The Neotropical *Lincus* Stål, 1867 includes 35 species, thirteen of which are known only from females. Several species are vectors of *Phytomonas staheli* McGhee & McGhee, 1979, a trypanosomatid parasitic in palm-trees in South America that causes hart-rot, sudden and slow wilt diseases. The hitherto unknown males of *L. singularis* Rolston, 1983 ("swollen head" species group found in the oil palm *Elaeis guineensis* Jacq.), and *L. incisus* Rolston, 1983 ("hatchet-lobed" species group; found in the coconut tree *Cocos nucifera* L.), are described with emphasis on the morphology of the genitalia, and taxonomic remarks are provided. Males of *L. singularis* can be distinguished from other species included in "swollen head" group by their pronotal lobes with anterior and posterior margins subparallel and projected laterally from the eye margin, while males of *L. incisus* can be distinguished from the species of the "swollen head" group by an obtuse projection with a deepest incision and several additional diagnostic characters of the genitalia.

KEY WORDS. Genitalia; morphology; Ochlerini; stink bugs; taxonomy.

Lincus Stål, 1867 is the richest genus of Ochlerini, comprising 35 species (CAMPOS & GRAZIA 2006). Even though the genus was described in the 19th century (STÅL 1867), most of its 25 species were described in the late 20th century (ROLSTON 1983, 1989, DOLLING 1984), and 13 of them are known only from one sex (ROLSTON 1983). Species of *Lincus* are found mostly in the Amazon region. There are a few exceptions to this, for instance *Lincus lobuliger* Breddin, 1908, recorded from the Brazilian Atlantic Forest and *Lincus anulatus* Rolston, 1983 and *Lincus discessus* (Distant, 1900) from Central America (ROLSTON 1983). Several species are sympatric in different countries. Geographic records of *Lincus* are particularly rich in Peru due to the extensive surveys on native palms carried out during the 1980's (COUTURIER & KAHN 1989, 1992, LLOSA et al. 1990). Sixteen species occur in that country, including *Lincus singularis* Rolston, 1983, although it has never been collected on *Elaeis guineensis* Jacq. Suriname comes next in terms of species richness in the Amazon region (ROLSTON 1983, 1989, DOLLING 1984), with six species, including *Lincus incisus* Rolston, 1983.

The association of pentatomids with the transmission of *Phytomonas staheli* McGhee & McGhee, 1979, a trypanosomatid parasitic in plants, has been known for a long time and is well documented (for a review see CAMARGO 1999 and MITCHELL 2004). Several species of *Lincus* play a major role as vectors of hart-rot, and of sudden and slow wilt (also called Marchitez sorpresiva in

Spanish-speaking countries) diseases in palm trees (Arecaceae) in South America, being of economic interest in crops of *E. guineensis* (African oil palm) and *Cocos nucifera* L. (coconut) (DESMIER DE CHENON 1984, COUTURIER & KAHN 1989, PERTHUIS et al. 1985, PANIZZI et al. 2000, DI LUCCA et al. 2013; for a review see HOWARD 2001). Although eleven species of *Lincus* have been reported on palm trees (HOWARD 2001), the genus was not listed as a possible vector of oil palm diseases until the 1980's (COUTURIER & KAHN 1992). Furthermore, transmission of *Phytomonas* trypanosomatids to palms has been documented in only six species, four of which transmit the parasite to *E. guineensis*: *Lincus lethifer* Dolling, 1984, *L. lobuliger*, *Lincus tumidifrons* Rolston, 1983, and *Lincus spurcus* Rolston, 1983 (CAMARGO 1999, DI LUCCA et al. 2013).

In 2009, the corresponding author received, for identification, specimens of *Lincus* collected from *E. guineensis* palm trees from Palmas del Espino S.A., Peru. These specimens were identified as *L. spurcus* and *L. singularis*, and included the only known males of the latter. Moreover, during the course of this study, we located males of *L. incisus* among specimens of Ochlerini received during the 1990's, two of which from *C. nucifera* crops cultivated by Sococo S.A., Moju, Pará State, Brazil. For the first time, *L. singularis* and *L. incisus* are reported from oil palm and coconut trees, respectively, and their males are described and illustrated for the first time, with emphasis on the morphology of genitalia.

MATERIAL AND METHODS

Five males and one female of *L. singularis* and three males and three females of *L. incisus* were examined in this study. The species were identified based in a revision by Rolston (1983). Observation of specimens, dissection and preservation followed Garbelotto et al. (2013). Measurements are in millimeters (mm) and follow mainly Garbelotto et al. (2013) and Rolston (1983) for: length and width of eye and pronotal lobe, and interocellar distance. The terminology of Baker (1931), Dupuis (1970), Campos & Grazia (2006) and Garbelotto et al. (2013) were adopted for genitalic structures. Photographs were taken using a Nikon AZ100M stereomicroscope and NIS-Elements Advanced Research software. Drawings were made under a stereomicroscope Leica MZ12 coupled with camera lucida and were vectored using Adobe Illustrator. Whenever possible, collection data were georeferenced following Garbelotto et al. (2013); coordinates are in decimal degrees.

Collections' acronyms follow Evenhuis (2014). Voucher specimens are deposited in the entomological collection of the Departamento de Zoologia at Universidade Federal do Rio Grande do Sul (UFRG), Porto Alegre, RS, Brazil.

TAXONOMY

Lincus singularis Rolston, 1983
Figs. 1-9

Lincus singularis Rolston, 1983: 1, 4, 5, 18-20, Figs. 34-35 (female holotype from Chauchamayo, Peru, deposited in USNM 76690, not examined, no paratypes); Couturier & Khan, 1992: 719 (map); Campos & Grazia, 2006: 153 (list).

Description of the male. The color of males is dark brown to fuscous and the general morphology is similar to that described for females by Rolston (1983) (Fig. 1). Genitalia. Pygophore oval, opening of genital cup narrow. Dorsal rim uniformly concave (Fig. 2, dr), bearing 1+1 tufts of setae lateral to segment X. Posterolateral angles rounded, projected distinctly beyond the ventral rim, depressed dorsally (Fig. 2, pa). Basal 1/3 of segment X membranous, lateral margins sinuous tapering to apex (Fig. 2, X). Ventral rim V-shaped, with setae along margin (Fig. 3, vr). Ventral surface tumescent on disc, with 1+1 lateral sulci following ventral rim (Fig. 3, t); ventral surface of posterolateral angles tumescent (Fig. 3). Parameres inconspicuous and covered by segment X, attached to the articulatory apparatus of phallus, subtriangular in lateral view, bearing a dorsal dense tuft of setae on apex (Figs. 4-6). Phallus. Phallotheca globose (Figs. 7-9, ph), strongly sclerotized. Vesica longer than the combined lengths of phallotheca and ductus seminis distalis (Figs. 7-9, v, ds), bearing an dorsal subtriangular process posteriorly directed (Figs. 7-9, dp), and 1+1 lateral processes short and truncate (Figs. 7-9, lp). Free portion of ductus seminis distalis very short, about half the length of the inner portion, projecting ventrad of vesica before the lateral processes (Figs. 7-9, ds).

Male. Measurements (n = 5). Total length 10.75 ± 0.29 (10.37-11.00); width of abdomen 6.62 ± 0.36 (6.12-7.00); head length 1.67 ± 0.08 (1.57-1.76); head width 2.24 ± 0.10 (2.14-2.39); eye length 0.50 ± 0.03 (0.47-0.55); eye width 0.55 ± 0.02 (0.52-0.57); interocellar distance 1.20 ± 0.03 (1.17-1.2); interocular distance 1.21 ± 0.06 (1.13-1.26); pronotum length 2.17 ± 0.14 (1.95-2.27); pronotum width 5.66 ± 0.19 (5.42-5.90); length of pronotal lobe 0.23 ± 0.03 (0.20-0.27); pronotal lobe width 0.17 ± 0.02 (0.15-0.20); scutellum length 4.27 ± 0.25 (3.91-4.60); scutellum width 3.55 ± 0.13 (3.39-3.72); length of antennomers: I 0.77 ± 0.03 (0.75-0.80); II 0.82 ± 0.03 (0.77-0.85); III 1.01 ± 0.04 (0.97-1.07); IV 1.42 ± 0.04 (1.37-1.45); V 1.81 ± 0.11 (1.62-1.92); length of labial segments: I 1.28 ± 0.11 (1.12-1.37); II 2.36 ± 0.09 (2.25-2.37); III 1.87 ± 0.03 (1.82-1.9); IV 1.79 ± 0.06 (1.75-1.90).

Material examined. Peru, *Tocache*: 5 males and 1 female, San Martin (Palmas del Espino S.A., Cultivo Palma Aceitera, parcela A11a [-8.41; -76.41] 500 m a.s.l.), 2009, E. Trindad leg.

Distribution. Peru, Cusco and San Martín regions.

Remarks. Although no phylogenetic hypothesis has been advanced for species of *Lincus*, the genus was recovered in the *Herrichella* Distant, 1911 clade in a cladistic analysis of the Ochlerini (Campos & Grazia 2006). The relationship between *Lincus* and the other members of the clade, however, remained unresolved. More recently, the genus (represented by *L. lobuliger*) was recovered as the sister group of the remaining taxa of the *Herrichella* clade in the phylogenetic analysis of Garbelotto et al. (2013). The monophyly of the genus, however, remains to be tested. Several species of *Lincus* are recognizable by their well-developed pronotal lobes, and all known males have tubular proctiger and reduced parameres (Rolston 1983, 1992). These characters were not used in the phylogenetic studies mentioned above. Regarding the phylogenetic relationships among the species of *Lincus*, Rolston (1983) placed *L. singularis* along with *Lincus parvulus* (Ruckes, 1958) and *L. tumidifrons* in the "swollen head" informal group of species ("species group of convenience" sensu Rolston 1983). This group was characterized by having a tumid vertex. Some features of the pygophore of *L. singularis* are consistent with Rolston's proposal to place the species in it, e.g. the 'V' shape of the ventral rim of the pygophore; subrectangular proctiger with acute apex; and globose phallotheca, the latter also observed in *L. tumidifrons*. *Lincus singularis* can be differentiated from the other species in the "swollen head" group by having the anterior and posterior margins of the pronotal lobes subparallel and each lobe projected laterad of its corresponding eye; the vertex of head not as tumid as in *L. parvulus* and *L. tumidifrons* (Fig. 1; for *L. parvulus* and *L. tumidifrons* see Rolston 1983, Figs. 30 and 36); and the ventral opening of the pygophore is narrower than in those species (Fig. 3, vr; for *L. parvulus* and *L. tumidifrons* see Rolston 1983, Figs. 32 and 41).

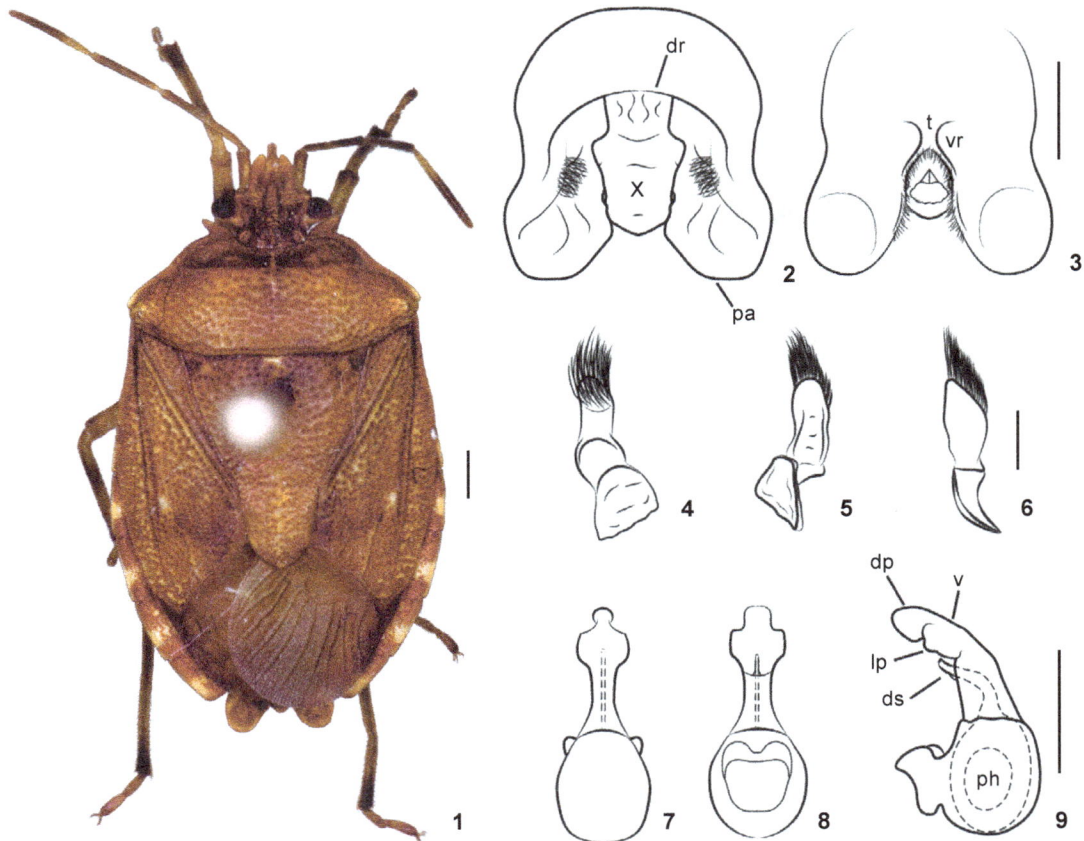

Figures 1-9. Male of *Lincus singularis*: (1) habitus in dorsal view; (2-3) pygophore: (2) dorsal view; (3) ventral view; (4-6) left paramere: (4) dorsal view; (5) ventral view; (6) lateral view; (7-9) phallus: (7) anterior view; (8) posterior view; (9) lateral view. (dp) Dorsal projections, (dr) dorsal rim, (ds) ductus seminis distalis, (lp) lateral projection, (pa) posterolateral angles, (ph) phallotheca, (t) tumescent area, (v) vesica, (vr) ventral rim, (X) segment X. Scale bars: 1-3 = 1 mm, 4-9 = 0.5 mm.

Lincus incisus Rolston, 1983
Figs. 10-18

Lincus incisus Rolston, 1983: 1, 3, 4, 9-10, Figs. 8-9 (female holotype from De Mapane, Suriname, deposited in RMNH, not examined, no paratypes); Campos & Grazia, 2006: 153 (list).

Description of the male. The fuscous general color of male and its general morphology, including the anterolateral margins of pronotum expanded in obtuse angle, posterior to pronotal lobes, are as described for females by ROLSTON (1983) (Fig. 10). Genitalia. Pygophore subrectangular. Surface with short setae. Dorsal rim concave, bearing setae lateral to segment X (Fig. 11, dr). Posterolateral angles obtuse (Fig. 11, pa), depressed, with 1+1 median projections (Fig. 11, mp). Segment X sclerotized, ventrally directed; apex expanded and flattened (Fig. 11, X); anal opening circular, and genital opening in longitudinal slit, both at ventral surface (Fig. 12). Ventral rim concave, with setae along the margin, medially carinated (Fig. 12, vr). Ventral

surface tumescent on disc, with 1+1 lateral sulci following ventral rim (Fig. 12, t). Parameres inconspicuous, attached to the articulatory apparatus of phallus, subrectangular and with an apical tuft of setae (Figs. 13-15). Phallus. Phallotheca globose, strongly sclerotized (Figs. 16-18, ph). Vesica elongated, medially narrowed, longer than the combined lengths of phallotheca and ductus seminis distalis (Figs. 16-18, v); with one globose dorsal projection posteriorly directed (Figs. 16-18, dp); 1+1 lateral globose projections, posteriorly directed (Figs. 16-18, lp); posterior projection truncated, bearing ductus seminis distalis (Figs. 16-18, pp). Ductus seminis distalis antero-dorsally arched toward the projections of vesica (Fig. 18, ds).

Measurements (n = 3). Total length 12.55 ± 0.79 (11.86-13.42); width of abdomen 6.77 ± 0.32 (6.46-7.10); head length 2.18 ± 0.11 (2.06-2.28); head width 2.63 ± 0.06 (2.56-2.68); eye length 0,60 ± 0.03 (0.57-0.63); eye width 0.78 ± 0.03 (0.75-0.82); interocellar distance 0.70 ± 0.01 (0.69-0.71); interocular distance 1.12 ± 0.05 (1.07-1.15); pronotum length 3.47 ± 0.18

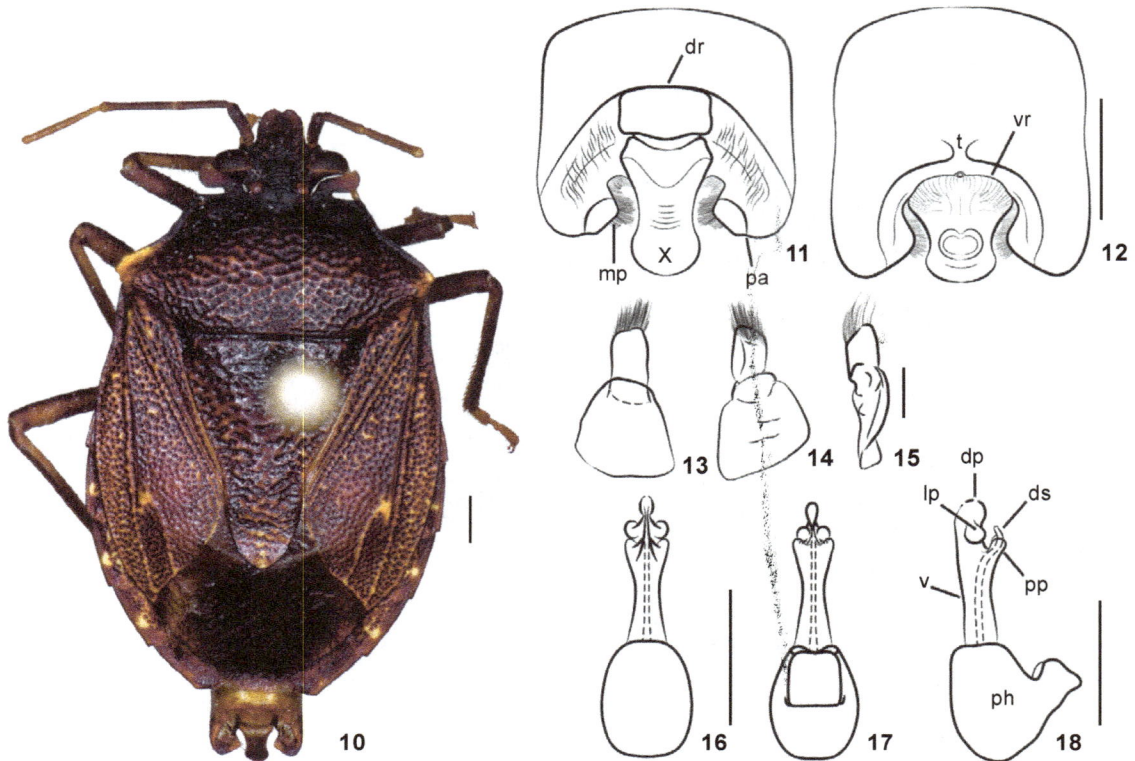

Figures 10-18. Male of *Lincus incisus*: (10) habitus in dorsal view; (11-12) pygophore: (11) dorsal view; (12) ventral view; (13-15) right paramere: (13) dorsal view; (14) ventral view; (15) lateral view; (16-18) phallus: (16) anterior view; (17) posterior view; (18) lateral view. (dp) Dorsal projections, (dr) dorsal rim, (ds) ductus seminis distalis, (lp) lateral projection, (mp) median projection, (pa) posterolateral angles, (ph) phallotheca, (t) tumescent area, (v) vesica, (pp) posterior projection, (vr) ventral rim, (X) segment X. Scale bars: 10-12 = 1 mm, 13-18 = 0.5 mm.

(3.27-3.59); pronotum width 6.01 ± 0.18 (5.82-6.17); length of pronotal lobe 0.41 ± 0.02 (0.40-0.44); pronotal lobe width 0.85 ± 0.02 (0.80-0.84); scutellum length 4.45 ± 0.30 (4.12-4.70); scutellum width 3.65 ± 0.08 (3.55-3.70); length of antennomers: I 0.97 ± 0.04 (0.92-1.00); II 1.12 ± 0.08 (1.07-1.21); III 1.42 ± 0.07 (1.35-1.50); IV 1.38 ± 0.20 (1.2-1.56); V 1.78 ± 0.00 (1.78-1.78); length of labial segments: I 1.35 ± 0.70 (1.28-1.42); II 2.43 ± 0.14 (2.34-2.60); III 1.97 ± 0.14 (1.85-2.13); IV 1.66 ± 0.05 (1.63-1.72).

Material examined. BRAZIL, *Amazonas*: 3 females, São Miguel da Cachoeira (Cachoeira do Tucano – Pico da Neblina), X.2007, Nogueira & Candiani leg.; *Pará*: 2 males, Moju (Fazenda Sococo) [-2.11; -48.00], 01.XII.1995, P. Lins leg.; 1 male, Tucuruí (Rio Tocantins) [-3.7; -49.7], 20.VII.1984, W. França leg.

Distribution: Suriname, Brazil (Amazonas and Pará States).

Remarks. *Lincus incisus* was placed, along with eight other species (*Lincus convexus* Rolston, 1983, *Lincus croupius* Rolston, 1983, *Lincus fatigus* Rolston, 1983, *Lincus operosus* Rolston, 1983, *Lincus securiger* Breddin, 1904, *Lincus sinuosus* Rolston, 1983, *Lincus spathuliger* Breddin 1908 and *Lincus vandoesburgi* Rolston,

1983), in the "hatchet-lobed" informal group of species (ROLSTON 1983). This placement was justified in view of the anterior pronotal angles resembling a hatchet blade in *Lincus incisus*. Males of *L. incisus* share some genitalic characters with males of *L. convexus*, *L. securiger*, *L. sinuosus* and *L. vandoesburgi*, such as the presence of 1+1 median projections at posterolateral angles, ventrally directed; and an elongated phallus bearing apical projections. The ductus seminis distalis bent toward the projections of the vesica is also observed in *L. vandoesburgi*. Within the hatchet-lobed group, *L. incisus* and *L. vandoesburgi* share the pronotal margin posterior to the lobes expanded on each side into an obtuse projection. The incision between each lobe and the anterolateral margin of the pronotum is deepest in *L. incisus*, reaching half the width of an eye (Fig. 10; ROLSTON 1983, Figs. 1, 8). *Lincus incisus* can also be distinguished from *L. vandoesburgi* by the more convex apical margin of the posterolateral angles of the pygophore, and by a more developed median projection (Figs. 11-12, mp; for *L. vandoesbugi* see ROLSTON 1983, Fig. 2). Among the other species in the hatchet-lobed group with known males, *L. incisus* differs from *L. sinuosus* by the median projection below the apical margin of the posterolateral angles of the pygophore (Figs. 11-12, mp; for *L. sinuosus* see Rolston 1983, Fig. 17); from *L. convexus* and *L. securiger* by the more developed median projection and the ventral opening of the pygophore broader and shallower (Figs. 11-12, mp, vr; for *L. convexus* and *L. securiger* see Rolston 1983, Figs. 23 and 26). Notwithstanding the placement of *L. incisus* within the "hatchet-lobed" group, it is noticeable that the shape of segment X, with an expanded and flattened apex, is also a feature of some species of the "big-eyed" group (Rolston 1983), such as *Lincus lethifer* Dolling, 1984, *Lincus substyliger* Rolston, 1983 and *Lincus subuliger* Breddin, 1908. *Lincus incisus* is recorded for the first time in Brazil.

ACKNOWLEDGMENTS

We thank the curators of the scientific collections for the loan of specimens. We also thank Conselho Nacional de Desenvolvimento Científico e Tecnológico (CNPq) for the scholarships granted to T.A. Garbelotto (process 142448/2011-7), Coordenação de Aperfeiçoamento de Pessoal de Nível Superior (CAPES) to T. Roell, and Universidade Federal do Rio Grande do Sul (UFRGS) to I.C. Winter and the funding from CNPq (process 305367/2012-9) as fellowship grant to L.A. Campos.

REFERENCES

Baker AD (1931) A study of the male genitalia of Canadian species of Pentatomidae. **Canadian Journal of Research** 4(3): 148-220. doi: 10.1139/cjr31-013

Camargo EP (1999) *Phytomonas* and other trypanosomatid parasites of plants and fruit. **Advances in Parasitology 42:** 29-112. doi: 10.1016/S0065-308X(08)60148-7

Campos LA, Grazia J (2006) Análise cladística e biogeografia de Ochlerini (Heteroptera, Pentatomidae, Discocephalinae). **Iheringia, Série Zoologia 96**(2): 147-163. doi: 10.1590/S0073-47212006000200004

Couturier G, Kahn F (1989) Bugs of *Lincus* spp. vectors of Marchitez and Hartrot (oil palm and coconut diseases) on *Astrocaryum* spp., Amazonian native palms. **Principes 33**(1): 19-20.

Couturier G, Kahn F (1992) Notes on the insect fauna on two species of *Astrocaryum* (Palmae, Cocoeae, Bactridinae) in Peruvian Amazonia, with emphasis on potential pests of cultivated palms. **Bulletin de l'Institut Français d'Études Andines 21**(2): 715-725.

Desmier de Chenon R (1984) Recherches sur le genre *Lincus* Stål, Hemiptera Pentatomidae Discocephalinae, et son role éventuel dans la transmission de la Machitez du palmier à huile et du Hart-Rot du cocotier. **Oléagineux 39**(1): 1-6.

Di Lucca AGT, Chipana EFT, Albújar MJT, Peralta WT, Piedra YCM, Zelada JLA (2013) Slow wilt: another form of Marchitez in oil palm associated with trypanosomatids in Peru. **Tropical Plant Pathology 38**(6): 522-533. doi: 10.1590/S1982-56762013000600008.

Dolling WR (1984) Pentatomid bugs (Hemiptera) that transmit a flagellate disease of cultivated palms in South America. **Bulletin of Entomological Research 74**(3): 473-476. doi: 10.1017/S000748530001573X

Dupuis C (1970) Heteroptera, p. 190-208. In: Tuxen SL (Ed.). **Taxonomist's glossary of genitalia of insects.** Copenhagen, Munksgaard, 359p.

Evenhuis NL (2014) **Abbreviations for insect and spider collections of the world.** Available at: http://hbs.bishopmuseum.org/codens/codens-inst.html [Accessed: 12 November 2014]

Garbelotto TA, Campos LA, Grazia J (2013) Cladistics and revision of *Alitocoris* with considerations on the phylogeny of the *Herrichella* clade (Hemiptera, Pentatomidae, Discocephalinae, Ochlerini). **Zoological Journal of the Linnean Society 168**(3): 452-472. doi: 10.1111/zoj.12032

Howard FW (2001) Sap-feeders on palms, p. 109-232. In: Howard FW, Moore D, Giblin-Davis RM, Abad RG (Eds). **Insects on palms.** New York, CABI Publishing, XIII+403p. doi: 10.1079/9780851993263.0109

Llosa JF, Couturier G, Kahn F (1990) Notes on the ecology of *Lincus spurcus* and *L. malevolus* (Heteroptera: Pentatomidae: Discocephalinae) on Palmae in forests of Peruvian Amazonia. **Annales de laSociété Entomologique de France (Nouvelle Série) 26**(2): 249-254.

Mitchell PL (2004) Heteroptera as vectors of plant pathogens. **Neotropical Entomology 33**(5): 519-545. doi: 10.1590/S1519-566X2004000500001

Panizzi AR, McPherson JE, Javahery JM, McPherson RM (2000) Stink Bugs (Pentatomidae), p. 421-474. In: Schaefer CW, Panizzi AR (Eds). **Heteroptera of economic importance.** Boca Raton, CRC Press, 856p.

Perthuis B, Desmier de Chenon R, Merland E (1985) Mise en évidence du vecteur de la Marchitez sorpresiva du palmier à huile, la punaise *Lincus lethifer* Dolling (Hemiptera Pentatomidae Discocephalinae). **Oléagineux 40**(10): 473-475.

Rolston LH (1983) A revision of the genus *Lincus* Stål (Hemiptera: Pentatomidae: Discocephalinae: Ochlerini). **Journal of the New York Entomological Society 91**(1): 1-47.

Rolston LH (1989) Three new species of *Lincus* (Hemiptera: Pentatomidae) from palms. **Journal of the New York Entomological Society 97**(3): 271-276.

Rolston LH (1992) Key and dignoses for the genera of Ochlerini (Hemiptera: Pentatomidae: Discocephalinae). **Journal of the New York Entomological Society 100**(1): 1-41.

Stål C (1867) Bildrag till Hemipterernas systematik. Conspectus generum Pentatomidum Americae. **Öfversigt af Kongliga Vetenskaps-Akademiens Förhandlingar 24**(7): 522-532.

12

Use of space by the Neotropical caviomorph rodent *Thrichomys apereoides* (Rodentia: Echimyidae)

Alex José de Almeida[1], Melina Maciel F. Freitas[1] & Sônia A. Talamoni[1,2]

[1] *Programa de Pós-graduação em Zoologia de Vertebrados, Pontifícia Universidade Católica de Minas Gerais. Avenida Dom José Gaspar 500, 30535-610 Belo Horizonte, MG, Brasil.*
[2] *Corresponding author. E-mail: talamoni@pucminas.br*

ABSTRACT. The objective of this study was to investigate some parameters of the space use by individuals in a population of the hystricognath rodent *Thrichomys apereoides* (Lund, 1839), using the spool-and-line tracking technique. This technique is useful for investigating characteristics of habitat use by individuals since it allows the mapping of the places where the individuals move. We evaluated three parameters of space use by 34 individuals of *T. apereoides*: 1) The daily home range (DHR) or the area used by individuals in their daily activities, 2) the distance moved on the leaf litter, and 3) the distance moved above ground using twigs, logs and rocks. The analysis of space use on such a small scale allows a better understanding of how individuals perceive and use the available space. The significant effect of age on DHR and the effect of the sex on the movements above ground were observed. Adult males had larger DHRs than adult females and subadults, and adult females showed the lowest displacement above ground. A statistically significant effect of the sex and seasonal period and the interaction between them was also observed on the size of DHRs of adults. During the dry season, females had lower DHRs than males and both females and males moved less on leaf litter in this season. There was no seasonal effect on the movement of males and females above ground, as well as no significant effect of age and sex on the movement of the individuals on leaf litter. We found that individuals responded differently to some aspects of the habitat structure and concluded that the pattern of movement is influenced by the sex and the age of the individuals and may vary according to ecological conditions.

KEY WORDS. Caviomorph rodent; daily home range; spacing pattern; spool-and-line.

A commonly studied parameter regarding space use by a population is the variation in size of the home ranges of different individuals or classes of individuals within that population. A number of life history factors have been used to explain this variation among individuals including age, sex, season, breeding season, and population density (MAGNUSSON *et al*. 1995, KELT & VAN VUREN 2001, ENDRIES & ADLER 2005). It has also been shown that the spacing pattern is associated with the social mating system used (ADLER 2011, MAHER & BURGER 2011), with the dispersion in the environment of individuals being strongly influenced by the way they behave in accordance to their needs and their role in their particular mating system (CLUTTON-BROCK & HARVEY 1978, GITTLEMAN & THOMPSON 1988).

Moreover, several models have been developed to explain the influence of ecological conditions on the social organization of mammals (EMLEN & ORING 1977, OSTFELD 1985, JOHNSON *et al*. 2002, MACDONALD 1983). It has been shown that different populations of the same species, or even the same population at different times, may vary in their social systems and spacing pattern due to ecological factors (see MAHER & BURGER 2011). Among caviomorph rodents, for example, distribution and availability of food resources, climatic conditions, and preda-

tion risk have been correlated with, and used to explain, intraspecific variation in the characteristics of home range and use of space (ADLER 1998, 2011, ADRIAN & SACHSER 2011, MAHER & BURGER 2011).

Capture-mark-recapture is a classical method employed in studies on space use (DUESER & SHUGART 1979, LACHER & ALHO 1989, JORGENSEN & DEMARAIS 1999). However, this technique provides little information about the movements of individuals (PREVEDELLO *et al*. 2010), which have their movements interrupted by their capture in live traps. There have been a number of studies on the use of space by several species using the spool-and-line tracking technique (PREVEDELLO *et al*. 2008, CERBONCINI *et al*. 2011). This technique allows the calculation of total home range of individuals in the traditional sense (CERBONCINI *et al*. 2011) as well the calculation of the "daily home range" (DHR), resulting from the movement of animals during their normal activities in one night (CUNHA & VIEIRA 2002, MENDEL & VIEIRA 2003, LORETTO & VIEIRA 2005, DELCIELLOS *et al*. 2006, ALMEIDA *et al*. 2008, PREVEDELLO *et al*. 2010). This technique is also useful for investigating other characteristics of habitat use by individuals since it allows the mapping of the places where the individuals move.

In this study, using the spool-and-line tracking technique, we investigate the influence of sex, age and seasonal period on the space use by individuals of a population of *Thrichomys apereoides* (Lund, 1839) specifically related with: 1) the area used by individuals in their daily activities (DHR), 2) the distance moved on the leaf litter, and 3) the distance moved above ground using twigs, logs and rocks. The analysis of space use on such a small scale allows a better understanding of how individuals perceive and use the available space.

MATERIAL AND METHODS

The study was conducted at Cauaia farm (19°28'57"S, 44°00'50"W), in an area (364 ha) of semideciduous forest, located in the protected area (Área de Proteção Ambiental) of Lagoa Santa Karst, in the municipality of Matozinhos, state of Minas Gerais, Brazil. The Cauaia farm itself has 20% of its area under preservation and consists of rocky outcrops covered with semideciduous and deciduous forests, and with cerrado in the immediate surroundings. The region is under the phytogeographic domains of the Atlantic Forest and Cerrado (Savannas) (IBGE 2004) and the climate of the region is classified as humid tropical savannah, with a rainy season from October to March and a dry season from April to September (SÁ JR et al. 2012). The average air temperature in the region is around 23°C (BERBERT-BORN 2002).

The species studied was the New World hystricognath rodent *T. apereoides*, an echimyid that occurs in areas of savanna and forests, where it uses crevices in the rocks for shelter, protection, and nesting; it is terrestrial and semi-arboreal, diurnal and nocturnal; with a diet that includes leaves, twigs, berries, seeds, fruits and insects (STREILEIN 1982a, REIS & PESSÔA 2004). The species occurs in the states of Minas Gerais, Goiás and Bahia (BONVICINO et al. 2008).

We captured animals for six consecutive days each month from August, 2008 to July, 2009. One hundred forty-three traps (15 x 15 x 30 cm) were set in a grid with traps 20 m apart and arranged in 15 parallel rows, covering an area of 4.7 ha. The total capture effort was 8,580 traps-nights. Traps were baited with a mixture of peanut butter, sardines, and bananas. Each point of capture was received in a trap on the ground. The traps were checked in the morning and the bait was changed when necessary. The information about captured individuals including sex, approximate age, and body mass were recorded monthly. Each captured animal was individually marked using numbered ear tags (National Band Tag and Company). Thirty-four animals were tagged, sexed, and placed in one of two age classes; either subadult and sexually immature (150-180 g), or adult and probably sexually mature (> 180-350 g) (THOMPSON 1985, ROBERTS et al. 1988, TEIXEIRA et al. 2005).

Each marked individual received a tracking spool (4 g, line length of 400 m) (BOONSTRA & CRAINE 1986), which was wrapped in plastic wrap and tape, then glued to the back of the animal with cyanoacrylate glue (Superbonder®) (RYAN et al. 1993, KEY & WOODS 1996, CUNHA & VIEIRA 2002). The line of the spool was tied to an object at the point of release of each animal. After the animal was left to move freely, the line was mapped by measuring the linear distance between two turning points using a measuring tape, and the azimuth was obtained for all the turning points of the line formed by the animal's movements. Only spools with more than 35 m line mapped were considered for analysis (LORETTO & VIEIRA 2005). For data analysis the initial 20 m were excluded which presumably reflected escape behavior (PIZZUTO et al. 2007).

The values of the azimuths and the distances between the turning points for each tracked animal were transformed into Cartesian coordinates (x, y), which were used to calculate the daily home range (DHR), using 100% of the minimum convex polygon (MCP) (MOHR 1947, STICKEL 1954), and to quantify the distance moved on the litter (LIU) and above ground (walking over logs, branches, roots, rocks – AGR). The vertical displacements also were included in this measure. The MCP method has been criticized as an estimate of the size of an animal's home range (BÖRGER et al. 2006), but it is simple to use and is not constrained by underlying statistical assumptions (HARLESS et al. 2010). The layout of each route and the corresponding area of the polygon were performed using the software 1.0.4 Biotas *alpha*.

General Linear Models (GLM) were used for the analysis of the intra-population variation of the log-transformed variables DHR, LIU and AGR as functions of the factors of sex, age class, seasonal period, and their interaction (LINDSEY 1997). Tukey's test for *post hoc* comparisons was used when needed. We performed the models: 1) analysis of the effects of sex, age class, and interaction between them, on DHR, on the movement on litter (LIU), and on the movement above ground (AGR); 2) analysis of the effects sex, seasonal period and interaction between them on DHR, on the movement on litter (LIU), and on the movement above ground (AGR).

All of the parameters analyzed correlated significantly with the total line-distance mapped. Thus, the total line-distance mapped was initially included in the models as an additional variable (covariate) (DRAPER & SMITH 1998). In the absence of a significant interaction between the covariate and the factors to be tested, the effect of total line-distance mapped was disregarded and the adjustment of the model was executed again (LORETTO & VIEIRA 2005). Many individuals were tracked during both seasons, thus, to analyze the effect of seasonality on the sizes of the DHR, these individuals were represented by the data from one spool obtained in each climatic season. For seasonality analysis, data obtained from adults only was used in order to avoid any influence of age in this analysis due to a possible seasonal recruitment of subadults. In the analysis of movement on leaf litter and above ground, we used data from the spools used in the first capture of individuals in each monthly session of sampling, thus, there were monthly repetitions for some individuals.

RESULTS

All of the parameters analyzed correlated significantly with the total line-distance mapped (Pearson correlation, DRH: $r = 0.735$, $p < 0.05$; LIU: $r = 0,820$, $p < 0.05$ AGR: $r = 0,665$, $p < 0.05$). Therefore, the total line-distance mapped was included in the models (GLM analysis) as an additional variable (covariate); in this model there was an absence of a significant result ($p > 0.05$), and the effect of total line-distance mapped on the factors tested was disregarded.

The general linear models revealed a significant effect of age on DHR as well as the effect of the sex on AGR (Table I). Adult males had greater DHRs than adult females and subadults (Fig. 1), and adult females showed the lowest displacement above ground (Fig. 5). There was no significant effect of the age and sex on the movement of individuals on leaf litter (Table I, Fig. 3).

There was a statistically significant effect of the sex and seasonal period and the interaction between them on the size of DHRs of adults (Table I). A seasonal effect was also observed on LIU of adults (Table I). Females had lower DHRs than males during dry season (Fig. 2), and both females and males moved less on leaf litter in the dry season (Fig. 4). There was no seasonal significant difference between the movement of males and females above ground (Fig. 6).

DISCUSSION

In this study, differences in the size of DHRs due to age were observed. Adult males had larger DHRs than subadults. This kind of difference has been attributed to the differing energy requirements among different individuals (MARQUET *et al.* 2005, ALMEIDA *et al.* 2008, VIEIRA & CUNHA 2008). However, the DHR expresses the spacing between the individuals and their daily movements, which can be viewed primarily as a result of hierarchical, demographic and economic processes (PRIOTTO *et al.* 2004, REHMEIER *et al.* 2004, SKVARLA *et al.* 2004, MITCHELL & POWELL 2012).

Females had smaller DHRs and movement on the litter during the dry season, as well as the lowest movements above ground. Data from animals kept in captivity (TEIXEIRA *et al.* 2005), and evidence obtained in the field for this population in another study, in which there was a greater recruitment of subadult individuals in the rainy season (A.J. Almeida, unpub. data), indicate that *T. apereoides* reproduce mainly during the dry season. If this is indeed the reproductive pattern, presumably then smaller DHRs in the dry season, as well as other parameters, might be a consequence of female reproductive activities, because more energy would need to be allocated to reproduction (mating, feeding and care of pups) (GITTLEMAN & THOMPSON 1988), at the expense of energy expended in movements. Therefore, it is expected that there is some relationship

Table I. Effect of sex, age class, seasonal period, and total amount of mapped line on the daily home range size (DHR), and movements (LIU, AGR) of *Thrichomys apereoides*. ([a]) Including the interactions between the factors and the probable effects of the interaction between these factors and the amount of mapped line, ([b]) including the interactions between the factors and excluding interactions not found to be significant with the amount of mapped line. (DHR) Daily home range, (LIU) Movements on leaf litter, (AGR) movements above ground, (r^2) coefficient obtained for adjusted model.

Factors	DHR[a] F	DHR[a] p	DHR[b] ($r^2 = 0.60$) F	DHR[b] ($r^2 = 0.60$) p	LIU[a] F	LIU[a] p	LIU[b] ($r^2 = 0.67$) F	LIU[b] ($r^2 = 0.67$) p	AGR[a] F	AGR[a] P	AGR[b] ($r^2 = 0.57$) F	AGR[b] ($r^2 = 0.57$) p
Sex	0.020	0.873	0.02	0.870	0.070	0.783	1.310	0.250	0.100	0.113	4.420	**0.003**
Age class	0.480	0.340	11.65	**0.009**	2.450	0.121	1.179	0.280	0.430	0.511	0.150	0.695
Mapped line	52.030	0.000	128.09	0.000	91.550	0.000	178.000	0.000	49.030	0.000	70.150	0.000
Sex*Age class	0.270	0.600	0.03	0.858	0.000	0.928	0.107	0.744	0.000	0.960	0.630	0.426
Sex*Mapped line	0.110	0.735	–	–	0.940	0.334	–	–	2.050	0.155	–	–
Age class*Mapped line	1.180	0.888	–	–	1.700	0.194	–	–	1.020	0.315	–	–
Sex*Age class*Mapped line	0.540	0.462	–	–	0.020	0.887	–	–	0.030	0.858	–	–
Sex	1.200	0.277	6.42	**0.013**	0.592	0.444	0.498	0.482	0.162	0.687	0.973	0.327
Seasonal period	0.167	0.683	15.24	**0.000**	0.841	0.362	4.008	**0.049**	1.107	0.296	0.443	0.507
Mapped line	68.970	0.000	142.78	0.000	66.68	0.000	90.690	0.000	18.310	0.000	43.285	0.000
Sex*Seasonal period	0.209	0.648	7.66	**0.007**	0.006	0.814	0.623	0.432	2.439	0.123	0.085	0.770
Sex*Mapped line	1.376	0.245	–	–	0.049	0.824	–	–	1.886	0.174	–	–
Seasonal period*Mapped line	0.111	0.739	–	–	1.753	0.190	–	–	0.487	0.487	–	–
Sex*Seasonal period*Mapped line	2.429	0.124	–	–	0.116	0.733	–	–	3.290	0.070	–	–

* Interaction between factors.

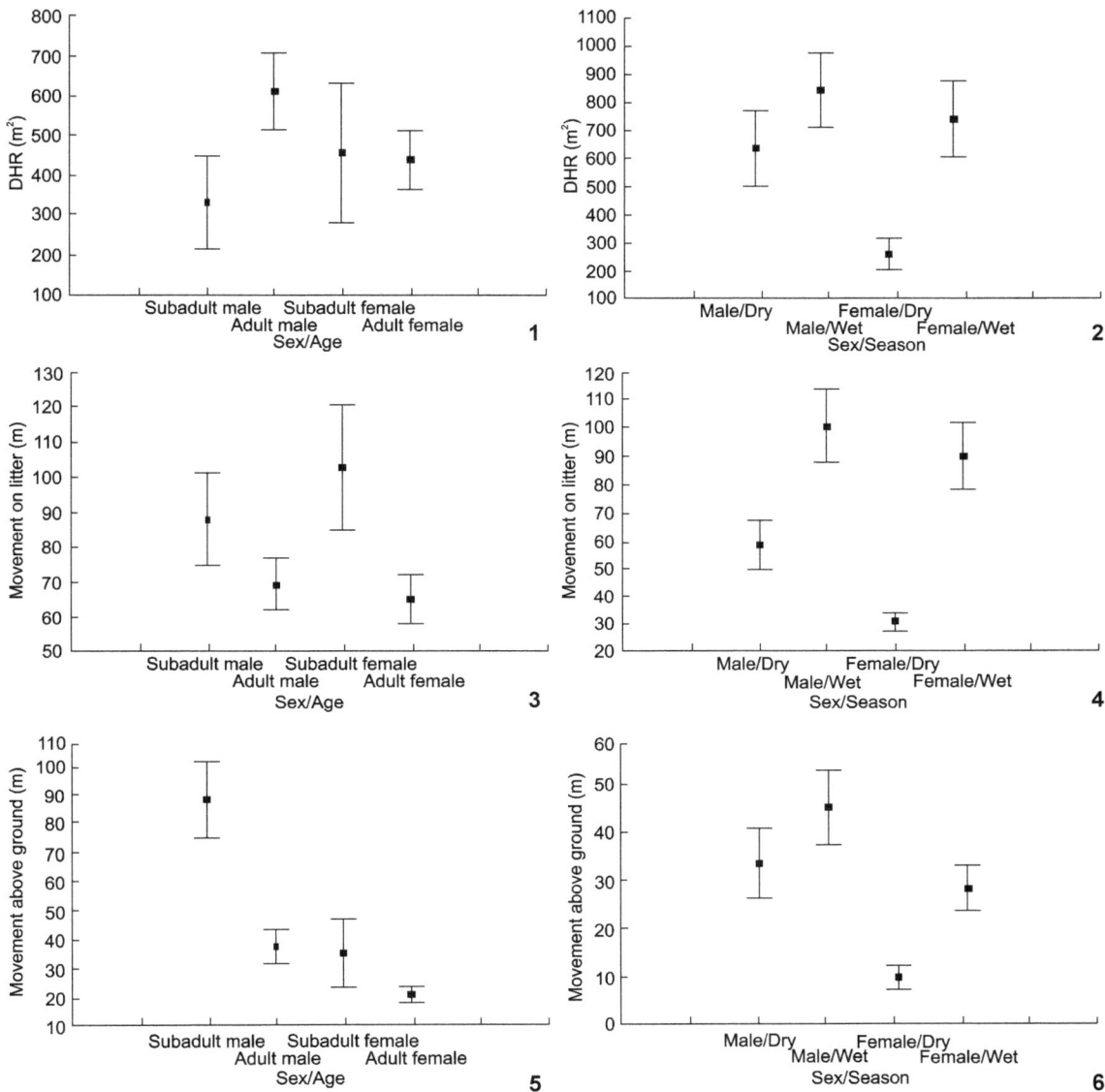

Figures 1-6. Mean values and respectively standard error of parameters of space use by *Thrichomys apereoides* obtained with the spool-and-line tracking technique for individuals living in the forest of Cauaia farm between the months August 2008 and July 2009. Daily Home Range (DHR) presented by (1) sex and age class, and (2) sex and seasonal period. Movements on leaf litter (LIU) presented by (3) sex and age class, and (4) sex and seasonal period. Movements above ground (AGR) presented by (5) sex and age class, and (6) sex and seasonal period.

between the activities of females with the availability of food resources in this period.

The availability of food resources has been considered as one of the most important factors that influence the pattern of use of space by mammals (ADLER 2011, ADRIAN & SACHSER 2011, MAHER & BURGER 2011). Depending on the spatial and temporal distribution of resources (ADLER 2011), in order

to maximize their fitness, female rodents tend to be territorial when food resources are scattered with lower renewal rates, while the males are distributed according to the space used by the females (OSTFELD 1985). Consequently, the spacing pattern of individuals has often been investigated in order to determinate the mating system of species (EMLEN & ORING 1977, OSTFELD 1985, GAULIN & FITZGERALD 1988, MAHER & BURGER 2011).

A pattern that has been found for some species of rodents based on space use is the polygyny (Gaulin & Fitzgerald 1988, Ostfeld 1990). This mating system was attributed to *T. apereoides* (Streilein 1982b, Thompson 1985), and for some hystricognath (Bergallo 1995, Bergallo & Magnuson 1999) and Sigmodontinae (Bonaventura *et al.* 1992, Heinemann *et al.* 1995, Gentile *et al.* 1997, Bergallo & Magnuson 1999, Pires *et al.* 2010) rodents. In this system, the males tend to travel greater areas in their search for females during the reproductive period, and the spacing patterns of males will be determined directly by the spacing of females rather than by resource distribution; in turn the females depending on resources distributed in patches tend to exhibit territorial tendencies and sometimes they are faced with the need to defend resources (Ostfeld 1990).

Territorial trends for females were found for other hystricognath species such as *Trinomys iheringi* (Bergallo 1995) and *T. moojeni* (Cordeiro Jr & Talamoni 2006). In addition, reproductive activity was observed during the dry season for these two *Trinomys* species. It is important to note that the studied echimyid occurs in area with pronounced seasonality, particularly related to the amount of rainfall (Sá Jr *et al.* 2012). The rain scarcity normally leads to a lower availability of fruit and insects during the dry season (Smithe 1982, Charles-Dominique 1983, Bergallo & Magnusson 1999, Stenvenson *et al.* 2000). Therefore, it is expected that food resources that commonly are locally distributed in temporal and spatial patches (Bergallo & Magnusson 1999, Stenvenson *et al.* 2000) can cause some influence over the movements of *T. apereoides* in this season. It was observed that for species that reproduce in a time of lower food availability, females may show a pattern of smaller displacement. In mammals, the greater parental investment lies with the females (Trivers 1972, Gittleman & Thompson 1988); therefore, the females tend to conserve energy expended in movement, establishing lower home ranges (Loretto & Vieira 2005). In Panamá, for the spiny rat *Proechimys semispinosus* (Tomes, 1860), Endries & Adler (2005) observed that individuals of both the sexes had larger home ranges during the rainy season. Furthermore, there was a positive relationship between home-range size and the number of fruiting trees within a home range. In contrast, no female spiny rats and only two of seven males captured during the dry season were in reproduction condition.

Alternatively, it has sometimes been observed that food shortages may force the individuals to expand their foraging areas and their home ranges (Slade & Swihart 1983, Gaulin & Fitzgerald 1988, Passamani 2000). We did not observe this phenomenon in this study, since both males and females did not significantly expanded their areas during the dry season, instead females showed a reduction in their movements, and the males showed DHRs always larger than females, this being the expected result assuming that *T. apereoides* is a polygynous species (Streilein 1982b).

In addition to the search for food and reproductive strategies, the predation pressure is recognized as having a strong influence on the characteristics of space use by various species of small mammals, directing their path choices during their movements (Holbrook 1979, Garshelis 2000, Getz *et al.* 2005). The risk of predation can limit the distance traveled by prey, thus reducing the home range of individuals as well as the pattern of overlapping home ranges among individuals (Yunger 2004, Petorelli *et al.* 2011). In this study, the adults of both sexes showed a significantly lower displacement on the leaf litter during the dry season. Assuming that differences occur in the sounds produced by movements of individuals on the leaf litter during the dry and rainy seasons, these animals might be avoiding movement on this type of substrate during the dry season because the sound produced could attract the attention of predators. However, only studies designed with the specific objective to test these hypotheses could show conclusively the influence of predation on the distance traveled by the individuals on the litter and the choice of specific paths by individuals.

It was observed that individuals of all age classes presented movement above ground but subadult males were more frequent. This result may indicate that subadult males utilize smaller areas, but more intensively, incorporating the use of the vertical stratum in their activities, as observed for some subadult marsupials (Cunha & Vieira 2005, Almeida *et al.* 2008). Subadult females use larger areas, but less intensively, moving predominantly on the litter.

In this study, we concluded that the pattern of movement is influenced by the sex and age of the individuals and that it may vary according to ecological conditions, especially those related to seasonal variations. Further studies should be directed to test separately the effect of the sex and the age and the environmental conditions on the DHR and movements of *T. apereoides*.

ACKNOWLEDGMENTS

We thank Marcus V. Vieira, Adriano P. Paglia and Gisele Lessa for their suggestions, Miguel Assis for the support and information he exchanged, and Andrew Linghorn for the English revision. Our gratitude goes to José Hein and all the residents and staff of the fazenda Cauaia. We thank CAPES for scholarship to A.J.A, and FIP PUC Minas for financial support. This study had the capture license 10807-2 from Chico Mendes Institute for Biodiversity Conservation (ICMBio).

REFERENCES

Adler, G.H. 1998. Impacts of resource abundance on populations of a tropical forest rodent. **Ecology** 79: 242-254.

Adler, G.H. 2011. Spacing patterns and social mating systems of echimyid rodents. **Journal of Mammalogy** 92: 31-38.

Adrian, O. & N. Sachser. 2011. Diversity of social and mating systems in cavies: a review. **Journal of Mammalogy** 92: 39-53.

ALMEIDA, A.J.; C.G. TORQUETTI & S.A. TALAMONI. 2008. Use of space by neotropical marsupial *Didelphis albiventris* (Didelphimorphia: Didelphidae) in an urban forest fragment. **Revista Brasileira de Zoologia 25**: 214-219.

BERBERT-BORN, M. 2002. Carste de Lagoa Santa, MG: berço da paleontologia e da espeleologia brasileira, p. 415-430. *In*: C. SCHOBBENHAUS; D.A. CAMPOS; E.T. QUEIROZ; M. WINGE & M. BERBERT-BORN (Eds). **Sítios geológicos e paleontológicos do Brasil.** Rio de Janeiro, Departamento Nacional de Produção Mineral (DNPM), Serviço Geológico do Brasil (CPRM).

BERGALLO, H.G. 1995. Comparative life-history characteristics of two species of rats, *Proechimys iheringi* and *Oryzomys intermedius*, in an Atlantic forest of Brasil. **Mammalia 59**: 51-64.

BERGALLO, H.G. & W.E. MAGNUSSON. 1999. Effects of climate and food availability on four rodent species in southeastern Brazil. **Journal of Mammalogy 80**: 472-486.

BONAVENTURA, S.M.; F.O. KRAVETZ & O.V. SUAREZ. 1992. The relationship between food availability, space use and territoriality in *Akodon azarae* (Rodentia, Cricetidae). **Mammalia 56**: 407-417.

BONVICINO, C.R.; J.A. DE OLIVEIRA & P.S. D'ANDREA. 2008. **Guia dos roedores do Brasil, com chaves para gêneros baseadas em caracteres externos.** Rio de Janeiro, Centro Pan-americano de Febre Aftosa – OPAS/OMS, 120p.

BOONSTRA, R. & T.M. CRAINE. 1986. Natal nest location and small mammal tracking with a spool and line technique. **Canadian Journal of Zoology 64**: 1034-1036.

BÖRGER L.; N. FRANCONI; G. DE MICHELE; A. GANTZ; F. MESCHI; A. MANICA; S. LOVARI & T. COULSON. 2006. Effects of sampling regime on the mean and variance of home range size estimates. **Journal of Ecology 75**: 1393-1405.

CERBONCINI, R.A.S.; M. PASSAMANI & T.V. BRAGA. 2011. Use of space by the black-eared opossum *Didelphis aurita* in a rural area in southeastern Brazil. **Mammalia 75**: 287-290.

CHARLES-DOMINIQUE, P. 1983. Ecology and social adaptations in didelphid marsupials: comparisons with eutherian of similar ecology. p. 395-420. *In*: J.F. EISENBERG & D.G. KLEIMAN (Eds). **Advances in the study of Mammalian Behavior.** Shippennshurg, Special Publication #7, American Society of Mammalogists.

CLUTTON-BROCK, T.H. & P.H. HARVEY. 1978. Mammals, resources and reproductive strategies. **Nature 273**: 191-195.

CORDEIRO-JÚNIOR, D. A. & S. A. TALAMONI. 2006. New data on the life history and occurrence of spiny rats *Trinomys moojeni* (Rodentia: Echimyidae), in Southeastern Brazil. **Acta Theriologica 51**: 1-6.

CUNHA, A & M.V. VIEIRA. 2002. Support diameter, incline, and vertical movements of four didelphid marsupials in the Atlantic Forest of Brazil. **Journal of Zoology 258**: 419-426.

CUNHA, A. & M.V. VIEIRA. 2005. Age, season, and arboreal movements of the opossum *Didelphis aurita* in an Atlantic rain forest of Brazil. **Acta Theriologica 50**: 551-560.

DELCIELLOS, A.C.; D. LORETTO & M.V. VIEIRA. 2006. Novos métodos no estudo da estratificação vertical de marsupiais neotropicais. **Oecologia Brasiliensis 10**: 135-153.

DRAPER, N.R. & H.A. SMITH. 1998. **Applied Regression Analysis.** New York, John Wiley & Sons Incorporation, III + 705 p.

DUESER, R.D. & H.H. SHUGART. 1979. Niche Pattern in Forest-floor in Small Mammal Fauna. **Ecology 60**: 108-118.

EMLEN, S.T. & L.W. ORING. 1977. Ecology, sexual selection and the evolution of mating systems. **Science 197**: 215-23.

ENDRIES, M.J. & G.H. ADLER. 2005. Spacing patterns of a tropical forest rodent, the spiny rat (*Proechimys semispinosus*), in Panamá. **Journal of Zoology 265**: 147-155.

GARSHELIS, D.L. 2000. Delusions in habitat evaluation: measuring use, selection, and importance, p.111-164. *In*: L. BOITANI & T.K. FULLER (Eds). **Research Techniques in Animal Ecology: Controversies and Consequences.** New York, Columba University Press.

GAULIN, J.C. & R.W. FITZGERALD. 1988. Home range size as a predictor of mating systems in *Microtus*. **Journal of Mammalogy 69**: 311-319.

GENTILE, R.; P.S. D'ANDREA & R. CERQUEIRA. 1997. Home ranges of *Philander frenata* and *Akodon cursor* in a Brazilian restinga (coastal shrubland). **Mastozoologia Neotropical 4**: 105-112.

GETZ, L.L.; M.K. OLI; J.E. HOFMANN; B. MCGUIRE & A. OZGUL. 2005. Factors influencing movement distances of two species of sympatric voles. **Journal of Mammalogy 86**: 647-654.

GITTLEMAN, J.L. & S.D. THOMPSON. 1988. Energy allocation in mammalian reproduction. **American Zoologist 28**: 863-875.

HARLESS, M.L.; A.D. WALDE; D.K. DELANEY; L.L. PATER & W.K. HAYES. 2010. Sampling considerations for improving home range Estimates of desert tortoises: effects of estimator, Sampling regime, and sex. **Herpetological Conservation and Biology 5** (3): 374-387.

HEINEMANN, K.M.; N. GUTHMANN; M. LOZADA & J.A. MONJEAU. 1995. Area de actividad de *Abrothrix xantorhinus* (Muridae, Sigmodontinae) e implicancias para su estratégia reproductiva. **Mastozoologia Neotropical 2**: 23-30.

HOLBROOK, S. 1979. Vegetation affinities, arboreal activity, and coexistence of three species of rodents. **Journal of Mammalogy 60**: 528-542.

IBGE. 2004. **Mapa de vegetação do Brasil.** Rio de Janeiro, Instituto Brasileiro de Geografia e Estatística, escala 1:5000000.

JOHNSON, D.D.P; R. KAYS; P.G. BLACKWELL & D.W. MACDONALD. 2002. Does the resource dispersion hypothesis explain group living? **Trends in Ecology & Evolution 17**: 563-570.

JORGENSEN, E.E. & S. DEMARAIS. 1999. Spatial scale dependence of rodent habitat use. **Journal of Mammalogy 80**: 421-429.

KELT, D.A. & D.H. VAN VUREN. 2001. The ecology and macroecology of mammalian home range area. **American Naturalist 157**: 637-45.

KEY, G.E. & R.D. WOODS. 1996. Spool-and-line studies on the behavioural ecology of rats (*Rattus* spp.) in the Galápagos Islands. **Canadian Journal of Zoology 74**: 733-737.

LACHER JR, T.E. & C.J.R. ALHO. 1989. Microhabitat use among small mammals in the Brazilian Pantanal. **Journal of Mammalogy 70**: 396-401.

LINDSEY, J.K. 1997. **Applying Generalized Linear Models.** New York, Springer-Verlag, 255p.

LORETTO, D. & M.V. VIEIRA. 2005. The effects of reproductive and climatic seasons on movements in the black-eared opossum (*Didelphis aurita* Wied-Neuwied, 1826). **Journal of Mammalogy 86**: 287-293.

MACDONALD, K. 1983. Production, social controls, and ideology: toward a sociobiology of the phenotype. **Journal of Social and Biological Structures 6**: 297-317.

MAGNUSSON, W.E.; A.L. FRANCISCO & T.M. SANAIOTTI. 1995. Home range size and territoriality in *Bolomys lasiurus* (Rodentia: Muridae) in an Amazonian savanna. **Journal of Tropical Ecology 11**: 179-188.

MAHER, C.R. & J.R. BURGER. 2011. Intraspecific variation in space use, group size, and mating systems of caviomorph rodents. **Journal of Mammalogy 92**: 54-64.

MARQUET, P.A.; R.A. QUIÑONES; A.S. ABADES; F. LABRA; M. TOGNELLI; M. ARIM & M. RIVADENEIRA. 2005. Scaling and Power-laws in ecological systems. **The Journal of Experimental Biology 208**: 1749-1969.

MENDEL, S.M. & M.V. VIEIRA. 2003. Movement distances and density estimation of small mammals using the spool-and-line technique. **Acta Theriologica 48**: 298-300.

MITCHELL, M.S. & R.A. POWELL. 2012. Foraging optimally for home ranges. **Journal of Mammalogy 93**: 917-928.

MOHR, C.O. 1947. Table of equivalent populations of North American small mammals. **American Midland Naturalist 37**: 223-249.

OSTFELD, R.S. 1985. Limiting resources and territoriality in microtine rodents. **American Naturalist 126**: 1-15.

OSTFELD, R.S. 1990. The ecology of territoriality in small mammals. **Tends in Ecology & Evolution 5**: 411-415.

PASSAMANI, M. 2000. Análise da comunidade de marsupiais em Mata Atlântica de Santa Teresa, Espírito Santo. **Boletim do Museu de Biologia Mello Leitão 11**: 215-228.

PETTORELLI, N.; T. COULSON; S.M. DURANT & J.M. GAILLARD. 2011. Predation, individual variability and vertebrate population dynamics. **Oecologia 167**: 305-314.

PIRES, A.S.; F.A.S. FERNANDES; B.R. FELICIANO & D. FREITAS. 2010. Use of space by *Necromys lasiurus* (Rodentia, Sigmodontinae) in a grassland among Atlantic Forest fragments. **Mammalian Biology 75**: 270-276.

PIZZUTO, T.A.; G.R. FINLAYSON; M.S. CROWTHER & C.R. DICKMAN. 2007. Microhabitat use by the brush-tailed bettong (Bettongia penicillata) and burrowing bettong (B. lesueur) in semiarid New South Wales: implications for reintroduction programs. **Wildlife Research 34**: 271-279.

PREVEDELLO, J.A.; A.F. MENDONÇA & M.V. VIEIRA. 2008. Uso do espaço por pequenos mamíferos: uma análise dos estudos realizados no Brasil. **Oecologia Brasiliensis 12**: 610-625.

PREVEDELLO, J.A.; R.G. RODRIGUES & E.L.A. MONTEIRO-FILHO. 2010. Habitat selection by two species of small mammals in the Atlantic Forest Brazil: Comparing results from live trapping and spool-and-line tracking. **Mammalian Biology 75**: 106-114.

PRIOTTO, J.; A. STEINMANN; C. PROVENSAL & J. POLOP. 2004. Juvenile dispersal in *Calomys venustus* (Muridae: Sigmodontinae). **Acta Oecologica 25**: 205-210.

REIS, S.F. & L.M. PESSÔA. 2004. *Thrichomys apereoides*. **Mammalian Species 727**: 1-8.

REHMEIER, R.L.; G.A. KAUFMAN & D.W. KAUFMAN. 2004. Long-distance movements of the deer mouse in tall grass prairie. **Journal of Mammalogy 85**: 562-568.

ROBERTS, M.S.; K.V. THOMPSON & J.A. CRANFORD. 1988. Reproduction and growth in captive punaré (*Thrichomys apereoides* Rodentia: Echimyidae) of the Brazilian Caatinga with reference to the reproductive strategies of the Echimyidae. **Journal of Mammalogy 69**: 542-551.

RYAN, J.M.; G.K. CREIGHTON & L.H. EMMONS. 1993. Activity patterns of two species of *Nesomys* (Muridae: Nesomyinae) in a Madagascar rain forest. **Journal of Tropical Ecology 9**: 101-107.

SÁ JR, A.; L.G. CARVALHO; F.F. SILVA & M.C. ALVES. 2012. Application of the Köppen classification for climatic zoning in the state of Minas Gerais, Brazil. **Theoretical and Applied Climatology 108**: 1-7.

SKVARLA, J.L.; J.D. NICHOLS; J.E. HINES & P.M. WASER. 2004. Modeling interpopulation dispersal by banner-tailed kangaroo rats. **Ecology 85**: 2737-2746.

SLADE, N.A. & R.K. SWIHART. 1983. Home range indices for the hispid cotton rat (Sigmodon hispidus) in northeastern Kansas. **Journal of Mammalogy 64**: 580-590.

SMITHE, N. 1982. The seasonal abundance of night-flying insects in a neotropical forest, p. 319-318. *In*: E.G. LEIGH, A.S. RAND & D.M. WINDSOR (Eds). **The ecology of a tropical Forest: Seasonal Rhythms and Long Term Changes.** Washington, D.C., Smithsonian Institute Press.

STENVENSON, P.R.; M.J. QUIÑONES & J.A. AHUMADA. 2000. Influence of fruit availability on ecological overlap among four Neotropical primates at Tinigua National Park, Colombia. **Biotropica 32**: 533-544.

STICKEL, L.F. 1954. A comparison of certain methods of measuring ranges of small mammals. **Journal of Mammalogy 35**: 1-15.

STREILEIN, K.E. 1982a. Ecology of small mammals in the semiarid Brazilian Caatinga: I. Climate and faunal composition. **Annals of Carnegie Museum 51**: 79-107.

STREILEIN, K.E. 1982b. Ecology of small mammals in the semiarid Brazilian Caatinga: V. Agonistic behavior and overview. **Annals of Carnegie Museum 51**: 345-369.

TEIXEIRA, B.R.; A.L.R. ROQUE; S.C. BARREIROS-GÓMEZ; P.M. BORODIN; A.M. JANSEN & P.S. D'ANDREA. 2005. Maintenance and breeding of *Thrichomys* (Trouessart, 1880) (Rodentia: Echimyidae) in captivity. **Memórias do Instituto Oswaldo Cruz 100**: 627-630.

THOMPSON, K.T. 1985. **Social play in the South American punaré (Thrichomys apereoides): a test of play function hypotheses.** Blacksburg, Virginia Polytechnic Institute and State University, 120p.

TRIVERS, R.L. 1972. Parental investment and sexual selection, p. 136-179. *In*: B. CAMPBELL (Ed). **Sexual selection and the descent of man 1871-1971.** Chicago, Aldine Publishing Company.

VIEIRA, M.V. & A.A. CUNHA. 2008. Scaling body mass and use of space in three species of marsupials in the Atlantic Forest of Brazil. **Austral Ecology 33:** 8720-879.

YUNGER, J.A. 2004. Movement and spatial organization of small mammals following vertebrate predator exclusion. **Oecologia 139:** 647-654.

13

Ecomorphology of *Astyanax* species in streams with different substrates

Marcela A. Souza[1], Daniela C. Fagundes[2], Cecília G. Leal[2,3] & Paulo S. Pompeu[3]

[1] *Laboratório de Ecologia de Peixes, Universidade Federal de Lavras. Campus Universitário, 37200-000 Lavras, MG, Brazil.*
E-mail: marcelaalvesdesouza@gmail.com
[2] *Programa de Pós-Graduação em Ecologia Aplicada, Universidade Federal de Lavras. Campus Universitário,*
37200-000 Lavras, MG, Brazil. E-mail: danielafagundes87@gmail.com, c.gontijoleal@gmail.com
[3] *Departamento de Biologia, Universidade Federal de Lavras. Campus Universitário, 37200-000 Lavras, MG, Brazil.*
E-mail: pompeu@ufla.br

ABSTRACT. In the present study, we assessed the ecomorphology of two species of *Astyanax* in streams with different substrates found in the Rio São Francisco Basin. The dominant substrate of each stream was defined as either "fine" (0 to 2 mm), "gravel" (2 to 250 mm), "rock" (> 250 mm), or "leaf bank". We analyzed a total of 22 ecomorphological attributes of *Astyanax intermedius* Eigenmann, 1908 (127 individuals) and *Astyanax rivularis* (Lütken, 1875) (238 individuals) adults. We detected significant ecomorphological differences between the populations of *A. rivularis* and *A. intermedius* from habitats with different types of substrates. However, the two species did not show the same morphological differences depending on the type of substrate. These results confirmed the hypothesis that individuals from environments with different characteristics may have different ecomorphological patterns. Knowing that morphology is associated with habitat use and available resources, the loss of a resource or a modification in the environment may directly affect the permanence of a species, leading to a loss of morphologic diversity.

KEY WORDS. Convergence; divergence; ecomorphological attributes; freshwater fish; habitat diversity; intraspecific differences.

The relationship between the body shape of organisms and the environments they inhabit was suggested by Aristotle in the fourth century BC. However, it was only in 1859 with the publication of The Origin of Species by Charles Darwin that the theoretical base of this relationship was consolidated (FREIRE & AGOSTINHO 2001).

Ecomorphology is an area of evolutionary biology that defines relationships between morphology and ecology by studying the relationship between body shape and organisms' ecological characteristics (GATZ 1979, WATSON & BALON 1984, WIKRAMANAYAKE 1990, WINEMILLER 1991, DOUGLAS & MATTHEWS 1992). The concept is based on the assumption that a phenotype provides useful information about the relationship between form and function, i.e., ecology and morphology are alternative expressions of the same ecological and evolutionary adjustments between phenotype and environmental conditions (WAINWRIGHT 1994). Body shape is the primary phenotypical component of an organism and influences foraging activities, locomotion, reproduction, and predator escape (GUILL *et al.* 2003). Species specialization is often associated with improved efficiency in the exploitation of the resources of a given habitat (PERES-NETO 1999). Morphological differences between species can result from different selective expressions, and species morphology should, therefore, reflect their habits and adaptations to their environment (GATZ 1979, WATSON & BALON 1984, WIKRAMANAYAKE 1990, WINEMILLER 1991, DOUGLAS & MATTHEWS 1992, BEAUMORD & PETRERE 1994, LANGERHANS *et al.* 2003, CASATTI & CASTRO 2006).

Adaptation is the most important concept for understanding the relationship between morphology and ecology, and it is described as structures that increase fitness (ELDREDGE 1989). This understanding states that adaptations can be detected by relating physiology to environmental variables (PERES-NETO 1999). A key feature is that the variation among individuals or species may be related to differences in performance and, ultimately, a change in resource use and fitness (WAINWRIGHT 1994). Ecomorphology is thus an alternative method of studying adaptation (MOTTA *et al.* 1995).

Intra and interspecific ecomorphological differences may be related to different environmental and/or biological pressures as well as phylogenetic factors (LANGERHANS *et al.* 2003, CASSATTI & CASTRO 2006, LEAL *et al.* 2011, 2013). Elements of the structural complexity of habitats, including substrates of different types and sizes and the presence of sites with different water velocities and depths, such as pools and rapids, are important in this regard (LEAL *et al.* 2011, 2013). Other factors may also influence habitat complexity and species distribution, including physical and chemical water parameters, as well as species and individuals interactions (JACKSON *et al.* 2001). Different habitats lead to divergent intraspecific diversification, but the degree of divergence is constrained by the mixture of individuals from alternative environments (LANGERHANS *et al.* 2003).

Conversely, significant relationships between morphology and ecology can occur due to phylogenetic factors. The integration of phylogenetic information and ecomorphological analyses is the only way to objectively identify cases of morphological and adaptive convergence and divergence (Casatti & Castro 2006). Importantly, this fusion allows the comprehension of larger evolutionary patterns within the communities (Peres-Neto 1999).

Astyanax encompasses morphologically diverse species that are highly abundant and widely distributed in the São Francisco River Basin. For this reason, these species can be good models for ecomorphological studies. Comparing the ecomorphology of populations of two species of *Astyanax* inhabiting streams characterized by different substrate types, this study addressed the following questions: I) are there intraspecific, differences in morphology related to the different stream substrate types?; and II) do different species show the same substrate related morphological differences?

MATERIAL AND METHODS

We conducted the present study in 12 1st-3rd order tributaries (Fig. 1) of the reservoir of the Três Marias hydroelectric dam (UHE Três Marias) located in the Upper Rio São Francisco Basin, in the Brazilian cerrado (Table I). The study area is characterized mainly by pasture and small family farms as well as small cities and towns (Ligeiro et al. 2013).

Table I. Location, dominant substrate, and number of individuals of *Astyanax intermedius* (Aint) and *Astyanax rivularis* (Ariv) of each sampled stream.

Stream Code	Coordinates (UTM)	Dominant substrate	Species (N)	
			Aint	Ariv
TM0027	23K 501037 E/7958924 S	Rock	10	–
TM0028	23K 426418 E/7900361 S	Gravel	–	4
TM0033	23K 460983 E/7872903 S	Rock	40	9
TM 0043	23K 501072 E/7965010 S	Leafbank	–	14
TM0058	23K 482680 E/7890720 S	Fine	51	100
TM0072	23K 429115 E/7961018 S	Gravel	–	9
TM0091	23K 498017 E/7952669 S	Fine	–	14
TM0106	23K 484778 E/7887307 S	Fine	30	–
TM0134	23K 498271 E/7897727 S	Gravel	–	40
TM0187	23K 487317 E/7961001 S	Rock	–	19
TM0220	23K 432976 E/7904242 S	Rock	–	4
TM3962	23K 504547 E/7940578 S	Fine	–	27

We collected the data during September 2010, and each stream was sampled once. We collected fish using seines and kick nets, labeled them and fixed them in a 10% formaldehyde solution.

The substrate characterization of each stream was conducted through a visual estimation along a stretch of 40 times the stream width, with the total length divided into 10 sections. Within each section, 15 equidistant measurements of the following substrate classes was made: "fine" (silt, clay, and/or sand, from 0 to 2 mm), "gravel" (fine gravel, coarse gravel, and/or block, from 2 to 250 mm), "rock" (boulders, wide boulders, and smooth and/or rough rocks, > 250 mm), and "leaf bank". The dominant substrate of each stream was defined as the one that occurred in 50% or more of the positions within that stream.

Morphometric analyses were obtained from 131 adult individuals of *Astyanax intermedius* Eigenmann, 1908 and 240 of *Astyanax rivularis* (Lütken, 1875). Specimens of both species are deposited in the Coleção Ictiológica da UFLA at the Universidade Federal de Lavras (CIUFLA 307 and 308). We obtained a total of 21 linear and six area measurements for each specimen and selected these measurements based on Winemiller (1991), Casatti & Castro (2006), Oliveira et al. (2010), and Leal et al. (2011). Linear measurements were made using a digital caliper (0.01 mm precision), and area measurements were obtained from drawings of the contours of the body and fins of the individuals. The drawings were then scanned and their areas calculated using the program ImageJ. From these measurements, 22 ecomorphological attributes were calculated: **1) Compression index (CI):** the maximum body depth divided by the maximum body width (Gatz 1979). High CI values indicate laterally compressed fishes inhabiting environments with slow-flowing waters (Watson & Balon 1984). **2) Relative depth (RD):** the maximum body depth divided by the standard body length (Gatz 1979). RD is inversely related to water velocity and directly related to the fish's ability to perform vertical movements (Gatz 1979). **3) Index of ventral flattening (IVF):** the average body height (vertical distance from the midline to the ventrum, with the midline defined as an imaginary line crossing the eye pupil towards the center of the ultimate vertebra) divided by the maximum body depth (Gatz 1979). Low IVF values indicate fishes living in environments with fast-flowing waters; these characteristics enable the fish to stay in position with no swimming (Hora 1930). **4) Mouth orientation (MO):** the angle formed between the tangential plane to both lips and the perpendicular plane to the longitudinal axis of the body when the mouth is open (Gatz 1979). MO is characterized as follows: superior = between 10° and 80°; terminal = 90°; inferior = between 100°; and 170° and ventral = 180° (Freire & Agostinho 2001). Degree values were converted to a decimal scale where 1° = 60 (Cassati & Castro 2006). **5) Relative eye position (REP):** the depth of the eye at the midline divided by the head depth (Watson & Balon 1984). REP is assumed to be related to vertical habitat preference (Gatz 1979), and high values correspond to benthic fishes with dorsally located eyes (Watson & Balon 1984). **6) Relative eye area (REA):** this index is related to food detection and provides information on the use of vision in predation activities (Pankhurst 1989, Pouilly et al. 2003). **7) Relative height of head (RHH):** the head depth divided by

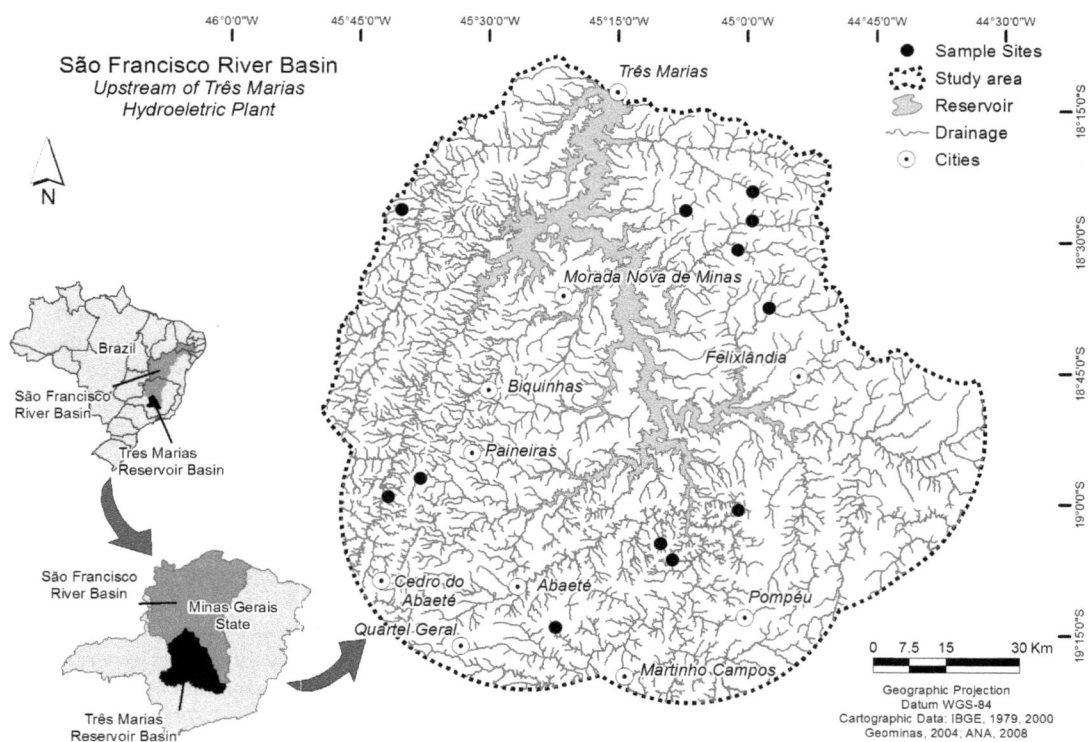

Figure 1. UHE Três Marias area consisting of 12 streams that were 1st, 2nd, and 3rd order tributaries of this reservoir.

the body depth (OLIVEIRA *et al.* 2010). Larger relative values of head height are found in fishes that feed on larger prey. Larger values for this index are expected for piscivores (WINEMILLER 1991, WILLIS *et al.* 2005). **8) Relative width of head (RWH):** the head width divided by the body width (OLIVEIRA *et al.* 2010). Larger relative values of head width are found in fishes that feed on larger prey. Larger values for this index are expected for piscivores (WINEMILLER 1991, WILLIS *et al.* 2005). **9) Relative head length (RHL):** the head length divided by the standard length (GATZ 1979). Larger relative values of head length are found in fishes that feed on larger prey. This index should be larger for piscivores (WATSON & BALON 1984, WINEMILLER 1991, POUILLY *et al.* 2003, WILLIS *et al.* 2005). **10) Relative caudal peduncle length (RCPL):** the caudal peduncle length (distance from a vertical line at the level of the posterior margin of the base of the most posterior median fin to the terminus of the vertebral column) divided by the standard length (GATZ 1979). Fishes with long caudal peduncles are goods swimmers. However, fishes adapted to rapid water flow, but not necessarily nektonic as armored catfishes, also presented long caudal peduncules for propulsion over short distances (HORA 1930, WATSON & BALON 1984, WINEMILLER 1991). **11) Caudal peduncle compression index (CPCI):** the depth of the caudal peduncle divided by the width of the caudal peduncle taken at the narrowest section (GATZ 1979). High values indicate compressed

peduncles, and this feature is assumed to be typical of less active swimmers (GATZ 1979). **12) Relative height of caudal peduncle (RHCP):** the peduncle height divided by the body depth (OLIVEIRA *et al.* 2010). Lower values indicate greater maneuverability potential (WINEMILLER 1991). **13) Relative width of caudal peduncle (RWCP):** the peduncle width divided by the body width (OLIVEIRA *et al.* 2010). Higher relative values indicate better continuous swimmers (WINEMILLER 1991). **14) Caudal fin aspect ratio (CFAR):** the square caudal fin depth divided by the caudal fin area (GATZ 1979). This index is directly proportional to the amount of swimming by a fish (GATZ 1979). **15) Pectoral fin aspect ratio (PcFAR):** the pectoral fin length divided by its width (GATZ 1979). High values suggest long fins typical of fishes with remarkable swimming ability (WATSON & BALON 1984). **16) Pelvic fin aspect ratio (PlFAR):** the pelvic fin length divided by its width (GATZ 1979). Low values are observed in fishes that use their pelvic fins for braking and swimming forward, while high values are observed in fishes that use their pelvic fins to swim backwards and to maintain their position in the water column (GATZ 1979). **17) Relative pectoral fin length (RPcFL):** the pectoral fin length divided by the SL (GATZ 1979). High values correspond to fishes that are able to perform many slow maneuvers and inhabit slow-flowing waters (GATZ 1979). **18) Relative pelvic fin length (RPlFL):** the pelvic fin length divided by the SL (GATZ 1979).

This index relates to the choice of habitat and is longer in rocky habitat-dwelling species and shorter in nektonic species (Gatz 1979). **19) Relative caudal fin area (RCFA):** the caudal fin area divided by the total body area (Watson & Balon 1984). High values indicate fins capable of quick propelling movements typical of benthic fishes (Watson & Balon 1984). **20) Relative pectoral fin area (RPcFA):** the pectoral fin area divided by the total body area (Gatz 1979). High values indicate slow swimmers using their fins to maneuver, although high values may also indicate fishes inhabiting fast-flowing waters that use their fins as water deflection surfaces to help maintain their body close to the substrate (Watson & Balon 1984). **21) Relative pelvic fin area (RPlFA):** the pelvic fin area divided by the total body area (Gatz 1979). Benthic fishes have relatively large RPlFAs (Gatz 1979). **22) Relative dorsal fin area (RDFA):** the dorsal fin area divided by the total body area (Casatti & Castro 2006). The dorsal fin provides fishes with stability, and small areas are assumed to be more efficient in fast-flowing waters (Gosline 1971).

We performed a principal components analysis (PCA) based on the morphological attributes to observe specimen distribution in morphological space. This ordination technique enables the simultaneous evaluation of several variables. The first two axes, representing the higher values of explained variance, were retained for biological interpretation.

We performed discriminant function analyses (DFA) for each species separately using the default method of Statistica 7.0 (Statsoft 2004). We used this approach to search for ecomorphological differences between individuals from different substrates. In DFA, the squared Mahalanobis distances are used to measure the distances between the centroids of each group. The farther one group is from another group, the greater confidence is that they are different (Statsoft 2004). Afterwards, we tested the attributes selected by the DFA for differences by analysis of variance (ANOVA) or by Kruskal-Waliis when assumptions of normality and homogeneity of variance were not met. The Tukey test or Fisher LDS (Post-Hoc) were subsequently applied to identify groups of streams responsible for the differences. The analyses were performed using Statistica 7.0 (Statsoft 2004).

RESULTS

The first two PCA axes for *A. intermedius* accounted for 29.0% of the variance; the first axis represented 18.1%, and the second axis 10.9% (Fig. 2). The contributions of the variables to the PCA axes (Table II) showed that the pelvic fin aspect ratio and the pectoral fin aspect ratio negatively influenced the first axis. Moreover, the relative pelvic fin area, the relative pectoral fin area, and the relative dorsal fin area positively influenced the first axis. The second axis had a positive contribution from the relative width of the caudal peduncle and a negative contribution from the caudal peduncle compression index.

Table II. Contribution of ecomorphological attributes to the first two axes of the principal component analysis for *Astyanax intermedius* and *Astyanax rivularis* collected in Rio São Francisco streams in September 2010.

Ecomorphological attribute	A. intermedius		A. rivularis	
	PCA 1	PCA 2	PCA 1	PCA 2
Compression index	0.26	0.29	0.21	-0.41
Relative depth	0.05	0.14	0.33	**-0.81**
Index of ventral flattening	-0.02	0.00	-0.06	-0.29
Mouth orientation	0.11	0.25	-0.21	0.30
Relative eye position	0.30	0.03	0.39	-0.37
Relative head height	0.10	-0.15	-0.35	**0.59**
Relative head width	0.13	0.40	-0.07	-0.03
Relative head length	-0.01	-0.27	0.04	**-0.64**
Relative caudal peduncle length	0.23	0.44	-0.32	-0.50
Caudal peduncle compression index	0.16	**-0.65**	-0.52	0.02
Relative caudal peduncle height	0.16	0.28	**-0.62**	0.33
Relative caudal peduncle width	-0.05	**0.85**	0.34	-0.04
Caudal fin aspect ratio	-0.52	-0.37	0.47	0.21
Pectoral fin aspect ratio	**-0.73**	0.39	**0.58**	0.05
Pelvic fin aspect ratio	**-0.84**	0.05	0.52	0.01
Relative pectoral fin length	-0.21	0.17	-0.01	-0.28
Relative pelvic fin length	-0.22	0.27	-0.22	-0.47
Relative caudal fin area	0.53	0.27	-0.53	-0.50
Relative pectoral fin area	**0.73**	-0.25	**-0.58**	-0.23
Relative pelvic fin area	**0.82**	0.07	**-0.60**	-0.35
Relative dorsal fin area	**0.68**	0.13	-0.05	0.15
Relative eye area	-0.35	0.10	-0.06	-0.20
Eigenvalues	3.99	2.40	3.22	3.08
Variance explained (%)	18.13	10.91	14.65	14.02
Cumulative variance (%)		29.05		28.68

The ecomorphological gradient indicates that on the first axis there was high segregation among individuals present in the streams with a fine, rocky substrate, while they overlap on the second axis. The ecomorphological attributes that most contributed to the observed differences were as follows: the compression index (CI), the relative width of the caudal peduncle (RWCP), the relative width of the head (RWH), and the relative dorsal fin area (RDFA) (Table III). The percentage of individuals within each group was 88.3% for the fine substrate and 80% for the rocky substrate.

We chose the first two PCA axes for the interpretation of the *A. rivularis* data. The first axis explained 14.6% of the variance, and the second axis accounted for 14.0% (Fig. 3). The contributions of the variables to the PCA showed that the first axis was negatively influenced by the relative height of the caudal peduncle, the relative pelvic fin area, and the relative pectoral fin area, and it was positively influenced by the pectoral fin aspect ratio (Table II). In the second axis, relative head length had a positive contribution, and relative caudal peduncle

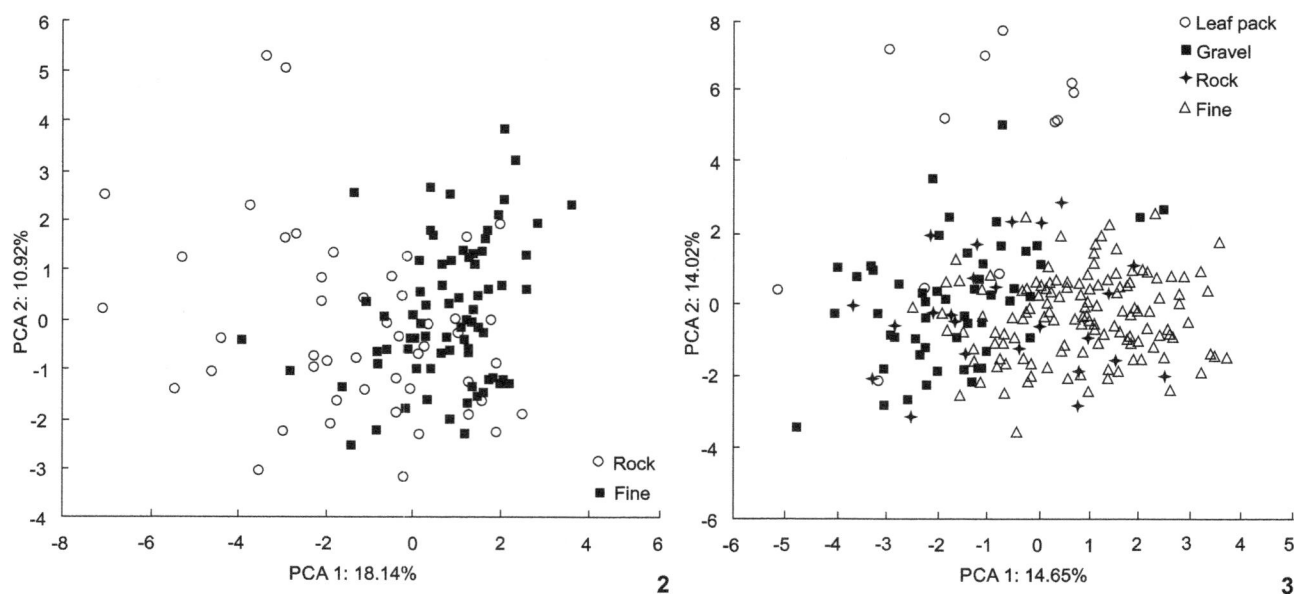

Figures 2-3. Projection of *Astyanax intermedius* (2) and (3) *Astyanax rivularis* individuals recorded in streams with different substrate types along the first two PCA axes. The axes represent the ecomorphological attributes of individuals collected in Rio São Francisco streams in September 2010.

Table III. F values to enter and exit the model and tolerance of the discriminant function analysis for *Astyanax intermedius* and *Astyanax rivularis* collected in Rio São Francisco streams in September 2010.

Variable retained in the model	Astyanax intermedius			Astyanax rivularis		
	F-exit	Tolerance	p	F-value-exit	Tolerance	p
Compression index	12.56	0.27	0.00	4.56	0.22	0.00
Relative depth	0.87	0.00	0.35	0.99	0.01	0.43
Index of ventral flattening	0.13	0.00	0.66	0.38	0.01	0.34
Mouth orientation	0.54	0.81	0.48	5.01	0.85	0.00
Relative eye position	0.35	0.57	0.55	5.71	0.61	0.00
Relative head height	0.43	0.39	0.49	2.59	0.31	0.05
Relative head width	10.34	0.53	0.00	12.82	0.43	0.00
Relative head length	1.07	0.68	0.30	7.79	0.53	0.00
Relative caudal peduncle length	1.13	0.00	0.29	1.37	0.01	0.28
Caudal peduncle compression index	3.37	0.09	0.07	0.23	0.06	0.87
Relative caudal peduncle height	0.55	0.00	0.46	1.60	0.01	0.21
Relative caudal peduncle width	4.51	0.07	0.00	0.29	0.06	0.84
Caudal fin aspect ratio	0.02	0.31	0.92	7.69	0.38	0.00
Pectoral fin aspect ratio	1.18	0.19	0.29	5.29	0.21	0.00
Pelvic fin aspect ratio	1.94	0.14	0.17	0.14	0.17	0.92
Relative pectoral fin length	0.03	0.46	0.87	0.83	0.56	0.51
Relative pelvic fin length	0.81	0.36	0.37	0.28	0.30	0.84
Relative caudal fin area	0.76	0.30	0.41	14.06	0.29	0.00
Relative pectoral fin area	2.55	0.22	0.11	0.63	0.18	0.52
Relative pelvic fin area	1.06	0.19	0.32	2.87	0.21	0.04
Relative dorsal fin area	15.31	0.63	0.00	0.81	0.89	0.48
Relative eye area	0.07	0.67	0.80	1.53	0.91	0.18

length and relative depth made negative contributions. It can be observed that for *A. rivularis*, the population from streams with abundant leaf banks appeared to be more segregated from the others in relation to the second axis of the PCA gradient, but the discriminant analysis still showed that all substrates were different.

The discriminant function analysis showed significant differences among the ecomorphological attributes of populations of *A. rivularis* present in different substrate types (Wilk's λ = 0.182, $F_{66,633}$ = 7.39, p < 0.01). The attributes that contributed most to this result were the compression index, the relative head length, the relative width of head, the relative eye position, the caudal fin aspect ratio, the pectoral fin aspect ratio, the relative caudal fin area, the relative pectoral fin area, and the mouth orientation (Table III). All Mahalanobis distances were significantly different (p < 0.05) for the substrate pairs. The percentage of individuals within each group was 95.7% for the fine substrate, 51.6% for the rocky substrate, 83% for the gravel, and 76.9% for the leaf bank. When separately evaluating the ecomorphological attributes of *A. intermedius* by discriminant analysis, streams with rocks showed lower values for the compression index (F (1, 125) = 12.6, p < 0.01) and the relative dorsal fin area (F (1,125) = 53.5,

p < 0.01). However, no differences were found among the substrates for the values of the relative width of the caudal peduncle (F (1,125) = 0.08, p = 0.76) or for the values of the relative head width (F (1, 125) = 1.71, p = 0.19) (Fig. 4).

For *A. rivularis*, the ecomorphological attributes indicated by the discriminant analysis differed primarily for the leaf bank substrate when compared with the other substrates. In streams with a leaf bank substrate, we observed lower values for the relative head length (RHL) and the relative eye position (REP) as well as greater values for the mouth orientation (MO). Populations from streams with a predominantly gravel substrate showed increased values for the relative caudal fin area (RCFA). For the other attributes, differences occurred between pairs of substrates, but there were no clear patterns (Fig. 5).

DISCUSSION

Ecomorphology studies the relationship between morphology and ecology across individuals, populations, and communities, and it examines the evolutionary consequences of these relationships, which are primarily adaptive convergence and divergence (GATZ 1979, WINEMILLER 1991). Comparisons

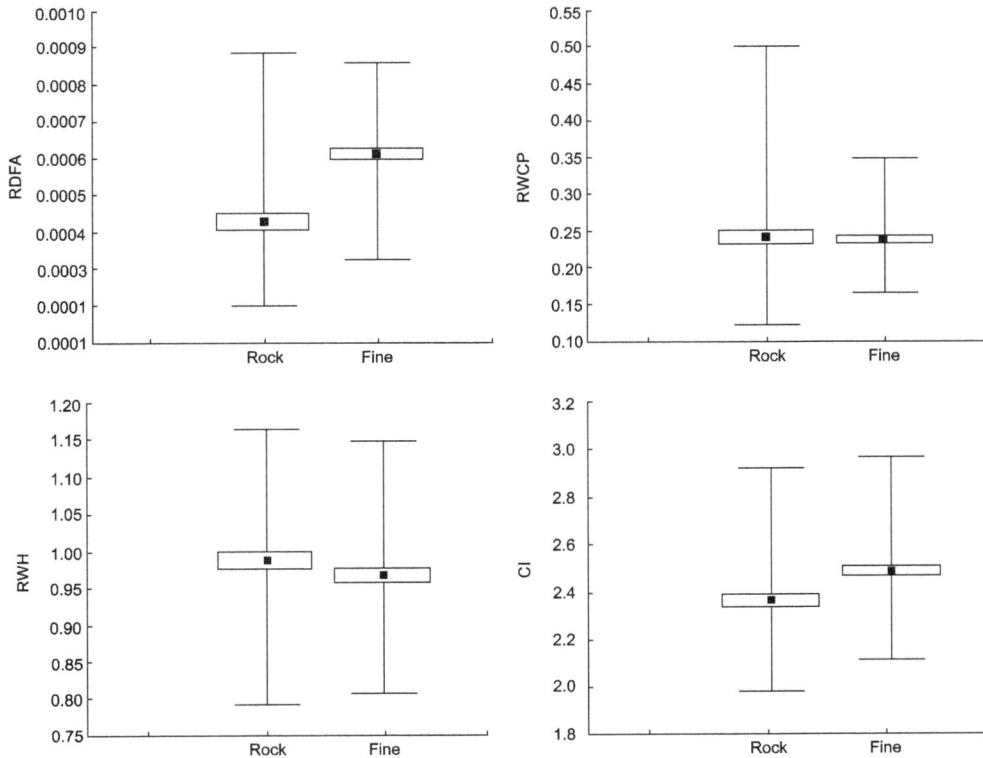

Figure 4. The means, standard deviations, and minimum and maximum values for the ecomorphological attributes of *Astyanax intermedius* present in streams with different substrates collected from Rio São Francisco streams in September 2010. CI (compression index), RWCP (relative width of caudal peduncle), RWH (relative width of head), and RDFA (relative dorsal fin area).

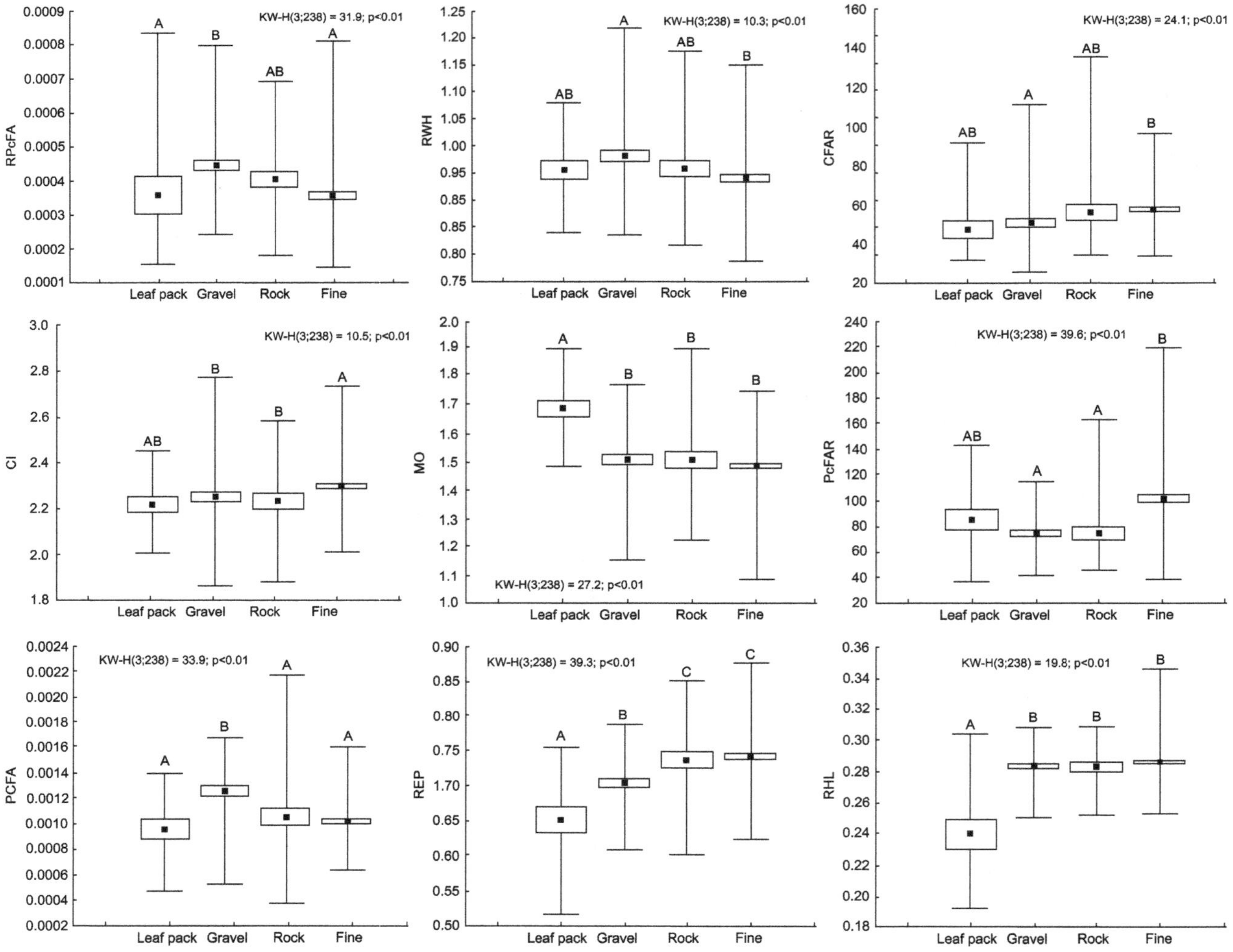

Figure 5. The means, standard deviations, and minimum and maximum values for the ecomorphological attributes of *Astyanax rivularis* present in streams with different substrates collected in Rio São Francisco streams in September 2010. CI (compression index), RHL (relative head length), RWH (relative width of head), RPE (relative eye position), CFAR (caudal fin aspect ratio), PcFAR (pectoral fin aspect ratio), RCFA (relative caudal fin area), RPcFA (relative pectoral fin area), MO (mouth orientation).

between related organisms occupying different habitats have been used as evidence of the role the environment plays in evolution and of the number of optimal adaptive solutions to a given habitat (CODY & MONEY 1978). Adaptive characteristics are those for which adaptive convergence occurs between different groups or for which adaptive divergence occurs within a group. Thus, these adaptations occur as evolutionary novelties for individuals and require minimal knowledge of the phylogenetic relationships between groups. The integration of phylogenetic information and ecomorphological analysis is the only way to identify cases of morphological and adaptive convergence and divergence, allowing the understanding of evolutionary patterns within communities (PERES-NETO 1999).

In the present study, we detected significant morphological differences among populations of *A. rivularis* and *A. intermedius* present in streams with different types of substrates. This finding confirmed the hypothesis that individuals from environments with different characteristics (i.e., velocity, depth, and substrate type) may have different ecomorphological patterns (LANGERHANS *et al.* 2003, NEVES & MONTEIRO 2003, LEAL *et al.* 2011, 2013). Our findings also support the idea that the morphology of individuals is shaped by different environments with different selection pressures, resulting in adaptations in body shape (LANGERHANS *et al.* 2003, JACKSON *et al.* 2001). However, in this study, the two species did not exhibit the same differences in morphological attributes across substrates.

Our study did detect ecomorphological differences between fish from rocky streams and fish from other substrates, with lower values observed for the compression index and the relative dorsal fin area for *A. intermedius* in rocky streams. These differences may be related to the faster water velocity in these streams. According to WATSON & BALON (1984), greater compression index values indicate laterally compressed fishes, which tend to reside in locations with a slow water velocity. The fin is responsible for the stability of the fish, and it is thought that dorsal fins with small areas are more effective in faster waters (GOSLINE 1971).

For *A. rivularis*, the attributes differed primarily between fish from the leaf bank substrate and fish from the other substrate types. Fish from the leaf bank substrate exhibited reduced values for the relative head length and the relative eye position as well as greater values for mouth orientation. The relative head length value is directly related to the size of the prey; therefore, larger values suggest predatory species of relatively large prey (GATZ 1979). Mouth orientation can reflect a species' foraging behavior, and the greater values here likely related to increased resource supply within the leaf bank, as many invertebrates are associated with this type of substrate. The importance of the substrate type for the composition of the stream benthic macroinvertebrate community is widely recognized in the literature and results from the differential availability of food resources and refuge from predation or flow disturbances (LIGEIRO *et al.* 2010). The importance of riparian vegetation inputs for stream substrates has also been emphasized by ANGERMEIER & KARR (1984) and JACKSON *et al.* (2001). These substrates locally modify the water flow and attract fish to these lower energy areas, in addition to being suitable areas for foraging, providing protection from predators, and serving as a substrate for aquatic invertebrates. Through these studies, accompanied by the significant results observed for the leaf bank substrate in our study, it is possible to highlight the ecological importance of the organic matter input in streams.

CASSATTI & CASTRO (2006) documented both a case of ecomorphological divergence between phylogenetically close Loricariidae species, indicating that the divergence occurred before their differentiation, and a case of ecomorphological convergence among Characidiinae and Parodontidae, which are phylogenetically distant groups. In our study, the two phylogenetically close species showed different ecomorphological attributes and thus represent adaptive divergence.

Considering that species morphology is associated with habitat use and resource availability (GATZ 1979), the loss and modification of habitats could directly affect the permanence of a species (CUNICO & AGOSTINHO 2006). Therefore, several aspects of physical habitat (i.e., stream size, channel gradient, habitat complexity and cover, channel morphology, substrate, and channel-riparian interactions) should be emphasized, as they might be impacted by human activities in different ways, resulting in habitat loss or homogenization (KAUFMANN *et al.*

1999). The loss of natural flow regimes have different impacts on the distribution of biodiversity in space and time, as well as on the functions and services provided by aquatic ecosystems (NILSSON *et al.* 2005).

We detected significant ecomorphological differences among the populations of *A. rivularis* and *A. intermedius* from habitats with different types of substrates. However, the two species did not show the same morphological differences depending on the type of substrate. These results confirmed the hypothesis that individuals from environments with different characteristics may have different ecomorphological patterns. Knowing that species morphology is associated with habitat use and available resources, the loss and modification of these factors may directly affect the permanence of a species and can cause a loss of morphological diversity.

ACKNOWLEDGMENTS

We thank our colleagues from Laboratório de Ecologia de Peixes (UFLA), NuVelhas (UFMG) and CEFET-MG for their help during fieldwork, Miriam A. de Castro for lab work and species identification, Diego Macedo for the map, Daniel Tregidgo, Hannah Griffiths, and Natalie Swan for English editing, IBAMA for collecting permits, and landowners for allowing access to their properties. Anonymous reviewers gave us important suggestions to improve the manuscript. This study was made possible by funding from Cemig, Fapemig, Capes, and CNPq. MAS and DCF received undergraduate scholarships, CGL received a doctoral fellowship, and PSP received a research fellowship from Fapemig and CNPq.

REFERENCES

ANGERMEIER, P. L. & J.R. KARR. 1984. Relationships between woody debris and fish habitat in a small warmwater stream. **Transactions of the American Fisheries Society 113**: 716-726.

BEAUMORD, A.C.; M. PETRERE. 1994. Fish communities of Manso River, Chapada dos Guimarães, MT, Brazil. **Acta Biologica Venezuelica152**: 21-35.

CASATTI, L.& R.M.C. CASTRO. 2006. Testing the ecomorphological hypothesis in a headwater riffles fish assemblage of the rio São Francisco, southeastern Brazil. **Neotropical Ichthyology 4** (2): 203-214.

CODY, M.L. & H.A. MOONEY. 1978. Convergence versus nonconvergence in Mediterranean-climate ecosystems. **Annual Review of Ecology and Systematics 9**: 265-321.

CUNICO, A.M. & A.A. AGOSTINHO. 2006. Morphological patterns of fish and their relationships with reservoirs hydrodynamics. **Brazilian Archives of Biology and Technology 49** (1): 125-134.

DOUGLAS, M.E.; W.J. MATTHEWS. 1992. Does morphology predict ecology? Hypothesis testing within a freshwater fish assemblage. **Oikos 65**: 213-224.

ELDREDGE, N. 1989. **Macroevolutionary Dynamics.** New York, McGraw-Hill, 226p.

FREIRE, A.G. & A.A. AGOSTINHO. 2001. Ecomorfologia de oito espécies dominantes da ictiofauna do reservatório de Itaipu (Paraná/Brasil). **Acta Limnologica Brasiliensia 13** (1): 1-9.

GATZ JÚNIOR, A.J. 1979. Ecological morphology of freshwater stream fishes. **Tulane Studies in Zoology and Botany 21:** 91-124.

GOSLINE, W.A. 1971. **Functional morphology and classification of teleostean fishes.** Honolulu, University of Hawaii, 208p.

GUILL, J.M.; C.S. HOOD & D.C. HEINS. 2003. Body shape variation within and among three species of darters (Perciformes: Percidae). **Ecology of Freshwater Fish 12:** 134-140.

HORA, S.L.1930. Ecology, bionomics and evolution of the torrential fauna, with special reference to the organs of attachment. **Philosophical Transactions of the Royal Society of London 28:** 171-282.

JACKSON, D.A.; P.R. PERES-NETO & J.D. OLDEN. 2001. What controls who is where in freshwater fish communities – the roles of biotic, abiotic, and spatial factors. **Canadian Journal of Fisheries and Aquatic Sciences 58:** 157-170.

KAUFMANN, P.R; P. LEVINE; E.G. ROBISON; C. SEELIGER & D.V. PECK. 1999. **Quantifying Physical Habitat in Wadeable Streams.** Washington, D.C., U.S. Environmental Protection Agency, 149p.

LANGERHANS, R.B.; C.A. LAYMAN; A.K. LANGERHANS & T.J. DEWITT. 2003. Habitat associated morphological divergence in two neotropical fish species. **Biological Journal of the Linnean Society 80:** 689-698.

LEAL, C.G.; N.T. JUNQUEIRA & P.S. POMPEU 2011. Morphology and habitat use by fishes of the Rio das Velhas basin in southeastern Brazil. **Environmental Biology of Fishes 90:** 143-157.

LEAL, C.G.; N.T. JUNQUEIRA; H.A. SANTOS & P.S. POMPEU. 2013. Variações ecomorfológicas e de uso de habitat em *Piabina argentea* (Characiformes, Characidae) da bacia do rio das Velhas, Minas Gerais, Brasil. **Iheringia, Série Zoologia, 103** (3): 222-231.

LIGEIRO, R.; A.S. MELO & M. CALLISTO. 2010. Spatial scale and the diversity of macroinvertebrates in a Neotropical catchment. **Freshwater Biology 55** (2): 424-435.

LIGEIRO, R.; R.M. HUGHES; P.R. KAUFMANN; D.R. MACEDO; K.R. FIRMIANO; W.R. FERREIRA; D. OLIVEIRA; A.S. MELO & M. CALLISTO. 2013. Defining quantitative stream disturbance gradients and the additive role of habitat variation to explain macroinvertebrate taxa richness. **Ecological Indicators 25:** 45-57.

MOTTA, P.J.; S.F. NORTON & J.J. LUCZKOVICH. 1995. Perspectives on the ecomorphology of bony fishes. **Environmental Biology of Fishes 44** (1-3): 11-20.

NEVES, F.M. & L.R. MONTEIRO. 2003. Body shape and size divergence among populations of Poecilia vivipara in coastal lagoons of south-eastern Brazil. **Journal of Fish Biology 63:** 928-941.

NILSSON, C.; C.A. REIDY; M. DYNESIUS & C. REVENGA. 2005. Fragmentation and flow regulation of the world's large river systems. **Science 308** (5720): 405-408.

OLIVEIRA, E.F.; E. GOULART; L. BREDA;C.V. MINTE VERA; L.R.S. PAIVA & M.R. VISMARA. 2010. Ecomorphological patterns of the fish assemblage in a tropical floodplain: effects of trophic, spatial and phylogenetic structures. **Neotropical Ichthyology 8** (3): 569-586.

PANKHURST, N.W.1989. The relationship of ocular morphology to feeding modes and activity periods in shallow marine teleosts from New Zealand. **Environmental Biology of Fishes 26:** 201-211.

PERES-NETO, P.R. 1999. Alguns métodos e estudos em Ecomorfologia de peixes de riacho, p. 209-236. *In*: E.P. CARAMASCHI; R. MAZZONI & P.R. PERES-NETO. **Ecologia de Peixes de Riachos.** Rio de Janeiro, PPGE/UFRJ, vol. 4.

POUILLY, M.; F. LINO; J.G. BRETENOUX & C. ROSALES. 2003. Dietary morphological relationships in a fish assemblage of the Bolivian Amazonian floodplain. **Journal of Fish Biology 62:** 1137-1158.

STATSOFT INC. 2004. STATISTICA: data analysis software system, version 10. Available online at: www.statsoft.com [Accessed: 17.VI.2013]

WAINWRIGHT, P.C. 1994. Functional morphology as a tool in ecological research, p. 42-59. *In*: P.C. WAINWRIGHT & S.M. REILLY (Eds). **Ecological morphology: integrative organismal biology.** Chicago, University of Chicago Press.

WATSON, D.J. & E. BALON. 1984. Ecomorphological analysis of taxocenes in rainforest streams of northern Borneo. **Journal of Fish Biology 25:** 371-384.

WILLIS, S.C.; K.O. WINEMILLER & H. LOPEZ-FERNANDES. 2005. Habitat structural complexity and morphological diversity of fish assemblages in a Neotropical floodplain river. **Oecologia 142:** 284-295.

WIKRAMANAYAKE, E.D. 1990. Ecomorphology and biogeography of a tropical stream fish assemblage: evolution of assemblage structure. **Ecology 71:** 1756-1764.

WINEMILLER, K.O. 1991. Ecomorphological diversification in lowland freshwater fish assemblages from five biotic regions. **Ecological Monographs 61** (4): 343-365.

14

Gross morphology and ultrastructure of the female reproductive system of *Diaphorina citri* (Hemiptera: Liviidae)

Fabio Cleisto Alda Dossi & Fernando Luis Cônsoli

Laboratório de Interações em Insetos, Departamento de Entomologia e Acarologia, Escola Superior de Agricultura "Luiz de Queiroz", Universidade de São Paulo, Avenida Pádua Dias 11, 13418-900 Piracicaba, SP, Brazil.
E-mail: fdossi@usp.br; fconsoli@usp.br

ABSTRACT. The morphological traits of the female reproductive system of *Diaphorina citri* were examined in detail. *Diaphorina citri* has telotrophic ovaries with ovarioles organized as a "bouquet", displaying a rudimentary terminal filament and a syncytial tropharium. The vitellarium carries a single growing oocyte at each maturation cycle, which is connected with the tropharium by a nutritive cord. Morpho-functional changes occur during oocyte development, mainly during mid to late vitellogenesis. Morphological events such as the patency of the follicular cells and the intense traffic of vesicles through para- or intracellular processes, suggest a possible route for endosymbiont invasion of *D. citri* reproductive tissues. Similar events have been demonstrated to be involved in the process of ovariole invasion by endosymbionts in other sternorrhynchans that share reproductive traits with psyllids.

KEY WORDS. Accessory structures; Asian citrus psyllid; oogenesis; ovary.

Insect ovaries are differentiated as panoistic and meroistic (BÜNING 1994). In the former all of the oogonia differentiate as oocytes, while in the latter part of the germline cells develop into trophocytes. Trophocytes are specialized cells that supply oocytes with macromolecules and organelles during the previtellogenic growth stage (KING & BÜNING 1985, BÜNING 2006). A typical insect ovariole is subdivided from apex to base into a fine terminal filament, tropharium (trophic chamber), vitellarium and a pedicel (ovariolar stalk) that opens into the calyx of a lateral oviduct (BÜNING 1994). The number of ovarioles per ovary varies considerably among different taxa, ranging from a single ovariole in aphids (BÜNING 1985, 1994) to thousands in termite queens and coccids (BÜNING 1994, GILLOT 2005). Auchenorrhyncha may have from one to 15 ovarioles per ovary, whereas members of Heteroptera commonly have seven ovarioles per ovary (BÜNING 1994, LALITHA *et al.* 1997, HODIN 2009). Plasticity in the number of ovarioles has been documented for aphids and scale insects in Sternorrhyncha, and the ovary of psyllids can contain up to 100 ovarioles (BÜNING 1994, HODIN 2009). The structure of the reproductive apparatus and the process of oogenesis in Heteroptera (Hemiptera) have been extensively studied (LALITHA et al. 1997, CAPERUCCI & CAMARGO-MATHIAS 2006), but information for the remaining suborders is scarce (BUNING 1985, SZKLARZEWICZ et al. 2008, 2013). In order to increase the overall knowledge on the reproductive system of sternorrhynchans, we characterized the morphology and ultrastructure of the female reproductive apparatus of *Diaphorina citri* Kuwayama, 1908 during oogenesis.

MATERIAL AND METHODS

Diaphorina citri was maintained at controlled conditions ($28 \pm 2°C$, $60 \pm 10\%$ UR, photophase 14 hours) using orange jasmine seedlings, *Murraya exotica* Linnaeus, 1771 (Sapindales: Rutaceae), as a host (TSAI & LIU 2000, NAVA *et al.* 2007). Fifth instars and adult females were sampled at different developmental stages for dissection of their reproductive apparatus.

Last instar, newly-emerged, and reproductively active females of *D. citri* (WENNINGER & HALL 2007), at the previtellogenic and vitellogenic stages (DOSSI & CÔNSOLI 2010), were dissected in insect saline (3 mM $CaCl_2.2H_2O$, 182 mM KCl, 46 mM NaCl, 10 mM Tris base, pH 7.2) (CSH PROTOCOLS 2007). Ovaries were immediately transferred to the proper fixative solution for further processing.

Dissected ovaries were fixed in AFATD solution (75 mL 96% ethanol, 10 mL 40% formaldehyde, 5 mL acetic acid, 10 mL dimethylsulfoxide, 1 g trichloroacetic acid) (MARTÍNEZ 2002) for two hours at room temperature, hydrated in a decreasing series of ethanol (96, 80, 70, 50 and 30% – five minutes each) and distilled water (five minutes), hydrolyzed in 2.5 N hydrochloric acid (five minutes), stained with Schiff reagent (30 minutes), rinsed in distilled water (five minutes) and counterstained with light green (one minute) (MARTÍNEZ 2002). Afterwards, samples were rinsed in distilled water (five minutes), absolute ethanol (2x – 10 minutes/each), diaphanized in xylol (2x – 10 minutes/each), and mounted in Entelan® (Merck). Samples were examined with a Zeiss Axiostar Plus light microscope.

Histological analysis. Dissected ovaries were fixed in PFA-PBTw (5% paraformaldehyde, 1% Tween20 in 0.1% phosphate buffer solution (w/v), pH 7.2) for 24 hours, rinsed in the same buffer (3x – 10 minutes) and dehydrated in a graded series of ethanol (1x – 30, 50, 70 and 90%; 3x – 100% – 10 minutes/each). Samples were then embedded in ethanol:historesin (Leica Historesin) solution (1:1) for 24 hours, followed by 24 hours at 4°C in pure historesin, and polymerized at room temperature for 48 hours. Semithin sections (1-2 µm) were stained in 1% toluidine blue + 1% sodium borate solution or Azan Heidenhain-Mallory (0.5% aniline blue, 1% orange G, 3% acid fuchsin, 1% phosphotungstic acid, w/v) (Behmer et al. 2003) for one minute. After staining, sections were quickly rinsed in distilled water, and hot plate dried at 45°C for 20 minutes followed by mounting with Entelan® (Merck) and examination with a Zeiss Axioskop 2 light microscope.

Fluorescence microscopy. For fluorescence microscopy, the dissected ovaries were fixed in PFA-PBTw solution and embedded in historesin as described before. Semithin sections (1 µm) were stained with 1.5 µg DAPI (4', 6'-diamidine-2-phenylindole) in mounting medium Vectashield® (DAPI, Sigma Chemical Co.) for 15 minutes to detect DNA, followed by 15 minutes incubation in rhodamine and phalloidin (Sigma Chemical Co.) for the detection of the actin filaments. Samples were observed under an Olympus BX51 epifluorescence microscope.

Transmission electron microscopy (TEM). Ovaries fixed in Karnovski fixative (3% glutaraldehyde, 3% paraformaldehyde, in 50 mM cacodylate buffer, 5 mM CaCl₂, pH 7.2) for 24 h, were rinsed in the same buffer (3x – 10 minutes), post-fixed in 1% osmium tetroxide (OsO₄) in 50 mM cacodylate buffer (60 minutes) and counterstained in bloco with 0.5% uranyl acetate for 12 hours. Samples were then dehydrated in a graded series of acetone, followed by embedding (EMBed 812, Electron Microscopy Sciences) (24 hours) and polymerization at 60°C for 24 hours. Ultrathin sections (50-70 nm) were mounted on copper grids, stained in 3% uranyl acetate followed by 1% lead citrate, and analyzed with a Zeiss EM900 transmission electron microscope.

Scanning electron microscopy (SEM). Ovaries fixed as described for TEM were dehydrated in graded series of acetone and critical point dried (CPD-030 Balzers, BAL-TEC). Afterwards, ovaries were mounted on stubs and gold sputter-coated (SCD-050 Sputter Coater, BAL-TEC). Image acquisition and analysis were carried out with a Zeiss LEO 435 VP scanning electron microscope.

RESULTS

Diaphorina citri has meroistic telotrophic ovaries formed by nearly 50 ovarioles arranged in a "bouquet" (Figs 1, 2, and 4). Ovaries are located ventro-laterally in the median region of the abdomen, just below the bacteriome, an organ that har-

bors the symbiotic associated bacteria (not shown) (for review, see Baumann 2005). They are surrounded by fat body tissue and a dense network of tracheae in both immature and adult stages. In newly-emerged adults, ovaries are small and all oocytes are at the previtellogenic stage (Fig. 2). But in mated females, they are fully developed and have oocytes at all stages of maturation (Figs 3 and 4).

The pedicel of the ovarioles opens into the apical bulb of their respective lateral oviducts (Figs 1, 3, and 4), which will merge forming the common oviduct (Fig. 1). The oviducts are formed by a single epithelium of columnar cells underlying a cuticular intima, which is surrounded by a thick basal lamina and muscle bundles (Fig. 22).

Three accessory structures are laterally connected to the common oviduct: a paired accessory gland, a spermatheca and a colleterial gland (Fig. 1).

A pair of tubular accessory glands is placed on opposite sides of the upper part of the common oviduct, near to the lateral oviducts. Glands are slightly dilated at their distal region, but bear a small sac-like structure at their base (Figs 1, 5, and 6). These glands are composed of large secretory cells, which contain a nucleus at the base of the cell and a conspicuous nucleolus. Mitochondria are densely distributed, mainly in the distal region of the cell, which has long microvilli (Fig. 7).

The sacciform reservoir of the spermatheca is connected to the common oviduct at a more distal position than the accessory glands (Figs 1 and 8). Both are formed by a single layer of epithelial cells, but only the cells of the reservoir have secretory activity. Their cytoplasm is filled with endoplasmic reticulum, mitochondria, and microvilli facing the gland lumen. These cells are surrounded by a basal lamina and a spermathecal sheath, a single layer of dome-shaped cells filled with muscle fibers and mitochondria (Fig. 9).

The colleterial gland has a sac-like shape and is located at the base of the ovipositor sclerites and above the accessory glands (Fig. 10). It is connected to the vagina by a long trachea-shaped duct made up of cuticular rings inserted into a single-layer epithelium. The colleterial gland has a single layer of secretory cells characterized by the presence of endoplasmic reticulum and vesicles that temporarily stores secreted molecules. The vesicle contents are unloaded into a network of collection channels that opens in the glandular lumen (Fig. 11).

Four distinct regions were identified in the ovarioles of D. citri (Fig. 12): I) a rudimentary terminal filament, II) an ovoid trophic chamber (Figs 15 and 16) that communicates with the growing oocyte into the III) vitellarium (Figs 15 and 16) by means of a cytoplasmic projection, the nutritive cord (Figs 16 and 20). At the base of the vitellarium, there is a IV) ovariolar stalk (pedicel) of a tubular aspect (Figs 3, 12, and 13) that allows the passage of the egg downwards to the calyx of the adjacent lateral oviduct (Fig. 22) after the completion of egg development. Development of the D. citri ovarioles is metachronic, with those at the margins maturing before those located internally (Fig. 14). The vitellarium has only one oocyte developing per maturation cycle, but as soon as the basal oo-

cyte choriogenesis has been completed, a new oocyte at a very early previtellogenic stage can be seen at the base of the trophic chamber (Fig. 14, inset).

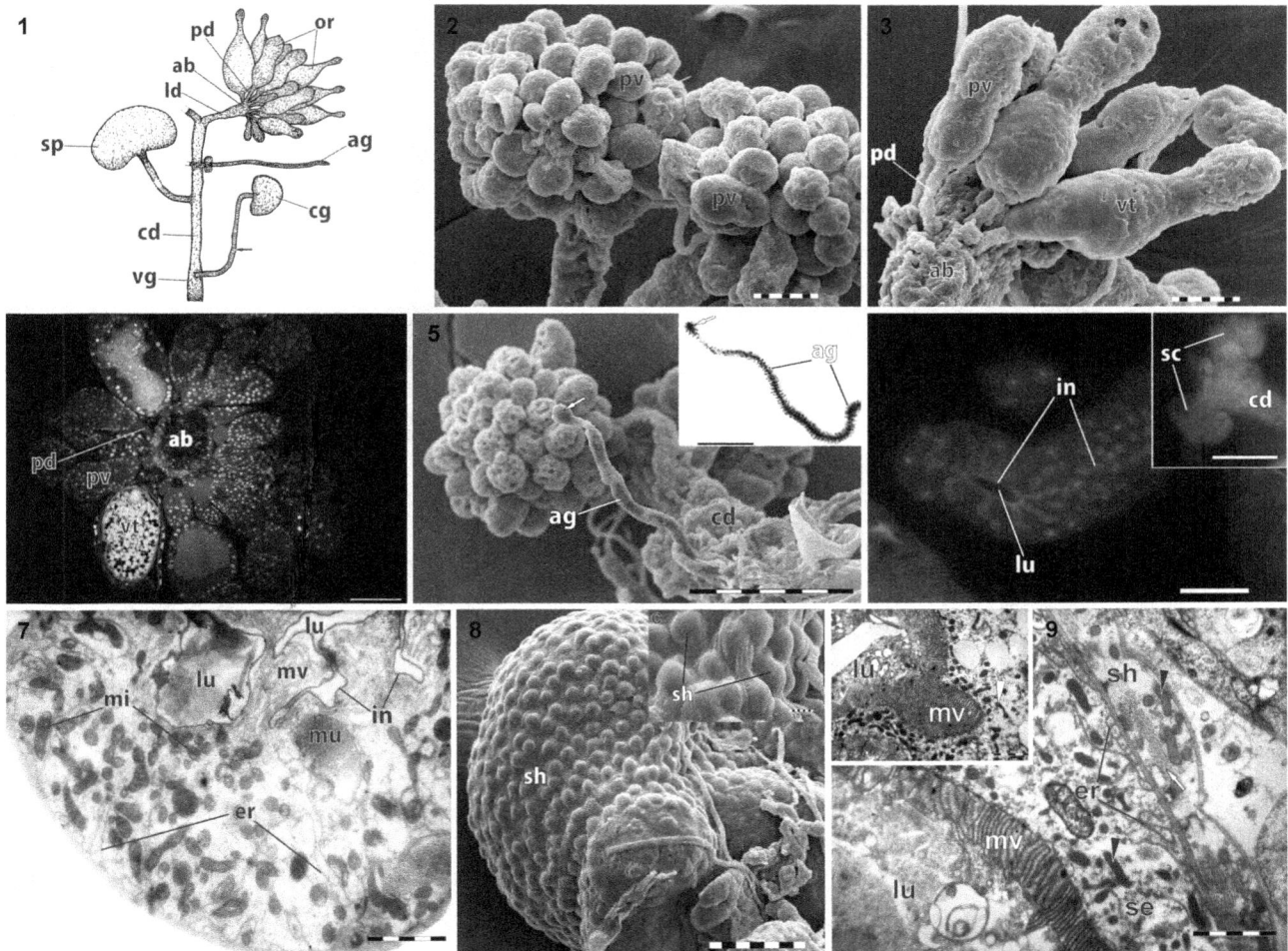

Figures 1-9. *Diaphorina citri*. (1-4) Gross morphology of the female reproductive system: (1) schematic diagram showing the ovary and its ovarioles (or) arranged in a bouquet, inserted by its pedicel (pd) in the apical bulb (ab) at the tip of the corresponding lateral oviduct (ld) which converges with the common oviduct (cd). A pair of accessory glands (ag) a spermatheca (sp) and a colleterial gland (cg) are observed. The colleterial gland (cg) is attached to the common oviduct by a long duct (setae). (2) Previtellogenic (pv) ovarioles of immature ovaries. (3-4) Vitellogenic (vt) and previtellogenic (pv) ovarioles of a mature ovary. (2-3) Scanning electron microscopy; (4) cross-section of ovary at the apical bulb region stained with DAPI (4′-6-diamidino-2-phenylindole), epifluorescence microscopy. (5-7) Accessory gland of *Diaphorina citri*: (5) view of the rounded tip (setae) of the accessory gland (ag) and the intricate network of channels (inset) as seen by whole-mount preparation. (6) Gland lumen (lu) and the intricate network of folded channels underlying a cuticular intima (in) in a tangential section. Whole-mount view (inset) of the sac-like structures (sc) at the base of the gland, placed in apposition to the common oviduct (cd). (7) Cytoplasm filled with mitochondria (mi), endoplasmic reticulum (er), muscle fibers (mu), and microvilli (mv) underlying a cuticular intima (in) at the lumen (lu) interface in a cross section. (5) scanning electron microscopy; (5-inset and 6) autofluorescence as seen by epifluorescence microscopy; (7) transmission electron microscopy. (8-9) Spermatheca: (8) Surface view of the spermathecal sheath (sh) represented in a section in (9) as a layer of cells (sh) containing bundles of muscle fibrils (setae) and mitochondria (arrowhead). Spermathecal sheath is covering a secretory epithelium (se) (er, endoplasmic reticulum; mv, folded network of microvilli; lu, lumen). (8) scanning electron microscopy; (9) transmission electron microscopy. Scale bars: 2 = 30 μm; 3 = 20 μm; 4 = 100 μm; 5 = 100 μm, inset = 50 μm; 6 = 20 μm, inset = 20 μm; 7 = 2 μm; 8 = 100 μm, inset = 10 μm; 9 = 2 μm, inset = 2 μm.

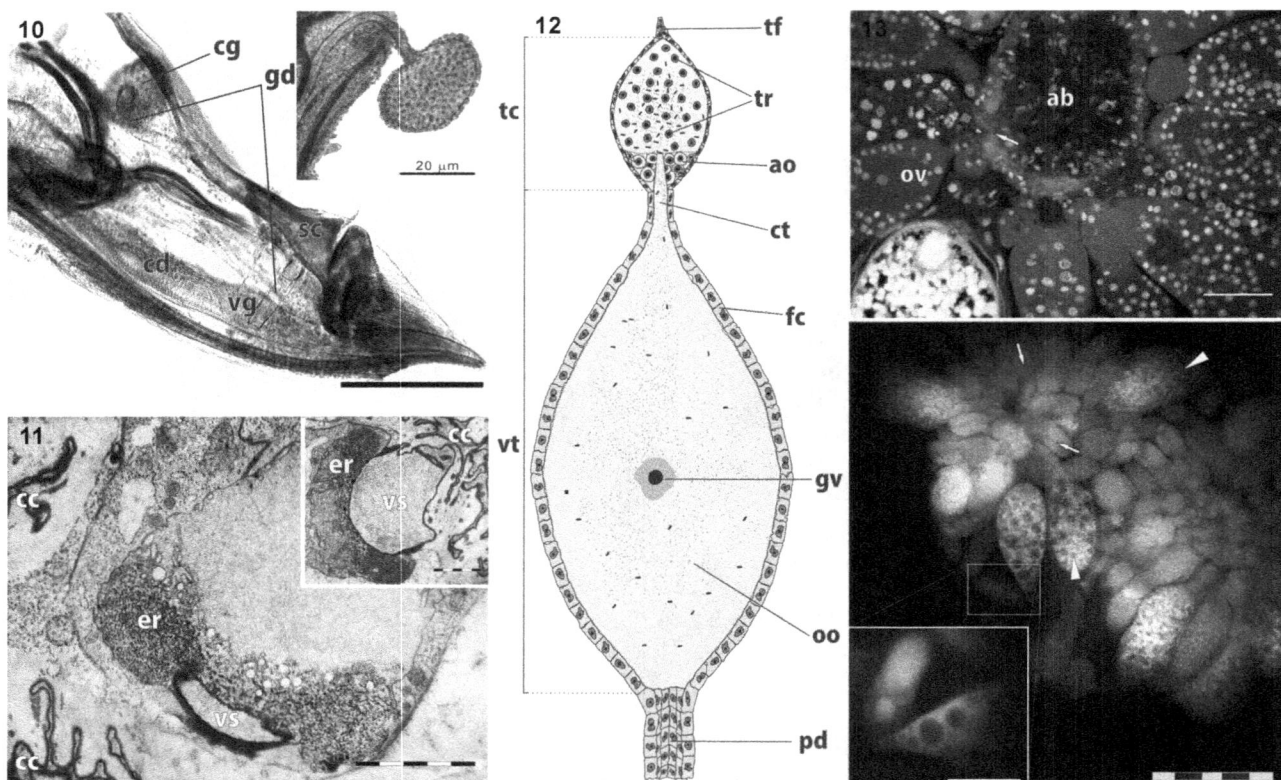

Figures 10-14. *Diaphorina citri*. (10-11) Colleterial gland: (10) general view of the colleterial gland (cg) placed at the distal region of the ovipositor sclerites (sc) and connected through its long and trachea-like duct (gd) to the vagina (vg) (cd, common oviduct); (11) secretory cell filled with endoplasmic reticulum (er) unloading its vesicle (vs) contents (see inset) in an intricate network of cuticular collector channels (cc). (10) bright field microscopy; (11) transmission electron microscopy. (12-14) Ovariole: (12) Schematic view of an ovariole: terminal filament (tf); trophic chamber (tc) containing trophocyte nuclei (tr) inserted into the syncytial tropharium and arrested oocytes (ao); vitellarium (vt) with follicular cells (fc) covering the growing oocyte (oo) that presents a conspicuous germinal vesicle (gv) and communicates with the tropharium through the trophic chord (ct). The pedicel (pd) is placed at the base of the ovariole; (13) Pedicel is connected (setae) to the apical bulb (ab); (14) Metachronic development of ovarioles, with well-developed maturing oocytes located at the periphery of the ovaries (arrowhead), while poor-developed maturing oocytes are internally located (setae). (13-14) epifluorescence microscopy. Scale bars: 10 = 20 μm, inset = 20 μm; 11 = 2 μm, inset = 2 μm; 13 = 50 μm, 14 = 300 μm, inset = 50 μm.

During the previtellogenic stage the follicular cells are columnar or cuboidal, changing their shape during ongoing maturation, becoming rectangular during vitellogenesis and elongated during the late choriogenic stage (Figs 17-18 and 21). The increase in the number of cytoplasmic organelles in the follicular cells and the accumulation of yolk was noticeable once vitellogenin uptake started through a paracellular route. Additionally, the large number of vesicles within the cell cytoplasm is also suggestive of the acquisition and transport of nutrients by pinocytosis (Fig. 18).

The trophic chamber of *D. citri* is surrounded by a single layer of flattened somatic cells which is covered by a tunica propria. The thickness of this layer of cells is variable along the periphery of the tropharium as the cells become very elongated and thin on the edges away from the nucleus (Fig. 19).

The trophocytes become devoid of plasmic membrane before the previtellogenic stage, originating a syncytial tropharium. There is an accumulation of electron-dense vesicles and multivesicular bodies, mainly in the central region of the trophic chamber. The previtellogenic oocyte accumulates the cytoplasmic components produced in the tropharium, transferred through the nutritive cord (Fig. 20), in which we did not observed microfilaments.

There are about 10 quiescent oocytes (i.e. oocytic cells) located at the base of the trophic chamber in each ovariole, but only one matures at a time (Fig. 15). During the previtellogenesis, the oocyte becomes enlarged due to the incorporation of cytoplasmic components (e.g., organelles, proteins and RNA) produced in the tropharium. The mid-previtellogenic stage is marked by the moderate secretory activity of the follicular cells, characterized by the presence of small vesicles between the plasmatic membrane and the cytoplasm of the oocyte. The follicular cells reach their maximal activity during the mid to late vitellogenesis. During this stage of oocyte development, their secretory activity is characterized by the abundance of vesicles of different sizes, probably due to the processing of macromolecules (mainly proteins and lipids) from hemolymph. Glycogen deposits were detected

by transmission electron microscopy at the perivitelline space, suggesting that such elements are thus added to the oocyte. The oocyte reaches its maximum size at the end of the vitellogenic stage, when the surrounding follicular cells become elongated and thin.

During late choriogenesis, the chorion (Fig. 21) is observed as a result of the accumulation of precursors secreted by the follicular cells into the perivitelline space.

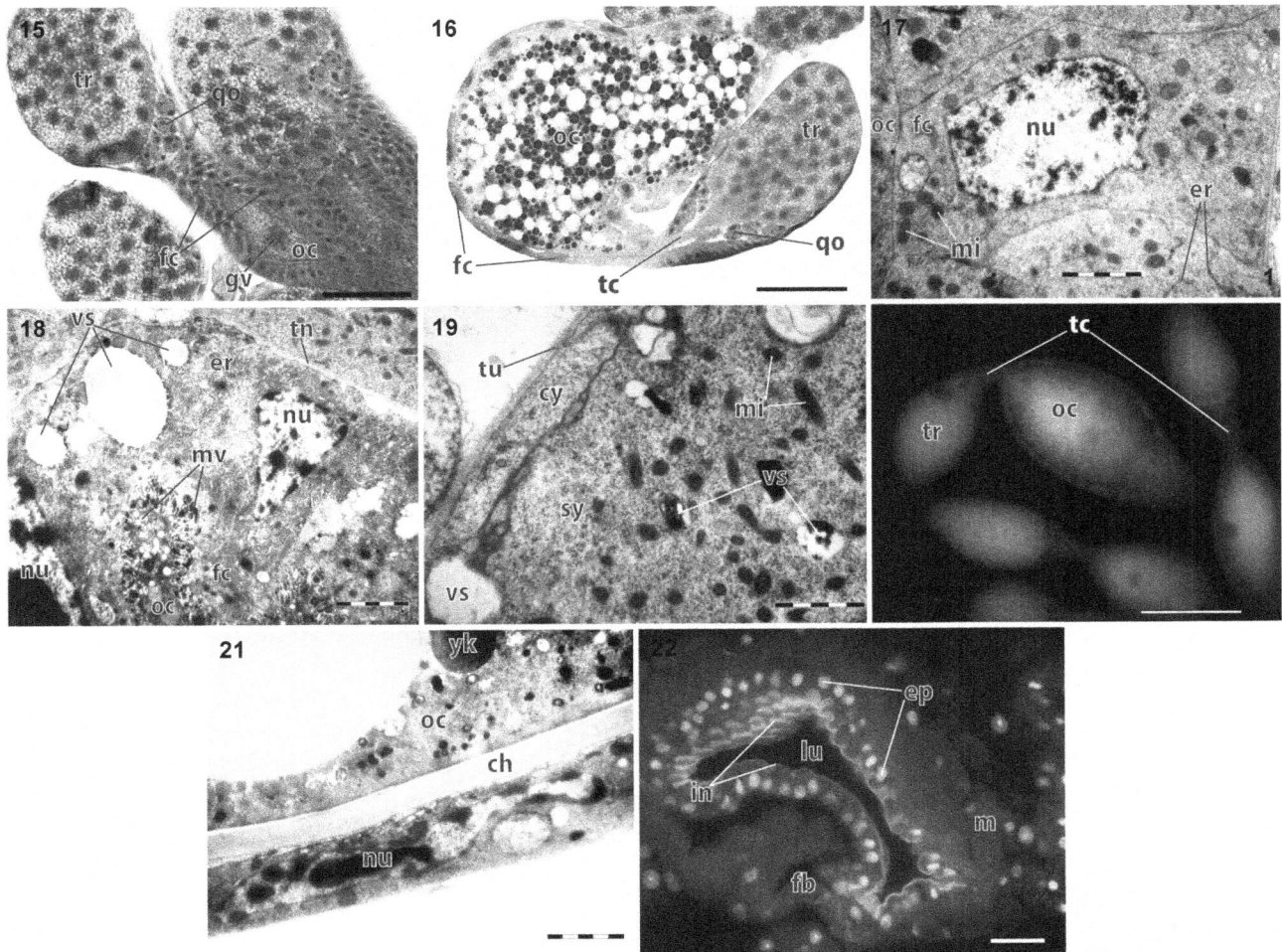

Figures 15-22. Ovariole maturation in *Diaphorina citri*: (15) previtellogenic and (16) vitellogenic ovarioles – tropharium (tr), quiescent oocytes (qo), nutritive cord (tc), growing oocyte (oc), germinal vesicle (gv) and follicle cells (fc). (17-18) follicle cells from previtellogenic (17) and vitellogenic (18) ovarioles. The distinct metabolic condition of the cells in each stage is revealed by the difference in the abundance of organelles, such as mitochondria (mi), endoplasmic reticulum (er) and secretory vesicles (vs) at the interface (mv) between follicular cells (fc) and oocyte (oc) during the accumulation of yolk. Note the presence of the tunica propria (tn) at the distal region of the follicular cell. (19) partial view of tropharium showing the thin, elongated epithelial cells (tu, tunica propria, vs, vesicle, mi, mitochondria, sy, syncytium). (20) ovarioles at different stages of maturation showing the nutritive cord (tc) (oc, oocyte, tr, tropharium). (21) mature oocyte revealing the elongated format of the follicle cell (nu, nucleus, ch, chorion, oc, oocyte cytoplasm, yk, yolk granules). (22) cross-section of the common oviduct showing the epithelium (ep) underlying a folded cuticular intima (in) (lu, lumen, fb, dense fibrous layer, m, muscle bundles), (15-16), bright field microscopy; (17-19 and 21) transmission electron microscopy; (20 and 22) epifluorescence microscopy. Scale bars: 15 = 30 µm; 16, 20 = 50 µm; 17, 18, 21 = 2 µm; 19 = 1 µm; 22 = 20 µm.

DISCUSSION

The meroistic telotrophic ovaries of *D. citri* are common to other Sternorrhyncha (Büning 1994, Michalik *et al.* 2013). The mature ovaries of this species, in which the ovarioles lack a functional terminal filament, are similar to those of some Aphidomorpha and Aleyrodomorpha (Büning 1985, Szklarzewicz & Moskal 2001). However, the types and localization of the accessory structures are variable from previously studied species (Lococo & Hubner 1980, Stacconi & Romani 2011, Sturm 2012, Ma *et al.* 2013). The occurrence and location of accessory structures may differ among species as they are highly specialized secretory organs involved in the synthesis of molecules related to several aspects of reproduction (sperm storage and nourishment, egg laying, among others). The synthesis and release of their contents are often linked to time-dependent physiological events during the reproductive stage (Sturm & Pohlhammer 2000, Gillott 2002, Klowden 2007). Although the accessory structures are frequently linked to the basal region of the common oviduct (Büning 1994), they are placed in different regions of the common oviduct in *D. citri*. One exception is the duct of the colleterial gland, which is linked to the distal part of the vagina.

The number of ovarioles of *D. citri* is much greater than in aphids (1-11) (Couchman & King 1979, Büning 1985) and aleyrodids (5-15) (Szklarzewicz & Moskal 2001), indicating that this trait is highly variable among Sternorrhyncha. The metachronic development observed in the ovarioles of *D. citri* has also been described for aphids, aleyrodids, scale insects (Couchman & King 1979, Büning 1985, 1994), and some Diptera and Psocoptera (Büning 1994).

The structure of the ovarioles of *D. citri* is very similar to that of aleyrodids (Büning 1994). The arrangement of the germline cells (cystocytes) in rosettes during the development of the trophic chamber is a relatively common trait in psyllids, aphids, aleyrodids (King & Büning 1985, Büning 1994), and scale insects (Szklarzewicz 1997).

We did not observe bundles of microfilaments into the nutritive cord of *D. citri*, as reported for other psyllids (Büning 1994), but mechanisms based on ionic gradients and on osmotic pressure are also known to be involved in the translocation of cytoplasmic components from the trophic chamber to the growing oocyte (Huebner & Diehl-Jones 1993, Telfer & Woodruff 2002).

The ultrastructural characteristics of the inner epithelial sheath of the trophic chamber of *D. citri* resemble those in aphids (Michalik *et al.* 2013). However, the syncytium of the tropharium of *D. citri* lacks a trophic core, differently from what is observed in aphids and heteropterans (Brough & Dixon 1989, Szklarzewics *et al.* 2000).

In *D. citri*, the activity of the nuclei in the tropharium is evident by the presence of "*nuage*-like", electron-dense agglomerates facing the nuclear envelope (as visualized by transmission electron microscopy), and is an indication of the nucleo-cytoplasmic transference of substances (Davenport 1976, Huebner & Diehl-Jones 1993). The accumulation of protein-filled vesicles, commonly associated with the Golgi complex and mitochondria in the tropharium, points to the selective uptake of molecules from hemolymph, and synthesis of morphogens (Couchman & King 1979). The *nuage* can be a ribonucleoprotein complex, as it shares similar morphology with different organisms. Nuages are frequently seen as small patches of dense granular material. They can be associated with mitochondrial clusters or lie adjacent to the nuclear envelope. Larger accumulations of dense material can also occur free in the cytoplasm (Eddy 1976). The major function of nuage is to maintain genome stability by repressing the expression of selfish genetic elements via small interference RNA-mediated gene silencing (for review, see Eddy 1976, Kim Lim & Kai 2007, Kim Lim *et al.* 2013). Furthermore, the presence of multivesicular structures in *D. citri*, which are similar to the cytolysomes dispersed in the syncytium, suggests that the tropharium may play a role in molecule processing and organelle resorption. The basal location of the arrested oocytes in the trophic chamber of *D. citri* follows the common pattern described for telotrophic ovaries (Büning 1994, Michalik *et al.* 2013), with only one oocyte developing per reproductive cycle. This reproductive strategy is also observed in *Megoura viciae* Buckton, 1876 (Homoptera: Aphididae) (Brough & Dixon 1989) and *Adelges laricis* Vallot, 1836 (Hemiptera: Adelgidae) (Szklarzewicz *et al.* 2000), and may be an adaptation to space or nutritional limitations, given that other insects have the capacity to produce several eggs per cycle (Lalitha *et al.* 1997, Szklarzewicz *et al.* 2008, Winnick *et al.* 2009).

The follicular cells surrounding the growing oocytes in the telotrophic ovarioles of *D. citri* have an important role during vitellogenesis, as they participate in the transport of molecules from hemolymph, and actively synthesize and incorporate macromolecules in the oocyte (Huebner & Anderson 1972, Raikhel & Dhadialla 1992). The occurrence of pinocytic vesicles in the follicular cells of *D. citri* was observed even during choriogenesis. Nevertheless, the late occurrence of molecule transport has been seldom indicated (e.g., Beams & Kessel 1969, Cruickshank 1972).

The accessory glands of the reproductive apparatus of insects are associated with the secretion of substances that participate in fertilization and oviposition (Wheeler 2003). The release of the electron-lucent secretions from the accessory gland of *D. citri*, may be linked to time-dependent physiological mechanisms, such as egg laying and fixation onto the substrate, as reported elsewhere (Gillott 2002, Klowden 2007, De Santis *et al.* 2008).

We provided here the first overview of the ultramorphology of the female reproductive system of *D. citri*. We believe that this information will be useful in furthering our understanding of the reproductive biology of this psyllid. Our data will also support future investigations on other reproductive traits of the Asian citrus psyllid and on the vertical transmission of symbiotic bacteria associated with this insect during its reproductive period.

ACKNOWLEDGEMENTS

The authors are grateful to the Centro de Microscopia Eletrônica Aplicada à Agricultura (NAP/MEPA), of the Escola Superior de Agricultura Luiz de Queiroz – USP, where ultrastructural studies were carried out. Authors thank FAPESP for providing a fellowship to FCAD (Processo FAPESP 2006/59300-0) and research funds to FLC (Processo FAPESP 2004/14215-0 and 2006/54792-1). FLC also thanks FUNDECITRUS for additional funding.

REFERENCES

BAUMANN, P. 2005. Biology of bacteriocyte-associated endosymbionts of plant sap-sucking insects. **Annual Review of Microbiology** 59: 155-189. doi: 10.1146/annurev.micro.59.030804.121041.

BEAMS, H.W. & R.G. KESSEL. 1969. Synthesis and deposition of oocyte envelopes (vitelline membrane, chorion) and the uptake of yolk in the dragonfly (Odonata: Aeschnidae). **Journal of Cell Science** 4: 241-264.

BEHMER, O.A.; E.M.C. TOLOSA; A.G. FREITAS NETO & C.J. RODRIGUES. 2003. **Manual de técnicas para histologia normal e patológica**. Barueri, Ed. Manole.

BROUGH, C.N. & A.F.G. DIXON. 1989. Follicular sheath (ovarian sheath) structure in virginoparae of the vetch aphid, *Megoura viciae* BUCKTON (Homoptera: Aphididae). **International Journal of Insect Morphology and Embryology** 18: 217-226. doi: 10.1016/0020-7322(89)90029-9.

BÜNING, J. 1985. Morphology, ultrastructure, and germ cell cluster formation in ovarioles of aphids. **Journal of Morphology** 186: 209-221. doi: 10.1002/jmor.1051860206.

BÜNING, J. 1994. The ovary of Ectognatha, the Insecta s.str., p. 281-299. *In*: J. BÜNING (Ed.). **The insect ovary: ultrastructure, previtellogenic growth and evolution**. London, Chapman & Hall.

BÜNING, J. 2006. Ovariole structure supports sistergroup relationship of Neuropterida and Coleoptera. **Arthropod Systematics and Phylogeny** 64: 115-126.

CAPERUCCI, D.; M.I. CAMARGO-MATHIAS. 2006. Ultrastructural study of the ovary of the sugarcane spittlebug *Mahanarva fimbriolata* (Hemiptera). **Micron** 37: 633-639. doi: 10.1016/j.micron.2006.02.002.

COLD SPRING HABOUR PROTOCOLS 2007. *Drosophila* **ringer's solution**. Available online at: http://cshprotocols.cshlp.org/cgi/content/full/2007/7/pdb.rec10919 [Accessed: 16/II/2014].

COUCHMAN, J.R. & P.E. KING 1979. Germarial structure and oogenesis in *Brevicoryne brassicae* (L.) (Hemiptera: Aphidae).

International Journal of Insect Morphology and Embryology 8: 1-10. doi: 10.1016/0020-7322(79)90002-3.

CRUICKSHANK, W.J. 1972. Ultrastructural modifications in the follicle cells and egg membranes during development of flour moth oocytes. **Journal of Insect Physiology** 18: 485-498.

DAVENPORT, R. 1976. Transport of ribosomal RNA into the oocytes of the milkweed bug, *Oncopeltus fasciatus*. **Journal of Insect Physiology** 22: 925-926. doi: 10.1016/0022-1910(76)90072-X.

DE SANTIS, F.; E. CONTI; R. ROMANI; G. SALERNO; F. PARILLO & F. BIN. 2008. Colleterial glands of *Sesamia nonagrioides* as a source of the host-recognition kairomone for the egg parasitoid *Telenomus busseolae*. **Physiological Entomology** 33: 7-16. doi: doi:10.1111/j.1365-3032.2007.00593.x.

DOSSI, F.C.A. & F.L. CÔNSOLI. 2010. Desenvolvimento ovariano e influência da cópula na maturação dos ovários de *Diaphorina citri* Kuwayama (Hemiptera: Psyllidae). **Neotropical Entomology** 39: 414-419. doi: 10.1590/S1519-566X2010000300015.

EDDY, E.M. 1976. Germ plasm and the differentiation of the germ cell line. **International Review of Cytology** 43: 229-280.

GILLOTT, C. 2002. Insect accessory reproductive glands: Key players in production and protection of eggs, p. 32-59. *In*: M. HILKER & T. MEINERS (Eds). **Chemoecology of insect eggs and egg deposition**. Berlin, Blackwell Verlag.

GILLOT, C. 2005. **Entomology**. Berlin, Springer, 3rd ed., 835p.

HODIN, J. 2009. She shapes events as they come: plasticity in female insect reproduction, p. 423-521. *In*: D.W. WHITMAN & T.N ANATHAKRISHNAN (Eds). **Phenotypic plasticity of insects: mechanisms and consequences**. Enfield, Science Publishers.

HUEBNER, E. & E. ANDERSON. 1972. A cytological study of the ovary of *Rhodnius prolixus* II. Oocyte differentiation. **Journal of Morphology** 137: 385-416. doi: 10.1002/jmor.1051370402.

HUEBNER, E. & W. DIEHL-JONES. 1993. Nurse cell-oocyte interaction in the telotrophic ovary. **International Journal of Insect Morphology and Embryology** 22: 369-387. doi: 10.1016/0020-7322(93)90020-2.

KIM LIM, A. & T. KAI. 2007. Unique germ-line organelle, nuage, functions to repress selfish genetic elements in *Drosophila melanogaster*. **Proceedings of the National Academy of Sciences** 104: 6714-6719. doi: 10.1073/pnas.0701920104.

KHIM LIM, A.; C. LORTHONGPANICH; T.G. CHEW; C.W.G. TAN; Y.T. SHUE; S. BALU; N. GOUNKO; S. KURAMOCHI-MIYAGAWA; M.M. MATZUK; S. CHUMA; D.M. MESSERSCHMIDT; D. SOLTER & B.B. KNOWLES. 2013. The nuage mediates retrotransposon silencing in mouse primordial ovarian follicles. **Development** 140: 3819-3825. doi: 10.1242/dev.099184.

KING, R.C. & J. BÜNING. 1985. The origin and functioning of insect oocytes and nurse cells, p. 37-82. *In*: G.A. KERKUT & L.I. GILBERT (Eds). **Comprehensive insect physiology, biochemistry and pharmacology: embryogenesis and reproduction**. Oxford, Pergamon Press, vol. 1.

KLOWDEN, M.J. 2007. **Physiological systems in insects.** San Diego, Academic Press, 661p.

LALITHA, T.G.; K. SHYAMASUNDARI & K.H. RAO. 1997. Morphology and histology of the female reproductive system of *Abedus ovatus* STAL (Belostomatidae: Hemiptera: Insecta). **Memórias do Instituto Oswaldo Cruz 92:** 129-135. doi: 10.1590/S0074-02761997000100028.

LOCOCO, D. & E. HUEBNER. 1980. The ultrastructure of the female accessory gland, the cement gland, in the insect *Rhodnius prolixus.* **Tissue Cell 12**(3): 557-580.

MA, N.; M. WANG & B. HUA. 2013. Ultrastructure of female accessory glands in the scorpionfly *Panorpa sexspinosa* Cheng (Mecoptera, Parnorpidae). **Tissue & Cell 45:** 107-114. doi: http://dx.doi.org/10.1016/j.tice.2012.09.010.

MARTÍNEZ, I.M. 2002. Técnicas básicas de anatomia microscópica y de morfometría para estudiar los insectos. **Boletín Sociedad Entomologica Aragonesa 30:** 187-195. Available online at: http://entomologia.rediris.es/aracnet/9/metodologias/tecnicas/index.htm. [Accessed: 16/II/2014].

MICHALIK, A.; T. SZKLARZWWICS; P. WEGIEREK & K. WIECZOREK. 2013. The ovaries of aphids (Hemiptera, Sternorrhyncha, Aphidoidea): morphology and phylogenetic implications. **Invertebrate Biology 132:** 226-240. doi: 10.1111/ivb.12026.

NAVA, D.E.; M.L.G. TORRES; M.D.L. RODRIGUES; J.M.S. BENTO & J.R.P. PARRA 2007. Biology of *Diphorina citri* (Hemiptera, Psyllidae) on different hosts and at different temperatures. **Journal of Applied Entomology 131:** 709-715. doi: 10.1111/j.1439-0418.2007.01230.x.

RAIKHEL, A.S. & T.S. DHADIALLA. 1992. Accumulation of yolk proteins in insect oocytes. **Annual Review of Entomology 37:** 217-251. doi: 10.1146/annurev.en.37.010192.001245.

STACCONI, M.V.R. & R. ROMANI. 2011. Ultrastructure and functional aspects of the spermatheca in the American Harlequin Bug *Murgatia histronica* (Hemiptera, Pentatomidae). **Neotropical Entomology 40:** 222-230. doi: 10.1590/S1519-566X2011000200011.

STURM, R. 2012. Morphology and ultrastructure of the accessory glands in the female genital tract of the house cricket, *Acheta domesticus.* **Journal of Insect Science 12:** 99. doi: 10.1673/031.012.9901

STURM, R. & K. POHLHAMMER. 2000. Morphology and development of the female accessory sex glands in the cricket *Teleogryllus commodus* (Saltatoria: Ensifera: Gryllidae). **Invertebr Repr Dev 38** (1): 13-21.

SZKLARZEWICZ, T. 1997. Structure and development of the telotrophic ovariole in ensign scale insects (Hemiptera, Coccomorpha: Ortheziidae). **Tissue & Cell 29:** 31-38. doi: 10.1016/S0040-8166(97)80069-9.

SZKLARZEWICZ, T. & A. MOSKAL. 2001. Ultrastructure, distribution, and transmission of endosymbionts in the whitefly *Aleurochiton aceris* MODEER (Insecta, Hemiptera, Aleyrodinea). **Protoplasma 218:** 45-53. doi: 10.1007/BF01288359.

SZKLARZEWICZ, T.; A. WNEK & S.M. BILÍNSKI. 2000. Structure of ovarioles in *Adelges laricis,* a representative of the primitive aphid family Adelgidae. **Acta Zoologica 81:** 307-313. doi: 10.1046/j.1463-6395.2000.00061.x.

SZKLARZEWICZ, T.; W. JANKOWSKA; K. WIECZOREK & P. WEGIEREK. 2008. Structure of the ovaries of the primitive aphids *Phylloxera coccinea* and *Phylloxera glabra* (Hemiptera, Aphidinea: Phylloxeridae). **Acta Zoologica 89:** 1-9. doi: 10.1111/j.1463-6395.2008.00335.x.

SZKLARZEWICZ, T.; M. KALANDYK-KOLODZIEJCZYK; M. KOT & A. MICHALIK. 2013. Ovary structure and transovarial transmission of microorganisms in *Marchalina hellenica* (Insect, Hemiptera, Coccomorpha: Marchalinidae). **Acta Zoologica 94:** 184-192. doi: 10.1111%2Fj.1463-6395.2011.00538.x.

TELFER, W.H. & R.I. WOODRUFF. 2002. Ion physiology of vitellogenic follicles. **Journal of Insect Physiology 48:** 915-923. doi: 10.1016/S0022-1910(02)00152-X.

TSAI, J.H. & Y.H. LIU. 2000. Biology of *Diaphorina citri* (Homoptera: Psyllidae) on four host plants. **Journal of Economic Entomology 93:** 1721-1725. doi: 10.1603/0022-0493-93.6.1721.

WENNINGER, E.J. & D.G. HALL. 2007. Daily timing of mating and age at reproductive maturity in *Diaphorina citri* (Hemiptera: Psyllidae). **Florida Entomology 90:** 715-722. doi: 10.1653/0015-4040(2007)90[715:DTOMAA]2.0.CO;2.

WHEELER, D.E. 2003. Accessory glands, p. 1-3. *In:* V.H. RESH & R.T. CARDÉ (Eds). **Encyclopedia of insects.** Hong Kong, Academic Press.

WINNICK, C.G.; G.I. HOLWELL & M.I. HERBERSTEIN. 2009. Internal reproductive anatomy of the praying mantid *Ciulfina klassi* (Mantodea: Liturgusidae). **Arthropod Structure & Development 38:** 60-69. doi: http://dx.doi.org/10.1016/j.asd.2008.07.002.

15

Morphology of the shell of *Happiella* cf. *insularis* (Gastropoda: Heterobranchia: Systrophiidae) from three forest areas on Ilha Grande, Southeast Brazil

Amilcar Brum Barbosa[1] & Sonia Barbosa dos Santos[1,2]

[1] *Laboratório de Malacologia Límnica e Terrestre, Departamento de Zoologia, Instituto de Biologia Roberto Alcantara Gomes, Universidade do Estado do Rio de Janeiro. Rua São Francisco Xavier 524, PHLC sala 525-2, 20550-900 Rio de Janeiro, RJ, Brazil.*
[2] *Corresponding author. E-mail: milkabrum@yahoo.com.br*

ABSTRACT. We conducted a study on shell morphology variation among three populations of *Happiella* cf. *insularis* (Boëttger, 1889) inhabiting different areas (Jararaca, Caxadaço, and Parnaioca trails) at Vila Dois Rios, Ilha Grande, Angra dos Reis, state of Rio de Janeiro, Brazil. Linear and angular measurements, shell indices representing shell shape, and whorl counts were obtained from images drawn using a stereomicroscope coupled with a camera lucida. The statistical analysis based on ANOVA (followed by Bonferroni's test), Pearson's correlation matrix, and discriminant analysis enabled discrimination among the populations studied. The variable that most contributed to discriminate among groups was shell height. Mean shell height was greatest for specimens collected from Jararaca, probably reflecting the better conservation status of that area. Good conservation is associated with enhanced shell growth. Mean measurements were smallest for specimens from Parnaioca, the most disturbed area surveyed. Mean aperture height was smallest for specimens from Parnaioca, which may represent a strategy to prevent excessive water loss. Discriminant analysis revealed that the snails from Jararaca differ the most from snails collected in the two other areas, reflecting the different conservation status of these areas: shells reach larger sizes in the localities where the humidity is higher. The similarities in shell morphology were greater between areas that are more similar environmentally (Caxadaço and Parnaioca), suggesting that conchological differences may correspond to adaptations to the environment.

KEY WORDS. Conchology; discriminant analysis; ecology; morphometry; threatened biome.

Land snails are exceptionally diverse in morphology, for instance they display great polymorphism in shell color and variations in shell dimensions. For this reason, they are a good subject for evolutionary biology studies (CLARKE *et al.* 1978). Differences in size, morphology and growth rates are associated with ecological conditions, natural selection, and phylogenetic history (VERMEIJ 1971, CLARKE *et al.* 1978, EMBERTON 1994, 1995b, COOK 1997, PARMAKELIS *et al.* 2003, TESHIMA *et al.* 2003). According to GOULD (1984), the low mobility of land snails influences character variability. The literature shows that habitat alterations, which result in fragmentation, are an important factor affecting shell morphological differentiation (COOK 1997, GOODFRIEND 1986, EMBERTON 1982, 1994), which can be accelerated in degraded environments (CHIBA 2004, CHIBA & DAVISON 2007).

Ilha Grande, a continental island in the southern portion of the state of Rio de Janeiro, harbors large, continuous and conserved fragments of Atlantic Forest (ROCHA *et al.* 2006), which is among the most threatened biomes in the world (MYERS *et al.* 2000). Over 50% of Ilha Grande is covered by ombrophilous dense forest, now at different levels of regeneration (ALHO *et al.* 2002, OLIVEIRA 2002, ALVES *et al.* 2005, CALLADO *et al.* 2009) from disturbances caused by a range of human activities over the past five decades, being now a natural laboratory to study shell morphological differentiation induced by in environment conditions.

The focus of this study was to investigate variations in the morphology of the shell of *Happiella* cf. *insularis* in three different environments (Table I). This species was described by BOËTTGER (1889) based on a single shell collected from the type locality, Ilha das Flores, São Gonçalo city, Rio de Janeiro, where additional specimens have not been found (SANTOS *et al.* 2010). BOËTTGER's (1889) description, which was not accompanied by illustrations, highlighted the following diagnostic features: maximum diameter with 5.25 mm, shell height 2 mm, large umbilicus, one-fourth the size of the shell base; shell pebble-shaped, thin, white, polished, spire apex slightly prominent, with ½ whorls, slightly convex; borders distinct, mildly striated, last border over the third, approximately as wide as shell, less arched at top than bottom, angled below central region;

suture deeply impressed. Aperture elliptical-lunular, with small slit, aperture height with 2 mm, aperture width with 2.25 mm, simple peristomatic edge, with curved, spherical, sub-angular syphunculus [sic] protruding to right side of base.

THIELE (1927), in addition to the type locality of *H.* cf. *insularis*, also listed it in Piracicaba (state of São Paulo), Blumenau (state of Santa Catarina) and Porto Alegre (state of Rio Grande do Sul); MORRETES (1949) also listed it only in Ilha das Flores and SIMONE (2007) to Xanxerê and São Carlos (state of Santa Catarina).

In the present study, we analyzed the shell morphology of three populations of *H.* cf. *insularis* subjected to different environmental conditions, with the goal to assess variability in shell morphology, as detailed morphology and range of variation can prove useful for refining species diagnoses.

MATERIAL AND METHODS

The specimens used in this study were collected from three areas, known as the Jararaca, Caxadaço, and Parnaioca trails, located in Vila Dois Rios, on the ocean side of Ilha Grande, Municipality of Angra dos Reis, southern region of the state of Rio de Janeiro (23°04'25" to 23°13'10"S, 44°05'35" to 44°22'50"W). In each collecting site (Fig. 1), a distinct level of forest regeneration (VERA-Y-CONDE & ROCHA 2006) can be found, making them suitable for investigations on the influence of environmental factors on shell morphology.

Table I contains a summary of the environmental parameters measured at the three areas studied.

We selected intact shells from 102 adults, grown to approximately three whorls and proportionally similar to each other. Thirty-three shells from the Jararaca Trail were selected, in addition to 34 and 35 shells from the Caxadaço and Parnaioca trails, respectively.

Material examined. *Happiella* cf. *insularis*. BRAZIL, *Rio de Janeiro*: Angra dos Reis, Ilha Grande, Vila Dois Rios, Trilha da Jararaca, 14.VI.1998, S.B. Santos *leg.* (Col. Mol. UERJ 942-1); 27.IX.1996, S.B. Santos *leg.* (Col. Mol. UERJ 977); 11.I.1996, V.C. Queiroz *leg.* (Col. Mol. UERJ 980-4 and 5); 12.I.1996, S.B. Santos *leg.* (Col. Mol. UERJ 990-1 and 2); 21.III.1997, S.B. Santos *leg.* (Col. Mol. UERJ 1132); 23.III.1997, S.B. Santos *leg.* (Col. Mol. UERJ 1133-1 and 2); 20.IX.1997, S.B. Santos *leg.* (Col. Mol.

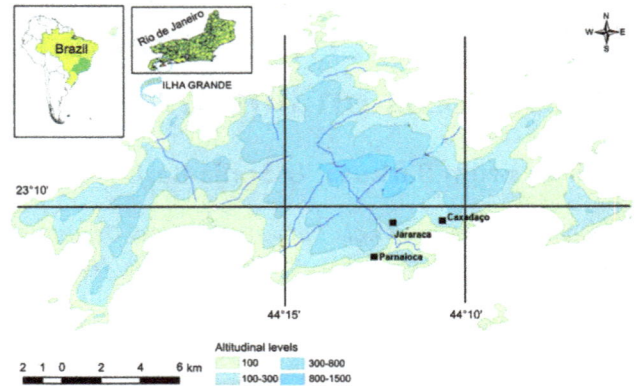

Figure 1. Location of Ilha Grande, in Angra dos Reis Municipality, state of Rio de Janeiro, Brazil, showing the Jararaca, Parnaioca, and Caxadaço trails. Map: Luiz E.M. Lacerda.

UERJ 1155); 30.XI.1997, D.P. Monteiro *leg.* (Col. Mol. UERJ 1168-1 and 2); *ditto*, 26.VI.1999, S.B. Santos *leg.* (Col. Mol. UERJ 1241-1 and 2); 21.III.1997, S.B. Santos *leg.* (Col. Mol. UERJ 1252-2); 14.I.1998, D.P. Monteiro *leg.* (Col. Mol. UERJ 1617-2); 17.II.1998, A.S. Alencar *leg.* (Col. Mol. UERJ 1618-2); 17.II.1998, D.P. Monteiro *leg.* (Col. Mol. UERJ 1646); 15.I.1998, S.B. Santos *leg.* (Col. Mol. UERJ 1647-2); 17.II.1998, M.A. Fernandez *leg.* (Col. Mol. UERJ 1650); 14.I.1998, A.S. Alencar *leg.* (Col. Mol. UERJ 1651); 14.I.1998, D.P. Monteiro *leg.* (Col. Mol. UERJ 1653-2); 17.I.1998, S.B. Santos *leg.* (Col. Mol. UERJ 1656-2 and 3); 17.II.1998, D.P. Monteiro *leg.* (Col. Mol. UERJ 1658-2, 3, 4 and 6); 17.II.1998, S.B. Santos *leg.* (Col. Mol. UERJ 1659-1, 2, and 3). Vila Dois Rios, Trilha do Caxadaço, 19.X.1995, V.C. Queiroz *leg.* (Col. Mol. UERJ 999-2, 3, 4, 5, and 6); 30.V.1997, S.B. Santos *leg.* (Col. Mol. UERJ 1064-7 and 3); 28.XI.1997, D.P. Monteiro *leg.* (Col. Mol. UERJ 1110-1, 2, 3, 4, and 5); 15.VIII.1996, S.B. Santos *leg.* (Col. Mol. UERJ 1114); 19.X.1995, V.C. Queiroz *leg.* (Col. Mol. UERJ 1144-1, 2, 3, and 4); 08.VIII.1999, M. Sttorti *leg.* (Col. Mol. UERJ 1310); 21.X.2000, D.P. Monteiro *leg.* (Col. Mol. UERJ 2061-1, 2, and 3); 15.III.2001, S.B. Santos *leg.* (Col. Mol. UERJ 2156-2, 3, 4, 5, and 6); 28.X.2001, C.C. Siqueira *leg.* (Col. Mol. UERJ 2225-2, 3, 4, 5, and 6); 2.VIII.2005, A.B Barbosa, Lacerda, L.E.M., T.A. Viana *leg.* (Col. Mol. UERJ 7445-1, 2, and 3). Vila Dois Rios, Trilha da Parnaioca), 28.V.1997, N. Salgado

Table I. Summary of local environmental parameters from three areas (Jararaca, Caxadaço and Parnaioca trails) at Ilha Grande.

Area	Mean ambient temperature (a) (°C)	Mean ground temperature (a) (°C)	Mean leaf litter depth layer (a) (cm)	Mean relative air humidity (a) (%)	Canopy height (m)	Canopy closure	Elevation	Degree of impact	Time of regeneration
Jararaca Trail	22.46 ± 3.42	20.81 ± 3.42	7.17 ± 2.68	84.81 ± 9.27	33 (c)	Greater	250 m asl	Advanced stage of ecological succession	At least 90 years-old (e)
Caxadaço Trail	29.95 ± 2.09	21.01 ± 2.15	4.11 ± 1.64	83.82 ± 8.98	15-20 (b)	Intermediate	180 m asl	Early stage of ecological succession	At least 50 years-old (c)
Parnaioca Trail	25.29 ± 3.12	22.04 ± 2.16	5.87 ± 2.66	80.93 ± 10.51	15 (c)	Lower	At sea level	Early stage of ecological succession	5 to 25 years-old (c)

a) D.P. Monteiro (unpubl. data), b) ALHO *et al.* (2002), c) VERA-Y-CONDE & ROCHA (2006), d) CALLADO *et al.* (2009), e) SLUYS *et al.* (2012).

leg. (Col. Mol. UERJ 1129-1); 13.VIII.1996, S.B. Santos *leg.* (Col. Mol. UERJ 1139); 13.VIII.1996, S.B. Santos *leg.* (Col. Mol. UERJ 1177-1); 13.VIII.1996, S.B. Santos *leg.* (Col. Mol. UERJ 1175); 13.VIII.1996, S.B. Santos *leg.* (Col. Mol. UERJ 1178-1 and 2); 16.VI.2002, S.B. Santos *leg.* (Col. Mol. UERJ 1827-2); 02.II.2002, D.P. Monteiro *leg.* (Col. Mol. UERJ 2989-1); 01.II.2000, D.P. Monteiro *leg.* (Col. Mol. UERJ 3005-1, 2, and 3); 31.I.2000, D.P. Monteiro *leg.* (Col. Mol. UERJ 3006-1, 2, 3, 4, 5, 6, 7, 8, 10, and 11); 31.I.2000, S.B. Santos *leg.* (Col. Mol. UERJ 3007-1, 2, 3, 4, 5, 6, 7, and 8); 31.I.2000, D.P. Monteiro *leg.* (Col. Mol. UERJ 3288-2); 28.I.2000, P. Coelho *leg.* (Col. Mol. UERJ 3289); 3.VIII.2005, A.B. Barbosa, Lacerda, L.E.M., T.A. Viana *leg.* (Col. Mol. UERJ 7444-1, 2, and 3).

Drawings of the shells in apical, umbilical, and lateral views were made with the aid of a camera lucida under an Olympus SZH10 stereomicroscope. The drawings were used to obtain angular and linear measurements, establish the number of whorls, and calculate the ratios between measurements, according to the criteria proposed by DIVER (1931), PARODIZ (1951), SOLEM & CLIMO (1985) and FONSECA & THOMÉ (1994). The following angular measurements were considered: maximum angle (MA), columellar angle (CA), sutural angle (SA), lower sutural angle (SS'), and spire angle (SPA) (Figs 2 and 3). The linear measurements taken were: shell height (h), aperture height (ah), aperture width (aw), spire height (sh) (Fig. 2), first whorl diameter (1wd), maximum diameter (D), smaller diameter (d), first whorl width (1ww), and second whorl width (2ww) (Fig. 4), diameter umbilical (ud) (Fig. 5). The following ratios were calculated: shell height/maximum diameter (h/D), maximum diameter/umbilical diameter (D/ud), umbilical diameter/ shell height (ud/h), aperture height/aperture width (ah/aw), aperture height/smaller diameter (ah/d), first whorl diameter/ maximum diameter (1wd/D), maximum diameter/total number of whorls (D/NW), and first whorl width/second whorl width (1ww/2ww) (SOLEM & CLIMO 1985, FONSECA & THOMÉ 1994, EMBERTON 1995a). The method proposed by DIVER (1931) was applied to obtain the number of protoconch whorls (pW), total number of whorls (NW), and total number of teleoconch whorls (TW) (Fig. 4).

The analysis of shell morphological variation followed VALORVITA & VÄISÄNEN (1986), with some modifications. Descriptive statistics were performed for each variable in each group, and normality was tested using the skewness test. In cases when a given variable had asymmetrical distribution, the following transformation procedures were applied to normalize it as appropriate: e-base logarithm of X (Neperian logarithm) [1wd/D, NW], square root of X [D/ud], sin X [MA, ah], cos X [SA, 1ww/2ww], tan X [ah/aw, ah/d], and reciprocal of X (i.e., 1/X) [1ww, 2ww, h/D, D/NW] (KREBS 1998, ZAR 1999), where X is the variable considered.

After normalization, each variable was standardized by reduction (SPIEGEL 1993) and compared using analysis of variance (ANOVA) followed by the Bonferroni's test. Differences

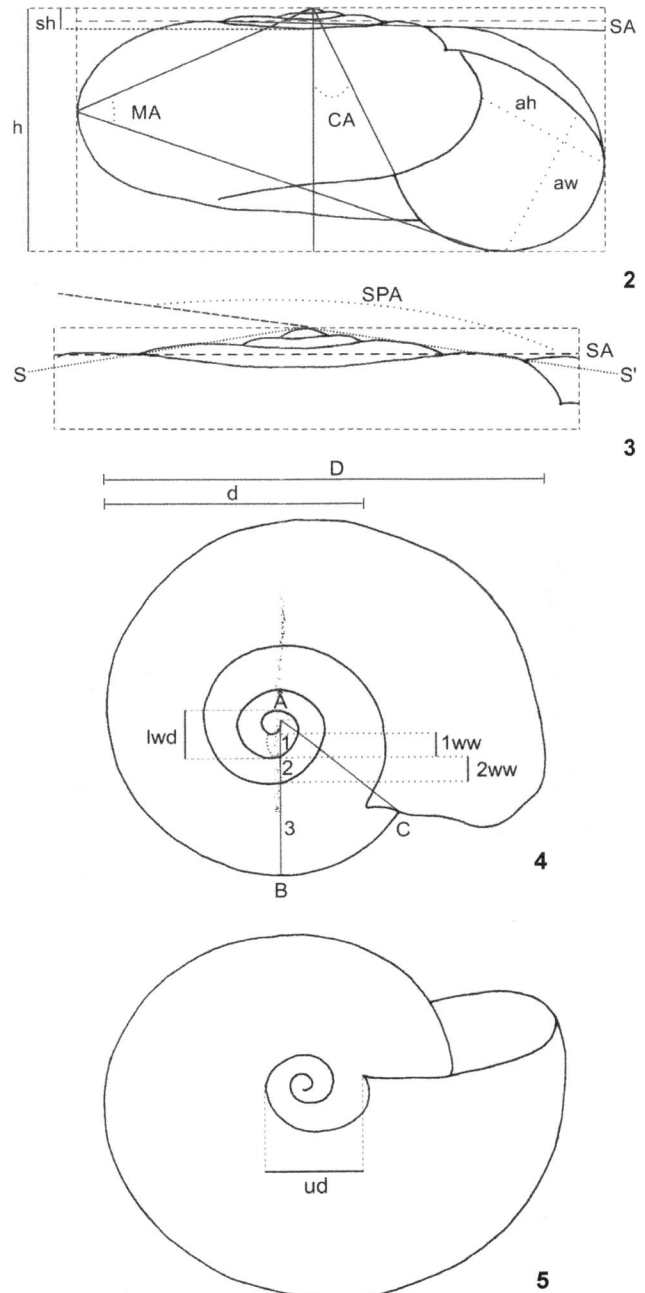

Figures 2-5. *Happiella* cf. *insularis*. Schematic depiction of shell, with angular, linear measurements, and number of whole whorls: (2) CA, columellar angle; MA, maximum angle; ah, aperture height; SA, sutural angle; sh, spire height; aw, aperture width. (3) SPA, spire angle; SA, sutural angle; SS', lower sutural angle. (4) D = maximum diameter; d, smaller diameter; 1ww, first whorl width; 2ww, second whorl width; 1wd, first whorl diameter; 1, 2, 3, number of whole whorls. (5) umbilical view: ud, umbilical diameter.

at p < 0.05 were considered statistically significant. The third decimal place was dropped and differences at p ≤ 0.05 were considered statistically significant.

After the exclusion of the highly correlated variables (KLECK 1982, ENGELMAN 1997), discriminant analysis was performed on 15 variables: MA, CA, SA, SS, SPA, 1wd,h, ah,aw,sh, 1ww, 2ww, pW, NW, and TW. Preliminary Pearson's correlation matrix revealed a high correlation (r ≥ 0.90) between the variables D, d, ud, and ah. These were removed from the analysis, and the variable h, representing all correlated variables excluded, was kept. Upon analysis, the variables MA, SA, sh, and 1ww were also removed, owing to their low contribution to group discrimination, as shown by the discriminant function coefficients. Statistical analyses were performed with the aid of the SYSTAT 7.0 statistical package (ENGELMAN 1997).

RESULTS

Shell morphology

Happiella cf. *insularis* (Fig. 6) has thin, translucent, depressed, shiny shells; periostracum color varies from yellow amber (when alive) to whitish yellow (when fixed in alcohol or in cases when only the shell is found *in situ*); spire slightly elevated (BOËTTGER 1889) (Fig. 9); under the optical microscope, the texture is smooth, with mildly marked growth lines (BOËTTGER 1889) (Fig. 7); under scanning microscopy, protoconch, more granular, and teleoconch, of rougher aspect (Fig. 10), umbilicus opened (Fig. 8) (BOËTTGER 1889); aperture wide, crescent-shaped, slightly oblique; peristome simple (BOËTTGER 1889), thin sharp edges, without teeth (Fig. 9) (THIELE 1931, ZILCH 1959, MONTEIRO & SANTOS 2001); suture not impressed (Fig. 7) (R.L. RAMÍREZ unpubl.

Figures 6-10. *Happiella* cf. *insularis*. (6) habitus; (7-10) specimen Col. Mol. UERJ 1653-2: (7) apical view; (8) umbilical view; (9) apertural view; (10) scanning electron microscopy of view's apical shell. Scale bar: 6-9 = 1 mm, 10 = 100 µm. Photos: 6 = Antônio C. de Freitas, 7-9 = Amilcar B. Barbosa, 10 = Alan C.N. de Moraes, LABMEL/UERJ.

data); body whorl rounded (Fig. 10) (R.L. Ramírez unpubl. data); rapid increment's shell growth (Fig. 7) (Emberton 1995a).

Shell morphometry

Table II shows the morphometric and meristic data of the 102 shells examined. The mean values of these features were lowest in specimens from the Parnaioca Trail.

The results of the ANOVA, distinguished among the three samples collected from Jararaca, Caxadaço, and Parnaioca, revealed significant differences in all linear and angular measure-

ments, except for the mean maximum angle. Specimens from the Jararaca and Parnaioca trails differed significantly in mean columellar and mean spire angles, but the differences in these measurements between samples from Jararaca and Caxadaço and from Parnaioca and Caxadaço were not statistically significant. The shells from Jararaca differed from those from the Caxadaço and Parnaioca trails in the mean sutural and lower sutural angles; shells from Caxadaço and Parnaioca, however, were statistically similar with regards to these two variables. The Bonferroni's test revealed differences between samples from Parnaioca and

Table II. Descriptive statistics of morphometric and meristic variables and ratios for *Happiella* cf. *insularis* collected from three areas on Ilha Grande. Linear measurements (cm): (D) maximum diameter, (d) smaller diameter, (ud) umbilical diameter, (1wd) first whorl diameter, (ah) aperture height, (sh) spire height, (h) shell height, (aw) aperture width, (1ww) first whorl width, (2ww) second whorl width. Angular measurements (degrees): (CA) columellar angle, (MA) maximum angle, (SPA) spire angle, (SA) sutural angle, (SS') lower sutural angle. Ratios: (D/ud) maximum diameter/umbilical diameter, (D/NW) maximum diameter/total number of whorls, (ud/h) maximum diameter/umbilical diameter, (1wd/D) first whorl diameter/maximum diameter, (h/D) shell height/maximum diameter, (aw/d) aperture height/smaller diameter, (ah/aw) aperture height/aperture width. Number of whorls: (pW) number of protoconch whorls, (TW) number of teleoconch whorls, (NW) total number of whorls. (N) sample size, (SD) standard deviation, (VAR) variance.

	Jararaca (N = 33)					Caxadaço (N = 34)					Parnaioca (N = 35)				
	Min.	Mean	Max.	SD	VAR	Min.	Mean	Max.	SD	VAR	Min.	Mean	Max.	SD	VAR
Linear measurements															
D	0.500	0.739	0.875	0.090	0.008	0.371	0.591	0.806	0.169	0.028	0.319	0.476	0.833	0.125	0.015
d	0.325	0.494	0.606	0.071	0.005	0.227	0.405	0.575	0.110	0.012	0.221	0.331	0.558	0.079	0.006
ud	0.090	0.161	0.206	0.030	0.000	0.059	0.124	0.241	0.054	0.002	0.044	0.089	0.193	0.037	0.001
1wd	0.038	0.068	0.086	0.011	0.000	0.031	0.056	0.080	0.012	0.000	0.030	0.056	0.083	0.012	0.000
sh	0.011	0.019	0.033	0.004	0.000	0.015	0.021	0.027	0.003	0.000	0.010	0.019	0.027	0.004	0.000
h	0.245	0.355	0.413	0.037	0.001	0.189	0.284	0.400	0.076	0.005	0.157	0.228	0.366	0.056	0.003
ah	0.127	0.191	0.240	0.026	0.000	0.100	0.161	0.225	0.046	0.002	0.082	0.127	0.233	0.031	0.000
aw	0.066	0.179	0.300	0.045	0.002	0.098	0.167	0.253	0.046	0.002	0.072	0.125	0.220	0.036	0.001
1ww	0.025	0.052	0.433	0.068	0.047	0.018	0.032	0.046	0.007	0.000	0.022	0.033	0.052	0.007	0.000
2ww	0.033	0.051	0.104	0.012	0.000	0.029	0.055	0.073	0.078	0.006	0.026	0.043	0.080	0.011	0.000
Angular measurements															
MA	44.0	47.772	55	2.446	5.985	42	46.426	51	2.074	4.305	42	47.200	59	3.595	12.924
CA	15.0	20.803	28	3.107	9.655	10	19.970	29	4.344	18.877	10	17.942	25	4.129	17.055
SA	0.5	1.590	5	0.930	0.866	1	2.500	4	0.904	0.818	1	2.242	5	1.017	1.034
SS'	151.0	161.742	171	5.026	25.267	147	156.264	167	5.029	25.291	147	154.742	166	4.767	22.726
SPA	158.0	168.151	175	3.700	13.757	160	166.588	172	3.210	10.310	160	165.200	172	3.332	11.105
Ratios															
h/D	0.445	0.481	0.533	0.021	0.000	0.428	0.483	0.581	0.029	0.000	0.439	0.480	0.519	0.020	0.000
D/ud	3.718	4.656	5.500	0.418	0.175	3.282	5.075	7.000	0.858	0.737	3.266	5.657	7.900	0.985	0.970
ud/h	0.361	0.450	0.583	0.054	0.003	0.238	0.419	0.658	0.085	0.007	0.276	0.380	0.632	0.079	0.006
ah/aw	0.643	1.129	2.606	0.340	0.115	0.735	0.965	1.234	0.132	0.017	0.650	1.037	1.388	0.165	0.027
ah/d	0.300	0.396	0.612	0.062	0.003	0.346	0.394	0.471	0.030	0.000	0.034	0.380	0.504	0.071	0.005
1wd/D	0.057	0.094	0.172	0.020	0.000	0.063	0.100	0.152	0.025	0.000	0.069	0.122	0.219	0.035	0.001
D/NW	0.115	0.192	0.224	0.020	0.000	0.103	0.163	0.573	0.081	0.006	0.098	0.128	0.215	0.029	0.000
1ww/2ww	0.521	0.826	1.200	0.154	0.023	0.066	0.758	1.000	0.178	0.031	0.519	0.781	1.000	0.104	0.010
Number of whorls															
NW	2.909	3.823	4.413	0.275	0.076	3.380	3.908	4.472	0.287	0.082	2.908	3.671	4.188	0.322	0.103
TW	1.433	2.538	3.587	0.410	0.168	1.794	2.376	3.166	0.316	0.099	1.720	2.191	2.936	0.334	0.112
pW	0.118	1.284	2.844	0.588	0.239	0.950	1.531	2.105	0.287	0.082	0.955	1.499	2.119	0.307	0.094

Jararaca in nine morphological features, between Jararaca and Caxadaço samples in eight features, and between samples from Caxadaço and Parnaioca in seven features – i.e., samples from Caxadaço and Parnaioca were less dissimilar to each other than to the sample from Jararaca (Table III).

ANOVA revealed significant differences in the D/ud, ud/h, ah/aw, 1wd/D, and D/NW ratios across samples. Bonferroni's test showed Parnaioca and Jararaca samples to differ in four of these ratios (D/ud, ud/h, 1wd/D, and D/NW), Caxadaço and Parnaioca samples to differ in three (D/ud, 1wd/D, and D/NW), and Jararaca and Caxadaço samples to differ on two of these ratios (ah/aw and D/NW) – i.e., differences in measurement

Table III. Results of ANOVA followed by Bonferroni's multiple comparison test, applied to linear and angular measurements, ratios, and number of whorls of *Happiella* cf. *insularis* specimens collected from the Jararaca (Jar), Caxadaço (Cax), and Parnaioca (Par) trails, Ilha Grande. Differences were considered statistically significant* at $p \leq 0.05$. For abreviations see Table II.

	p	Jar x Cax	Jar x Par	Cax x Par
Linear measurements				
D	0.000*	0.000*	0.000*	0.001*
d	0.000*	0.000*	0.000*	0.002*
ud	0.000*	0.002*	0.000*	0.002*
1wd	0.000*	0.000*	0.000*	1.000
sh	0.036*	0.099	1.000	0.061*
h	0.000*	0.000*	0.000*	0.000*
ah	0.000*	0.002*	0.000*	0.000*
aw	0.000*	0.753	0.000*	0.000*
1ww	0.000*	0.000*	0.001*	1.000
2ww	0.008*	0.033*	0.012*	1.000
Angular measurements				
MA	0.470	1.000	1.000	0.751
CA	0.010*	1.000	0.010*	0.101
SA	0.000*	0.000*	0.002*	0.187
SS'	0.000*	0.000*	0.000*	0.611
SPA	0.003*	0.193	0.002*	0.285
Ratios				
h/D	0.947	1.000	1.000	1.000
D/ud	0.000*	0.106	0.000*	0.012*
ud/h	0.000*	0.212	0.000*	0.082
ah/aw	0.024*	0.020*	0.559	0.427
ah/d	0.490	1.000	0.757	1.000
1wd/D	0.000*	1.000	0.000*	0.004*
D/NW	0.000*	0.000*	0.000*	0.003*
1ww/2ww	0.175	0.238	0.450	1.000
Number of whorls				
NW	0.005*	0.807	0.107	0.004*
TW	0.001*	0.192	0.000*	0.099
pW	0.015*	0.022*	0.056*	0.929

ratios were most pronounced between samples collected from the Jararaca and Parnaioca trails, as were differences in linear measurements (Table III).

The mean total number of whorls (NW) differed significantly between the samples from Caxadaço (greater mean) and Parnaioca (Table III). The mean total number of teleoconch whorls (TW) differed significantly between the Jararaca and Parnaioca samples. The mean number of protoconch whorls (pW) differed significantly not only between samples from Jararaca and Caxadaço, but also between samples from Jararaca and Parnaioca (Table III).

Discriminant analysis

The discriminant analysis (Fig. 11) allowed the distinction of all three samples (Wilks's Lambda = 0.300, F = 6.689, df = 22, p = 0.000), particularly with respect to the sample from Jararaca, which differed the most from the others. The samples from Caxadaço and Parnaioca were more similar to each other than each was to the sample from Jararaca (Fig. 11). This analysis correctly classified 67% of the specimens (Fig. 11), with 34 out of 102 being incorrectly classified.

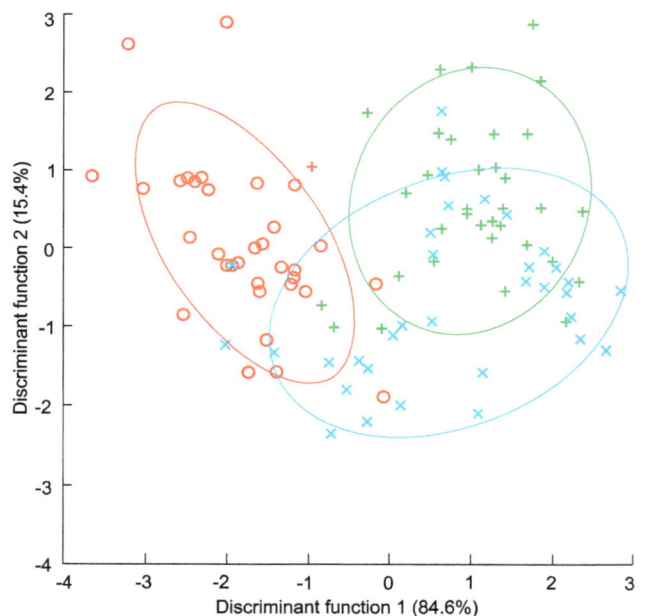

Figure 11. Graphic depiction of the discriminant analysis of morphometric variables of three *Happiella* cf. *insularis* samples collected from the Jararaca (○), Caxadaço (×), and Parnaioca (+) trail areas, Ilha Grande, RJ, Brazil.

The proportions of explanation were of 84.6% and 15.4% for the first and second discriminant functions, respectively. The coefficients also revealed the following variables to be major contributors to the degree of differentiation achieved with the first function: shell height (h), lower sutural angle (SS'), spire angle (SPA), aperture height (ah), number of protoconch whorls (pW) e number of teleoconch whorls (TW).

Discriminant function 1 = -2.263 h – 1.077 SS' + 0,964 SPA + 0.877 ah + 0.736 pW + 0.674 TW – 0.411 2ww + 0.296 aw + 0.294 + 0.113 1wd + 0.038 CA.

Discriminant function 2 = -1.046 NW + 0.992 SS' – 0.941 SPA + 0.900TW – 0.870 ah + 0.799 pW – 0.780 aw + 0.465 h + 0.396 CA + 0.249 1wd + 0.035 2ww.

DISCUSSION

The morphology of the shell of *H.* cf. *insularis* fits the description for Systrophiidae perfectly: shell thin, translucent, polished, generally smooth, spire apex slightly prominent, discoid, simple peristomatic edge. According to BAKER (1925), *Happiella* shells are characterized by a very low spire and an umbilicus normally reduced to a small perforation. In his original description of *H.* cf. *insularis*, BOETTGER (1889) reported a maximum diameter of 5.25 mm, very close to the mean values found for the Caxadaço and Parnaioca samples, and shell height of 2 mm, very close to the mean obtained for the sample from Parnaioca. Aperture height and aperture width were originally described as measuring 2 mm and 2.25 mm, respectively, with the latter measurement falling within the confidence interval of the sample from Jararaca. The original description of the number of whorls (3.5) is also within the confidence intervals of both Jararaca and Parnaioca samples. According to the original description, the umbilicus size is one-fourth the maximum diameter of the shell, a similar ratio to that found in our specimens.

Our results revealed significant differences among the three populations of *H.* cf. *insularis* examined, which may be explained by differences in the degree of forest conservation in each area surveyed. The original vegetation in the Caxadaço and Parnaioca areas has been disturbed, more dramatically so in the latter, where the vegetation was entirely slashed down in some areas to make way for the now disabled Vila Dois Rios-Jararaca Dam road and for a number of plantations that served the now defunct Cândido Mendes Penal Colony. Along the Caxadaço Trail, inhabited until the construction of the penal colony by local native fishermen, the environmental changes were less pronounced. The Jararaca Trail region, by contrast, is better preserved. It has a relatively undisturbed secondary forest in lower-altitude areas and primary forest in higher areas. This translates into a deeper leaf litter layer, a more closed canopy, and lower temperature (Table I). As in other instances (SHIMEK 1930, BOYCOTT 1934, CAIN 1977, CLARKE *et al.* 1978, TILLIER 1981, EMBERTON 1995b, COOK 1997, WELTER-SCHULTES 2000, TESHIMA *et al.* 2003), better conservation certainly influences the environmental conditions overall, including leaf litter quality, structure, humidity, and depth, which in turn influence mollusk morphology (GOULD 1968, PEAKE 1978, CIPRIANI 2007). Investi-

gating *Ainohelix editha* (Adams, 1868) in Hokkaido, Japan, TESHIMA *et al.* (2003) demonstrated that shell size and growth rates are adaptations to the environmental conditions; CHIBA (2004), investigating the genus *Mandarina* Pilsbry, 1894 on the Bonin Islands, found that differentiation in shell shape and dimensions are accelerated in degraded environments.

In the present study, the smallest mean values were obtained for specimens collected from the Parnaioca Trail area (Table II), the most disturbed of the three areas (shallower leaf litter, more open canopy, higher temperature). The lower capacity of the local leaf litter to retain water is likely responsible for the smaller size of the snails. The smaller mean aperture height of the shell may represent a strategy to prevent excessive water loss (GOODFRIEND 1986). MACHIN (1967), PEAKE (1978), EMBERTON (1982), and GOODFRIEND (1986), along with other investigators, have reported that smaller specimens are found in terrestrial gastropod populations living in dry areas with a strong incidence of sunlight.

The mean shell diameter and number of whorls were greatest in snails collected from Jararaca (Table II), consistent with the hypothesis that higher humidity and lower temperatures promote increased rates of shell growth in terrestrial gastropods (GOULD 1984, GOODFRIEND 1986, EMBERTON 1994). BAUR (1988) concluded that the size of the shell of *Chondrina clienta* (Westerlund, 1883) increases in higher temperatures and lower population densities, as verified by ANDERSON *et al.* (2007) for *Oreohelix cooperi* (Binney, 1838). BAUR (1988) commented that the phenotypic plasticity found in *C. clienta* may be adaptive, as the genetic makeup of snails allows for different shell growth patterns under different environmental conditions.

The populations of the most disturbed areas – the Caxadaço and Parnaioca trails – are more similar to each other than to the population of the Jararaca Trail, which is the best-preserved area (Fig. 11). The greater similarity observed between the Parnaioca and Caxadaço samples can be explained by the intermediate degree of conservation of the Caxadaço region, which is closer to the degree of conservation of the Parnaioca area than to that of the Jararaca region.

The findings of this study corroborate investigations conducted in other countries showing that morphological differentiation is a result of the isolation of populations in areas that are distinct in vegetation cover, dominant plant species, maximum altitude, and soil type – i.e., areas that offer different microhabitats. As CHIBA (2004) pointed out, degraded environments accelerate differentiation by eliciting new ecological interactions and new habitat conditions, thus subjecting species to a number of selective pressures.

We believe that a similar process has occurred in the areas investigated in the present study, where the different degrees of forest degradation, added to different degrees of moisture, contributed to the morphological differentiation of the three studied populations of *H.* cf. *insularis*. However, as Brazilian species of Systrophiidae are not yet well defined, we have decided not to treat the three populations as separate species, and we recommend that they continue to be identified as *H.* cf. *insularis* until ongoing anatomical studies are concluded, and a decision on their taxonimical status is reached.

ACKNOWLEDGEMENTS

This study is one of the results of the Project "Fauna malacológica aquática e terrestre da Ilha Grande", supported by research grants from the Fundação de Amparo à Pesquisa do Estado do Rio de Janeiro (FAPERJ) to the second author (APQ1 E-26/110.402/2010 and E-26/110.362/2012). We would like to express our gratitude to PEIG/INEA (Parque Estadual da Ilha Grande/Instituto Estadual do Ambiente) for license 18/2007; to IBAMA/Sisbio (Instituto Brasileiro do Meio Ambiente e dos Recursos Naturais Renováveis/Sistema de Autorização e Informação em Biodiversidade) for license 19836-1 to ABB and 10812-1 to SBS; to Capes (Coordenação de Aperfeiçoamento do Pessoal de Ensino Superior for a PhD scholarship to ABB; to the reviewers who gave valuable suggestions to improve the manuscript; to Cientifica Consultoria for review of English and to CEADS (Centro de Estudos Ambientais e Desenvolvimento Sustentável da UERJ) for logistic support.

REFERENCES

ALHO, C.J.R.; M. SCHNEIDER & L.A. VASCONCELLOS. 2002. Degree of threat to the biological diversity in the Ilha Grande State Park (RJ) and guidelines for conservation. **Brazilian Journal of Biology 62** (3): 375-385. doi: 10.1590/S1519-69842002000300001.

ALVES, S.L.; A.S. ZAÚ; R.R. OLIVEIRA; D.F. LIMA & C.J.R. MOURA. 2005. Sucessão florestal e grupos ecológicos em Floresta Atlântica de encosta, Ilha Grande, Angra dos Reis/RJ. **Revista Universidade Rural: Série Ciências da Vida 25** (1): 26-32.

ANDERSON, T.K.; K.F. WEAVER & R.P. GURALNICK. 2007. Variation in adult shell morphology and life-history traits in the land snail *Oreohelix cooperi* in relation to biotic and abiotic factors. **Journal of Molluscan Studies 73**: 129-137. doi: 10.1093/mollus/eym006.

BAKER, H.B. 1925. The Mollusca collected by the University of Michigan-Williamson expedition in Venezuela. Part III. **Occasional Papers of the Museum of Zoology 156**: 1-44.

BAUR, B. 1988. Microgeographical variation in shell size of the land snail *Chondrina clienta*. **Biological Journal of the Linnean Society 35**: 247-259. doi: 10.1111/j.1095-8312.1988.tb00469.x.

BOËTTGER, O. 1889. Bemerkung uber ein paar brasilianische Landschneken, nebst Beschreibung drein neuer Hyalinien von dort. Nachrichtsblatt der deutschen **Malakozoologischen 20** (1-2): 27-30.

BOYCOTT, A.E. 1934. The habitats of land mollusca in Britain. **Journal of Ecology 22**: 1-38.

CAIN, A.J. 1977. Variation in the spire index of some coiled gastropods shells, and its evolutionary significance.

Philosophical Transactions of the Royal Society B277: 377-428.

CALLADO, C.H.; A.A.M. BARROS; L.A. RIBAS; N. ALBARELLO; R. GAGLIARDI & C.E.S. JASCONE. 2009. Flora e cobertura vegetal, p. 91-162. *In*: M. BASTOS & C.H. CALLADO (Eds). **O Ambiente da Ilha Grande**. Rio de Janeiro, UERJ/CEADS, 562p.

CHIBA, S. 2004. Ecological and morphological patterns in communities of land snails of the genus *Mandarina* from the Bonin Islands. **Journal of Evolutionary Biology 17**: 131-143. doi: 10.1046/j.1420-9101.2004.00639.x.

CHIBA, S. & A. DAVISON. 2007. Shell shape and habitat use in the North-west Pacific land snail *Mandarina polita* from Hahajima, Ogasawara: current adaptation or ghost of species past? **Biological Journal of the Linnean Society 91**: 149-159. doi: 10.1111/j.1095-8312.2007.00790.x.

CIPRIANI, R. 2007. Modelando las conchas de los moluscos, o la búsqueda de la espiral perfecta, p. 3-11. *In*: S.B. SANTOS; A.D. PIMENTA; S.C. THIENGO; M.A. FERNANDEZ & R.S. ABSALÃO (Eds). **Tópicos em Malacologia – Ecos do XVIII Encontro Brasileiro de Malacologia**. Rio de Janeiro, Sociedade Brasileira de Malacologia, XIV+365p.

CLARKE, B.; W. ARTHUR; D.T. HORSLEY & D.T. PARKIN. 1978. Genetic variation and natural selection in pulmonate molluscs. 219-270. *In*: V. FRETTER & J. PEAKE (Eds). **Pulmonates: Systematics, Evolution and Ecology**. Londres, Academy Press, 540p.

COOK, L.M. 1997. Geographic and ecological patterns in Turkish land snails. **Journal of Biogeography 24**: 409-418. doi: 10.1111/j.1365-2699.1997.00139.x.

DIVER, C. 1931. A method to determining the number of whorls of a shell and its application to *Cepaea hortensis* Müll. **Proceedings of the Malacological Society of London 19**: 1931.

EMBERTON, K.C. 1982. Environment and shell shape in the Tahitian land snail *Partula otaheitana*. **Malacologia 23** (1): 23-35.

EMBERTON, K.C. 1994. Partitioning a morphology among its controlling factors. **Biological Journal of the Linnean Society 53**: 353-369.

EMBERTON, K.C. 1995a. Land-snail community morphologies of the highest-diversity sites of Madagascar, North America and New Zealand, with recommended alternatives to height-diameter plots. **Malacologia 36** (1-2): 43-66.

EMBERTON, K.C. 1995b. Sympatric convergence and environmental correlation between two land-snail species. **Evolution 3**: 469-475.

ENGELMAN, K. 1997. **SYSTAT 7.0**. Chicago, SPSS Inc press, 421p.

FONSECA, A.L.M. & J.W. THOMÉ. 1994. Conquiliomorfologia e anatomia dos sistemas excretor e reprodutor de *Radiodiscus thomei* Weirauch, 1965 (Gastropoda, Stylommatophora, Charopidae). **Biociências 2** (1): 163-188.

GOODFRIEND, G.A. 1986. Variation in land-snail shell form and size and its causes: a review. **Systematic Zoology 2**: 204-223.

GOULD, S.J. 1968. Ontogeny and the explanation of form: an allometric analysis. **Paleontological Society Memoirs 2**: 81-98.

GOULD, S.J. 1984. Covariance sets and ordered geographic variation in *Cerion* from from Aruba, Bonaire and Curaçao: a way of studying nonadaptation. **Systematic Zoology** 33 (2): 217-237.

KLECK, W. 1982. **Discriminant analysis.** Sage University Paper Series on Quantitative Applications in the Social Sciences, 07-0119. Beverly Hills, Sage Publications, 71p.

KREBS, J.C. 1998. **Ecological Methodology.** New York, Benjamin Cummings, XII + 620p.

LEVINE, D. M.; M.L. BERENSON & D. STEPHAN. 2000. **Estatística: teoria e aplicações.** Rio de Janeiro, LTC, 812 p.

MACHIN, J. 1967. Strutural adaptation for reducing water-loss in three species of terrestrial snail. **Journal of Zoology** 152: 55-65. doi: 10.1111/j.1469-7998.1967.tb01638.x.

MONTEIRO, D.P. & S.B. SANTOS. 2001. Conquiliomorfologia de *Tamayoa* (*Tamayops*) *banghaasi* (Thiele) (Gastropoda, Systrophiidae). **Revista Brasileira de Zoologia** 18 (4): 1049-1055. doi: 10.1590/S0101-81752001000400002.

MORRETES, F.L. 1949. Ensaio de catálogo dos moluscos do Brasil. **Arquivos do Museu Paranaense** 7: 1-216.

MYERS, N.; R.A. MITTERMEIER; C.G. MITTERMEIER; G.A.B. FONSECA & J. KENT. 2000. Biodiversity hotspots for conservation priorities. **Nature** 403: 853-858. doi: 10.1126/science.1067728.

OLIVEIRA, R.R. 2002. Ação antrópica e resultantes sobre a estrutura e composição da Mata Atlântica na Ilha Grande, RJ. **Rodriguésia** 53 (82): 33-58. doi: 10.1590/S1414-753X2007000200002.

PARMAKELIS, A.; E. SPANOS; G. PAPAGIANNAKIS; C. LOUIS & M. MYLONAS. 2003. Mitochondrial DNA phylogeny and morphological diversity in the genus *Mastus* (Beck, 1837): a study in recent (Holocene) island group (Koufonisi, south-east Crete). **Biological Journal of the Linnean Society** 78: 383-399. doi: 10.1046/j.1095-8312.2003.00152.x.

PARODIZ, J.J. 1951. Métodos de Conquiliometria. **Physis** 20 (38): 241-248.

PEAKE, J.F. 1978. Distribution and ecology of Stylommatophora, p. 429-526. *In*: V. FRETTER & J. PEAKE (Eds). **Pulmonates. Systematics, Evolution and Ecology.** New York, Academic Press, vol. 2A, 540p.

ROCHA, C.F.D.; H.G. BERGALLO; M.A.S. ALVES & M.V. SLUYS. 2003. **A biodiversidade nos grandes remanescentes florestais do Estado do Rio de Janeiro e nas restingas da Mata Atlântica.** São Carlos, Editora Rima, 160p.

SANTOS, S.B.; A.B. BARBOSA; R.M.R.B. BRAGA; J.L. OLIVEIRA & R.F. XIMENES. 2010. Moluscos da Ilha das Flores, São Gonçalo, Rio de Janeiro. **Informativo SBMa** 173: 10-14.

SHIMEK, B. 1930. Land snails as indicators of ecological conditions. **Ecology** 11 (4): 673-686. doi: 10.2307/1932328.

SIMONE, L.R.L. 2007. **Land and freshwater molluscs of Brazil.** São Paulo, EGB, Fapesp, 390p.

SLUYS, M.V.; R.V. MARRA; L. BOQUIMPANI-FREITAS; & C.F.D. ROCHA. 2012. Environmental factors affecting calling behavior of sympatric frog species at an Atlantic Rain Forest area, Southeastern Brazil. **Journal of Herpetology** 46 (1): 41-46.

SOLEM, A. & F.M. CLIMO. 1985. Structure and habitat correlations of sympatric New Zealand land snail species. **Malacologia** 26: 1-30.

SPIEGEL, M.R. 1993. **Estatística.** São Paulo, Makron Books, Coleção Schaum, 643p.

TESHIMA, H.; A. DA VISON; Y. KUWAHARA; J. YOKOHAMA; S. CHIBA; T. FUKUDA; H. OGIMURA & M. KAWATA. 2003. The evolution of extreme shell shape variation in the land snail *Ainohelix editha*: a phylogeny and hybrid zone analysis. **Molecular Ecology** 12: 1869-1878. doi: 10.1046/j.1365-294X.2003.01862.x.

THIELE, J. 1927. Über einige brasilianische Landschnecken. **Abhandlungen der Senckenbergischen Naturforschenden Gesellschaft** 40 (3): 307-329.

THIELE, J. 1931. **Handbuch der Systematischen Weichtierkunde.** Jena, Gustav Fischer, vol. 1, 778p.

TILLIER, S. 1981. Clines, convergence and character displacement in new Caledonian diplommatinids (land prosobranchs). **Malacologia** 21 (1-2): 177-208.

VALORVITA, I. & VÄISÄNEN, R. A. 1986. Multivariate morphological discrimination between *Vitrea contracta* (Westerlund) and *V. crystallina* (Müller)(Gastropoda, Zonitidae). **Journal Molluscan Studies** 52: 62-67. doi: 10.1093/mollus/52.1.62

VERMEIJ, G.J. 1971. Gastropod evolution and morphological diversity in relation to shell geometry. **Journal of Zoology** 163: 15-23. doi: 10.1111/j.1469-7998.1971.tb04522.x.

VERA-Y-CONDE, C.F. & C.F.D. ROCHA. 2006. Habitat disturbance and small mammal richness and diversity in an atlantic rainforest area in southeastern Brazil. **Brazilian Journal of Biology** 66 (4): 983-990. doi: 10.1590/S1519-69842006000600005.

WELTER-SCHULTES, F.W. 2000. Human-dispersed land snails in Crete, with special reference to *Albinaria* (Gastropoda: Clausiliidae). **Biologia Gallo-hellenica** 24: 83-106.

ZAR, J. H. 1999. **Biostatistical Analysis.** New Jersey, Prentice-Hall, 663p.

ZILCH, A. 1959. **Gastropoda: Euthyneura.** Berlim, Borträger, vol. 2, 834p.

16

Habitat use and movements of *Glossophaga soricina* and *Lonchophylla dekeyseri* (Chiroptera: Phyllostomidae) in a Neotropical Savannah

Ludmilla M.S. Aguiar[1,3], Enrico Bernard[2] & Ricardo B. Machado[1]

[1] *Departamento de Zoologia, Instituto de Ciências Biológicas, Universidade de Brasília, Campus Darcy Ribeiro, Asa Norte, 70910-900 Brasília, DF, Brazil.*
[2] *Departamento de Zoologia, Universidade Federal de Pernambuco. Rua Nelson Chaves, Cidade Universitária, 50670-420 Recife, PE, Brazil.*
[3] *Corresponding author. E-mail: ludmillaaguiar@unb.br*

ABSTRACT. The greatest current threat to terrestrial fauna is continuous and severe landscape modification that destroys and degrades animal habitats. This rapid and severe modification has threatened species, local biological communities, and the ecological services that they provide, such as seed dispersal, insect predation, and pollination. Bats are important pollinators of the Cerrado (woodland savanna) because of their role in the life cycles of many plant species. However, there is little information about how these bat species are being affected by habitat loss and fragmentation. We used radio-tracking to estimate the home ranges of *Glossophaga soricina* (Pallas, 1776) and *Lonchophylla dekeyseri* Taddei, Vizotto & Sazima, 1983. The home range of *G. soricina* varies from 430 to 890 ha. They combine short-range flights of up to 500 m to nearby areas with longer flights of 2 to 3 km that take them away from their core areas. The maximum flight distance tracked for *L. dekeyseri* was 3.8 km, and its home range varies from 564 to 640 ha. The average distance travelled by this species was 1.3 km. Our data suggest that *G. soricina* and *L. dekeyseri* are able to explore the fragmented landscape of the Central Brazilian Cerrado and that they are likely to survive in the short- to medium-term. The natural dispersal ability of these two species may enable them to compensate for continued human disturbance in the region.

KEY WORDS. Bats; Brazil; Cerrado biome; home range; radio-tracking.

Habitat destruction and the associated degradation are the greatest threats to terrestrial fauna. Habitat loss impacts 86% of threatened mammals (BAILLIE *et al.* 2004), and there is no indication that landscape transformation is diminishing, especially in the Cerrado, which is a Brazilian biome that currently has a high level of industrial agriculture.

The Cerrado of Central Brazil, the most biologically rich savannah in the world, has been under serious threat for the last 35 years. More than 50% of its land has been lost to pasture and cropland (KLINK & MACHADO 2005, BRASIL 2009a). This rapid and severe landscape modification has threatened many species, which in turn has affected local biological communities and the ecological services that they provide, including seed dispersal, insect predation, and pollination. Long-distance pollen dispersal is important for plants, and bat pollinators are particularly well suited to provide this service (FLEMING & SOSA 1994).

Bats are among the most important pollinators of the Cerrado because of their roles in the life cycles of many typical cerrado *sensu stricto* plant species, such as *Caryocar brasiliensis* Cambess., *Bauhinia holophylla* Steud., *Hymenaea stigonocarpa* Mart and *Luehea grandiflora* Mart. & Zucc. (BOBROWIEC & OLIVEIRA

2012). However, there is little knowledge about how these bat species are being affected by continued habitat loss and fragmentation in the region. There are also a number of gaps in our knowledge of the basic characteristics of the bat species of the Cerrado, including their diet, ecology, roosting requirements, behavior, and physiology (e.g., BERNARD *et al.* 2011).

Radio-tracking is an important tool in the exploration of how native species perceive the structure of natural or altered landscapes in their habitats. Bats are seemingly unaffected by small changes in landscape structure because of their ability to fly, but contradictory results have been reported in the literature (FENTON 1997, GORRENSEN *et al.* 2005). Part of this contradiction can be explained by the species-dependent nature of responses to landscape change (GORRENSEN *et al.* 2005). AVILA-CABADILLA *et al.* (2012), have shown that nectivorous bats are negatively associated with dry forest patches, while frugivorous bats are positively associated with riparian forest. Phyllostomid bats in savannah areas of the Brazilian Amazon were found by radio-tracking to have ranges of 0.5 to 2.5 km, which demonstrates interspecific differences in flight ability (BERNARD & FENTON 2003). In contrast, phyllostomids in French Guiana and Panama were reluctant to traverse open spaces that composed a water matrix (Cosson *et al.* 1999, Albrecht *et al.* 2007). Therefore, the type of matrix has an impact on flight behavior. The perception of landscape fragmentation may also vary among individuals and species (Gorrensen *et al.* 2005), especially in regions that are naturally composed of a mosaic of different phytophysiognomies (see Montiel *et al.* 2006, Loayza & Loiselle 2008), such as the Cerrado (woodland savannah) (Eiten 1972).

We do not know how bats are responding to the anthropogenic changes that are occurring in the Cerrado biome. The use of radio-tracking studies may contribute to the understanding of how pollinator bat species are affected. Data on the home range and habitat use of a given species can be applied to determine its dispersal capacity, which indicates the ability of that species to utilize and survive in a fragmented landscape (Lima *et al.* 2012). This information can be used to establish conservation strategies for bat species, such as defining minimum acceptable distances between roosts and feeding areas, or determining the optimal quality of the ecosystems that they inhabit. We therefore used radio-tracking to investigate habitat use and movements of two small, primarily nectivorous phyllostomid species, Dekeyser's nectar bat – *Lonchophylla dekeyseri* Taddei, Vizotto & Sazima, 1983 –, and Pallas' long-tongued bat – *Glossophaga soricina* (Pallas, 1776). The two species are similar in size and diet (both eat pollen, fruits and insects) but differ in abundance. We hypothesize that the home range of the less abundant species (*L. dekeyseri*) is more restricted than that of *G. soricina* because it avoids crossing altered areas. Thus, we expect that the home range and the mean flight distance of *G. soricina* is greater than that of *L. dekeyseri*.

MATERIAL AND METHODS

The study was conducted in the Roncador Ecological Reserve (RER – 15°56'S, 47°53'W), which is 35 km south of the capital Brasília, and near the Sal Fenda cave (SFC – 15°30'35"S, 48°09'59"W) in Brazlândia, which is 45 km east of Brasília (Fig. 1). Spaced thorny trees (3 to 8 m tall) that are surrounded by smaller bushes and grasslands known as the cerrado *sensu stricto* characterize the vegetation in both areas. (The cerrado is a phytophysiognomy and Cerrado is the biome). The rainy season lasts from October to April, and the dry season lasts from May to September. The average temperature is 22°C, and the average annual rainfall is approximately 1,500 mm (Eiten 1994).

Brazlândia is within the Cafuringa Area of Environmental Protection (APA Cafuringa, 46,000 ha), which contains some of the last remaining native grasslands in the region and several limestone caves in the Distrito Federal itself (Baptista 1998). Sal Fenda is the largest granite cave in the Distrito Federal and is located at 840 m above sea level. There is an extension of the cave that has an elevation of 865 m. The original cerrado vegetation that surrounded the cave has been heavily deforested and replaced by pastures and degraded semi-deciduous

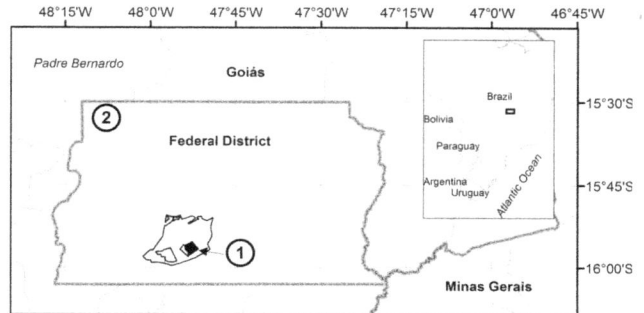

Figure 1. Location of the study sites in the Distrito Federal. The number "1" indicates the Reserva Ecológica do Roncador (RER) within the Área de Proteção Ambiental Gama-Cabeça do Veado, and the number "2" indicates the location of the Sal Fenda cave.

forest. The RER is surrounded by two other reserves, the Ecological Station of Brasília's Botanical Garden (JBB) on its northeastern border and Brasília University's Farm Água Limpa (FAL) on its southeastern border. These reserves together form a single large continuous tract of cerrado (*sensu stricto*). Nearly 75% of the RER's 1,350 ha are covered by this cerrado, with a heterogeneous landscape that is composed of smooth slopes and is crossed by small rivers and streams.

We defined a polygon of approximately 20 x 15 km (approximately 30,000 hectares) around each study site (RER and SFC) to assist in evaluating the status of natural coverage. According to maps prepared in 2009 by the Center for Remote Sensing of the Brazilian Institute of Environment and Renewable Natural Resources (Brasil 2009b), the remnants of native vegetation cover in the RER area represented 71.2% of this polygon. There are 15 fragments in the polygon. The average fragment size is of 1163.1 ha, and there is an average distance of 138 m between them. Native remnants represented 52% of the SEC area. It has 16 fragments, with an average fragment size of 790.4 ha. The average distance between fragments in this area was 395.5 m. We calculated all measurements that presented here with the extension Patch Analyst 5.0 (Rempel *et al.* 2012) and ArcGIS 10.0 software (ESRI 2010).

Lonchophylla dekeyseri is a small bat (10.7 ± 0.85 g; forearm length: 35.3 ± 8.5 mm) that is endemic to the Cerrado. It is threatened by extinction ("Endangered" according to the Brazilian Ministry of the Environment and "Near Threatened" according to the IUCN Red List) (Brasil 2003, IUCN 2011). This species is still relatively unknown and began to receive more attention only after its inclusion in the official list of Brazilian endangered species. In contrast, *G. soricina* (10.6 ± 1.7 g; forearm length: 35.4 ± 1.4 mm) is found from northern México to southern Argentina, including Jamaica and other islands near the northern coast of South America (Alvarez *et al.* 1991). This widespread species is not at risk of extinction and occupies a large variety of habitats, ranging from arid-subtropical thorn forest to tropical rainforests and savannahs, from sea level to 2,600 m elevation (Alvarez *et al.* 1991). Both species feed on insects, fruits and flower parts in addition to pollen and nectar. The known food sources of these bats include the pollen of several Cerrado plant species, including *Hymenaea stigonocarpa* Mart, ex Hayne, *Bauhinia brevipes* Vogel, *Bauhinia cupulata* Benth., *Bauhinia multinervia* (Kunth) D.C., *Bauhinia megalandra* Griseb., *Bauhinia pauletia* Pers., *Bauhinia ungulata* L., *Bauhinia rufa* (Bong.), and *Luehea speciosa* Wild. (Gibbs *et al.* 1999, Gribel & Hay 1993, Heithaus *et al.* 1975, Hokche & Ramírez 1990, Ramírez *et al.* 1984).

From 6-9 September 2005, four *G. soricina* individuals (one male, three females) were captured at RER, and from 10-12 October 2005, 10 *L. dekeyseri* individuals (five males, five females) were captured at the entrance of the Sal Fenda cave. All of the bats were captured with mist nets that were set around flowering pequi trees (*C. brasiliensis*). Radio transmitters (model LB-2, Holohil, Carp, Ontario, Canada) that were compliant with the 5% mass limit rule (Aldridge & Brigham 1988) were glued to the backs of the bats with Skin-Bond glue. Three pairs of researchers who were in radio contact with each other tracked the bats. At RER, where only *G. soricina* was tracked, two of the tracking teams were stationed at the tops of two existing 8- and 12-m towers, while the third team was instructed to move around to improve signal reception (White & Garrott 1990, Mason & Hope 2014). At Sal Fenda, where only *L. dekeyseri* was tracked, two 6-m aluminum pole structures were constructed to hold one antenna each, and the third was mobile (White & Garrott 1990, Mason & Hope 2014).

The location of each tracking station was marked using the average function of a GPS device (Garmin ETrex). The receivers that were used were TRX-1000 models (Wildlife Materials, Murphysboro, Illinois, USA) paired with Yagi Three-Element Antennas (Titley Scientific, Columbia, Missouri, USA). The teams attempted to locate each bat and collect a location point for each individual at five-minute intervals over the tracking period. Three bearings (one per team) were required for a location to be recorded (White & Garrott 1990). A bat was considered stationary when three consecutive bearing measurements indicated the same position for an individual. Data on the frequencies, date, time, and direction of the strongest signal intensities were taken as positional bearings using a compass and were recorded for each bat (White & Garrott 1990, Mason & Hope 2014).

The trajectory of displacement for each of the bats was represented by the sum of the sequential movements of each individual. A movement corresponded to the distance that was covered between two points as measured by triangulation of the bearings (White & Garrott 1990, Mason & Hope 2014). It was therefore possible to obtain the pattern of dispersion for each bat at the end of a tracking session.

The average distance covered and the flight directions taken by all of the tracked bats were recorded and plotted on a map. The bats were tracked until their signals were lost. *G. soricina* individuals were tracked from approximately 18:00 to 03:00 h the following day, and *L. dekeyseri* individuals were tracked from approximately 18:00 to 05:00 h.

The home range for each individual was calculated using the 'Animal Movement' extension (Hooge *et al.* 2001) of the ArcView 3.3 software (ESRI 2002) and was based on minimum convex polygons (MCP) and 95% kernel estimators (Hooge & Eichenlaub 1997, Jacob & Rudran 2003, White & Garrott 1990). To estimate the type and prevalence of the habitats that were within the calculated home ranges, the polygons were plotted on a 30-m resolution, classified Landsat ETM+ satellite image (taken on August 2002) that contained three native vegetation classes (cerrado, grassland, and forest) and one altered class (grouped anthropogenic areas, such as pasture, crop, urban and deforested areas).

RESULTS

Two of the four tracked *G. soricina* individuals (one male and one female) disappeared immediately following their release, and their signals were not detected for the remainder of the study. The two remaining individuals (females #58 and #69) were tracked for seven consecutive days, resulting in 105 location records (Table I). Based on their MCPs, their home ranges varied from 427 to 893.6 ha (Fig. 2). Both of the tracked individuals combined short-range flights of up to 500 m, possibly to scan nearby areas, with longer distance flights of 2 to 3 km to reach areas that were more distant from the core. The two bats used different parts of their home ranges to different degrees. On average, the main habitat types that they visited were cerrado vegetation (49.7%), grasslands (6%), and gallery forest areas (14.0%, used just twice) (Table II). The bats crossed disturbed areas (pasture and crops) to reach native areas that were outside of the RER. Based on 27 movements, the maximum flight distance tracked was 3.8 km and the average distance was 1.5 km (SD = 0.9 km).

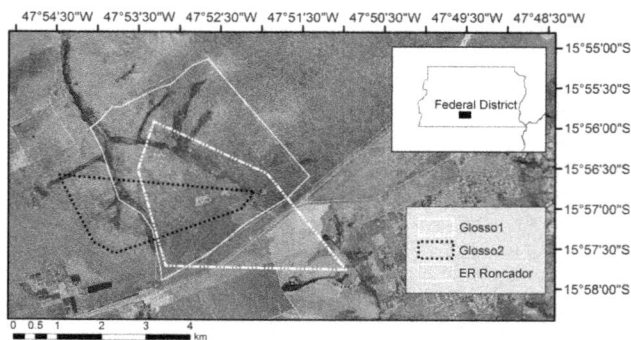

Figure 2. Home ranges of the two radio-tracked *Glossophaga soricina* bats (white polygons) in the Reserva Ecológica do Roncador (RER) (black polygon).

Table I. Basic data from each transmitter used to evaluate home range and movements of *Glossophaga soricina* and *Lonchophylla dekeyseri* in the Cerrado of Brasília, Brazil.

Species	Indiv.	Date	Time (min)	Valid points	Frequency
Glossophaga soricina	#58	6-8/Sep	580	85	151.140
Glossophaga soricina	#69	9/Sep	100	20	151.155
Lonchophylla dekeyseri	#59	10-Oct	600	15	151.161
Lonchophylla dekeyseri	#60	10-12/Oct	1800	35	151.180
Lonchophylla dekeyseri	#61	10-12/Oct	1800	15	151.198
Lonchophylla dekeyseri	#62	10-Oct	600	4	151.220

[1] The Time column indicates the total time that the individual was tracked in the study region.
[2] The Valid points column indicates the number of triangulated points that were used to calculate the individual's home range.

Table II. Composition of the native and anthropogenic environments within the home range of the tracked individuals of *Glossophaga soricina* at RER, Brasília, Brazil.

Coverage	Indv. 58		Indiv. 62		Mean	
	Area (ha)	%	Area (ha)	%	Area (ha)	%
Pasture/grass	17.30	4.1	152.6	17.1	161.3	14.6
Forest	72.90	17.1	118.9	13.3	155.3	14.0
Cerrado s.s.	214.50	50.2	443.3	49.6	550.5	49.7
Grassland	14.20	3.3	58.8	6.5	65.9	6.0
Others (nonnative)	108.09	25.3	120.0	13.4	174.0	15.7
Total	427.00		893.6		660.3	

Six of the 10 *L. dekeyseri* individuals that were equipped with transmitters disappeared immediately after release and their signals were not recorded for the remainder of the tracking period. Two of the four remaining bats (females #60 and #61) were tracked for three consecutive nights, while the others (females #59 and #62) were tracked for a single night before their signals disappeared. The tracking of these four bats resulted in 69 location records (Table I). Three of the four tracked bats provided enough readings to estimate their home ranges (Fig. 3). Their average home range was 640 ha (SD = 704 ha) based on the MCPs and 564 ha (SD = 265 ha) based on 95% kernel estimates. The bats explored both fragments and forests, with an average of 53% of their home ranges (approximately 343 ha) being composed of cerrado and the remaining 47% of pasture (Table III).

In general, these home ranges were composed of irregular terrain, with drier forests along the steepest parts of the area and rock outcrops. Most of the native areas were in poor conservation condition, and there were cattle in several areas. Consequently, the understory in those areas was more degraded and more open than expected. Like the *G. soricina* individuals, the tracked *L. dekeyseri* individuals visited different parts of their home ranges during different nights. For example, one female

(#60) visited the northeastern, northwestern, and northern parts of its home range and several points along the southern, southwestern and northeastern areas on four consecutive nights. The average movement distance of this species was 1.3 km (SD = 1.0 km), and the greatest distance flown was 3.8 km. The tracked bats exhibited two activity peaks, at approximately 19:00 and 02:00 h. The lowest activity occurred at approximately 21:00 and 04:00 h.

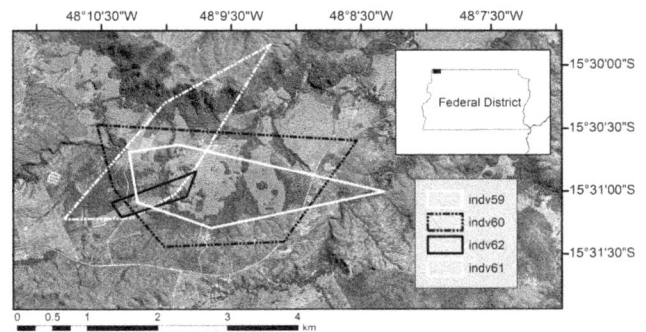

Figure 3. Home ranges of the four radio-tracked *Lonchophylla dekeyseri* bats in the northern region of the Distrito Federal.

Table III. Composition of the native and anthropogenic environments within the home range of the tracked individuals of *Lonchophylla dekeyseri* at Sal Fenda Cave, Brazlândia, Brazil.

Coverage	Indiv. 59		Indiv. 60		Indiv. 61		Mean	
	Area (ha)	%	Area (ha)	%	Area (ha)	%	Area (ha)	%
Pasture	105.8	44.7	700.2	48.2	73.7	32.0	293.2	45.8
Forest	93.1	39.3	274.3	18.9	99.5	43.2	155.6	24.3
Cerrado s.s.	24.0	10.2	335.8	23.1	30.9	13.4	130.3	20.3
Grassland	13.7	5.8	143.6	9.9	16.8	7.3	58.0	9.1
Others (nonnative)	0.0	0.0	0.0	0.0	9.3	4.0	3.1	0.5
Total	236.6		1453.9		230.2		640.3	

DISCUSSION

Although the sample sizes in the present study are small and the measurements were limited to the dry season, the data support the hypothesis that the home range of *G. soricina* is larger than that of *L. dekeyseri*. The sample sizes do not permit more powerful statistical comparisons. The flight distances that were recorded for both species were similar. We therefore cannot confirm our hypothesis regarding differences in flight distance. Both species showed the same ability to visit natural vegetation patches in the study areas.

The characteristics of the vegetation structure of the study areas did not impose any significant impediment to radio signal transmission. This allowed us to detect signals at distances of up to 5 km. This detection distance was even greater when the antennas were positioned on a 10- to 12-meter-high fire observation tower. Other studies (McGuire 2010) have reported detection of signals from bats that were equipped with such small radios at distances of up to 12 km in estuarine regions.

ROTHENWÖHRER et al. (2011) state that the spatial and temporal activity of *Glossophaga commissarisi* Gardner, 1962 are closely matched to the local resource landscape, with high resource density allowing smaller home ranges, lower flight duration and thus reduced foraging costs. *G. soricina* and *L. dekeyseri* were much more mobile than expected despite their small body sizes. Both species are able to consume pollen, fruits and insects, suggesting that their broad home ranges reflect the wide spatial distribution of food or food quality rather than food scarcity. Previously published data show that bats of the savannahs of Brazil, Bolivia and Africa have smaller home ranges and flight distances than those that were observed in the present study. For example, the home range of *G. commissarisi*, at La Selva (Costa Rica) ranges from 7.4 to 23.9 ha (ROTHENWÖHRER et al. 2011), and that of *Artibeus watsoni* Thomas, 1901 is 3.6 ha (ROTHENWÖHRER et al. 2011). *Carollia perspicillata* (Linnaeus, 1758), a larger but still small frugivore, was found to have a home range of 155 to 320 ha in the savannah of the Brazilian Amazon (BERNARD & FENTON 2003). Even when we compare data that were collected only from areas of open vegetation, the home ranges of our study exceed those that have been reported for larger species. In the Bolivian savannah, the home range of *Sturnira lilium* (E. Geoffroy, 1810) ranges from 36.5 to 190.7 ha (LOAYZA & LOISELLE 2008), and the home range of *Megaloglossus woermannii* Pagenstecher, 1885, (which weighs less than 20 g) in the Lama Forest Reserve, southern Benin, West Africa, varies from 99.8 to 146.8 ha (WEBER et al. 2009).

The larger home ranges that were observed in the present study for both *L. dekeyseri* and *G. soricina* might be explained by one of two different hypotheses. First, the ability to fly longer distances than those observed in other regions may be a consequence of the natural horizontal heterogeneity of the Cerrado. The Cerrado is a mosaic of different vegetation physiognomies, from grassland to forest, and it varies on the horizontal scale at the level of hundreds of meters (EITEN 1972, 1994). This characteristic of the Cerrado may encourage environmental plasticity of its bat species through the separation of habitat types. In this case, covering distances of 1 to 3 km to explore the surrounding area would be a natural circumstance for these animals, and may be sufficient to allow them to survive in landscapes with an intermediate level of fragmentation (such as the study site, where the fragments are separated by an average distance of 300-400 m).

Alternatively, these longer flight distances may be a result of the habitat fragmentation of the study areas. The bats may have to fly long distances to find suitable habitat fragments. The longer flight distances of both species are noteworthy regardless of the underlying cause.

Others have argued that the nectar-feeding phyllostomid bats of the subfamilies Lonchophyllinae and Glossophaginae are highly susceptible to extinction because many of these species are food and habitat specialists, roost in caves, show migratory behavior, and are already rare (ARITA & SANTOS-DEL-PRADO 1999). Others have found that these species are more resilient than other guilds to changes in land use (WILLIG et al. 2007). Our data indicate that *G. soricina* and *L. dekeyseri* are able to utilize the fragmented landscape of the Central Brazilian cerrado successfully. This ability may make them more likely to survive in the short run, but we still do not know how the quality of these different remnant fragments influences the long-term survival of both species. Nevertheless, populations of *L. dekeyseri* are declining in response to habitat loss and human disturbance (AGUIAR et al. 2010), and this species has already been declared as endangered (BRASIL 2003).

ACKNOWLEDGEMENTS

This study is part of the *L. dekeyseri* Action Plan. We thank PROBIO/FNMA/CNPq for funding this research. We thank field assistants Adriana Bochiglieri, Nicholas Camargo, William Camargo, Alexandre Portela, and Fábio Souza for their invaluable help in the fieldwork. We also thank Brock Fenton and three other anonymous reviewers for their invaluable comments. CNPq provides a research fellowship to RBM. English revision was done by American Jounal Expert Reviews (AJE).

REFERENCES

AGUIAR, L.M.S.; D. BRITO & R.B. MACHADO. 2010. Do current vampire bat (*Desmodus rotundus*) population control practices pose a threat to Dekeyser's nectar bat's (*Lonchophylla dekeyseri*) long-term persistence in the Cerrado? **Acta Chiropterologica** 12 (2): 275-282. doi: 10.3161/150811010X537855

ALBRECHT, L.; C.F.J. MEYER & E.K.V. KALKO. 2007. Differential mobility in two small phyllostomid bats, *Artibeus watsoni* and *Micronycteris microtis*, in a fragmented Neotropical landscape. **Acta Theriologica** 52 (2): 141-149.

ALDRIDGE, H.D.J.N. & R.M. BRIGHAM. 1988. Load carrying and manoeuvrability in an insectivorous bat: a test of the 5% rule of radio-telemetry. **Journal of Mammalogy** 69 (2): 379-382.

ALVAREZ, J.; M.R. WILLIG; J.K.J. JONES & W.D. WEBSTER. 1991. *Glossophaga soricina*. **Mammalian Species** 379: 1-7.

ARITA, H.T. & K. SANTOS-DEL-PRADO. 1999. Conservation biology of nectar-feeding bats in Mexico **Journal of Mammalogy** 80 (1): 31-41.

AVILA-CABADILLA, L.D.; G.A. SANCHEZ-AZOFEIFA; K.E. STONER; M.Y. ALVAREZ-AÑORVE; M. QUESADA & C.A. PORTILLO-QUINTERO. 2012. Local and landscape factors determining occurrence of Phyllostomid bats in tropical secondary forests. **PLoS ONE** 7: e35228. doi:10.1371/journal.pone.0035228

BAILLIE, J.E.M.; L.A. BENNUN; T.M. BROOKS; S.H.M. BUTCHART; J.S. CHANSON; Z. COKELISS; C. HILTON-TAYLOR; M. HOFFMANN; G. MACE & S.A. MAINKA. 2004. **IUCN Red List of Threatened Species – a global species assessment.** Cambridge, The IUCN Species Survival Commission.

BAPTISTA, G.M.M. 1998. Caracterização climatológica do Distrito Federal, p. 187-208. *In*: **Inventário hidrogeológico e dos recursos hídricos superficiais do Distrito Federal.** Brasília, IEMA/SEMATEC/UnB.

BERNARD, E. & M.B. FENTON. 2003. Bat mobility and roosts in a fragmented landscape in central Amazonia, Brazil. **Biotropica 35** (2): 262-277. doi: 10.1111/j.1744-7429.2003.tb00285.x

BERNARD, E.; L.M.S. AGUIAR & R.B. MACHADO. 2011. Discovering the Brazilian bat fauna: a task for two centuries? **Mammal Review 41** (1): 23-39. doi: 10.1111/j.1365-2907.2010.00164.x

BOBROWIEC, P.E.D. & P.E. OLIVEIRA. 2012. Removal effects on nectar production in bat-pollinated flowers of the Brazilian Cerrado. **Biotropica 44** (1): 1-5. doi: 10.1111/j.1744-7429.2011.00823.x

BRASIL. 2003. Espécies da fauna brasileira ameaçadas de extinção, p. 88-97. *In*: **Instruction 3 of 27 May 2003.** Brasília Ministério do Meio Ambiente, Published in the Official Gazette 101 of 28 May 2003, Section 1.

BRASIL. 2009a. **Technical Report on monitoring deforestation in the Cerrado biome, 2002 a 2008 revised data.** Brasília, MMA/IBAMA/CID.

BRASIL. 2009b. **Mapas de cobertura vegetal dos biomas brasileiros.** Brasília, Secretaria de Biodiversidade e Florestas, Ministério do Meio Ambiente-MMA.

COSSON, J.F.; S. RINGUET; O. CLAESSENS;, J.C. DE MASSARY; A. DALECKY; J.F. VILLIERS; L. GRANJON & J.M. PONS. 1999. Ecological changes in recent land-bridge islands in French Guiana, with emphasis on vertebrate communities. **Biological Conservation 91** (2-3): 213-222. doi: 10.1016/S0006-3207(99)00091-9

EITEN, G. 1972. The Cerrado vegetation of Brazil. **Botanical Review 38** (2): 201-341.

EITEN, G. 1994. CerradoVegetation. *In*: M. PINTO (Ed.). **Cerrado: characterization, occupation and perspectives.** Brasília, Editora Universidade de Brasília.

ESRI. 2002. **ArcView 3.3 – Geographical information system.** Readlands, Environment System Research Institute, Inc.

ESRI. 2010. **ArcGIS 10.0 – Geographical Information System.** Readlands, Environment System Research Institute, Inc.

FENTON, M.B. 1997. Science and the conservation of bats. **Journal of Mammalogy 78** (1): 1-1.

FLEMING, T.H. & V. SOSA. 1994. Effects of nectarivorous and frugivorous mammals on reproductive success of plants. **Journal of Mammalogy 75** (4): 845-851.

GIBBS, P.E.; P.E. OLIVEIRA & M.B. BIANCHI. 1999. Postzygotic control selfing in *Hymenaea stigonocarpa* (Leguminosae-Caesalpinioideae), a bat-pollinated tree of the Brazilian cerrados. **International Journal of Plant Science 160** (1): 72-78.

GORRENSEN, P.M.; M.R. WILLIG & R.E. STRAUSS. 2005. Multivariate analysis of scale-dependent associations between bats and landscape structure. **Ecological Applications 15** (6): 2126-2136. doi: 10.1890/04-0532

GRIBEL, R. & J. D. HAY. 1993. Pollination ecology of *Caryocar brasiliensis* (Caryocaraceae) in Central Brazil cerrado vegetation. **Journal of Tropical Ecology 9** (2): 199-211. doi: 10.1017/S0266467400007173

HEITHAUS, E.R.; T.H. FLEMING & P.A. OPLER. 1975. Foraging patterns and resource utilization in seven species of bats in a seasonal tropical forest. **Ecology 56** (4): 841-854.

HOKCHE, O. & N. RAMÍREZ. 1990. Pollination ecology of seven species of Bauhinia L. (Leguminosae: Caesalpinioideae). **Annals of Missouri Botanical Garden 77** (3): 559-572.

HOOGE, P.N. & W. EICHENLAUB. 1997. **Animal movement extension to arcview ver. 1.1.** Anchorage, Alaska Biological Science Center, U.S. Geological Survey.

HOOGE, P.N.; W. EICHENLAUB & E.R. HOOGE. 2001. **Animal movement 2.5.** Anchorage, US geological survey, Alaska Biological Science Center.

IUCN. 2011. **Iucn Red List of Threatened Species.** Version 2011.1.

JACOB, A.A. & R. RUDRAN. 2003. Radiotelemetry in population studies, p. 285-342. *In*: L.C. CULLEN JR; R. RUDRAN & C. VALLADARES-PÁDUA (Eds). **Study methods in conservation biology and wildlife management.** Curitiba, Paraná.

KLINK, C.A. & R.B. MACHADO. 2005. Conservation of the Brazilian Cerrado. **Conservation Biology 19** (3): 707-713. doi: 10.1111/j.1523-1739.2005.00702.x

LIMA, E.S.; K.E. DEMATTEO; R.S.P. JORGE; M.L.S.P. JORGE; J.C. DALPONTE; H.S. LIMA & S.A. KLORINE. 2012. First telemetry study of bush dogs: home range, activity and habitat selection. **Wildlife Research 39** (6): 512-519. doi: 10.1071/WR11176

LOAYZA, A. & B. A. LOISELLE. 2008. Composition and distribution of a bat assemblage during the dry season in a naturally fragmented landscape in Bolivia. **Journal of Mammalogy 90** (3): 732-742. doi: 10.1644/08-MAMM-A-213R.1

MCGUIRE, L. P. 2010. **Bat Migration Stopover Ecology.** Ontario, Ontario Ministry of Natural Resources, Technical Report.

MASON, V. & P.R. HOPE. 2014. Echoes in the dark: Technological encounters with bats. **Journal of Rural Studies 33**: 107-118. doi: 10.1016/j.jrurstud.2013.03.001

MONTIEL, S.; A. ESTRADA & P. LEÓN. 2006. Bat assemblages in a naturally fragmented ecosystem in the Yucatan Peninsula, Mexico: species richness, diversity and spatio-temporal dynamics. **Journal of Tropical Ecology 22**: 267-276. doi: 10.1017/S026646740500307X

RAMÍREZ, N.; C. SOBREVILA; N.X. ENRECH & T. RUIZ-ZAPATA. 1984. Floral biology and breeding system of *Bauhinia benthamiana* Taub. (Leguminosae), a bat-pollinated tree in Venezuelan Llanos. **American Journal of Botany 71** (2): 273-280.

REMPEL, R.S.; D. KAUKINEN & A.P. CARR. 2012. Patch Analyst and Patch Grid. Ontario, Ontario Ministry of Natural Resources, Centre for Northern Forest Ecosystem Research.

Rothenwöhrer, C.; N.I. Becker & M. Tschapka. 2011. Resource landscape and spatio-temporal activity patterns of a plant-visiting bat in a Costa Rican lowland rainforest. **Journal of Zoology 283** (2): 108-116. doi: 10.1111/j.1469-7998.2010.00748.x

White, G.C. & R.A. Garrott. 1990. **Analysis of wildlife radio-tracking data.** San Diego, Academic Press.

Willig, M.R.; S.J. Presley; C.P. Bloch; C.L. Hice; S.P. Yanoviak; M.M. Diaz; L.A. Chauca; V. Pacheco & S.C. Weaver. 2007. Phyllostomid bats of lowland Amazonia: effects of habitat alteration on abundance. **Biotropica 39** (6): 737-746. doi: 10.1111/j.1744-7429.2007.00322.x

Weber, N.; E.K.V. Kalko & J. Fahr. 2009. A First Assessment of Home Range and Foraging Behaviour of the African Long-Tongued Bat *Megaloglossus woermanni* (Chiroptera: Pteropodidae) in a Heterogeneous Landscape within the Lama Forest Reserve, Benin. **Acta Chiropterologica 11** (2): 317-329. doi: 10.3161/150811009X485558

17

Ecomorphological relationships among four Characiformes fish species in a tropical reservoir in Southeastern Brazil

Débora de S. Silva-Camacho[1,2], Joaquim N. de S. Santos[1], Rafaela de S. Gomes[1] & Francisco G. Araújo[1]

[1] Laboratório de Ecologia de Peixes, Universidade Federal Rural do Rio de Janeiro. Antiga Rodovia Rio-SP km 47, 23851-970 Seropédica, RJ, Brazil.
[2] Corresponding author. E-mail: debora_desouza@yahoo.com.br

ABSTRACT. The aim of this study was to assess the ecomorphological patterns and diet of four Characiformes fish species in a poorly physically structured tropical reservoir. We tested the hypothesis that body shape and diet are associated, because environmental pressure acts on the phenotype, selecting traits according to the available resources. Ten ecomorphological attributes of 45 individuals of each species – *Astyanax* cf. *bimaculatus* (Linnaeus, 1758), *Astyanax parahybae* Eigenmann, 1908, *Oligosarcus hepsetus* (Cuvier, 1829), and *Metynnis maculatus* (Kner, 1858) –, collected between February and November 2003, were analyzed, and the patterns were assessed using Principal Components Analysis (PCA). Diet similarity among fish species was assessed using cluster analysis on feeding index. The first two axes from PCA explained 61.73% of the total variance, with the first axis being positively correlated with the compression index and relative height, whereas the second axis was positively correlated with the pectoral fin aspect. Two well-defined trophic groups, one herbivorous/specialist (*M. maculatus*) and the other formed by two omnivorous/generalist (*A.* cf. *bimaculatus*, *A. parahybae*) and one insectivorous-piscivorous (*O. hepsetus*) were revealed by the cluster analysis. *Astyanax.* cf. *bimaculatus* and *A. parahybae* differed. The first has comparatively greater relative height, relative length of the caudal peduncle and lower caudal peduncle compression index. However, we did not detect a close correspondence between diet and body shape in the reservoir, and inferred that the ecomorphological hypothesis of a close relationship between body shape and diet in altered systems could be not effective.

KEY WORDS. Body shape; diet; freshwater fishes; morphological diversity; niche overlap.

Ecomorphological studies of fishes aim to understand the patterns of association between the morphology of these organisms and their resource use. A major focus of some of these studies is the relationship between morphological variables and feeding behavior, or habitat use (TEIXEIRA & BENNEMANN 2007, OLIVEIRA *et al.* 2010, FAYE *et al.* 2012). One of the first studies on the ecomorphological patterns of fishes was published by KEAST & WEEB (1966), who observed that specializations of the oral apparatus determined habitat preferences and contributed to lessening interspecific competition among the fish species of the Opinicon Lake, Canada. Coexistence among several species in fish communities are facilitated because morphological segregation enables spatial and feeding partitioning (WIKRAMANAYAKE 1990).

Morphological divergences among species can be assessed by certain indices that can be interpreted as indicators of life strategies for habitat colonization (FREITAS *et al.* 2005) and use of food resources (WAINGHWRITH & RICHARD 1995). Correlations between the diversity of morphological patterns and resource partitioning have been used to test ecomorphological hypotheses in fish communities (WIKRAMANAYAKE 1990, HUGENY & POUILLY 1999). However, in the current literature there is no consensus about the direct relationship between ecology and morphology. Some studies maintain that there is a strong relationship between the morphology of organisms and their use of resources (MOYLE & SENANAYAKE 1984, WIKRAMANAYAKE 1990), whereas others have not found support for this relationship (GROSSMAN 1986, MOTTA *et al.* 1995), or have found only a weak correlation (CLIFTON & MOTTA 1998).

The use of ecomorphological approaches in the Neotropical region may be particularly relevant to address questions on niches and shared resources, since the region is characterized by a high diversity of fish (WINEMILLER 1991). This approach has been taken to study fish species inhabiting reservoirs, and which possess morphological and reproductive plasticity to adapt to modified environmental conditions (DUARTE *et al.* 2011). Moreover, impoundments can cause changes in the feeding strategies of species (HAHN & FUGI 2007), through selection: traits that are more suited for the colonization and use of available resources in the altered environment will be selected (CUNICO &

AGOSTINHO 2006). This study describes the ecomorphological patterns of four Characiformes fish species in the Lajes Reservoir, and evaluates the relationship between their diet and ecomorphological variables associated with their feeding behavior, locomotion, and their use of the water column. We tested the hypothesis that the shape of the body and diet of fishes are closely associated, since the environmental pressure acts on the phenotype, selecting traits according to the available resources.

MATERIAL AND METHODS

Fishes were collected using gillnets between February and November 2003, from the Lajes Reservoir (22°42'-22°50'S, 43°53'-44°05'W) and were fixed in 10% formalin during 48 hours and transferred to 70% ethanol. Four abundant Characiformes species were analyzed and forty-five individuals of each species were selected: *Astyanax* cf. *bimaculatus* (Linnaeus, 1758), Standard Length average (SL) = 96.65 ± 9.69 mm standard deviation, *Astyanax parahybae* Eigenmann, 1908, SL = 104.37 ± 7.27 mm, *Oligosarcus hepsetus* (Cuvier, 1829), SL = 144.28 ± 23.5 mm, and *Metynnis maculatus* (Kner, 1858) SL = 110.11 ± 6,81 mm. Voucher specimens were deposited in the fish collection of the Laboratory of Fish Ecology, Universidade Federal Rural do Rio de Janeiro, Brazil under numbers 1636 to 1641.

The Lajes Reservoir is a major impoundment in the State of Rio de Janeiro and was built between 1905 and 1908 by damming small streams. Initially built to generate hydroelectric power, the reservoir today is a strategic water supply for the municipality of Rio de Janeiro because it provides good quality water. This reservoir, situated at 415 m a.s.l., has a poorly structured physical habitat, lacking routes for fish migration because the tributaries that it receives in the slopes of the Serra do Mar are small (ARAÚJO & SANTOS 2001). Water levels range between 5-9 m during the year (ARAÚJO & SANTOS 2001, SANTOS *et al.* 2004).

Thirteen morphometric measurements were taken for each individual: standard length, body height, body width, head height, head length, pectoral fin length, pectoral fin width, caudal peduncle length, caudal peduncle height, caudal peduncle width, eye height, mouth height and mouth width. Absolute measurements were taken on the left side of each specimen with a caliper accurate to 0.01 mm. Only adults above L_{50} (average length of first gonadal maturation, according to data from VAZZOLER 1996) were used in order to avoid eventual allometric effects. Ecomorphological attributes of compression index (CI), relative height (RH), relative length of the caudal peduncle (RLP), caudal peduncle compression index (CPC), pectoral fin aspect (PFA), eye position (EP), relative length of the head (RLH), relative width of the mouth (RWM), relative height of the mouth (RHM) and mouth aspect (MA) were calculated based on morphometric measurements interpreted as indicators of lifestyle or adjustments to the different habitats and diets, as described in several studies (GATZ 1979a, b, MAHON 1984, WATSON & BALON 1984, BALON *et al.* 1986, BARRELLA *et al.* 1994, FREIRE & AGOSTINHO 2001).

Diet analysis was based on stomach contents examined under a stereoscopic microscope and identified to the lowest taxonomic level. Food items were identified according to BRUSCA & BRUSCA (2007) and MUGNAI *et al.* (2010). Empty stomachs were excluded from the analyses. The feeding index (IAi according to KAWAKAMI & VAZZOLER 1980) was calculated to obtain the relative importance of each food item –, using the wet weight of each item.

Principal Components Analysis (PCA) was performed using a correlation matrix of the 10 ecomorphological attributes of the four species to characterize them according to their morphological characteristics. The broken-stick criterion (JACKSON 1993) was used to select the most significant axes in the PCA. This criterion selects axes with eigenvalues higher than those expected by chance (KING & JACKSON 1999). We performed a Discriminant Function Analysis (DFA) to identify the ecomorphological attributes that best discriminated morphological characteristics among *A.* cf. *bimaculatus* and *A. parahybae*. This procedure was performed using Statistica 7.0 software.

Fish diets were analyzed by cluster analysis on the feeding index based on wet weight of food items, using the UPGMA linkage method and the Euclidian Distance matrix. To perform this analysis we used the software Primer 6.0. The amplitude of the trophic niche of each species was estimated using the Shannon's index of niche breadth (H' $\log_2$). This index varies from 0 (only one kind food item is present in the species' diet) to 1 (similar quantities of many food items are present in the species' diet). Feeding overlap among species was assessed using PIANKA's (1973) index, with the help of the software EcoSim700. This index varies from zero (signifying no overlap) to one (complete overlap).

The relationships between diet and ecomorphological attributes among the fishes were explored using Canonical Correspondence Analysis (CCA), a direct gradient analysis technique that ordinates a set of observations (in this case species) by directly relating them to two series of associated variables (ecomorphological attributes and diet) (TER BRAAK 1986). CCA is a modification of Correspondence Analysis that adds a multiple regression step and simultaneously relates the primary set of variables (in this case diet) with the secondary variables (ecomorphological attributes), constraining the ordination in such a manner that scores represent the maximum correlation between diet and morphology. A permutation test was used to assess the statistical significance of the relationship. The analysis was performed on log10-transformed ecomorphological attributes and dietary data with the CANOCO software, version 4.5.

RESULTS

Principal Component Analysis of the ecomorphological attributes produced two axes with eigenvalues higher than those expected by chance, explaining 61.73% of the total variance (Table I). The first component (PC1) explained 42.37% of the variability and was positively correlated with the compression index and relative height, and negatively correlated with the relative length of the caudal peduncle, relative height of the mouth and aspect of the mouth. *Metynnis maculatus* was positively correlated with axis 1 and was characterized by a laterally compressed and high body, whereas *O. hepsetus* was characterized by large mouth opening and elongated caudal peduncle, being negatively correlated with axis 1. The species *A.* cf. *bimaculatus* and *A. parahybae* positioned close to the origin of the ordination diagram and were characterized by a fusiform body. Despite this similarity, *A.* cf. *bimaculatus* had a comparatively wider mouth and more compressed caudal peduncle, whereas *A. parahybae* had a longer caudal peduncle, higher head and higher and narrow mouth. The second component (PC2) explained 19.36% of the variability and was positively correlated with the aspect of the pectoral fin. This axis was negatively correlated with *A.* cf. *bimaculatus,* which had a shorter and wider pectoral fin and was positively correlated with *M. maculatus, O. hepsetus* and *A.* cf. *bimaculatus,* which had a longer and narrower pectoral fin (Table I, Fig. 1).

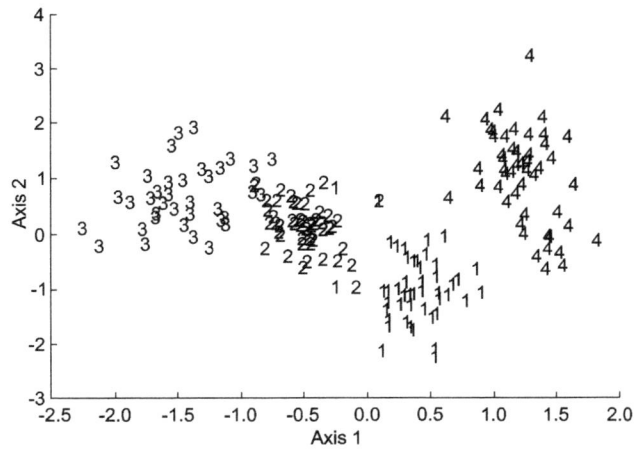

Figure 1. Ordination diagram from two principal components on ecomorphological attributes of four studied species: (1) *A.* cf. *bimaculatus;* (2) *A. parahybae;* (3) *O. hepsetus;* (4) *M. maculatus.*

Table I. Factor loads from principal components analysis on ecomorphological attributes of four examined Characiformes fish species in Lajes Reservoir, Rio de Janeiro, Brasil.

Ecomorphological attributes	Axis 1	Axis 2
Compression Index (CI)	**0.80**	0.43
Relative Height (RH)	**0.88**	0.33
Relative Length of the Caudal Peduncle (RLP)	**-0.77**	0.36
Caudal Peduncle Compression Index (CPC)	0.60	-0.53
Pectoral Fin Aspect (PFA)	0.00	**0.73**
Eye Position (EP)	-0.40	-0.40
Relative Length of the Head (RLH)	-0.14	0.51
Relative Width of the Mouth (RWM)	-0.01	-0.45
Relative Height of the Mouth (RHM)	**-0.93**	0.00
Mouth Aspect (MA)	**-0.92**	0.13
Percentage of explained variance (%)	42.37	19.36
Eigenvalues	4.24	1.98
Broken-stick eigenvalues	2.93	1.93

According to the Discriminant Function Analyses, there are significant morphological differences between *A. parahybae* and *A.* cf. *bimaculatus* (Wilks' λ = 0.147, p < 0.0001). The most important attributes for discriminating among the studied species were relative height, relative length of the caudal peduncle, and caudal peduncle compression index.

Astyanax parahybae (H' = 0.25), *A.* cf. *bimaculatus* (H' = 0.57) and *O. hepsetus* (H' = 0.31) clustered in a single group in the results of the cluster analysis of the diet (Index of Feeding Importance), whereas *M. maculatus* (H' = 0.03) formed a separated branch (Fig. 2, Table II). The diets of the former three species consist mainly of insects and the items in the diets of

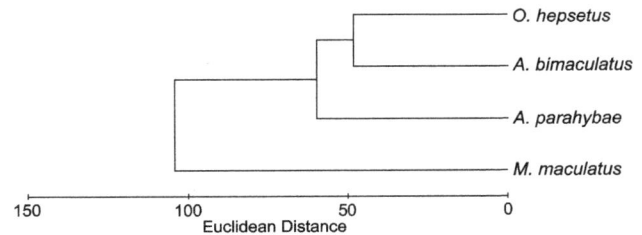

Figure 2. Dendrogram from cluster analysis on Index of Feeding Importance for the four examined Characiformes fish species in Lajes Reservoir.

Table II. Index of Feeding Importance for feeding item of the four examined Characiformes fish species in Lajes Reservoir. Number of examined individuals for each species = 45. Main items are show in bold.

Items	*A.* cf. *bimaculatus*	*A. parahybae*	*O. hepsetus*	*M. maculatus*
Hymenoptera	2.97	0.00	4.57	0.00
Coleoptera	0.83	0.00	0.00	0.00
Diptera	0.25	0.79	0.00	0.00
Odonata	0.00	0.79	5.14	0.22
Insects remains	**75.39**	**41.23**	**41.13**	0.20
Bivalvia	0.02	0.00	0.00	0.00
Nematoda	0.00	0.00	0.99	0.00
Fish	2.12	0.00	**31.07**	0.00
Algae	8.42	7.29	3.99	**89.54**
Seeds	2.21	**49.87**	0.00	10.02
Larval fish	0.38	0.00	0.00	0.00
Fish eggs	0.00	0.00	**12.62**	0.00
Vegetal remains	0.66	0.00	0.00	0.00
Detritus and Sediment	6.65	0.00	0.45	0.00

A. cf. *bimaculatus* and *O. hepsetus*, which have a wider trophic niche, overlap by 78% (Table III). Moreover, seeds are an important food item in the diet of *A. parahybae*, which have a comparatively narrower trophic niche than *A.* cf. *bimaculatus*. On the other hand, the diet of *M. maculatus* consists mainly on vegetal fragments (algae), and the trophic niche of this species is narrower.

The Canonical Correspondence Analysis did not show significant correlation between diet and morphology (p = 0.567, test with 1000 permutations). The sum of the eigenvalues for all axes defined by CCA was only 0.011, suggesting that differences in dietary composition were poorly explained by variations in morphological variables.

DISCUSSION

The four studied species in the Lajes Reservoir presented clear differentiated ecomorphological traits, suggesting that these species potentially use different resources. We found a lack of correspondence between diet and morphology, although the patterns of habitat use (locomotion and position in the water column) were consistent with the ecomorphological hypothesis. This finding suggests that the fish species were more influenced by spatial than by trophic structure.

Metynnis maculatus presented higher CI and RH than the other species examined. These attributes are associated with the use of structured habitats that have a variety of shelters and different types of substrate, where fish with high and laterally compressed bodies can perform maneuvers such as pitch or yaw (ALEXANDER 1967). Structured habitats are limited in the Lajes Reservoir, especially during the dry season, when most of the shoreline is exposed. Then, structured areas, which are favorable to the herbivorous feeding habits of *M. maculatus*, are restricted to some shallow bays, which house shrubs and grasses. Conversely, during the wet season, the submerged vegetation in the shoreline is well structured and should allow for the maneuvers of *M. maculatus*. The herbivorous habit of *M. maculatus* inhabiting reservoirs of the middle and lower Tiete River was described by SMITH *et al.* (2003). It is made possible by the small terminal mouth of this species, which enables the capture of plant fragments in the water column. Moreover, herbivorous feeding habits suit slow swimmers that have the ability to perform vertical movements in habitats where dynamism is low (BALON *et al.* 1986), such as reservoirs. Their results were corroborated by our cluster analysis, in which *M. maculatus* formed an isolated branch in the dendrogram based on its herbivorous feeding habits (mostly algae and seeds), thus characterizing this species as the most specialist among the species studied.

In contrast to *M. maculatus*, *O. hepsetus* had lower CI and RH, and higher RHM, MA and RLP, indicating a narrow mouth with large gap and a longer head and caudal peduncle. This set of attributes are typical of fishes that feed on large prey items that have good swimming capacity, for instance piscivorous

predators that inhabit benthic habits, dwelling in more dynamic environments (GATZ 1979a, b, WATSON & BALON 1984, BALON *et al.* 1986, FREIRE & AGOSTINHO 2001). Although fishes were the main prey of this species, insects and fish eggs were also consumed, suggesting that the trophic plasticity of *O. hepsetus* is high. The insectivorous-piscivorous habits of *O. hepsetus* at the Lajes Reservoir were described by ARAÚJO *et al.* (2005).

The morphological attributes related to body height and apparatuses to capture food of *Astyanax* cf. *bimaculatus* and *A. parahybae* were intermediate between *M. maculatus* and *O. hepsetus*. The scores obtained for individuals of *Astyanax*, close to the origin in the ordination diagram, indicate a more fusiform body, a characteristic that is associated with generalist feeding habits (OLIVEIRA *et al.* 2010). According to UIEDA (1984) and CASATTI *et al.* (2001), species of *Astyanax* have a continuous swimming behavior across different parts of the water column, which is favored by their laterally positioned eyes and small and relatively elongate pectoral fins. Such characteristics suggest a lack of specialization to explore a particular feeding resource. These species, which are similar in phenotypic traits, present similar adaptations to use resources and thus have a strong potential to compete with one another (WOOTTON 1990). However, in the tropics, competition seems to be reduced, owing to the feeding plasticity of most species, resource partitioning (ARAÚJO-LIMA *et al.* 1995) and resource availability (WINEMILLER & JEPSEN 1998), as well as the phenotypic characteristics of each species (WAINWRIGHT & RICHARD 1995, LABROPOULOU & ELEFTHERIOU 1997, BELLWOOD & WAINWRIGHT 2001). Contrarily to the classical results of the traditional niche theory, species that use similar resources may coexist if they are sufficiently similar in their skills to compete for the limiting resources (FAGERSTRÖM 1988). This pattern was confirmed for the Lajes Reservoir because of the high trophic niche overlap of *Astyanax* species, although their ecomorphological differences allow the exploration of differentiated resources that probably decrease competition.

In spite of exhibiting a more generalist body shape, the two *Astyanax* species have morphological differences that enable them to use resources differently. This pattern was also found by SANTOS *et al.* (2011) for the same species in the Funil Reservoir. *Astyanax parahybae* fish have phenotypic characteristics that enable them to capture larger prey items, or prey items that have faster movements, than the items eaten by *A.* cf. *bimaculatus*. This is possible because *A. parahybae* has narrower mouth, greater gap and longer caudal peduncle, whereas *A.* cf. *bimaculatus* has a higher RH, CI and CPC, and a wider mouth. *Astyanax* cf. *bimaculatus* is more frequent and abundant than *A. parahybae* in the Lajes Reservoir (ARAÚJO & SANTOS 2001), presenting wide niche breadth. We speculate that the generalist feeding habits and the morphological characteristics of *A.* cf. *bimaculatus* are more suitable for standing waters and enable them to be more successful exploring a wide range of trophic resources compared with *A. parahybae*.

According to the cluster analysis, A. cf. *bimaculatus* presented a diet more similar to *O. hepsetus* than to *A. parahybae* because they have a comparatively wider niche breadth and prey mainly on insects, whereas *A. parahybae* eats a great amount of seeds. Although the two species of *Astyanax* had high niche overlap, they seem to concentrate on different food items. While *A.* cf. *bimaculatus* use a variety of food items, *A. parahybae* concentrates on a few food items such as insects and seeds. These findings show the trophic plasticity of *A. parahybae*, since they use other kinds of food items in other systems (HIRT *et al.* 2011). Moreover, *A.* cf. *bimaculatus* seems to be better adapted to lentic environments (e.g. height body, compressed peduncle), which help this species to better explore different food items, while *A. parahybae* seems to be less efficient in obtaining access to a wide range of food resources, thus restricting its diet to a fewer number of items.

No significant association was found between the ecomorphological attributes and diet of these four Characiformes species in the reservoir. The lack of correspondence between morphology and diet can be related to the trophic plasticity of species such as *O. hepsetus*, *A.* cf. *bimaculatus* and *A.parahybae*. As these species present great flexibility in their feeding behavior, it is unlikely that we will observe a strong link between diet and morphology, because they belong to the same broad trophic category, as indicated by cluster analysis. Moreover, HUGUENY & POUILLY (1999) suggested that food availability may weaken the relationship between diet and morphology because stomach contents probably reflect food availability more than morphological adaptation.

Trophic plasticity can also be associated with morphological divergence of fish species in reservoirs. SANTOS *et al.* (2011) suggested that there is morphological divergence between *A.* cf. *bimaculaus* and *A. parahybae* in the Funil Reservoir, Southeastern Brazil, as consequence of impoundment. Similarly, FRANSSEN (2011), studying fish species of Central Plains of the USA, found that, although the components of body shape are plastic, anthropogenic habitat modifications may drive trait divergence in native populations in reservoir-altered habitats. According to LANGERHANS (2008), phenotypic variation can be adaptive in conditions of low flow and in habitats with high densities of predators, and these two factors could be driving the observed morphological shifts. Morphological divergence between phylogenetically related species has been reported to occur when morphological variations are found as responses to selective pressures of the environment (CASATTI & CASTRO 2006, OLIVEIRA *et al.* 2010). Thus, the mechanism involved in competition in altered environments may suggest an intensification of generalist habits, rather than the specialization predicted by the classical competition theory (GABLER & AMUNDSEN 2009). Therefore, altered systems like the Lajes Reservoir, which have a simple habitat structure, may favor trophic generalist fish species and contribute to the lack of correlation between morphology and diet.

We conclude that the ecomorphological hypothesis of a close relationship between fish body shape and diet was not corroborated in the present study for the altered system studied. However, the patterns of habitat use of the four species studied corresponded to the expectation of the ecomorphological approach. In altered systems, the patterns of resource use can be broad enough to allow fish species to change their choices and to respond to local biotic and/or abiotic conditions, contradicting the traditional ecological theories. Further studies are necessary to give a more detailed picture on the relationship between body shape and diet, especially in dammed rivers, because the species that persist in this new system had to adapt to new environmental constraints.

ACKNOWLEDGMENTS

Specimens of fish used in this study were obtained from the Project "Fish ecology in Lajes Reservoir", which is financially support by a Research & Development Program by ANNEEL/LIGHT/UFRRJ consortium. FAPERJ – Rio de Janeiro Carlos Chagas Filho Research Support Agency also supplied financial support for this project.

REFERENCES

ALEXANDER, R.M. N. 1967. **Functional design in fishes.** London, Hutchinson University Library, 160p.

ARAÚJO, F.G. & L.N. SANTOS. 2001. Distribution and composition of fish assemblages in Lajes Reservoir, Rio de Janeiro, Brazil. **Revista Brasileira de Biologia 61**: 563-576.

ARAÚJO, F.G.; C.C. ANDRADE; R.N. SANTOS; A.F.G.N. SANTOS & L.N. SANTOS. 2005. Spatial and seasonal changes in the diet of *Oligosarcus hepsetus* (Characiformes: Characidae) in a Brazilian reservoir. **Revista Brasileira de Biologia 65**: 1-8.

ARAÚJO-LIMA, C.A.R.M.; A.A. AGOSTINHO & N.N. FABRÉ. 1995. Trophic aspects of fish communities in Brazilian rivers and reservoirs, p. 105-136. *In*: J.G. TUNDISI; C.E.M. BICUDO & T. MATSUMURA-TUNDISI (Eds). **Limnology in Brazil.** Rio de Janeiro, ABC/SBL, 376p.

BALON, E.K.; S.S. CRAWFORD & A. LELEK. 1986. Fish communities of the upper Danube River (Germany, Austria) prior to the new Rhein-Main-Donau connection. **Environmental Biology of Fishes 15**: 243-271.

BARRELLA, W.; A.C. BEAUMORD & M. PETRERE-JR. 1994. Comparison between the fish communities of Manso river (MT) and Jacare Pepira river (SP), Brazil. **Acta BiologicaVenezuelica 15**: 11-20.

BELLWOOD, D.R. & P.C. WAINWRIGHT. 2001. Locomotion in labrid fishes: implications for habitat use and cross-shelf biogeography on the Great Barrier Reef. **Coral Reefs 20** (2): 139-150.

BRUSCA, R.C. & G.J. BRUSCA. 2007. **Invertebrados.** 2nd. ed. Rio de Janeiro, Guanabara Koogan, XXII+968p.

CASATTI, L. & R.M.C. CASTRO. 2006. Testing the ecomorphological hypothesis in a headwater riffles fish assemblage of the rio São Francisco, southeastern Brazil. **Neotropical Ichthyology** 4 (2): 203-2014.

CASATTI, L.; F. LANGEANI & R.M.C. CASTRO. 2001. Peixes de riacho do parque estadual morro do Diabo, bacia do alto rio Paraná, SP. **Biota Neotropica** 1 (1): 1-15.

CLIFTON K.B. & P.J. MOTTA. 1998. Feeding morphology, diet and ecomorphological relationships among five caribbean labrids (Teleostei, Labridae). **Copeia** (4): 953-966.

CUNICO, A.M. & A.A. AGOSTINHO. 2006. Morphological patterns of fish and their relationships with reservoirs hydrodynamics. **Brazilian Archives of Biology and Technology** 49 (1): 125-134.

DUARTE, S.; F.G. ARAÚJO; N. BAZZOLI. 2011. Reproductive plasticity of *Hypostomus affinis* (Siluriformes: Loricariidae) as a mechanism to adapt to a reservoir with poor habitat complexity. **Zoologia** 28 (5): 577-586.

FAGERSTRÖM, T. 1988. Lotteries in communities of sessile organisms. **Trends in Ecology and Evolution** 3: 303-306.

FAYE, D.; F. LE LOC'H; O.T. THIAW & L.T. MORAIS. 2012. Mechanisms of food partitioning and ecomorphological correlates in ten fish species from a tropical estuarine marine protected area (Bamboung, Senegal, West Africa). **African Journal of Agricultural Research** 7 (3): 443-455.

FRANSSEN, N.R. 2011. Anthropogenic habitat alteration induces rapid morphological divergence in a native stream fish. **Evolutionary Applications** 4 (6): 791-804.

FREIRE, A.G. & A.A. AGOSTINHO. 2001. Ecomorfologia de oito espécies dominantes da ictiofauna do reservatório de Itaipu (Paraná/Brasil). **Acta Limnologica Brasiliensia** 13 (1): 1-9.

FREITAS, C.E.C.; E.L. COSTA & M.G.M. SOARES. 2005. Ecomorphological correlates of thirteen dominant fish species of Amazonian floodplain lakes. **Acta Limnologica Brasiliensia** 17 (3): 339-347.

GATZ JR, A.J. 1979a. Ecological morphology of freshwater stream fishes. **Tulane Studies of Zoology and Botany** 21: 91-124.

GATZ JR, A.J. 1979b. Community organization in fishes as indicated by morphological features. **Ecology** 60: 711-718.

GABLER, H.M. & P.A. AMUNDSEN. 2009. Feeding strategies, resource utilisation and potential mechanisms for competitive coexistence of Atlantic salmon and alpine bullhead in a sub-Arctic river. **Aquatic ecology** 44 (2): 325-336.

GROSSMAN, G.D. 1986. Food resource partitioning in a rocky intertidal fish assemblage. **Journal of Zoology** 1: 317-355.

HAHN, N.S. & R. FUGI. 2007. Alimentação de peixes em reservatórios brasileiros: alterações e conseqüências nos estágios iniciais do represamento. **Oecologia Brasiliensis** 11 (4): 469-480.

HIRT, L.M.; P.R. ARAYALA & S.A. FLORES. 2011. Population structure, reproductive biology and feeding of *Astyanax fasciatus* (Cuvier, 1819) in an Upper Paraná River Tributary, Misiones, Argentina. **Acta Limnologica Brasiliensia** 23 (1): 1-12.

HUGUENY, B. & M. POUILLY. 1999. Morphological correlates of diet in an assemblage of West African freshwater fishes. **Journal of Fish Biology** 54 (6):1310-1325.

JACKSON, D.A. 1993. Stopping rules in principal components analysis: a comparison of heuristical and statistical approaches. **Ecology** 74: 2201-2214.

KAWAKAMI, E. & G. VAZZOLER. 1980. Método gráfico e estimativa de índice alimentar aplicado no estudo de alimentação de peixes. **Boletim do Instituto Oceanográfico** 29 (2): 205-207.

KEAST, A. & D. WEBB. 1966. Mouth and body form relative to feeding ecology in the fish fauna of a small lake, Lake Opinicon, Ontario. **Journal of the Fisheries Research Board of Canada** 23: 1846-1874.

KING, J.R. & D.A. JACKSON. 1999. Variable selection in large environmental data sets using principal components analysis. **Environmetrics** 10: 67-77.

LABROPOULOU, M. & A. ELEFTHERIOU. 1997. The foraging ecology of two pairs of congeneric demersal fish species: importance of morphological characteristics in prey selection. **Journal of Fish Biology** 50: 324-340.

LANGERHANS, R. B. 2008. Predictability of phenotypic differentiation across flow regimes in fishes. **Integrative and Comparative Biology** 48: 750-768.

MAHON, R. 1984. Divergent structure in fish taxocenes of north temperate stream. **Canadian Journal of Fisheries and Aquatic Sciences** 41: 330-350.

MOTTA, J.P.; K.B. CLIFTON; P. HERNANDEZ & B.T. EGGOLD. 1995. Ecomorphological correlates in ten species of subtropical seagrass fishes: diet and microhabitat utilization. **Environmental Biology of Fishes** 44: 37-60.

MOYLE, P.B. & F.R. SENANAYAKE. 1984. Resource partitioning among the fishes of rainforest streams in Sri Lanka. **Journal of Zoology** 202: 195-223.

MUGNAI, R.; J.L. NESSIMIAN & D.F. BAPTISTA. 2010. **Manual de identificação de macroinvertebrados aquáticos do Estado do Rio de Janeiro.** Rio de Janeiro, Technical Books, 174p.

OLIVEIRA, E.F.; E. GOULART; L. BREDA; C.V. MINTE-VERA, L.R.S. PAIVA & M.R. VISMARA. 2010. Ecomorphological patterns of the fish assemblage in a tropical floodplain: effects of trophic, spatial and phylogenetic structures. **Neotropical Ichthyology** 8 (3): 569-586.

PIANKA, E.R. 1973. The structure of lizard communities. **Annual Review of Ecology and Systematics** 4: 53-74.

SANTOS, A.B.I; F.L. CAMILO; R.J. ALBIERI & F.G. ARAÚJO. 2011. Morphological patterns of five fish species (four characiforms, one perciform) in relation to feeding habits in a tropical reservoir in south-eastern Brazil. **Journal of Applied Ichthyology** 27: 1360-1364.

SANTOS, A.F.G.N.; L.N. SANTOS & F.G. ARAÚJO. 2004. Water level influences on body condition of *Geophagus brasiliensis* (Perciformes, Cichlidae) in a Brazilian oligotrophic reservoir. **Neotropical Ichthyology** 2: 151-156.

Smith, W.S.; C.C.G. F. Pereira; E.L.G. Espíndola & O. Rocha. 2003. A importância da zona litoral para a disponibilidade de recursos alimentares à comunidade de peixes em reservatórios, p. 233-248. *In*: R. Henry (Ed.). **Nas interfaces dos ecossistemas aquáticos.** São Carlos, Rima, 350p.

Teixeira, I. & S.T. Bennemann. 2007. Ecomorfologia refletindo a dieta dos peixes em um reservatório no sul do Brasil. **Biota Neotropica 7** (2): 67-76.

ter Braak, C.J.F. 1986. Canonical correspondence analysis: a new eigenvector technique for multivariate gradient analysis. **Ecology 67**: 1167-1179.

Uieda, V.S. 1984. Ocorrência e distribuição dos peixes em um riacho de água doce. **Revista Brasileira de Biologia 44**: 203-213.

Vazzoler, A.M.A. de M. 1996. **Biologia da reprodução de peixes teleósteos: teoria e prática.** Maringá, EDUEM-SBI, 169p.

Wainwright, P.C. & B.A. Richard. 1995. Predicting patterns of prey use from morphology of fishes. **Environmental Biology of Fishes 44**: 97-113.

Watson, D.J. & E.K. Balon. 1984. Ecomorphological analysis of taxocenes in rainforest streams of northern Borneo. **Journal of Fish Biology 25**: 371-384.

Wikramanayake, E.D. 1990. Ecomorphology and biogeography of a tropical stream fish assemblage: evolution of assemblage structure. **Ecology 71**: 1756-1764.

Winemiller, K.O. 1991. Ecomorphological diversification in lowland freshwater fish assemblages from five biotic regions. **Ecological Monographs 61**: 343-365.

Winemiller, K. & D.B. Jepsen. 1998. Effects of seasonality and fish movement on tropical river food webs. **Journal of Fish Biology 53** (Suppl. A): 267-296.

Wootton, R.J. 1990. **Ecology of teleost fishes.** London, Chapman and Hall, 404p.

18

Assessing the efficacy of hair snares as a method for noninvasive sampling of Neotropical felids

Tatiana P. Portella[1,2,4], Diego R. Bilski[2], Fernando C. Passos[3] & Marcio R. Pie[1]

[1] *Laboratório de Dinâmica Evolutiva e Sistemas Complexos, Departamento de Zoologia, Universidade Federal do Paraná. Caixa Postal 19020, 81531-990 Curitiba, PR, Brazil.*
[2] *Programa de Pós-Graduação em Ecologia e Conservação, Universidade Federal do Paraná.*
[3] *Laboratório de Biodiversidade, Conservação e Ecologia de Animais Silvestres, Departamento de Zoologia, Universidade Federal do Paraná. Caixa Postal 19020, 81531-990 Curitiba, PR, Brazil.*
[4] *Corresponding author. E-mail: portellatp@yahoo.com.br*

ABSTRACT. Hair snares have been used in North and Central America for a long time in assessment and monitoring studies of several mammalian species. This method can provide a cheap, suitable, and efficient way to monitor mammals because it combines characteristics that are not present in most alternative techniques. However, despite their usefulness, hair snares are rarely used in other parts of the world. The aim of our study was to evaluate the effectiveness of hair snares and three scent lures (cinnamon, catnip, and vanilla) in the detection of felids in one of the largest remnants of the Brazilian Atlantic Forest. We performed tests with six captive felid species – *Panthera onca* (Linnaeus, 1758), *Leopardus pardalis* (Linnaeus, 1758), *L. tigrinus* (Schreber, 1775), *L. wiedii* (Schinz, 1821), *Puma concolor* (Linnaeus, 1771), and *P. yagouaroundi* (É. Geoffroy Saint-Hilaire, 1803) – to examine their responses to the attractants, and to correlate those with lure efficiency in the field. The field tests were conducted at the Parque Estadual Pico do Marumbi, state of Paraná, Brazil. Hair traps were placed on seven transects. There were equal numbers of traps with each scent lure, for a total of 1,551 trap-days. In captivity, vanilla provided the greatest response, yet no felids were detected in the field with any of the tested lures, although other species were recorded. Based on the sampling of non-target species, and the comparison with similar studies elsewhere, this study points to a possible caveat of this method when rare species or small populations are concerned. Meanwhile, we believe that improved hair snares could provide important results with several species in the location tested and others.

KEY WORDS. Atlantic Forest; detection technique; hair; scent lure; rubbing behavior.

Of all 37 recognized wild felid species, eight can be found in Brazil. All of them are placed in the threatened category in regional (BRESSAN *et al.* 2009, BRAGA & VIDOLIN 2010), national (MACHADO *et al.* 2008) or international (IUCN 2011) official lists. Habitat loss and poaching are considered to be the major threats to these animals, (MACHADO *et al.* 2008) and the lack of knowledge about their behavior, basic biology and distribution (OLIVEIRA 2006) is a threat to their conservation in the future.

Due to their secretive behavior, low densities and predominantly crepuscular and/or night habits, the observation and capture of wild felids can be very expensive and require a considerable effort (TOMAS *et al.* 2006, AGUIAR & MORO-RIOS 2009). As a consequence, non-invasive methods and indirect evidence such as scats, tracks, camera-traps, and hair-traps are often used to study felids (LONG *et al.* 2008). Feces can be used in dietary and molecular analyses (MIRANDA *et al.* 2005, LUDWIG *et al.* 2007, MIOTTO *et al.* 2007, HEINEMEYER *et al.* 2008, SILVA-PEREIRA *et al.* 2011). However, finding scats in the field can be harder in humid and

montane forests, where frequent rains over an irregular landscape can remove feces from trails, where they are more easily found (CRAWSHAW *et al.* 1997). Tracks can be used to monitor felids, but the correct discrimination between tracks of small felids is difficult (BECKER & DALPONTE 1999). Automatic camera-traps can provide reliable identifications, but these devices are relatively expensive for developing countries, and several models cannot handle the high humidity of tropical forests. Hair-traps, on the other hand, represent a cheaper alternative, and are less affected by the weather. The identification of the species can be done through microscopic analysis of the cuticle patterns of the guard hairs or through DNA analysis (WEAVER *et al.* 2005, KENDALL & MCKELVEY 2008). Moreover, passive methods such as camera-traps and hair snares can be installed in sites of difficult access, thus minimizing the possible bias in the area covered, a pitfall of other methods (WASSER *et al.* 2004).

There are several kinds of hair-traps, developed according to the target species or group (KENDALL & MCKELVEY 2008).

The most common snares used for felids consist of a rigid plate covered with short-napped carpet or hook-and-loop fasteners (e.g. Velcro™), and wires or nails attached to the carpet (Weaver et al. 2005). These plates are used by the felids in their natural cheek-rubbing behavior, during which some hairs are snagged. However, in order to work properly, this method requires scented lures that attract the species, and also elicit the rubbing behavior (McDaniel et al. 2000, Weaver et al. 2005, Kendall & McKelvey 2008). The effectiveness of lures and snares varies among species, individuals and even among study sites (Harrison 1997, Thomas et al. 2005, Castro-Arellano et al. 2008, Schlexer 2008). For this reason, controlled tests of their efficiency are recommended before they are used in field studies (Schlexer 2008). Multiple studies have evaluated the efficacy of hair-snares with several lures on the Nearctic region (Weaver et al. 2005, Bertrand et al. 2006, McKelvey et al. 2006, Long et al. 2007, Ruell & Crooks 2007) and in Central America (Harrison 1997, Downey et al. 2007, Castro-Arellano et al. 2008), but there are no published studies in South America. Therefore, the main objective of our study was to evaluate the effectiveness of three scent lures to attract six Neotropical felid species to hair snares, in one of the largest Atlantic Forest remnants in South America.

MATERIAL AND METHODS

Catnip (Nepeta cataria, Lamiaceae), cinnamon, and vanilla were chosen as potential attractant scents. Catnip is commonly employed in this type of study (Harrison 1997, McDaniel et al. 2000, Castro-Arellano et al. 2008, Schlexer 2008), and cinnamon was successfully used by Niara Martins (per. comm.) to detect pumas, Puma concolor (Linnaeus, 1771), in a cerrado region of the Estação Ecológica do Jataí, Brazil. We found no previous studies that employed vanilla as a scent lure, but it proved to be effective in stimulating rubbing behavior in tests that we performed with captive felids.

How efficient these scents are in stimulating the rubbing behavior, and the capacity of the hair-snare to remove hairs were evaluated with captive felids using a methodology similar to that of Harrison (1997). Using this test, we can assess the intensity of the interactions in captivity, and potentially associate them with the rate of detection of wild animals. The behavioral responses of ten ocelots, Leopardus pardalis (Linnaeus, 1758), seven margays, Leopardus wiedii (Schinz, 1821), six oncillas, Leopardus tigrinus (Schreber, 1775), six pumas, five jaguaroundis, Puma yagouaroundi (É Geoffroy Saint-Hilaire, 1803), and five jaguars, Panthera onca (Linnaeus, 1758) were evaluated, for all the three scents. To evaluate their responses, we provided, to each individual in an enclosure, wooden blocks sprayed with one of the lures, and then observed the animals for one hour, to record the occurrence and duration of the rubbing behavior. The total amount of time the animal spent rubbing on the block was used as a proxy for the efficiency of the lure. All scent lures were provided for all species and individu-

als, on different days, and the choice of the lure used was defined randomly. As a negative control, we repeated the tests providing only the wooden blocks with no lures. The non-parametric Friedman test was used to evaluate if there was a difference between interactions with each lure for each species. When significant differences occurred, Wilcoxon tests were used to compare the pairs.

Field tests were conducted in the Parque Estadual Pico do Marumbi (PEPM), state of Paraná, southern Brazil. This park has an area of more than 8,000 ha, located within the AEIT (Área de Especial Interesse Turístico) of Marumbi, an area of 66,732.99 ha of continuous Atlantic Ombrophylous Dense Forest, comprising low-slope, montane and high montane forests (Marques et al. 2011). In accord with Köppen's climate classification, the climate at the PEPM can be classified as Cfb – with cool summers and precipitation in all seasons (SEMA-IAP 1996). Field campaigns were conducted on July and August of 2009, and from April through July of 2010. We used hair-snare mobile stations like those described by Downey et al. (2007) in seven transects of varying lengths, with three, six, nine or 12 snares per transect, comprising a total effort of 1,551 trap-days. Along each transect we deployed the same number of snares with each scent lure, evenly spaced, and revisited these sites after five or six nights, thus relocating the snares. The seven transects were never sampled consecutively. Hairs recovered from the snares were mounted on glass slides according to Quadros & Monteiro-Filho (2006), and identified by comparison of the microscopic structure of the cuticular and medullar patterns with those described by Quadros & Monteiro-Filho (2010). All the species included in the captivity tests occur in the PEPM (Leite & Galvão 2002, Cáceres 2004).

RESULTS

In the tests performed in captivity, pumas did not interacted with any of the scent lures. The Friedman tests show significant differences between the scent lures and the negative control for the ocelot ($\chi^2 = 16.9$, p = 0.0007), margay ($\chi^2 = 9.76$, p = 0.020) and the oncilla ($\chi^2 = 8.37$, p = 0.038). Based on the Wilcoxon tests results (Table I), the ocelot interacted significantly more with cinnamon and vanilla than with catnip and the negative control, and the margay interacted more with vanilla than with catnip and the control, with no significant difference between the other scents. There were no differences between any pair of scents for the oncilla. Evaluating those species that interacted with at least one scent lure, vanilla can be considered the most effective substance in eliciting the rubbing behavior (Table I).

Hair snares yielded little success in the field tests, with no detection of felid species. Hairs of other mammals were recovered from the snares, including Lontra longicaudis (Olfers, 1818), Cerdocyon thous (Linnaeus, 1766), Procyon cancrivorus (G. Cuvier, 1798), Sapajus nigritus (Goldfuss, 1809), and one species that

Table I. Results of the Wilcoxon tests comparing pairs of stimulants for those species that had differences indicated by the Friedman test. Cinnamon (Cin), Vanilla (Van), Catnip (Cat), Control (Ctr). n = number of individuals.

	n	Cin/Van		Cin/Cat		Cin/Ctr		Van/Cat		Van/Ctr		Cat/Ctr	
		Z	p	Z	p	Z	p	Z	p	Z	p	Z	p
Ocelot	10	1.35	0.17	2.02	0.04	2.02	0.04	2.36	0.01	2.36	0.01	–	–
Margay	7	1.21	0.22	0.67	0.50	1.48	0.13	2.02	0.04	2.02	0.04	–	–
Oncilla	6	1.60	0.10	1.60	0.10	1.60	0.10	1.60	0.10	1.60	0.10	–	–
All species[1]	33	2.10	0.03	2.53	0.01	3.29	<0.001	3.54	<0.001	3.51	<0.001	1.27	0.20

[1] excluding *Puma concolor*.

could not be identified based on Quadros & Monteiro-Filho (2010). The respective attractants of these species are presented in Table II. Catnip and cinnamon were the only attractants in the field tests that had any success, while vanilla yielded no records.

Table II. Species recorded on the hair snares on Parque Estadual Pico do Marumbi, and the respective attractants. Cinnamon (Cin), Vanilla (Van), Catnip (Cat), (N) number of occurrences.

Species	Common name	N		
		Cin	Van	Cat
Primates				
Cebidae				
Sapajus nigritus	Capuchin monkey	–	–	1
Carnivora				
Canidae				
Cerdocyon thous	Crab-eating fox	–	–	1
Mustelidae				
Lontra longicaudis	Otter	1	–	–
Procyonidae				
Procyon cancrivorous	Crab-eating raccoon	1	–	–
Unidentified species		–	–	1
Total		2	–	3

DISCUSSION

Due to the fact that no felid species was detected during our field tests, we could not evaluate the correlation between the intensity of responses in captivity with the efficiency of the scent lures to attract wild felids. In a similar study, Harrison (1997) also could not make this kind of comparison due to the low success in felid detection. But considering all detected carnivorous species, Harrison (1997) found a positive association between the intensity of the responses of felids in captivity and the attraction of carnivorous species in the field. On the other hand, in our study vanilla was an ineffective attractant in the field, despite eliciting the strongest responses in the tests with captive felids.

The lack of response by the wild species to the scent lures, in contrast to the prominent interaction of the captive individuals, can be a side effect of the different environmental conditions to which these animals are exposed. According to Weller & Bennett (2001), captive animals are often sedentary and less engaged in exploratory behaviors than their wild counterparts, probably due to the lower complexity and higher previsibility of captivity. This can undermine the motivation, opportunity, or the necessity of certain behaviors (McPhee 2002). Because of this lack of stimuli, when new attractants are introduced into the captive environments, they can elicit stronger behavioral responses than those observed in wild animals (Tanás & Pisula 2011) or in more complex enclosures. In fact, in our study individuals maintained in less complex enclosures interacted more intensely with the scent lures than those kept in more enriched places (data not shown). By contrast, the captive pumas in our study (some of those maintained in poor enclosures) failed to respond to the stimulants. This is surprising given that this species was detected in the wild by other researchers using catnip (Weaver *et al.* 2005, Castro-Arellano *et al.* 2008).

The failure to detect wild felids through hair snares, where they are known to occur, is not uncommon (Thomas *et al.* 2005, Downey *et al.* 2007, Reed 2011). As in our study, several researchers had more success detecting generalist carnivores such as some canid, procyonid and ursid species (Harrison 1997, Thomas *et al.* 2005, Downey *et al.* 2007, Ruell & Crooks 2007). These groups are known to have better olfactive acuity than felids (Gittleman 1991), and therefore could be attracted to the snares sooner. According to Downey *et al.* (2007), felids may avoid rubbing in snares already marked by other species, particularly where these other species occur at high densities. In our study, however, non-target species were seldom detected, and therefore it is unlikely that other species inhibited felid rubbing on the snares. However, even the human odor possibly present on the snares can hinder the approximation of felids (Schlexer 2008). Contrary to felids, which are known to avoid human contact (Martins *et al.* 2008), most of the species detected in our study (with the exception of the otter) are generalist-opportunists, and are acknowledged as being well adapted to anthropic environments (Facure & Monteiro-Filho 1996, Sabbatini *et al.* 2008, Aguiar *et al.* 2011). Although we were careful when transporting the snares in the field, avoiding unnecessary human contact, we cannot dismiss the possibility of human scents on the snares, given that we did not use gloves or odor removers. Commercial odor removers can be useful to avoid human odor in this kind of study, but they are not necessarily odorless to carnivores, a point that needs to be evaluated prior to their use.

The profitability of hair snares can also be associated with the study site and density of target species. In a study similar to ours, at Estação Ecológica do Jataí, the use of hair snares, with catnip and cinnamon as attractants, recovered a high number of hairs of *P. concolor* (Niara Martins, pers. comm.). A study conducted by MIOTTO *et al.* (2011) points to a high density of pumas at the Estação Ecológica do Jataí and surrounding areas, and although no similar studies have been conducted on PEPM, Atlantic Ombrophylous Dense Forests are known to support lower densities of medium and large sized mammals than other formations of the Atlantic Forest (GALETTI *et al.* 2009). The fact that no felid hairs were recovered in our study can indicate a low density of these animals at the PEPM, and the inefficiency of hair snares in detecting rare species when used for short periods of time.

The low efficiency of this method in detecting species that occurs at low densities becomes evident when we compare the studies performed by DOWNEY *et al.* (2007) and CASTRO-ARELLANO *et al.* (2008), both conducted in El Cielo Biosphere Reserve, Mexico. DOWNEY *et al.* (2007) had an effort of 1,920 trap-days, and detected only six species, four of those domestic animals, and was unable to detect felids. Using the same transects, but with a total effort of 8,149 trap-days, CASTRO-ARELLANO *et al.* (2008) registered 14 wild species, including four felid species. However, despite their success in detecting felids, only the mountain lion was detected more than 10 times. This comparison points to the necessity of great efforts to reliably assess rare species using hair snares.

The design of the trap can equally affect detection success. We designed a hair trap to be disguised in the environment. Felids are more responsive to visual than to olfactive attractants (SCHELEXER 2008), and several studies already used CDs and aluminum plates near the hair snares to visually entice felids (McDANIEL *et al.* 2000, WEAVER *et al.* 2005, RUELL & CROOKS 2007). However, despite being frequently used, no study to date has evaluated the efficiency of visual attractants in improving the efficiency of hair snares, nor attempted to assess a possible negative impact of them. HARRISON (1997) and CASTRO-ARELLANO *et al.* (2008) had success using felt cloths hung above the snares, embedded with commercial trapping lures (Carman's Canine Call, Hawbaker's Wildcat lure #.1, Carman's Raccoon lure # 1; Minnesota Trapline Products). However, these commercial lures are sold only in North American countries, what hinders their use elsewhere.

Regardless of the lack of success of hair snares in the present study, and the different factors that can affect the outcomes of this method, such as interference from non-target species, low abundance, and difficult attraction of some groups to the snares, we believe that this method is still useful. This technique has been used in North and Central America for the last 20 years (KENDALL & McKELVEY 2008), and through continuous improvement it has contributed in wildlife assessments (CASTRO-ARELLANO *et al.* 2008), recording of species presence (BERTRAND *et al.* 2006) and population monitoring (MOWAT & PAETKAU 2002, DE BARBA *et al.* 2010). Therefore, we believe that, through proper improvements, this technique can prove to be useful for several kinds of studies on many species around the world.

ACKNOWLEDGEMENTS

We thank T. Margarido and N. Benavicius from Zoológico de Curitiba, P. Mangini and H. Chupil from Criadouro Conservacionista Onça-Pintada, and J. Pereira and C. Adania from Associação Mata Ciliar for their support to our study on their respective institutions. R. Miotto and E. Monteiro-Filho made substantial comments on an early draft of this manuscript, and N. Martins for her contribution. We are also thankful to Diego Astúa and three anonymous reviewers for their suggestions. This study was part of a master's thesis at the Programa de Pós-Graduação em Ecologia e Conservação, and received support from IAP and CNPq by means of the fellowship 135204/2009-7 to TPP. We also thank CAPES for the fellowship to DRB, and CNPq/MCT for provided funding to FCP (grant 300466/2009-9) and MRP (grant 571334/2008-3).

REFERENCES

AGUIAR, L.M. & R.F. MORO-RIOS. 2009. The direct observational method and possibilities for Neotropical Carnivores: an invitation for the rescue of a classical method spread over the Primatology. **Zoologia 26** (4): 587-593.

AGUIAR, L.M.; R.F. MORO-RIOS; T. SILVESTRE; J.E. SILVA-PEREIRA; D.R. BILSKI; F.C. PASSOS; M.L. SEKIAMA & V.J. ROCHA. 2011. Diet of brown-nosed coatis and crab-eating raccoons from a mosaic landscape with exotic plantations in southern Brazil. **Studies on Neotropical Fauna and Environment 46** (3): 153-161.

BECKER, M. & J.C. DALPONTE. 1999. **Rastros de mamíferos silvestres brasileiros.** Brasília, Universidade de Brasília, 180p.

BERTRAND, A-S.; S. KENN; D. GALLANT; E. TREMBLAY; L. VASSEUR & R. WISSINK. 2006. MtDNA analyses on hair samples confirm cougar, *Puma concolor*, presence in southern New Brunswick, Eastern Canada. **Canadian Field-Naturalist 120** (4): 438-442.

BRAGA, F.G. & G.P. VIDOLIN. 2010 **Mamíferos Ameaçados no Paraná.** Curitiba, IAP, 78p.

BRESSAN, M; M.K. KIERULFF & A.M. SUGIEDA. 2009. **Fauna Ameaçada de Extinção no Estado de São Paulo. Vertebrados.** São Paulo, Fundação Parque Zoológico de São Paulo, Secretaria do Meio Ambiente, 645p.

CÁCERES, N.C. 2004. Occurrence of *Conepatus chinga* (Molina) (Mammalia, Carnivora, Mustelidae) and other terrestrial mammals in the Serra do Mar, Paraná, Brazil. **Revista Brasileira de Zoologia 21** (3), 577-579.

CASTRO-ARELLANO, I.; C. MADRID-LUNA; T.E. LACHER & L. LÉON-PANIAGUA. 2008. Hair-Trap efficacy for detecting mammalian carnivores in the tropics. **Journal of Wildlife Management 72**: 1405-1412.

CRAWSHAW JR, P.G. 1997. Recomendações para um modelo de pesquisa sobre felídeos neotropicais, p. 70-94. *In*: C. VALLADARES-PÁDUA & R.E. BODMER (Eds). **Manejo e conservação de vida silvestre no Brasil.** Belém, MCT, CNPq, Sociedade Civil Mamirauá.

DE BARBA, M.; L.P. WAITS; E.O. GARTON; P. GENOVESI; E. RANDI; A. MUSTONI & C. GROFF. 2010. The power of genetic monitoring for studying demography, ecology and genetics of a reintroduced brown bear population. **Molecular Ecology 19**: 3938-3951.

DOWNEY, P.J.; E.C. HELLGREN; A. CASO; S. CARVAJAL & K. FRANGIOSO. 2007. Hair Snares for Noninvasive Sampling of Felids in North America: Do Gray Foxes Affect Success? **Journal of Wildlife Management 71** (6): 2090-2094.

FACURE, K.G. & E.L.A.MONTEIRO-FILHO. 1996. Feeding habits of crab-eating fox, *Cerdocyon thous* (Carnivora:Canidae), in a suburban area of southeastern Brazil. **Mammalia 60** (1): 147-149.

GALETTI, M.; H.C. GIACOMINI; R.S. BUENO; C.S.S. BERNARDO; R.M. MARQUES; R.S. BOVENDORP; C.E. STEFFLER; P. RUBIM; S.K. GOBBO & C.I. DONATTI. 2009. Priority areas for the conservation of Atlantic forest large mammals. **Biological Conservation 142**: 1229-1241.

GITTLEMAN, J.L. 1991. Carnivore olfactory bulb size: allometry, phylogeny and ecology. **Journal of Zoology 225** (2): 253-272.

HARRISON, R.L.1997. Chemical attractants for Central American felids. **Wildlife Society Bulletin 25**: 93-97.

HEINEMEYER, K.S.; T.J. ULIZIO & R.L. HARRISON. 2008. Natural sign: tracks and scats, p. 45-74. *In*: R. LONG; P. MACKAY; W. ZIELINSKI & J.C. RAY (Eds). **Noninvasive survey methods for carnivores.** Washington, DC, Island Press.

IUCN. 2010. **International Union for Conservation of Nature and Natural Resources.** Avaliable online at: http://www.iucnredlist.org. [Accessed: 12.II.2011].

KENDALL, K.C. & K.S. MCKELVEY. 2008. Hair collection, p. 135-176. *In*: R.A. LONG; P. MACKAY; W.J. ZIELINSKI & J.C. RAY (Eds). **Noninvasive survey methods for carnivores.** Washington, DC, Island Press.

LEITE, M.R.P. & F. GALVÃO. 2002. El jaguar, el puma y el hombre en tres áreas protegidas del bosque atlántico costero de Paraná, Brasil, p. 237-250. *In*: R.A. MEDELLIN; C. CHETKIEWICZ; A. RABINOWITZ; K.H. REDFORD; J.G. ROBINSON; E. SANDERSON & A. TABER (Eds). **El jaguar en el nuevo milenio.** Mexico, Universidad Nacional Autonoma de Mexico, Wildlife Conservation Society.

LONG, R.A.; T. DONOVAN; P. MACKAY; W.J. ZIELINSKI & J.S. BUZAS. 2007. Comparing scat detection dogs, cameras, and hair snares for surveying carnivores. **Journal of Wildlife Management 71** (6): 2018-2025.

LONG, R.A.; P. MACKAY; W.J. ZIELISNKI & J.C. RAY 2008. **Noninvasive survey methods for carnivores.** Washington, DC, Island Press. 386p.

LUDWIG, G; MALANSKI; M.M. SHIOZAWA; C.L.S. HILST; I.T. NAVARRO & F.C. PASSOS. 2007. Cougar predation on Black-and-Gold Howlers on Mutum Island, Southern Brazil. International **Journal of Primatology 28** (1): 39-46.

MACHADO, A.B.M.; G.M. DRUMMOND & A.P. PAGLIA. 2008. **Livro Vermelho da Fauna Brasileira Ameaçada de Extinção.** Belo Horizonte, Fundação Biodiversitas.

MARQUES, M.C.M.; M.D. SWAINE & D. LIEBSCH. 2011. Diversity distribution and floristic differentiation of the coastal lowland vegetation: implications for the conservation of the Brazilian Atlantic Forest. **Biodiversity and Conservation 20** (1): 153-168.

MARTINS, R.; J. QUADROS & M. MAZZOLLI. 2008. Hábito alimentar e interferência antrópica na atividade de marcação territorial do *Puma concolor* e *Leopardus pardalis* (Carnivora:Felidae) e outros carnívoros na Estação Ecológica de Juréia-Itatins, São Paulo, Brasil. **Revista Brasileira de Zoologia 25** (3): 427-435.

MCDANIEL, G.W.; K.S. MCKELVEY; J.R. SQUIRES & L.F. RUGGIERO. 2000. Efficacy of lures and hair snares to detect lynx. **Wildlife Society Bulletin 28** (1): 119-123.

MCKELVEY, K.S.; J.V. KIENAST; K.B. AUBRY; G.M. KOEHLER; B.T. MALETZKE; J.R. SQUIRES; E.L. LINQUIST; S. LOCH & M.K. SCHWARTZ. 2006. DNA analysis of hair and scat collected along snow tracks to document the presence of Canada lynx (*Lynx canadensis*). **Wildlife Society Bulletin 34**: 451-455.

MCPHEE, M.E. 2002. Intact carcasses as enrichment for large felids: effects on on-and off exhibit behaviors. **Zoo Biology 21**: 37-47.

MIOTTO, R.A.; G. CIOCHETI; F.P. RODRIGUES & P.M. GALETTI Jr. 2007. Identification of pumas (*Puma concolor* (Linnaeus, 1771)) through faeces: a comparison between morphological and molecular methods. **Brazilian Journal of Biology 67** (4): 963-965.

MIOTTO, R.A.; M. CERVINI; R.A. BEGOTTI; P.M & GALETTI JR. 2011. Monitoring a Puma (*Puma concolor*) Population in a Fragmented Landscape in Southeast Brazil. **Biotropica 44** (1): 98-104. doi: 10.1111/j.1744-7429.2011.00772.x

MIRANDA, J.M.D.; I.P. BERNARDI; K.C. ABREU & F.C. PASSOS. 2005. Predation on *Alouatta guariba clamitans* Cabrera (Primates, Atelidae) by *Leopardus pardalis* (Linnaeus) (Carnivora, Felidae). **Revista Brasileira de Zoologia 22** (3): 793-795.

MOWAT, G. & D. PEATKAU. 2002. Estimating marten *Martes americana* population size using hair capture and genetic tagging. **Wildlife Biology 8** (3): 201-209

OLIVEIRA, T.G. 2006. Research in terrestrial Carnivora from Brazil: current knowledge and priorities for the new Millennium,

p. 39-45. *In*: R.G. Morato; F.H.G. Rodrigues; E. Eizirik; E. Mangini (Eds). **Manejo e Conservação de Carnívoros Neotropicais.** São Paulo, Ibama.

Quadros, J. & E.L.A. Monteiro-Filho. 2006. Coleta e preparação de pêlos de mamíferos para identificação em microscopia óptica. **Revista Brasileira de Zoologia 23** (1): 274-278.

Quadros, J. & E.L.A. Monteiro-Filho. 2010. Identificação dos mamíferos de uma área de Floresta Atlântica utilizando a microestrutura de pelos-guarda de predadores e presas. **Arquivos do Museu Nacional 68** (1, 2): 47-66.

Reed, S.E. 2011. Non-invasive Methods to Assess Co-Occurrence of Mammalian Canivores. **The Southwestern Naturalist 56** (2): 231-240.

Ruell, W.W. & K.R. Crooks. 2007. Evaluation of non-invasive genetic sampling methods for felid and canid populations. **Journal of Wildlife Management 71**: 1690-1694.

Sabbatini, G.; M. Stammati; M.C.H. Tavares & E. Visalberghi. 2008. Behavioral flexibility of a group of bearded capuchin monkeys (Cebus libidinosus) in the National Park of Brasília (Brazil): consequences of cohabitation with visitors. **Brazilian Journal of Biology 68** (4): 685-693.

Schlexer, F.V. 2008. Attracting Animals to Detection Devices, p. 263-292. *In*: R.A. Long; P. Mackay; W.J. Zielinski & J.C. Ray (Eds). **Noninvasive survey methods for carnivores.** Washington, DC, Island Press.

SEMA-IAP. 1996. **Plano de Manejo do Parque Estadual do Pico do Marumbi – PR.** Curitiba, Secretaria de Estado do Meio Ambiente-Instituto Ambiental do Paraná,114p.

Silva-Pereira, J.E.; R.F. Moro-Rios; D.R. Bilski & F.C. Passos. 2011. Diets of three sympatric Neotropical small cats: Food niche overlap and interspecies differences in prey consumption. **Mammalian Biology 76** (3): 308-312.

Tanás, L. & W. Pisula. 2011. Response to novel object in Wistar and Wild-type (WWCPS) rats. **Behavioural Processes 86** (2011): 279-283.

Thomas, P.; G. Balme; L. Hunter & J. McCabe-Parodi. 2005. Using scent attractants to non-invasively collect hair samples from cheetahs, leopards and lions. **Animal Keeper's Forum 7** (8): 342-384.

Tomas, W.M.; F.H.G. Rodrigues & R. Fusco-Costa. 2006. Levantamento e monitoramento de populações de carnívoros, p. 145-167. *In*: R.G. Morato; F.H.G. Rodrigues; E. Eizirik; P.R. Mangini; F.C.C. Azevedo & J. Marinho-Filho (Eds). **Manejo e Conservação de Carnívoros Neotropicais.** Brasília, IBAMA.

Wasser, S.K.; B.E.R. Davenport; K. Ramage, E. Hunt; M. Parker, C. Clarke & G. Stenhouse. 2004. Scat detection dogs in wildlife research and management: application to grizzly and black bears in the Yellowhead Ecosystem, Alberta, Canada. **Canadian Journal of Zoology 82** (3): 475-492.

Weaver, J.L.; P. Wood; D. Paetkau & L.L. Laack. 2005. Use of scented hair snares to detect ocelots. **Wildlife Society Bulletin 33**: 1384-1391.

Weller, S.H. & C.L. Bennett. 2001. Twenty-four hour activity budgets and patterns of behaviour in captive ocelots (*Leopardus pardalis*) **Applied Animal Behaviour Science 71**: 67-69.

19

The spatial distribution of the subtidal benthic macrofauna and its relationship with environmental factors using geostatistical tools

Fernanda M. de Souza[1,2], Eliandro R. Gilbert[1], Maurício G. de Camargo[1] & Wagner W. Pieper[1]

[1] Centro de Estudos do Mar, Universidade Federal do Paraná. Caixa Postal 50002, 83255-976 Pontal do Paraná, PR, Brazil.
[2] Corresponding author. E-mail: fernanda.cem@gmail.com

ABSTRACT. Modeling the distribution patterns of the estuarine macrobenthic community has revealed itself as a difficult task due to spatio-temporal heterogeneity. This study uses ordinary kriging and Poisson modeling to generate distribution maps of the subtidal benthic macrofauna in the Trapandé Bay (southeastern Brazil). Samples were taken in duplicate from 36 locations distributed along nine transects perpendicular to the main estuarine axis in October 2006 and March 2007. One-hundred and seventy taxa belonging to 12 phyla, were identified, with dominance of Annelida Polychaeta. Distribution maps were prepared to illustrate the total density, the number of species and the six most numerous taxa, as well as abiotic parameters. The general distribution pattern has revealed that the greatest number of species and the highest densities are at the estuary mouth, decreasing towards its inner areas. However the temporal and spatial changes observed at the estuary mouth have clearly shown the impact of environmental variations such as nutrients and freshwater input, attributed to increased rainfall in March. The increased flow in the Cananeia Sea, coming from the drainage basin, produces major changes in sediment and faunal composition. Ordinary kriging associated with Poisson modeling has proved to be a powerful and promising tool for modeling the macrofauna, despite the fact that it is not frequently used due to the scarcity of appropriate software.

KEY WORDS. Cananeia; kriging; macrobenthic.

The distribution of benthic organisms in unconsolidated sediments is determined both by abiotic factors, such as sedimentary and physicochemical characteristics, and by biological factors, such as predation, competition and bioturbation (DAUER 1993, LEVINTON 1995, SNELGROVE 1998, THRUSH & DAYTON 2002, MAIRE et al. 2010, FANJUL et al. 2011). The active interdependence among multiple potentially important factors causes the distribution of benthic associations to be frequently asymmetrical, non-linear and heterocedastic (ANDERSON 2008). Therefore, the distribution of benthic associations resulting from these complex interactions is very heterogeneous (SCHNEIDER 1994), which renders the recognition of spatial patterns difficult. In addition to the heterogeneity in species distribution, another complicating factor for modeling efforts is that species variability is not constant along environmental gradients, as it happens in estuaries, but is instead dependent upon the average abundances of species (ELLIOT & WHITFIELD 2011).

Several studies have shown that the spatial variation in the structure of benthic communities and the nature of marine sediments occur at various scales (e.g., THRUSH et al. 1997, BELL et al. 1999, ANDERSON 2008, CHAPMAN et al. 2010, RODRIGUES et al. 2011), contributing to a better understanding of the ecological relationships between physical processes and species distribution (ELLIOT & McLUSKY 2002). Thus, this work describes the spatial distribution of subtidal benthic macrofaunal associations and the environmental factors in Trapandé Bay, a subtropical estuarine system in southeastern Brazil; compares the changes of these variables between two seasons and; evaluates the relationship between environmental variables along an estuarine gradient and the benthic macrofauna. We hypothesize that the response of the benthic macrofauna to environmental variables may be visually detected in spatial distribution maps, using appropriate geostatistical methods.

MATERIAL AND METHODS

The Trapandé Bay is located in the southern sector of the Lagoon-Estuarine Complex of Cananeia-Iguape, southern coast of São Paulo, Brazil. The estuary is geologically constructed by a series of bars, which form a complex system of channels between Comprida, Cananeia and Cardoso islands. The main islands are separated by channels (Fig. 1) that connect to the ocean through the Cananeia Inlet, in the central sector of the estuary, the Icapara Inlet, in its northern sector, and the Ararapira Inlet, in its southern sector (BERNARDES & MIRANDA 2001).

The main channels that connect with the Trapandé Bay are the Cubatão Sea in the Northwest, and the Cananeia Sea in the Northeast, which have average depths of 6 and 10 m, respectively. The opening to the Southwest is approximately 10 m deep.

The regional drainage basin has variable seasonal discharges and may carry high concentrations of suspended sediments (Miranda *et al.* 2002). About 60% of the Ribeira River discharge flows by the internal channels of the Cananeia-Iguape estuary, carrying suspended sediments and nutrients to the Trapandé Bay through the Cubatão Sea and the Cananeia Sea (Saito *et al.* 2006). The rainy season extends from December to April, and the dry season from May to November (Silva 1984).

The tidal regimen is semi-diurnal, with inequalities during a diurnal cycle (Miyao *et al.* 1986). Currents are more intense on the southern bank of the Trapandé Bay, near Cardoso Island, and less intense on the northern bank, near the Cananeia Island (Tessler & Souza 1998).

Sampling was carried out in October 2006, during the dry period, and in March 2007, during the period of highest average annual rainfall. In each campaign, nine transects were sampled at the main axis, covering the entire extent of the bay, and four sampling points were established on each transect (Fig. 2). Transects were divided into inner (1, 2 and 3), intermediate (4, 5 and 6) and outer (7, 8 and 9). At each point, two

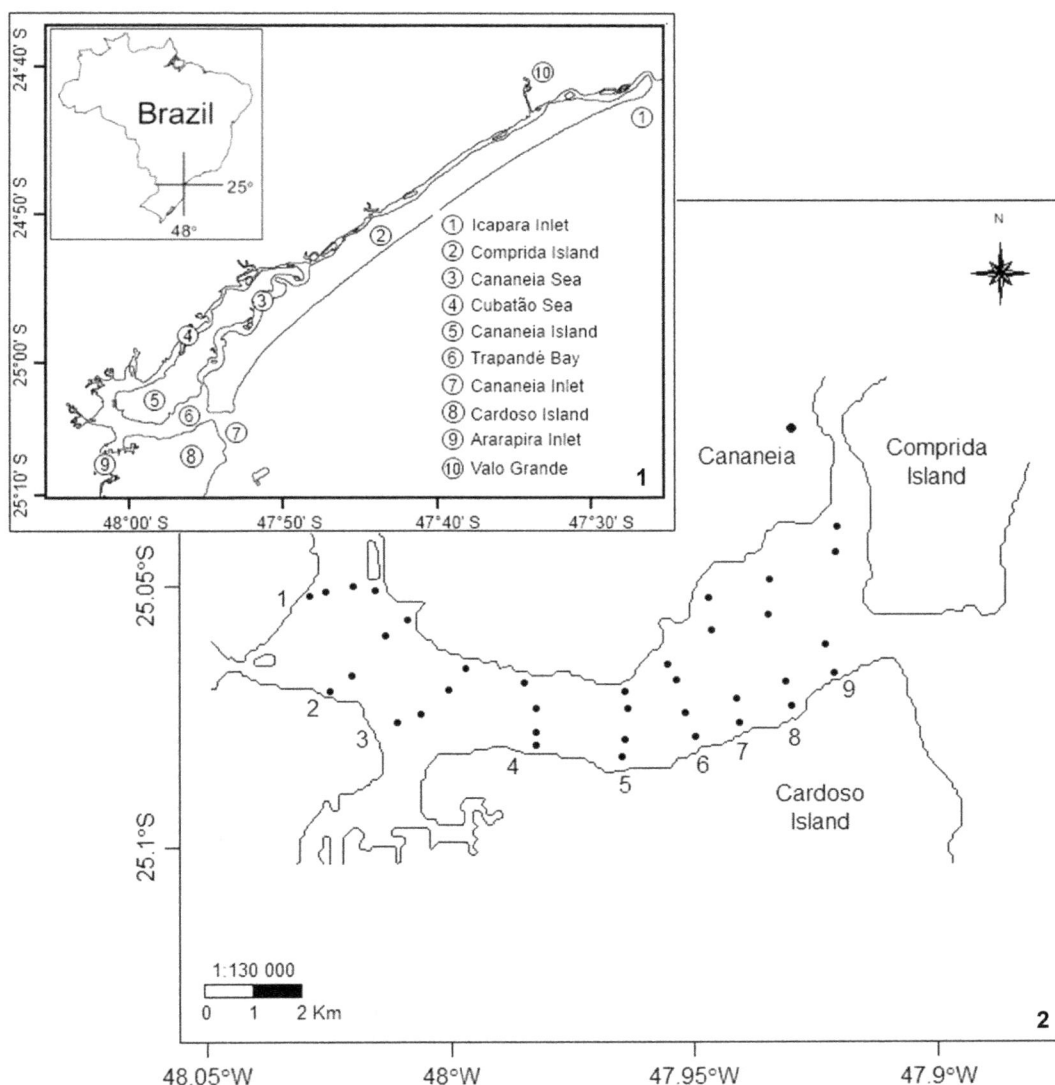

Figures 1-2. A map illustrating the study area, Trapandé Bay and the nine sampled transects (labeled 1 to 9). Sampling was performed at four points in each transect.

samples were taken for benthic macrofauna and one sample
for granulometric and chemical analysis using a Van Veen grab
with an area of 0.065 m². The faunal samples were fixed with
5% formaldehyde and washed on 0.5 mm mesh sieves. Animals were counted and identified to the lowest possible taxonomic level. The replicates were pooled and the data were
converted into individuals per square meter (ind/m²).
Granulometric measures were performed through the method
described by Suguio (1973) and the data was analyzed using
the package rysgran (Gilbert *et al.* 2012) within the R environment (R Development Core Team 2012). The sedimentary organic
matter content was determined after 1 hour at 550°C. Total
nitrogen and phosphorus content were determined following
Grasshoff *et al.* (1983).

Geostatistics was originally developed by Matheron (1971)
and Journel & Huijbregts (1978) in mining geology and is currently employed in a variety of scientific fields, such as soil
sciences, meteorology, hydrology and ecology (Axis-Arroyo &
Mateu 2004). Although it is most frequently applied to marine
fisheries (Petitgas 2001, Maynou *et al.* 1998, Simard *et al.* 1992,
Rufino *et al.* 2004), it has also been used to assess the patterns
of spatial distribution of the benthic macrofauna (Bergstrom *et
al.* 2002, Cole *et al.* 2001).

The geostatistical Gaussian models do not properly estimate the response patterns of benthic species counting due to
the existence of a correlation between a species' average density at a given point and its variance (Anderson 2008). Whenever such a correlation exists, Poisson models are more adequate
and generate results that are far more realistic. Here,
geostatistical maps were generated by ordinary kriging, and
adjusted to the Poisson model for densities of the most representative species (in terms of species abundance and occurrence), the total number of species, and the total species
abundance. Maps were generated using the R package geoRglm
(Christensen & Ribeiro Jr 2002).

RESULTS

In transects 3, 4 and 5, sediments were predominantly
silty. Mean grain size was coarser towards the inner bay and
the Cananeia Inlet, with higher values around 2 Φ (fine sand)
(Fig. 3).

Total nitrogen (TN) ranged between 0 and 3.5 mg/g, with
higher values in the inner and intermediate regions, decreasing towards the Cananeia Inlet in both campaigns (Figs 4 and
5). Total phosphorus contents (TP) were higher in the October
campaign with maximum values of 3.0 mg/g in the outer region and lowest values in transects 4 and 5. In March, TP did
not exceed 0.7 mg/g near the Cananeia Inlet (Figs 6 and 7).

Organic matter concentrations (OM) were higher in the
northern sector of the bay in October, especially in transect 4,
with values up to 20%. The lowest OM values occurred in the
southern sector of the outer transects. In March, the values of

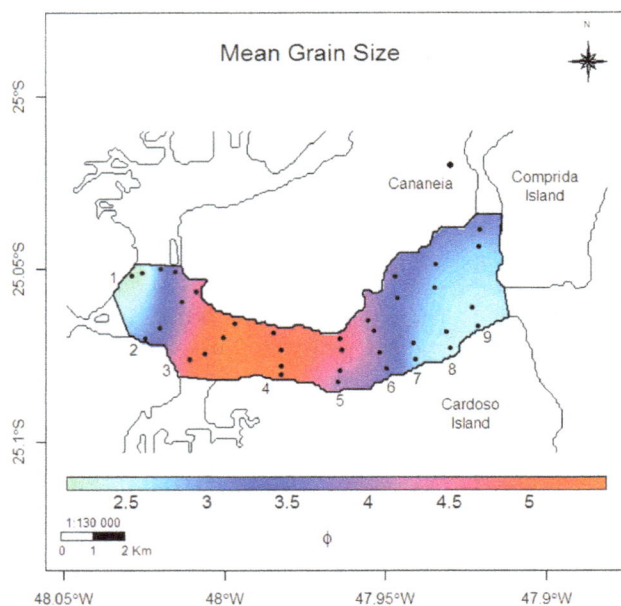

Figure 3. A map of Trapandé Bay representing the mean grain size
in phi intervals (Φ) observer in sampled points.

OM were homogeneous along transects 1 to 8, despite the high
value observed in the southern sector of transect 3. The lowest
concentrations were observed in transect 9 (Figs 8 and 9).

A total of 5,531 organisms were identified, distributed
among 12 phyla and 170 taxa. Annelids were numerically dominant, totaling 87.8% of the overall abundance, together with
molluscs and crustaceans. The most representative taxa in terms
of abundance and occurrence were *Aricidea* sp., Capitellidae,
Magelonidae, Sipuncula, *Scoletoma tetraura* (Schmarda, 1861)
and Ophiuroidea, which together totaled 2,154 organisms and
comprised 39% of the total abundance (Table I).

In October, species numbers gradually increased towards
the mouth of the bay and were greatest in transect 9. In March,
species numbers were greatest in transects 5, 6 and 7, where
approximately 40 taxa per m² were observed (Figs 10 and 11).

Littoridina australis (Orbigny, 1835) was the most abundant taxon due to its extremely high density at a single sampled
point (691 individuals in one sample alone in March). However, aside from this extreme value, the abundances of *L. australis* were low in all the other samples in both campaigns.
Likewise, *Bulla striata* (Bruguière, 1792) showed high densities
in specific occurrences in transect 2. Both taxa were removed
from the analysis.

The total density maps indicated higher densities in the
outer transects in October, whereas in March the highest densities were concentrated in the intermediate transects and decreased in transect 9 (Figs 12 and 13). In March, transect 5 had
the highest total density on its northern sector, where the to-

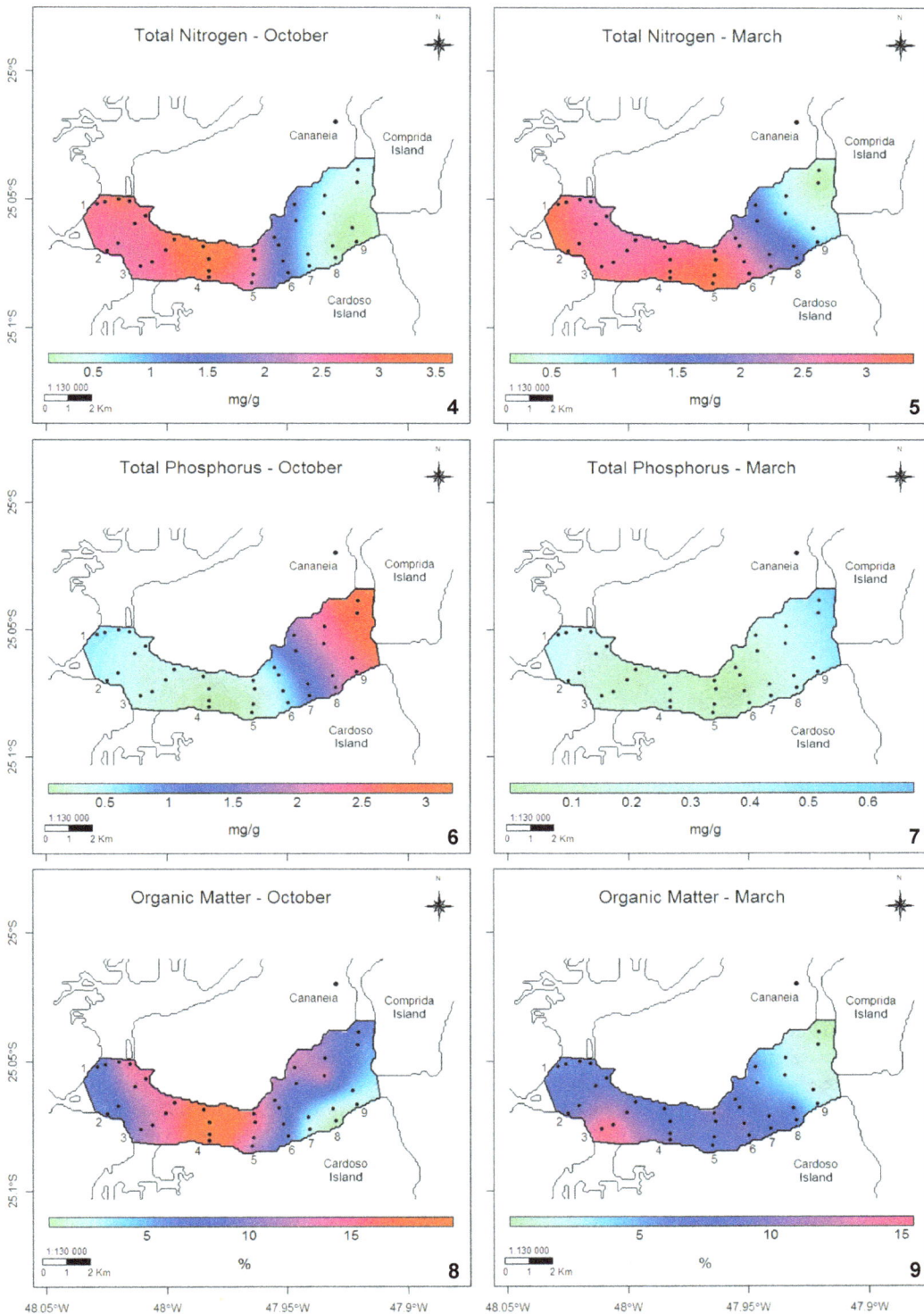

Figures 4-9. A map of Trapandé Bay representing the concentrations of Total Nitrogen, Total Phosphorus and Organic Matter for the October 2006 and March 2007 campaigns.

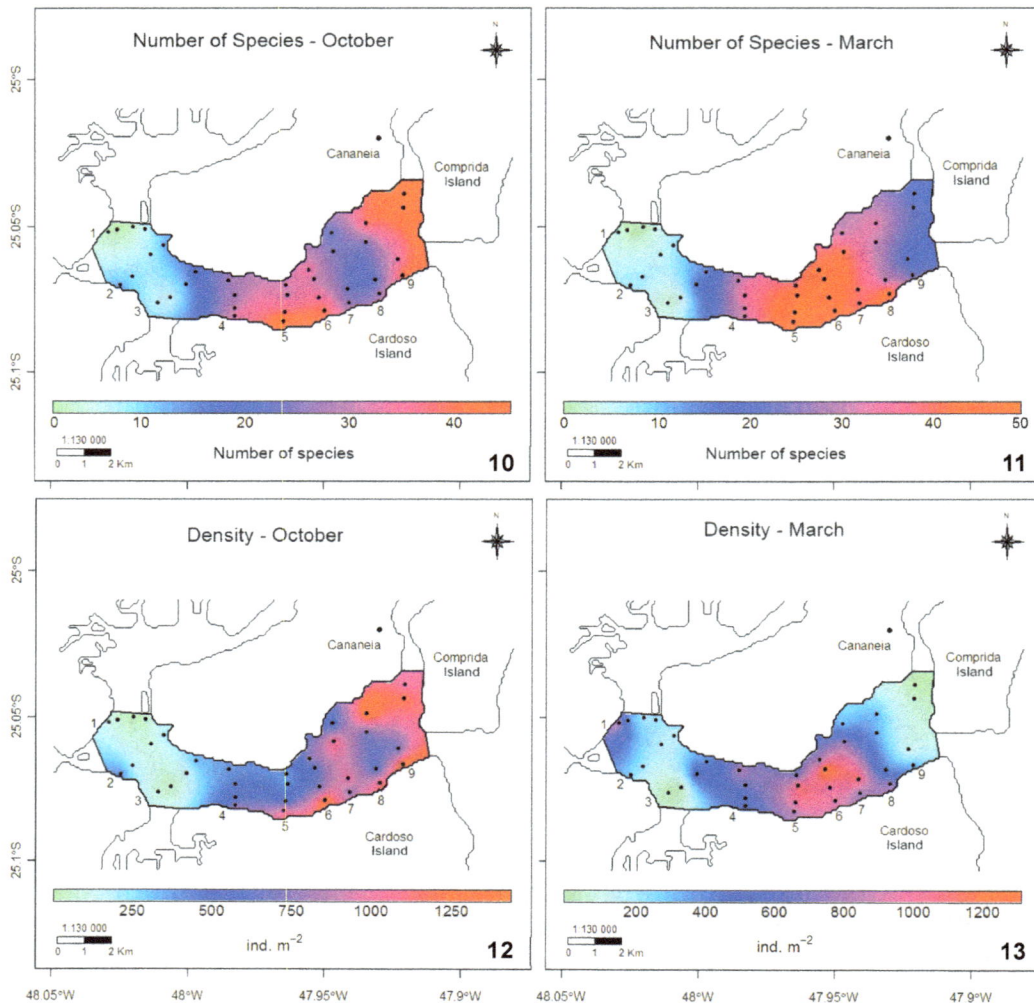

Figures 10-13. A map of Trapandé Bay representing the number of species and density for the October 2006 and March 2007 campaigns.

tal density reached 5,000 ind/m². In October, the densities were highest in the intermediate and outer regions of the estuary and on the estuarine margins.

Aricidea sp. had the lowest densities in transects inside the bay, not exceeding 50 ind/m² per sample in October and 20 ind/m² in March (Figs 14 and 15). There was a gradual increase in its density towards the northern sectors of the outer transects, where the highest densities observed ranged between 400 and 600 ind/m² in October. However, in March, the average densities in the outer areas were reduced to 70 ind/m².

Members of Capitellidae were more abundant in the outer areas, particularly in the sampling points at the south shore, and displayed variance between the two campaigns only in transect 9, which presented lower densities in March than in October (Figs 16 and 17). Members of the Magelonidae were first found in high densities in transect 4, and tended to increase in density in the outermost transects. In March, the highest densities of Magelonidae were restricted to the intermediate transects (4, 5 and 6). A maximum of 250 ind/m² was observed in the northern sectors of transects 5 and 6, and decreasing densities were noted at the mouth of the bay (Figs 18 and 19).

In October, Ophiuroidea increased in density towards the mouth of the bay, with higher density values on the southern banks of transects 8 and 9, where up to 200 ind/m² were recorded. In March, the highest densities of Ophiuroidea were recorded in transects 5, 6 and 7, with a maximum of 50 ind/m² on the southern bank sampling points (Figs 20 and 21).

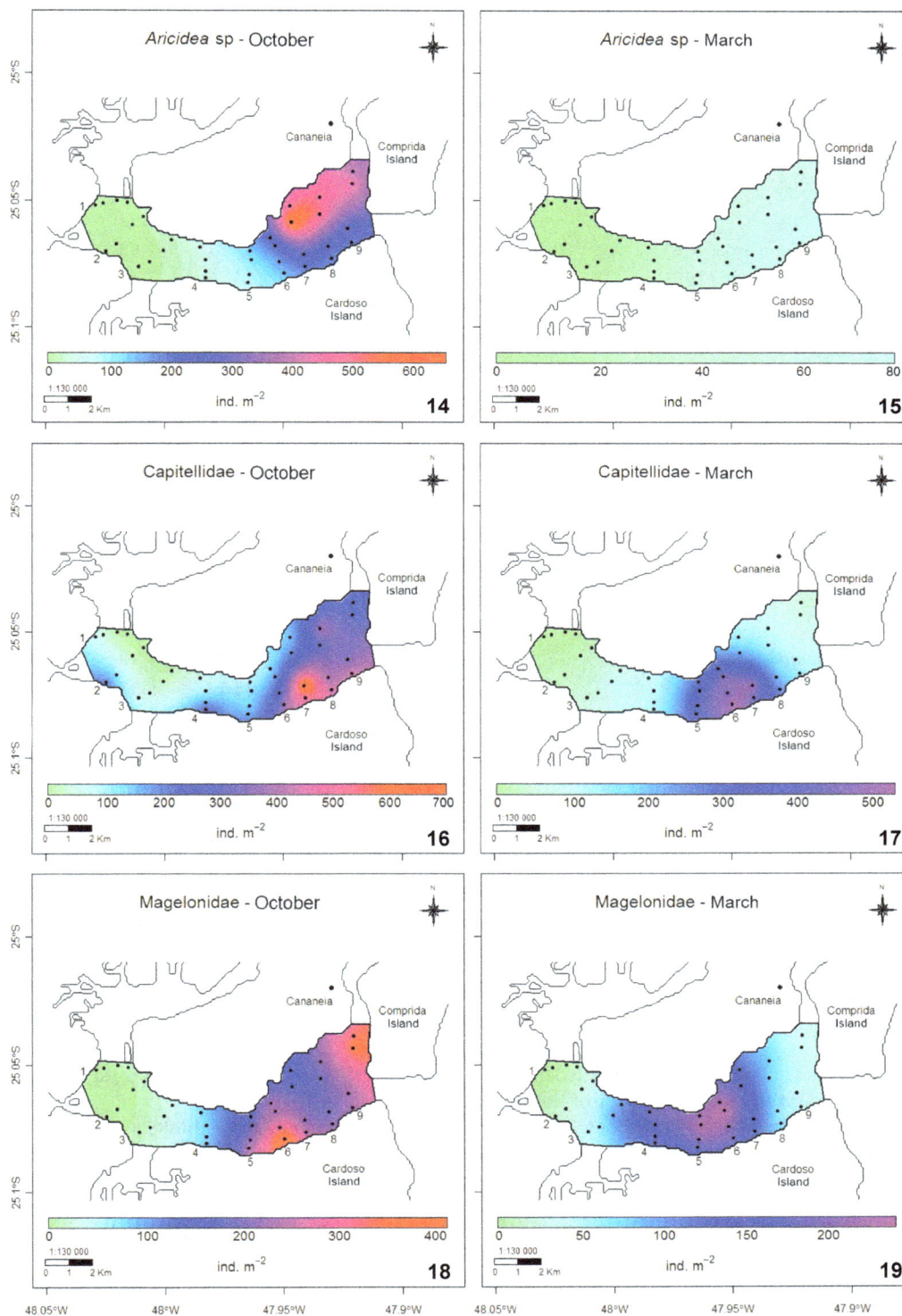

Figures 14-19. A map of Trapandé Bay representing the densities of *Aricidea* sp., Capitellidae and Magelonidae for the October 2006 and March 2007 campaigns.

Table I. The mean density (ind/m²) and relative frequency (%) of the most representative groups and species in the October 2006 and March 2007 sampling campaigns.

Taxa	October		March		Mean	
	Ind/m²	%	Ind/m²	%	Ind/m²	%
Polychaetes	392.52	67.38	300.43	50.12	346.47	58.63
Capitellidae	89.53	15.46	77.35	13.00	83.44	14.21
Magelonidae	44.66	7.71	55.13	9.26	49.89	8.50
Aricidea sp.	49.57	8.56	11.75	1.97	30.66	5.22
Scoletoma tetraura (Schmarda, 1861)	15.81	2.73	15.81	2.66	15.81	2.69
Paraprionospio pinnata (Ehlers, 1901)	4.91	0.85	14.10	2.37	9.50	1.62
Paraprionospio sp.	11.54	1.99	3.63	0.61	7.58	1.29
Parandalia tricuspis (Müller, 1858)	9.62	1.66	5.13	0.86	7.37	1.26
Sigambra sp.	10.47	1.81	3.63	0.61	7.05	1.20
Neanthes bruaca Lana & Sovierzovsky, 1987	5.56	0.96	8.12	1.36	6.83	1.16
Sthenelais limicola (Ehlers, 1864)	11.97	2.07	1.28	0.22	6.62	1.13
Clymenella dalesi Mangum, 1966	8.33	1.44	4.49	0.75	6.41	1.09
Mollusca	82.26	14.12	197.44	32.94	139.85	23.67
Littoridina australis	12.39	2.14	155.56	26.14	83.97	14.30
Bulla striata (Bruguière, 1792)	41.88	7.23	5.77	0.97	23.82	4.06
Angulus versicolor (De Kay, 1843)	10.90	1.88	10.26	1.72	10.57	1.80
Tagelus sp.	1.07	0.18	12.18	2.05	6.62	1.13
Arthropoda	20.09	3.44	45.09	7.52	32.58	5.51
Kalliapseudes schubarti Mane-garzon, 1969	7.69	1.33	11.32	1.90	9.50	1.62
Sipuncula	45.51	7.86	33.12	5.57	39.31	6.70
Echinodermata	10.90	1.87	11.11	1.85	11.00	1.86
Ophiuroidea	10.90	1.54	11.11	1.82	11.00	1.67
Echiura	13.89	2.38	0.00	0.00	6.94	2.38
Cnidaria	8.12	1.39	5.34	0.89	6.73	1.86
Phoronida	7.26	1.25	3.21	0.53	5.23	0.89
Porifera	0.00	0.00	3.63	0.61	1.81	0.31
Platyhelminthes	1.07	0.18	0.00	0.00	0.53	0.09
Chordata	0.64	0.11	0.00	0.00	0.32	0.05
Nemertea	0.21	0.04	0.00	0.00	0.10	0.02
Total	582.48		599.36		590.92	

The highest densities of S. tetraura were recorded in October, mainly concentrated on the southern sector of transect 5 and the northern sector of transect 8, with about 200 ind/m² (Fig. 22). In March the highest densities of S. tetraura occurred in transects 5, 6 and 7. In these locations, average densities varied between 80 and 100 ind/m² (Fig. 23). Inner transects showed the lowest densities of S. tetraura in both sampling campaigns.

Sipunculans were distributed in a similar pattern in both campaigns, with higher density values consistently observed in the intermediate transects. In October, their highest densities occurred in transects 4 and 5, shifting to transects 5 and 6 in March (Figs 24 and 25).

DISCUSSION

The Trapandé Bay exhibits a well-defined environmental gradient with population densities and number of taxa decreasing towards the inner region. The intermediated sector, which corresponds to transects 3 to 7, contained a higher number and density of species than the inner region, particularly on the north bank of the estuary, which is shallow. The outer region presented the highest faunal differences between campaigns, presumably due to the powerful hydrodynamic effects generated by interactions with the Cananeia Sea and the Icapará Inlet. This region has strong, dynamic currents near its bottom

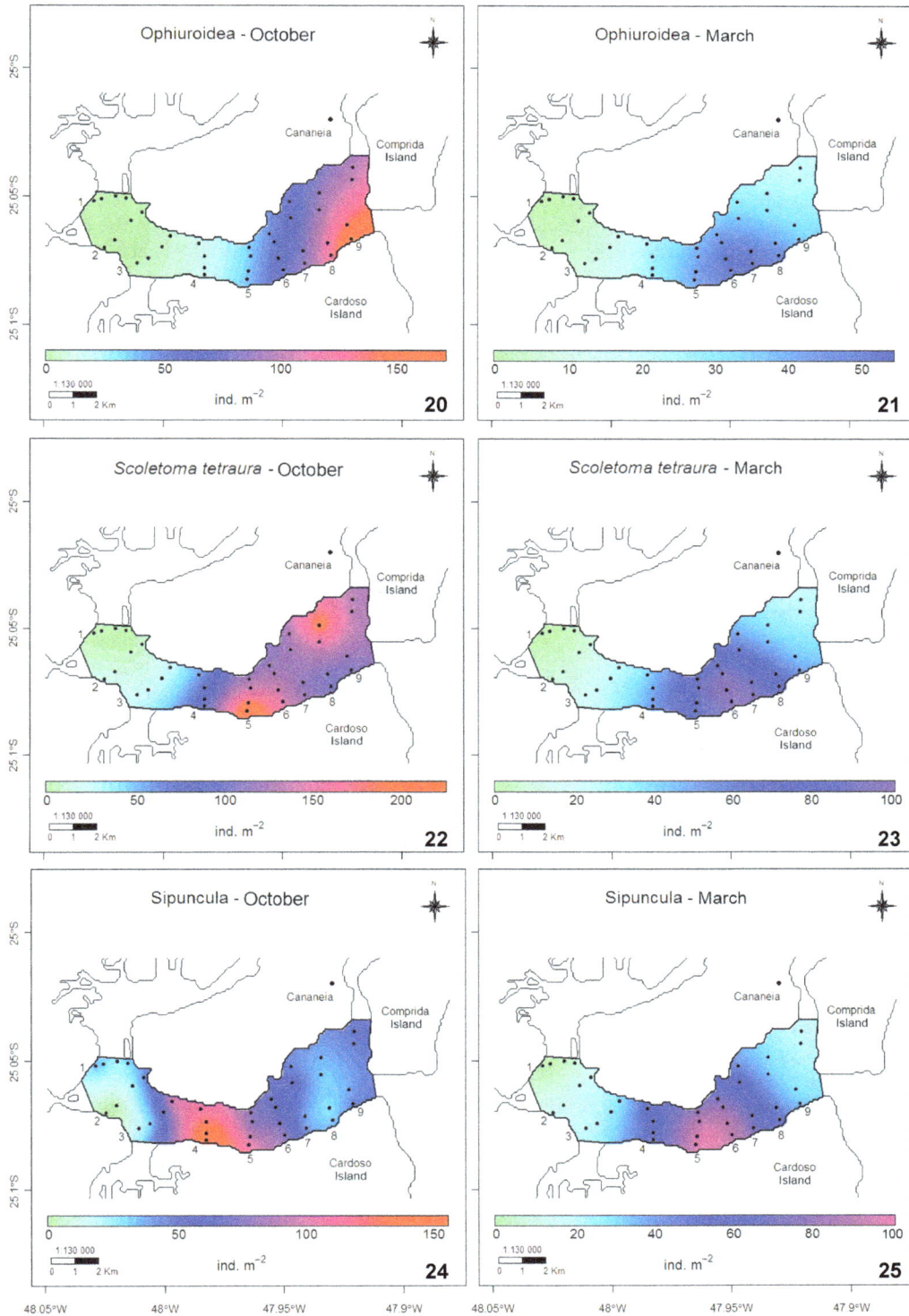

Figures 20-25. A map of Trapandé Bay representing densities of Ophiuroidea, *Scoletoma tetraura* and Sipuncula for the October 2006 and March 2007 campaigns.

(>1.0 m/s) due to the influence of the mouth of the Cananeia Sea and the Icapará Inlet (MIYAO et al. 1986), as well as the constriction of the lagoon channels (KUTNER 1962, TESSLER & SOUZA 1998). The friction stress caused by the bottom currents tends to resuspend fine sediments, nutrients and organic matter.

The Cananeia Sea influences the outer region of the bay due to the freshwater and nutrient input from the Ribeira do Iguape River through the Valo Grande Channel. Although located 40 km from Trapandé Bay, the Valo Grande Channel influences nutrient cycling in the southern sector of the Cananeia-Iguape system. MIYAO et al. (1986) observed higher concentrations of phosphates in the mouth region of the Trapandé Bay that were caused by the resuspension of sediments at the tide entrance. JORCIN (2000) found higher concentrations of total phosphorus and nitrogen in the Cananeia region during the summer, in areas coinciding with the region of transects 8 and 9 that were associated with a seasonal increase in the levels of organic matter.

The circulation channel of Trapandé Bay is located next to the Cardoso Island and there is a transition within the bay from sandy sediments, associated with the channel, to sediments with a high mud content (silt and clay), which are found on the northern sector of the region, near the Cananeia Island (SAITO et al. 2006). The northern margin of this area is less deep and suffers less hydrodynamic influence.

The distribution pattern of benthic organisms observed on estuaries typically fit the ecocline model. The community progressively changes following the gradual variation in at least one major environmental variable. In estuaries, this variable usually is salinity, but hydrodynamics, organic matter and mud content play important roles. Furthermore, estuaries differ from previously defined terrestrial ecoclines because they present two overlapping gradients in the major stressor: from river to mid-estuary for freshwater species and from sea to mid-estuary for marine species (ATTRILL & RUNDLE 2002).

Sediments with higher organic matter contents are likely to present higher abundances of deposit-feeder polychaetes and further opportunists or early colonizing species (PEARSON & ROSENBERG 1978). Capitellid polychaetes are a good example of opportunistic species, characteristically found in areas with heavy input of organic matter (SCOTT et al. 1987, RIZZO & AMARAL 2001).

Individuals of Paraonidae family are considered non-selective surface diggers (FAUCHALD & JUMARS 1979). In the October campaign, Aricidea sp. was predominantly found at the shallow waters near the Cananeia Sea, associated with higher values of TP and OM. However, in March, the densities of Aricidea sp. decreased, most likely due to the increased rainfall and hydrodynamic flows at the estuary's bottom, which leaded to a decrease in TP and OM contents.

Magelonid polychaetes, a pooling of Magelona papillicornis (Müller, 1858), Magelona posterelongata (Bolivar & Lana, 1986) and Magelona variolamellata (Bolivar & Lana, 1986), were present in the intermediate and outer zone of the Trapandé Bay, not

well correlated with the observed peaks of organic matter content. Higher densities of magelonids were usually found in sediments with about 5-10% of organic matter. Magelonids are mainly non-selective surface deposit feeders, but some species may alternate to suspension feeding (FAUCHALD & JUMARS 1979). This alteration in feeding strategies may make it possible for magelonids to live both in muddy and sandy sediments (ROUSE & PLEIJEL 2001).

The distribution pattern of the omnivorous S. tetraura was intrinsically linked to the total density and number of species. This specie may be more influenced by interspecific relationships than by environmental variables. Ophiuroids are typically marine and are thus restricted to the outer estuarine area in the dry season (October). In March, most likely due to the freshwater inflow that originates from the North through the Cananeia Sea, their occurrence was restricted to the southern bank of the outer area.

The general benthic community pattern observed in the Trapandé Bay, in which highest densities and species numbers were recorded at the estuary mouth and decreased towards the inner areas, is recurrent in other estuaries from southern and southeastern Brazil, as shown in Cananeia (TOMMASI 1970), Paranaguá Bay (LANA 1986) and Patos Lagoon (BEMVENUTI & NETTO 1998). The temporal and spatial variation in faunal parameters clearly demonstrated the impact of environmental variations in either chemical composition or particle size on the distribution of the macrofauna. The marked temporal variations may be mainly attributed to increased rainfall in March, when the increased flow in the Cananeia Sea coming from the Ribeira do Iguape River affects sediment properties.

Interpolation procedures represent a gain in information over unsampled areas, but with the restriction that the results contain some degree of uncertainty (JEROSCH et al. 2007).

Several statistical/mathematical methodologies exist on how to derive spatially continuous distributions from available localized point data (LI et al. 2011 for a review), for instance decision tree models (PESCH et al. 2011), machine learning methods (LI & HEAP 2008) and combined methods that use ordinary kriging with Gaussian linear models (STELZENMÜLLER et al. 2010) or non-Gaussian generalized linear models (GML) (GOTWAY & STROUP 1997, MAXWELL et al. 2009). GLM is more adequate than linear models to analyze spatial structure of the benthic macrofauna because it can fit Poisson data, which is much more suitable for counting data like abundance or richness. Most available software for analyzing spatial data, however, works only with Gaussian methods and that is the reason why the method is rarely used in modeling benthic macrofauna.

REFERENCES

ANDERSON, M.J. 2008. Animal-sediment relationships re-visited: characterizing species' distributions along an environmental gradient using canonical analysis and quantile regression splines. **Journal of Experimental Marine Biology and Ecology** 366 (1/2): 16-27. doi: 10.1016/j.jembe.2008.07.006.

ATTRILL, M. & S.D. RUNDLE. 2002. Ecotone or ecocline: ecological boundaries in estuaries. **Estuarine, Coastal and Shelf Science 55**: 929-936. doi: 10.1006/ecss.2002.1036.

AXIS-ARROYO, J. & J. MATEU. 2004. Spatio-temporal modeling of benthic biological species. **Journal of Environmental Management 71**: 67-77.

BELL, S.S.; B.D. ROBBINS & S.L. JENSEN. 1999. Gap dynamics in a seagrass landscape. **Ecosystems 2** (6): 493-504.

BEMVENUTI, C.E. & S.A. NETTO. 1998. Distribution and seasonal patterns of the sublittoral benthic macrofauna of Patos Lagoon (South Brazil). **Revista Brasileira de Biologia 58** (2): 211-221.

BERGSTROM, U.; G. ENGLUND & E. BONSDORFF. 2002. Small-scale spatial structure of Baltic Sea zoobenthos – inferring processes from patterns. **Journal of Experimental Marine Biology and Ecology 281**: 123-136.

BERNARDES, M.E.C. & L.B. MIRANDA. 2001. Circulação estacionária e estratificação de sal em canais estuarinos: simulação com modelos analíticos. **Revista Brasileira de Oceanografia 49** (1/2): 115-132.

CHAPMAN, M.G.; T.J. TOLHURST; R.J. MURPHY & A.J. UNDERWOOD. 2010. Complex and inconsistent patterns of variation in benthos, micro-algae and sediment over multiple spatial scales. **Marine Ecology Progress Series 398**: 33-47. doi:10.3354/meps08328.

CHRISTENSEN, O.F. & P.J. RIBEIRO JR. 2002. geoRglm: A package for generalized linear spatial models. **R-NEWS 2** (2): 26-28.

COLE, R.G.; T.R. HEALY; M.L. WOOD & D.M. FOSTER. 2001. Statistical analysis of spatial pattern: a comparison of grid and hierarchical sampling approaches. **Environmental Monitoring and Assessment 69**: 85-91.

DAUER, D.M. 1993. Biological criteria, environmental health and estuarine macrobenthic community structure. **Marine Pollution Bulletin 26** (5): 249-257. doi:10.1016/0025-326X(93)90063-P.

ELLIOTT, M. & D.S. MCLUSKY. 2002. The need for definitions in understanding estuaries. **Estuarine, Coastal and Shelf Science 55** (6): 815-827. doi: 10.1006/ecss.2002.1031.

ELLIOTT, M. & A.K. WHITFIELD. 2011. Challenging paradigms in estuarine ecology and management. **Estuarine, Coastal and Shelf Science 94**: 306-314. doi: 10.1016/j.ecss.2011.06.016.

FANJUL, E.; M.C. BAZTERRICA; M. ESCAPA; M.A. GRELA & O. IRIBARNE. 2011. Impact of crab bioturbation on benthic ûux and nitrogen dynamics of Southwest Atlantic intertidal marshes and mudûats. **Estuarine, Coastal and Shelf Science 92**: 629-638.

FAUCHALD, K. & P. A. JUMARS. 1979. The diet of worms: a study of polychaete feeding guilds. **Oceanography and Marine Biology, an Annual Review 17**: 193-284.

GILBERT, E.R.; M.G. CAMARGO & L. SANDRINI-NETO. 2012. **rysgran: Grain size analysis, textural classifications and distribution of unconsolidated sediments.** R package version 2.0.

GOTWAY, C.A. & W.W. STROUP. 1997. A generalized linear model approach to spatial data analysis and prediction. **Journal of Agricultural, Biological, and Environmental Statistics 2** (2): 157-178.

GRASSHOFF, K.; M. EHRHARDT & K. KREMLING. 1983, **Methods of seawater analysis.** Weinhein, Verlag Chemie, 2nd ed., 419p.

JEROSCH, K; M. SCHLÜTER; J.P. FOUCHER; A.J. ALLAIS; M. KLAGES & C. EDY. 2007. Spatial distribution of mud flows, chemoautotrophic communities, and biogeochemical habitats at Håkon Mosby Mud Volcano. **Marine Geology 243**: 1-17. doi:10.1016/j.margeo.2007.03.010.

JORCIN, A. 2000. Physical and chemical characteristics of the sediment in the estuarine region of Cananéia (SP), Brazil. **Hydrobiologia 431** (1): 59-67. doi: 10.1023/A:1004054305496.

JOURNEL, A. & C.H. HUIJBREGTS. 1978. **Mining geostatistics.** London, Academic Press, 600p.

KUTNER, A.S. 1962. Granulometria dos sedimentos de fundo da região de Cananéia (SP). **Boletim da Sociedade Brasileira de Geologia 11** (2): 41-54.

LANA, P.C. 1986. Macrofauna bêntica de fundos sublitorais não consolidados da Baía de Paranaguá (Paraná). **Nerítica 1** (3): 79-89.

LEVINTON, J. 1995. Bioturbators as ecosystem engineers: control of the sediment fabric, interindividual interactions, and material fluxes, p. 29-36. *In*: C.G. JONES & J.H. LAWTON (Eds). **Linking Species and Ecoystems.** New York, Chapman and Hall, 387p.

LI, J.; A.D. HEAP. 2008. **A Review of Spatial Interpolation Methods for Environmental Scientists.** Canberra, Geoscience Australia Record, 137p.

LI, J.; A.D. HEAP; A. POTTER; Z. HUANG & J.J. DANIELL. 2011. Can we improve the spatial predictions of seabed sediments? A case study of spatial interpolation of mud content across the southwest Australian margin. **Continental Shelf Research 31**: 1365-1376. doi: 10.1016/j.csr.2011.05.015.

MAIRE, O.; J.N. MERCHANT; M. bulling; L.R. TEAL; A. GRÉMARE; J.C. DUCHÊNE & M. SOLAN. 2010. **Journal of Experimental Marine Biology and Ecology 395** (1-2): 30-36.

MATHERON, G. 1971. **The theory of regionalized variables and its applications.** Paris, Ecole Nationale Supérieure de Mines, 209p.

MAYNOU, F.; F. SARDA & G.Y. CONAN. 1998. Assessment of the spatial structure and biomass evaluation of *Nephrops norvegicus* (L.) populations in the northwestern Mediterranean by geostatistics. **ICES Journal of Marine Science 55**: 102-120.

MAXWELL, D.L.; V. STELZENMÜLLER; P.D. EASTWOOD & S.I. ROGERS. 2009. Modelling the spatial distribution of plaice (Pleuronectes platessa), sole (Solea solea) and thornback ray (Raja clavata) in UK waters for marine management and planning. **Journal of Sea Research 61** (4): 258-267. doi: 10.1016/j.seares.2008.11.008.

MIRANDA, L.B; B.M. CASTRO & B. KJERFVE. 2002. **Princípios de Oceanografia Física em estuários.** São Paulo, EDUSP, 414p.

MIYAO, S.Y; L. NISHIHARA & C.C. SARTI. 1986. Características físicas e químicas do sistema estuarino-lagunar de Cananéia – Iguape. **Boletim do Instituto Oceanográfico 34** (1): 23-36.

PEARSON, T.H & R. ROSENBERG. 1978. A Macrobenthic succession in relation to organic enrichment and pollution of the marine environment. **Oceanography and Marine Biology: An Annual Review 16**: 229-311.

PESCH, R.; G. SCHMIDT; W. SCHROEDER & I. WEUSTERMANN. 2011. Application of CART in ecological landscape mapping: Two case studies. **Ecological Indicators 11**(1): 115-122. doi: 10.1016/j.ecolind.2009.07.003.

PETITGAS, P. 2001. Geostatistics in fisheries survey design and stock assessment: models, variances and applications. **Fish and Fisheries 2** (3): 231-249. doi: 10.1046/j.1467-2960.2001.00047.x.

R DEVELOPMENT CORE TEAM. 2012. **R: A language and environment for statistical computing.** Vienna, R Foundation for Statistical Computing. Available online at: http://www.R-project.org [Accessed: 15.X.2012].

RIZZO, A.E. & A.C.Z. AMARAL. 2001. Spatial distribution of annelids in the intertidal zone in São Sebastião Channel, Brazil. **Scientia Marina 65** (4): 323-331. doi:10.3989/scimar.2001.65n4323.

RODRIGUES, A.M.; V. QUINTINO; L. SAMPAIO; R. FREITAS & R. NEVES. 2011. Benthic biodiversity patterns in Ria de Aveiro, Western Portugal: Environmental-biological relationships. **Estuarine, Costal and Shelf Science 95**: 338-348.

ROUSE, G.W. & F. PLEIJEL. 2001. **Polychaetes.** New York, Oxford University Press, 354p.

RUFINO, M.M.; F. MAYNOU; P. ABELLÓ & A.B. YULE. 2004. Small-scale non-linear geostatistical analysis of *Liocarcinus depurator* (Crustacea: Brachyura) abundance and size structure in a western Mediterranean population. **Marine Ecology Progress Series 276**: 223-235. doi: 10.3354/meps276223.

SAITO, R.T.; R.C.L. FIGUEIRA; M.G. TESSLER & I.I.L. CUNHA. 2006. A model of recent sedimentation in the Cananéia-Iguape Estuary, Brazil. **Journal of Radioanalytical and Nuclear Chemistry 8**: 419-430.

SCHNEIDER, D.C. 1994. **Quantitative ecology: spatial and temporal scaling.** San Diego, Academic Press, 395p.

SCOTT, J.; D. RHOADS; J. ROSEN; S. PRATT & J. GENTILE. 1987. **Impact of open-water disposal of Black Rock Harbor dredged material on benthic recolonization at the FVP Site.** Technical Report D-87-4, Vicksburg, Army Engineer Waterway Experiment Station, 65p.

SILVA, J.F. 1984. Dados Climatológicos de Cananéia e Ubatuba (Estado de São Paulo). **Boletim Climatológico 5**: 1-18.

SIMARD, Y.; P. LEGENDRE; G. LAVOIE & D. MARCOTTE. 1992. Mapping, estimating biomass, and optimizing sampling programs for spatially autocorrelated data: case study of the northern shrimp (*Pandalus borealis*). **Canadian Journal of Fisheries and Aquatic Science 49** (1): 32-45. doi: 10.1139/f92-004.

SNELGROVE, P.V.R. 1998. The biodiversity of macrofaunal organisms in marine sediments. **Biodiversity and Conservation 7** (9): 1123-1132. doi: 10.1023/A:1008867313340.

STELZENMÜLLER, V.; J.R. ELLIS & S.I. ROGERS. 2010. Towards a spatially explicit risk assessment for marine management: Assessing the vulnerability of fish to aggregate extraction. **Biological Conservation 143** (1): 230-238. doi: 10.1016/j.biocon.2009.10.007.

SUGUIO, K. 1973. **Introdução à sedimentologia.** São Paulo, Edgard Blücher/EDUSP, 317p.

TESSLER, M.G. & L.A.P. SOUZA. 1998. Dinâmica sedimentar e feições sedimentares identificadas na superfície de fundo do sistema Cananéia – Iguape, SP. **Brazilian Journal of Oceanography 46** (1): 69-83. doi: 10.1590/S1679-87591998000100006.

THRUSH, S.F. & P.K. DAYTON. 2002. Disturbance to marine benthic habitats by trawling and dredging: implications for marine biodiversity. **Annual Review of Ecology and Systematics 33** (1): 449-473. doi: 10.1146/annurev.ecolsys.33.010802.150515.

THRUSH, S.F.; D.C. SCHNEIDER; P. LEGENDRE; R.B. WHITLATCH; P.K. DAYTON; J.E. HEWITT; A.H. HINES; V.J. CUMMINGS; S.M. LAWRIE; J. GRANT; R.D. PRIDMORE; S.J. TURNER & B.H. MCARDLE. 1997. Scaling-up from experiments to complex ecological systems: where to next? **Journal of Experimental Marine Biology and Ecology 216** (1): 243-254.

TOMMASI, L.R. 1970. Observações sobre a fauna bêntica do Complexo Estuarino-Lagunar de Cananéia (SP). **Boletim do Instituto Oceanográfico 19**: 43-56. doi: 10.1590/S1679-87591970000100003.

20

Dimorphism and allometry of *Systaltocerus platyrhinus* and *Hypselotropis prasinata* (Coleoptera: Anthribidae)

Ingrid Mattos[1], José Ricardo M. Mermudes[1,3] & Mauricio O. Moura[2]

[1] *Laboratório de Entomologia, Departamento de Zoologia, Universidade Federal do Rio de Janeiro. Caixa Postal 68044, 21941-971 Rio de Janeiro, RJ, Brazil.*
[2] *Departamento de Zoologia, Universidade Federal do Paraná. Caixa Postal 19020, 81531-980 Curitiba, PR, Brazil.*
[3] *Corresponding author. E-mail: jrmermudes@gmail.com*

ABSTRACT. Males of sexually dimorphic anthribid species display structural modifications that suggest sexual selection. Polyphenism, which is expressed through morphological and behavioral novelties, is an important component of the evolutionary process of these beetles. In this study, we endeavored to ascertain the presence of variations in selected monomorphic traits, polyphenism in males, and variation in structures associated with sexual dimorphism and allometric patterns in two species: *Systaltocerus platyrhynus* Labram & Imhoff, 1840 and *Hypselotropis prasinata* (Fahraeus, 1839). To that end, we used Principal Components Analysis (PCA) and Canonical Variate analysis (CVA) to statistically analyze 26 measurements of 91 specimens. The PCA discriminated three groups (females, major, and minor males) for *S. platyrhinus*, but only two groups (males and females) for *H. prasinata*. The same groups discriminated by the PCA for *Systaltocerus* were confirmed by the CVA analysis, indicating a highly significant variation separating the three groups. We also analyzed positive allometry with respect to prothorax length – independent variable by Reduced Major Axis (RMA). The allometric pattern indicated by most of the linear measurements was strong and corroborates a possible relationship between male polyphenism and the reproductive behavior of major and minor males. We believe that these patterns, in species that show both sexual dimorphism and male polyphenism, are associated with the behavior of defending the female during oviposition, performed by major males.

KEY WORDS. Anthribinae; morphometry; polyphenism; sexual dimorphism.

Male sexual dimorphism and polyphenism are ubiquitous in several species of Coleoptera (EMLEN *et al.* 2005, KAWANO 2006). These phenotypic differences are thought to be linked to fitness, since they influence reproductive success (EBERHARD & GUTIEREZ 1991, EMLEN & NOJHOUT 2000, EMLEN 1994, 1996, 2008, EMLEN *et al.* 2005, 2007, KAWANO 2006). In insects, body size is an important phenotypic trait which often corresponds to adaptations (POSSADAS *et al.* 2007). Some species of Coleoptera, for instance beetles with horns (e.g., Scarabaeidae, Dynastinae) and those with oversized mandibles (Cerambycidae, Prioninae, and Lucanidae) are model systems for studies on the evolution of sexual dimorphism and polyphenism (EBERHARD & GUTIEREZ 1991, KAWANO 2000, SHIOKAWA & IWAHASHI 2000). Moreover, Anthribidae species show both sexual dimorphism and polyphenism (MERMUDES 2002, YOSHITAKE & KAWASHIMA 2004).

Fungus weevils (Anthribidae: Curculionoidea) comprise about 370 genera and at least 3,900 species (SLIPINSK *et al.* 2011). Most species of Anthribinae have remarkable sexual dimorphism, particularly with respect to the size of the rostrum and antennae (HOLLOWAY 1982, MERMUDES 2002, 2005, MERMUDES & NAPP 2006). Anthribidae females have toothed sclerotized plates at the apex of the ovipositor, which bear conchoidal projections that are used to excavate plant tissues for oviposition. This behavior is unique and distinct among Curculionoidea, which use only the rostrum to dig plant tissues (HOWDEN 1995).

Although sexual dimorphism in size and polyphenism in male size are widespread in Anthribinae (MERMUDES 2002, 2005, MERMUDES & NAPP 2006, MERMUDES & MATTOS 2010), detailed information about it is only available for a few species (HOLLOWAY 1982). YOSHITAKE & KAWASHIMA (2004) and MATSUO (2005) demonstrated that in large, intermediate, and small males of the Japanese fungus weevil *Exechesops leucopis* Jordan, 1928 the length of the eyestalks, which are associated with the agonistic behavior males use to protect females against other males on fruits of *Styrax japonica* Siebold & Zuccarini (Styracaceae) differs. Large males that have more developed cephalic eyestalks win the disputes, indicating that sexual dimorphism and polyphenism in males are under sexual selection. However, smaller males (without developed eyestalks) can copulate in the absence of competition when females are not accompanied by larger males, which may partly explain the sneaky behavior of small males described by YOSHITAKE & KAWASHIMA (2004).

Agonist behavior in Anthribidae was also observed by THOMPSON (1963) and HOWDEN (1992). Thompson reported that guarding males of *Deuterocrates longicornis* (Fabricius, 1781), a species from West Africa, defend females and engage in fights with other males using their mandibles. HOWDEN (1992) recorded that males of *Ptychoderes rugicollis* Jordan, 1895, a Neotropical species, use their antennae and rostrum to protect females while they lay eggs on dead trees.

Considering the past detection of polyphenism in size in two species of Neotropical Anthribinae, *Systaltocerus platyrhinus* Labram & Imhoff, 1840 (variations in the length and shape of the rostrum; MERMUDES 2002) and *Hypselotropis prasinata* (Fahraeus, 1839) (different length of rostrum and antennae; MERMUDES 2005, MERMUDES & RODRIGUES 2010), we endeavored to determine whether there is variation in monomorphic characters (such as eyes, prothorax, and elytra), polyphenism in males, variation in sexually dimorphic structures (rostrum, antennae, and ventrites) and allometric patterns. This study contributes to the understanding of patterns of dimorphism and polyphenism in Anthribidae and evaluates structures that are likely to interfere with body size and/or with the relative size of other structures in the two species. However, whether agonistic interactions occur between males in those species remains unknown.

MATERIAL AND METHODS

In this study, we used a sample of 34 specimens (25 males and 9 females) of *S. platyrhinus* and 57 specimens (32 males and 25 females) of *H. prasinata* loaned from three collections (curators between parenthesis): MNRJ, Museu Nacional, Universidade Federal do Rio de Janeiro, Rio de Janeiro (M. Monné); AMCT, American Coleoptera Museum, San Antonio, Texas (J. Wappes); and DZUP, Coleção Padre Jesus S. Moure, Departamento de Zoologia, Universidade Federal do Paraná, Curitiba (L. Marinoni).

All individuals were measured using the standard image-analysis software Moticam 1000, or in the case of elytral length, a digital caliper. Before each trait was measured, the specimen was oriented so that the trait of interest was as closely parallel to the plane of the objective lens as possible. The anatomical landmarks measured follow MERMUDES & NAPP (2006) with some modifications. These modifications, defined in Table I, are based on characters that display variation among males and the sexes, independently of geographical locality. The 26 traits (measurements in millimeters) used were log-transformed (Table I and Figs 1-7).

Linear models and cluster analysis were performed in PAST version 2.0 (HAMMER *et al.* 2001). Multivariate analyses (PCA and CVA) were run in vegan (OKSANEN *et al.* 2013) and Morph (SCHLAGER 2013). Both packages were implemented in R (R CORE TEAM 2013).

Variations in phenotypic traits between and within sexes were accessed through the coefficient of variation (CV).

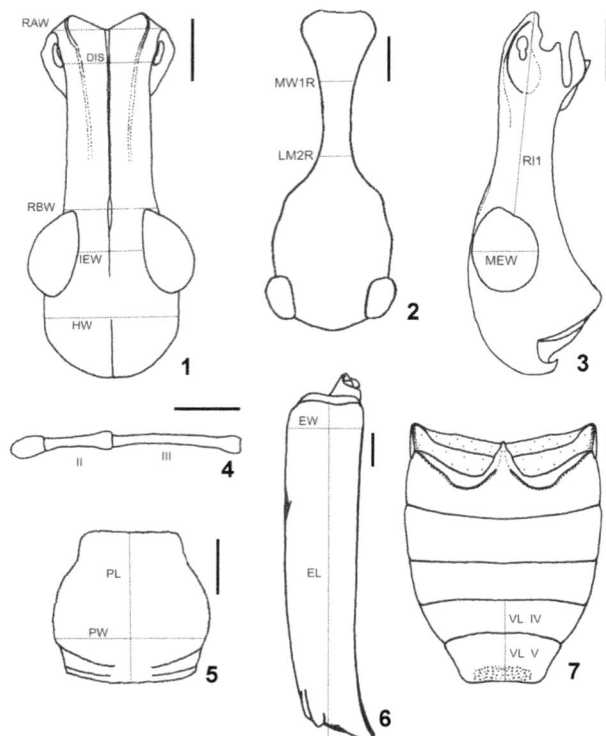

Figures 1-7. Diagram of the morphological traits measured: (1) *Hypselotropis prasinata*, head, dorsal; (2) *Systaltocerus platyrhinus*, head, frontal; (3-7) *H. prasinata*: (4) antennal segments I-III; (5) prothorax, dorsal; (6) elytron, dorsal; (7) abdomen, ventral. For abbreviations see Material and methods. Scale bars: 1 mm.

A cluster analysis with Ward's methods (based on Euclidean distance) was carried out with 1,000 Bootstrap replicates (VALENTIN 2000). In this analysis, missing data were replaced by the column average. Additionally, a Principal Components Analysis (PCA) of the covariance-variance matrix of all variables was performed to reduce the dimension of the data matrix and to visualize possible differences among groups and characters that contributed the most to these differences. The first two component axes were then used as variables in a Canonical Variate Analysis (CVA) to test morphometric differences among groups.

The analyses were designed to test the relationship between body size (prothorax length = PL) and all other variables. For this reason we used the allometric function $y = ax^b$ (HUXLEY 1932, 1950). However, the data was log-transformed and expressed by: $\log y = \log a + b (\log x)$, to fit a straight line (GOULD 1966).

Body size (prothorax length = PL) was used as a predictor variable and all other measurements were considered as response variables. However, in allometric studies, no variable

Table I. Measurements obtained from each part of the body.

Measures and abbreviation	Description
Rostral length 1 (RL 1)	laterally between the anterior margin of the eye and the apex of the rostrum
Apical width of rostrum (RAW)	dorsally at the apical margin of the rostrum
Basal width of rostrum (RBW)	measured dorsally at the base of the rostrum
Medial width 1 of rostrum (MW1R)	dorsally in the rostrum, only in Systaltocerus platyrhinus (modified from Mermudes 2002)
Medial width 2 of rostrum (MW2R)	dorsally in the rostrum, only in Systaltocerus platyrhinus (modified from Mermudes 2002)
Head width (HW)	dorsally between the lateral margins of the head
Antennal segments, length = seven variables (II, III, IV, V, VI, VII, VIII)	along the midline of each segment
Antennal segments of club, length = three variables (IX, X, XI)	along the midline
Inter-eye width (IEW)	maximum distance measured between the inner eye margins
Maximum eye width (MEW)	laterally between the outer eye margins
Inter-scrobal distance (DIS)	maximum width between the inner margins
Prothorax length (PL)	dorsally along the midline between the anterior and posterior margins
Prothorax width (PW)	dorsally near the antebasal carina (Fig. 5)
Elytra length (EL)	dorsally between the anterior margin and the apical margin
Elytra width (EW)	dorsally across the humeri
Total body length (TL)	sum of PL, EL, and RL 1
Ventrite length IV (VL IV)	along the midline
Ventrite length V (VL V)	along the midline

can be considered independent (Gould 1966). Therefore, we decided to fit a model II regression, or Reduced Major Axis regression (RMA). This allows the combined variation of the two variables to be better described because there are associated errors in both.

The slope (b) of the model II regression is the allometric constant that expresses the relationship between two variables and it has been used as an indication of the allometric pattern (Emden 2008). Therefore, when b equals 0 there is no allometric relationship. However, when b = 1 the relationship is isometric, b < 1 determines a negative allometry, and b > 1 describes positive allometry. The level of statistical significance was set at 0.05 in all analyzes.

RESULTS

The mean and standard deviation of all measurements were given in the Appendixes 1 and 2. The amplitude of total body length (TL) and the coefficient of variation (CV) for *S. platyrhinus* and *H. prasinata* were summarized in Table II.

Sexual dimorphism. Males of *H. prasinata* (Fig. 8) are relatively larger than females (Fig. 9). Major males of *S. platyrhinus* (Fig. 10) are similar to females in size, whereas minor males (Fig. 11) of this species are smaller than their female counterparts (Fig. 12). Males and females of *H. prasinata* and *S. platyrhinus* did not differ in the following variables that correspond to monomorphic characters in both species: apical width of rostrum (RAW), basal width of rostrum (RBW), head width (HW), prothorax length (PL), prothorax width (PW), elytra

Table II. Amplitude of the total length (mm) for males and females of *S. platyrhinus* and *H. prasinata* (n = 34, males = 25 and females = 9).

Species	Groups	CV	TL
S. platyrhinus	Males	0.17	6.37-12.81
	Females	0.12	8.22-11.36
S. platyrhinus	Major males		10.32-12.95
	Minor males		6.37-9.51
H. prasinata	Males	0.15	10.86-19.60
	Females	0.16	8.51-17.96

length (EL), elytra width (EW), inter-scrobal distance (DIS), and inter-eye width (IEW), as detailed in Appendix 1.

The independent t test for sexual dimorphism of all variables is shown for the two species analyzed (Appendix 1). Males and females of the two species did not differ only in the maximum eye width (MEW). Based on the RMA results for *S. platyrhinus* (Table III), the elytral length and width did not show allometry. These results showed that all other structures are indicative of sexual dimorphism, as previously suggested by Holloway (1982) and Mermudes (2002).

Polyphenism in males. In *S. platyrhynus*, major and minor males differ significantly in almost all variables, with the exception of antennomeres VII, VIII, and IX and ventrite V. The result of the independent t test for the polyphenism in males of the two species analyzed is shown in Table III. The presence of two groups of males in *S. platyrhinus*, relatively discrete in size, indicates size polyphenism (Table III and Ap-

Figures 8-12. Dorsal habitus. (8-9) *Hypselotropis prasinata*: (8) male; (9) female. (10-12) *Systaltocerus platyrhynus:* (10) major male; (11) minor male; (12) female. Scale bar = 2 mm.

pendix 2). In *H. prasinata*, although there is no evidence of major and minor males, we found intermediate males, suggesting a continuous variation in size (Table IV). Therefore, there were no discrete groups, rejecting the hypothesis of size polyphenism for *H. prasinata* males.

Multivariate analysis. Cluster Analyses with the Ward's Method, considered very efficient (VALENTIN 2000), identified different groups for each species analyzed. Bootstrap support values for these groups are shown within parentheses: three groups found for *S. platyrhinus* (Fig. 13): major males (76), minor males (75), and females (99); and two for *H. prasinata* (Fig. 14): males (respectively 29, 23) and females (74).

The Principal Components Analysis (PCA) indicated that size has a greater influence on the identification of groups (major males, minor males, and females) of *S. platyrhinus* (Table V and Fig. 15). The separation of the groups was evident by the analysis of the axes of components 1 and 2, which explain more than 80% of total variance. In the first axis (PC1), two groups were identified: males and females. The second axis (PC2) shows the separation between major and minor males. For *H. prasinata*, the principal components analysis indicated that size contributes to the differentiation of groups (Table V). However, there is no evidence of polyphenism in males (Fig. 16). The first and second components explained 87% of total evidence.

Canonical Variate Analysis (CVA), together with MANOVA, confirmed that there are three different morphotypes in *S. platyrhinus* (MANOVA CVA: Wilks' Lambda = 0.000194; df1 = 50; df2 = 14; F = 19.82; p < 0.0001) with correct allocation of

specimens exceeding 90%. The separation of groups in *S. platyrhynus* (Fig. 17) was evident through the first two axes, of which the first CV provided information for the separation of males and females and the second CV distinguished major and minor males. This separation is obtained essentially by a size contrast among head width (HW), prothorax width (PW), elytra width (EW), and length of antennal segment VII. CVA was not undertaken for *H. prasinata* because it is only recommended when there are more than two groups (HAMMER 2002).

Allometry and sexual dimorphism. Results of analysis by the RMA in *S. platyrhinus* males, without separating major and minor groups (Table IV), showed positive allometry between the independent variable PL (prothorax length) and each of the six variables connected with the rostrum (rostral length 1, apical width of rostrum, medial width 1 and 2 of rostrum, basal width of rostrum, and inter-scrobal distance). Even within the analysis of males, only one variable of the head, inter-eye width (IEW), and three antennal segments (the proximal III-V), did not fit an allometric pattern, differing from females of *S. platyrhinus* in this respect (Table IV).

Differing from the results above, evidence of sexual dimorphism with allometric patterns was confirmed only for females of *H. prasinata* in the following characters: width of head, prothorax, and elytra. Males of *H. prasinata* (Table IV) showed positive allometry for only one trait in the antennae (segment III). Males and females of this species, however, showed positive allometry in thirteen measurements, whereas females showed exclusive positive allometry in five traits.

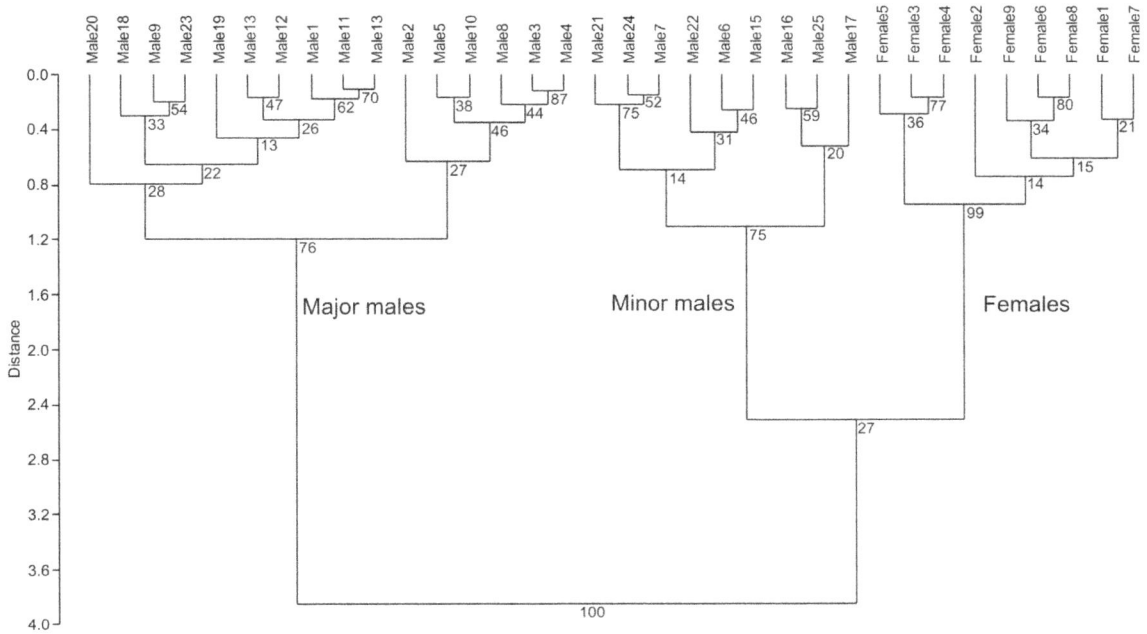

Figure 13. Dendogram obtained with Ward's Cluster Analysis methods for *S. platyrhinus.* 1,000 bootstrap replicates.

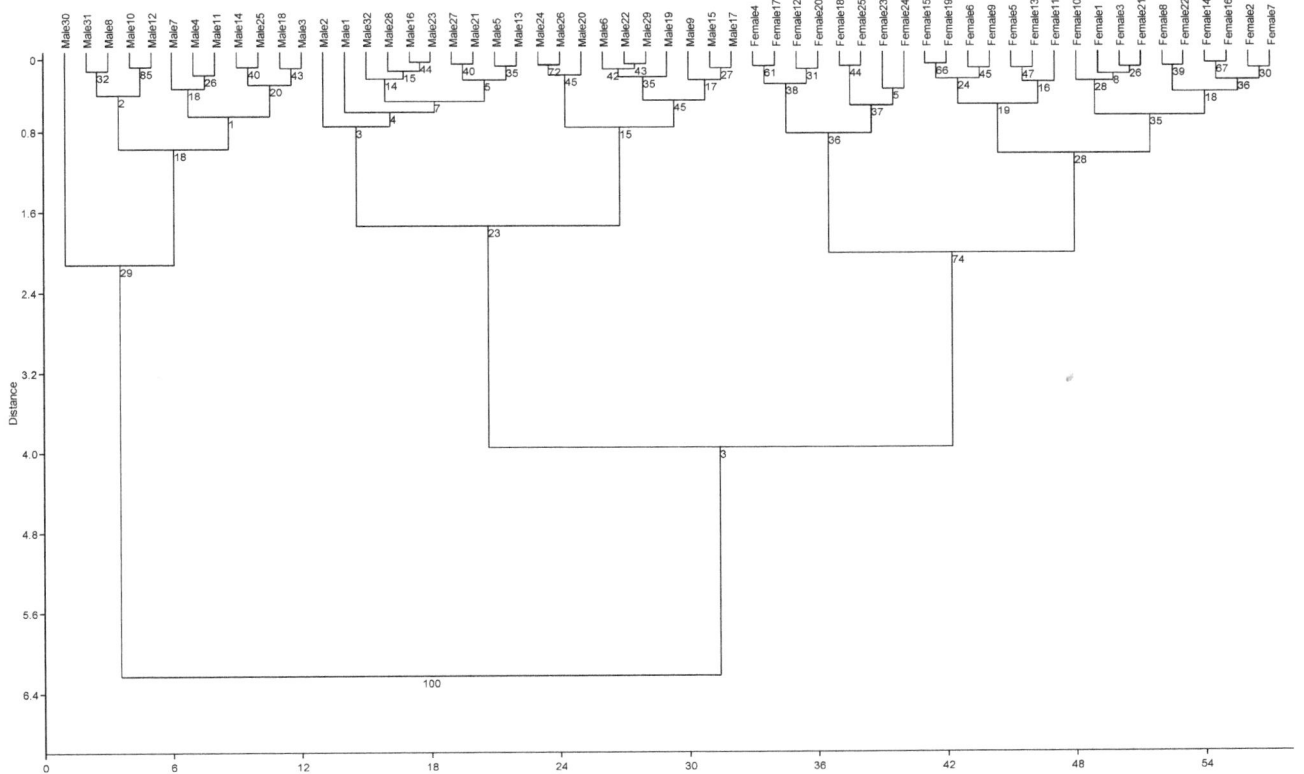

Figure 14. Dendogram obtained with Ward's Cluster Analysis methods for *H. prasinata.* 1,000 bootstrap replicates.

Allometry and polyphenism in males. In reviewing the evidence of allometry between major and minor males of *S. platyrhynus* (Table III), five variables from the rostrum (RAW, MW1R, MW2R, RBW, and DIS), two of antennomeres (III) and only one from the ventrite (VL IV) have positive allometry only in major males. On the other hand, minor males showed positive allometry in prothorax and elytral width, and length of rostrum and antennomeres II, III, and VI. This demonstrates a clear morphological plasticity between males.

DISCUSSION

According to Kawano (2006), body size is the most appropriate morphological trait for allometric analyses because it depends on the quality of the nutrition received by an individual during growth. However, other traits have been used as allometric predictors, for instance elytra length (Clark 1977, Goldsmith 1985), elytra width (Eberhard 1980, Cook 1987), and pronotum width (Emlem 1994, 1996, Eberhard *et al.* 1998, Emlen *et al.* 2005, Tomkins *et al.* 2005). Kawano (2006) stated that in sexually dimorphic beetles characters such as elytra length (EL), elytra width (EW), and prothorax width (PW) are not adequate allometric predictors because they do not represent a true measurement of body size (see Moczek *et al.* 2002 for a different view). Notably, we know that nutritional and environmental factors influence body size and also dimorphic structures, and that body size, as a dimorphic structure, is determined by endocrine mechanisms regulating development (Emlen *et al.* 2005).

We did not use the TL as an allometric predictor because it is a composite measure formed by the sum of rostral length (RL1), prothorax length (PL), and elytra length (EL). Moreover, all these traits exhibited positive allometry (see discussion below), which could lead, through high correlations, to an indirect correlation. In addition, the elytra width (EW) showed positive allometry for males of *S. platyrhinus* and *H. prasinata*, and females of *H. prasinata*. Thus, only the prothorax length (PL) was used as a proxy for the true size of the body in the subsequent allometric analysis.

The results on variation in monomorphic characters also corroborated the work of Holloway (1982), who postulated that the inter-scrobal distance (DIS) is a diagnostic character for genera in Anthribidae. Our results also showed that the DIS did not differ between sexes because it is a monomorphic character. Additionally, the results showed that, in a supposedly dimorphic structure such as the rostrum, there are measurements that do not vary between the sexes (monomorphic). For *H. prasinata* there were no differences between the sexes in rostral length (RL1). For *S. platyrhynus*, there were no differences in: (1) rostrum, the average width 1 (MW1B) and width 2 (MW2R); (2) antenna, length of antennomere II; and (3) abdomen, length of ventrite IV (VL IV).

Sexual dimorphism was discussed by Holloway (1982) and Mermudes & Napp (2006); it occurs in the relative length of the

Table III. The results based on regression. Reduced Major Axis regression (RMA) of pairs of morphological variables selected with positive allometry in *S. platyrhinus*. Prothorax length (PL) was used as a predictor variable, and all variables were log-transformed prior to the two male groups (major and minor). Values of positive allometry shown in bold, with slope >1 and p ≤ 0.05.

Measures	Males	Slope	r	p	95% IC
RL1	Major Males	-1.48550	-0.093925	0.7255	-2.2580; 1.7650
	Minor Males	**1.81560**	**0.758530**	**0.0181**	**1.0100; 2.8050**
RAW	Major Males	**1.22850**	**0.656570**	**0.0053**	**0.7506; 1.7230**
	Minor Males	0.96200	0.899870	0.0020	0.6779; 1.2760
MW1R	Major Males	**1.22850**	**0.656570**	**0.0053**	**0.7506; 1.7230**
	Minor Males	1.55180	0.218300	0.5583	-2.2640; 2.5990
MW2R	Major Males	**1.75670**	**0.762370**	**0.0005**	**1.0840; 2.4550**
	Minor Males	1.33680	0.802830	0.0143	0.8597; 1.9060
RBW	Major Males	**1.39400**	**0.652370**	**0.0054**	**0.6713; 1.9830**
	Minor Males	1.39230	0.815180	0.0089	0.8295; 1.9170
HW	Major Males	0.96648	0.565020	0.0229	0.6182; 1.3610
	Minor Males	0.85738	0.742430	0.0168	0.4798; 1.2800
PW	Major Males	0.93269	0.678800	0.0030	0.5844; 1.2920
	Minor Males	**1.08050**	**0.842200**	**0.0026**	**0.6262; 1.5120**
EL	Major Males	0.80780	0.619160	0.0099	0.5425; 1.0200
	Minor Males	0.77868	0.619390	0.0818	-0.6077; 1.1640
EW	Major Males	0.99183	0.657430	0.0059	0.6287; 1.3850
	Minor Males	**1.09810**	**0.819920**	**0.0034**	**0.6483; 1.7040**
MEW	Major Males	1.14760	0.363730	0.1763	-1.3830; 1.6280
	Minor Males	0.75328	0.862790	0.0029	0.5855; 0.9414
DIS	Major Males	**3.04740**	**0.530800**	**0.0355**	**1.4770; 4.4360**
	Minor Males	1.66940	0.474440	0.1888	-1.7580; 2.6060
IEW	Major Males	1.23250	0.534230	0.0283	0.5754; 1.9310
	Minor Males	1.15860	0.813750	0.0102	0.7183; 1.5460
II	Major Males	2.61660	0.277900	0.3112	-1.9190; 4.4160
	Minor Males	**1.40660**	**0.894920**	**0.0016**	**0.8673; 1.8060**
III	Major Males	**2.63410**	**0.577120**	**0.0139**	**1.1790; 4.4660**
	Minor Males	**1.23150**	**0.930700**	**0.0007**	**0.9571; 1.5440**
IV	Major Males	1.70980	0.494290	0.0512	0.9584; 2.6600
	Minor Males	1.90930	0.661310	0.0546	1.1760; 2.8730
V	Major Males	1.13960	0.434660	0.0868	-0.5834; 1.8540
	Minor Males	1.75840	0.644540	0.0652	0.9586; 2.5610
VI	Major Males	0.74156	0.419830	0.1016	-0.6722; 1.0240
	Minor Males	**1.88570**	**0.672400**	**0.0489**	**1.0130; 2.8120**
VII	Major Males	0.73473	0.286160	0.2817	-0.9168; 0.9983
	Minor Males	1.59160	0.596570	0.0893	-1.0050; 2.3940
VIII	Major Males	1.15560	0.397810	0.1253	-1.4790; 1.6420
	Minor Males	1.37890	0.456590	0.2185	-1.0770; 2.2130
IX	Major Males	1.60190	0.347030	0.1804	-1.9980; 2.3000
	Minor Males	1.19260	0.657950	0.0557	0.6023; 1.7510
X	Major Males	1.77750	0.281690	0.3055	-2.8070; 2.7750
	Minor Males	0.94794	0.432220	0.2407	-0.8429; 1.3880
XI	Major Males	1.37250	0.502660	0.0498	0.8856; 2.0770
	Minor Males	1.13450	0.572490	0.1132	-0.3999; 1.7300
VL IV	Major Males	**1.64070**	**0.701600**	**0.0017**	**0.9728; 2.4760**
	Minor Males	0.91300	0.015324	0.9692	-1.2600; 1.4550
VL V	Major Males	1.80020	0.219740	0.4421	-1.2380; 3.1420
	Minor Males	1.34300	0.466900	0.2049	-0.8419; 2.1480

Table IV. Reduced Major Axis regression (RMA) results between pairs of morphological variables selected with positive allometry in *S. platyrhinus* and *H. prasinata*. Prothorax length (PL) was used as a predictor variable. All variables were log-transformed prior to males and females. (***) for p ≤ 0.0001.

Measures	Sex	S. platyrhinus Slope b	r	p	I.C.	H. prasinata Slope b	r	p	I.C.
RL1	Males	1.90000	0.699050	***	1.4610; 2.3560	1.06040	0.931790	***	0.9302; 1.1910
	Females	0.59137	0.440290	0.2449	-0.4048; 1.800	1.10600	0.940410	***	0.9485; 1.3050
RAW	Males	1.16070	0.875300	***	1.0080; 1.3720	-4.99440	-0.058107	0.7606	-9.096; 4.3820
	Females	-0.41585	-0.078700	0.8496	-0.7777; 1.1800	0.98130	0.919110	***	0.8439; 1.2070
MW1R	Males	1.68740	0.549920	0.0064	1.0480; 2.5660	–	–	–	–
	Females	-0.32354	-0.012680	0.9765	-0.4807; 0.9039	–	–	–	–
MW2R	Males	1.60640	0.878990	***	1.3820; 1.9150	–	–	–	–
	Females	0.41981	0.038053	0.9181	-0.5954; 1.2100	–	–	–	–
RBW	Males	1.50130	0.863070	***	1.2590; 1.7940	1.17200	0.873160	***	0.8929; 1.4110
	Females	0.53790	0.031836	0.9298	-1.0410; 1.6160	1.02050	0.972850	***	0.9115; 1.1610
HW	Males	1.03900	0.823670	***	0.8540; 1.2510	0.91548	0.953520	***	0.8060; 1.0180
	Females	0.47593	0.043096	0.9097	-0.9619; 1.4070	1.02800	0.975950	***	0.9448; 1.1690
PW	Males	1.04600	0.875940	***	0.8642; 1.2310	0.99005	0.961220	***	0.8782; 1.0860
	Females	0.41758	0.088446	0.8306	-0.7940; 1.1870	1.10450	0.985070	***	1.0230; 1.2320
EL	Males	0.90022	0.807660	***	0.7675; 1.0580	0.98289	0.967430	***	0.8957; 1.0540
	Females	0.43514	0.348030	0.3544	-0.1564; 1.1820	1.05030	0.988480	***	0.9686; 1.1080
EW	Males	1.05590	0.860950	***	0.8797; 1.2260	0.98049	0.974080	***	0.8989; 1.0570
	Females	0.45187	0.155270	0.6926	-0.6374; 1.370	1.08220	0.983880	***	1.0170; 1.1800
MEW	Males	0.82335	0.695740	***	0.6704; 1.0870	0.78105	0.814940	***	0.6461; 0.9124
	Females	-0.34873	-0.10388	0.8103	-0.4492; 0.9072	0.87605	0.936000	***	0.7657; 1.0860
DIS	Males	2.07910	0.649830	0.0008	1.5690; 2.8290	1.12190	0.872020	***	0.8799; 1.3570
	Females	0.45967	0.465290	0.2183	0.1739; 1.2460	1.04790	0.970210	***	0.9672; 1.2050
IEW	Males	1.35670	0.835360	***	1.1270; 1.6580	1.15460	0.889160	***	0.9594; 1.3300
	Females	0.59589	0.116280	0.723	-0.8943; 1.8360	1.04290	0.957510	***	0.9386; 1.1890
II	Males	1.71650	0.620900	0.0005	0.9697; 2.6140	1.82900	0.917990	***	1.5820; 2.0320
	Females	-0.94552	-0.729110	0.0761	-1.1480; 1.3230	1.03330	0.908060	***	0.9029; 1.3000
III	Males	1.69210	0.740040	***	1.1920; 2.5400	1.92700	0.959180	***	1.7420; 2.1320
	Females	-1.07410	-0.089660	0.8043	-1.2830; 2.8000	1.01900	0.933010	***	0.9035; 1.2060
IV	Males	1.68500	0.749730	***	1.2730; 2.1020	2.24450	0.972940	***	2.0800; 2.4190
	Females	-0.44250	0.378060	0.3446	-1.2100; 0.8810	1.07030	0.939310	***	0.9352; 1.2770
V	Males	1.54270	0.761590	***	1.1060; 1.9080	6.15970	0.526750	***	2.5750; 10.8700
	Females	0.49559	0.074484	0.8421	-1.2560; 1.4010	1.07280	0.916670	***	0.9190; 1.2640
VI	Males	1.51620	0.764820	***	0.9724; 1.9400	2.99950	0.944720	***	2.8050; 3.2500
	Females	-0.45601	-0.077770	0.8245	-0.8710; 1.4290	1.02590	0.879440	***	0.8348; 1.2660
VII	Males	1.34630	0.727210	***	0.9090; 1.7510	2.83610	0.916910	***	2.5600; 3.1810
	Females	-0.44280	-0.023300	0.9504	-0.8681; 1.3760	0.95728	0.824990	***	0.8010; 1.2830
VIII	Males	1.27890	0.686200	0.0002	0.8871; 1.6360	2.56950	0.924410	***	2.248; 2.9630
	Females	0.56072	0.210630	0.5876	-0.9753; 1.6330	1.12210	0.613780	0.0013	0.8289; 1.7310
IX	Males	1.19600	0.641790	0.0005	0.8995; 1.5690	2.70990	0.836270	***	2.2630; 3.1940
	Females	-0.97553	-0.262960	0.406	-1.7080; 2.6140	1.02260	0.415650	0.0415	-0.9485; 1.4340
X	Males	1.07710	0.400380	0.0479	0.5625; 1.6440	1.66220	0.743030	***	1.3390; 2.0690
	Females	-0.96886	-0.548470	0.1422	-2.400; 1.5710	1.28970	0.746940	***	1.0500; 1.6130
XI	Males	1.06410	0.671810	0.0004	0.8033; 1.3920	2.27550	0.836560	***	1.8510; 2.7770
	Females	-0.86524	-0.41590	0.2525	-2.3730; 0.7534	1.52600	0.665790	0.0001	1.1680; 2.0830
VL IV	Males	1.24970	0.665430	***	0.9291; 1.7370	0.79050	0.705380	***	0.6296; 0.9581
	Females	-0.62952	-0.237740	0.546	-1.3870; 1.7320	0.95457	0.932760	***	0.8309; 1.0880
VL V	Males	1.26970	0.484400	0.0123	0.6120; 1.9140	0.88102	0.814830	***	0.6817; 1.0890
	Females	-1.07080	-0.285230	0.3561	-2.9930; 1.1460	0.85026	0.925360	***	0.7052; 0.9477

Table V. Loadings of the morphometric variables in the first two components of the Principal Components Analysis (PCA). Variables not measured marked with an asterisk.

Measures	S. platyrhinus		H. prasinata	
	PC1	PC2	PC1	PC2
RL1	-0.544	-0.114	0.135	-0.183
RAW	-0.150	-0.240	0.108	-0.182
MW1R	0.015	-0.318	*	*
MW2R	-0.353	-0.225	*	*
RBW	-0.375	-0.213	0.095	-0.214
HW	-0.156	-0.222	0.079	-0.200
PL	-0.098	-0.229	0.124	-0.178
PW	-0.153	-0.248	0.100	-0.206
EL	-0.127	-0.234	0.110	-0.191
EW	-0.156	-0.259	0.091	-0.207
MEW	-0.027	-0.126	0.073	-0.148
DIS	-0.295	-0.224	0.088	-0.210
IEW	-0.199	-0.204	0.115	-0.199
II	-0.076	-0.052	0.398	-0.096
III	-0.401	0.066	0.565	-0.009
IV	-0.551	0.158	0.613	-0.027
V	-0.600	0.174	1.018	0.087
VI	-0.583	0.187	0.845	0.048
VII	-0.561	0.211	0.864	0.104
VIII	-0.459	0.157	0.815	0.115
IX	-0.227	0.043	0.583	0.010
X	-0.102	0.018	0.308	-0.109
XI	-0.174	0.011	0.554	-0.045
VL IV	-0.030	0.117	-0.002	-0.199
VL V	0.037	-0.134	-0.056	-0.235

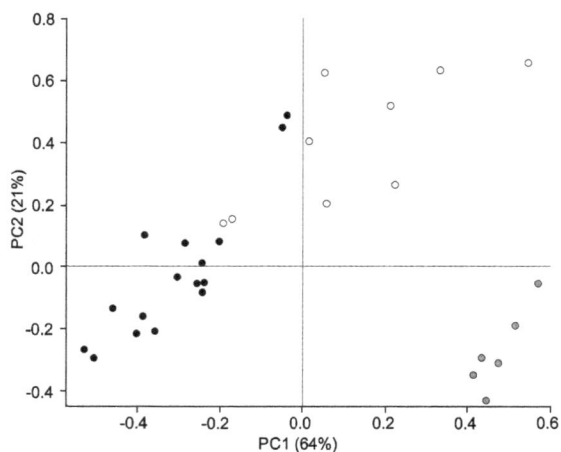

15

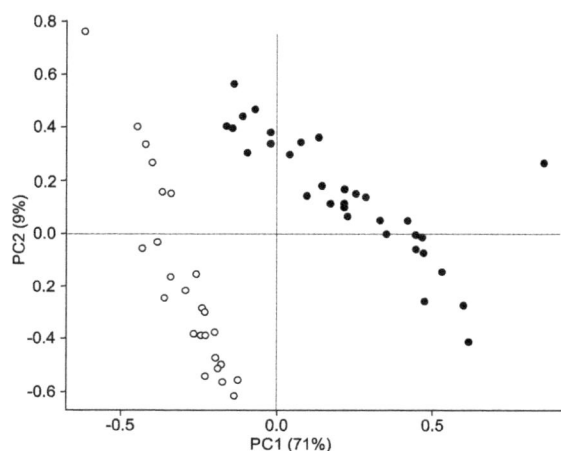

16

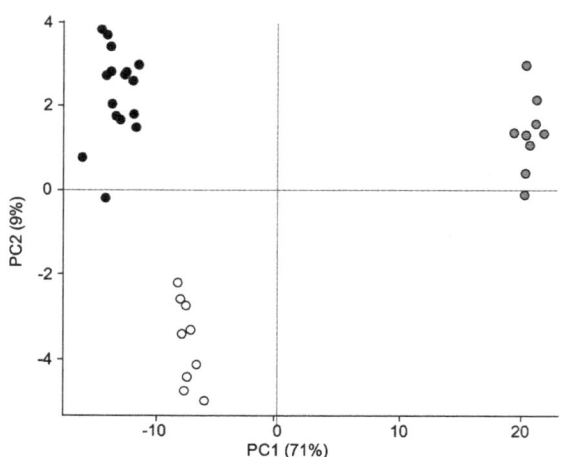

17

antenna of males of some Anthribinae. Also, MERMUDES (2005) used a relationship between the length of ventrites IV and V to distinguish between males and females of *Hypselotropis* Jekel, 1855. Our results here show that ventrite V is always longer than ventrite IV in both males and females of *H. prasinata*. This differs from the opinion of MERMUDES (2005) who believes that ventrite V in males is always slightly shorter than ventrite IV. However, our results confirmed this relationship for females. The large number of variables with values also tested by PCA and CVA (Figs 15-17) suggest that there is marked sexual dimorphism in some structural characters that had not been previously investigated.

The multivariate analysis (PCA) indicated that two relatively discrete groups of males of *S. platyrhinus* exist with respect to size (major males, minor males and females, Fig. 15), which was confirmed by CVA (Fig. 17), revealing the presence of polyphenism in males of this species. The analysis suggested that the allometric component contributes to the differentiation of groups, but there is no evidence of polyphenism in males of *H. prasinata* (Fig. 16), rejecting, at least in this analysis, the hypothesis of size polyphenism in this species.

Figures 15-17. (15-16) Principal Components Analysis (PCA) for: (15) *S. platyrhynus*: females (●), major males (●), and minor males (light gray); (16) *H. prasinata*: females (○) and males (●); (17) Canonical Variate Analysis (CVA) of *S. platyrhinus*: females (●), major males (●), and minor males (○).

Considering together all the results on allometry and sexual dimorphism, we conclude that, in the case of *S. platyrhinus*, the dimensions of the rostrum, antennal segments, and ventrites IV and V indicate that sexual dimorphism is in place, as previously suggested by MERMUDES (2002). Traits that exhibited positive allometry are a strong indication of sexual dimorphism (EMLEN 1996, MOCZEK *et al.* 2002, MATSUO 2005, MOCZEK 2006). It is worth noting that the variables with positive allometry in *S. platyrhinus* are at the anterior part of body (rostrum and frons). In this species, the rostrum and forehead are vertical (hypognathous), providing evidence that such structures are subject to sexual selection and are probably associated with male fighting, similar to the condition found in *Exechesops leucopis*.

Despite the fact that we have analyzed only two species, our results emphasize allometric patterns in structures with sexual dimorphism that can be highly variable within *S. platyrhynus* males. Almost all species of Anthribidae that show sexual dimorphism lack structures known as weapons (e.g., horns). Polyphenism in males was also found to be present in *S. platyrhinus*, making it possible to infer that many traits related to dimorphism could play a role in tactical alternatives that minor males developed when confronted with major males, as reported by HOWDEN (1992), YOSHITAKE & KAWASHIMA (2004), and MATSUO (2005).

Initially, the allometric variation could be derived from either behavioral differences between major or minor males or from a threshold size to developing weapons (horns or mandibles) with exaggerated sizes (MOCZEK & EMLEN 2000, MOCZEK *et al.* 2002, YOSHITAKE & KAWASHIMA 2004, MATSUO 2005). In males of some species of Anthribidae (which do not have horns = weapons), sexually dimorphic traits exhibit positive allometry with body size, whereas isometry or negative allometry is detected when sexually monomorphic traits are considered (or which are not associated with dimorphism) (MATSUO 2005).

The behavioral relationship involves male-male competition for females, but it does not eliminate the interactions between minor males when they meet, as well as alternative tactics developed by minor males to copulate (EMLEN 1994, MOCZEK *et al.* 2002, MATSUO 2005, TOMKINS & MOCZEK 2009).

Finally, it is possible that the morphological patterns of Anthribidae are linked to the protection of the female, which is secured by males during oviposition, and that a relationship between reproductive behavior and alternative morphologies exists (as noted by HOWDEN 1992). This behavioral pattern can be elucidated in further studies on *S. platyrhinus*.

ACKNOWLEDGMENTS

We are grateful to the curators who made material available and to two anonymous reviewers for comments. This research was partially supported by grants from CNPq (processes 470980/2011-7, 475461/2007, and 312357/2006), FAPERJ (processes 101.476/2010 and 100.927/2011), and Programa de Pós-Graduação em Biociências da Universidade do Estado do Rio de Janeiro.

REFERENCES

CLARK, J.T. 1977. Aspects of variation in the stag beetle *Lucanus cervus* (L.) (Coleoptera: Lucanidae). **Systematic Entomology** 2: 9-16. doi: 10.1111/j.1365-3113.1977.tb00350.x.

COOK, D. 1987. Sexual selection in dung beetles. I. A multivariate study of the morphological variation in two species of *Onthophagus* (Scarabaeidae: Onthophagini). **Australian Journal of Zoology 35**: 123-132. doi: 10.1071/ZO9870123.

EBERHARD, W.G. 1980. Horned beetles. **Scientific American 242**: 166-182.

EBERHARD, W.G. & E.E. GUTIÉRREZ. 1991. Morphometric Variability in Continental and Atlantic Island Populations of Chaffinches *Fringilla coelebs*. **Evolution 45** (1): 29-39.

EBERHARD, W.G.; B.A. HUBER; R.L. RODRIGUEZ; R.D. BRICENO; I. SALAS & V. RODRIGUEZ. 1998. One size fits all? Relationships between the size and degree of variation in genitalia and other body parts in twenty species of insects and spiders. **Evolution 52**: 415-431.

EMLEN, D.J. 1994. Environmental control of horn length dimorphism in the beetle *Onthophagus acuminatus* (Coleoptera: Scarabaeidae). **Proceedings of the Royal Society B. 256**: 131-136. doi: 10.1098/rspb.1994.0060.

EMLEN, D.J. 1996. Artificial selection on horn length-body size allometry in the horned beetle *Onthophagus acuminatus* (Coleoptera:Scarabaeidae). **Evolution 50**: 1219-1230.

EMLEN, D.J. 2008. The evolution of animal weapons. **Annual Review of Ecology and Systematics 39**: 387-413. doi: 10.1146/annurev.ecolsys.39.110707.173502.

EMLEN, D.J. & H.F. NIJHOUT. 2000. The Development and Evolution of Exaggerated Morphologies in Insects. **Annual Review of Entomology 45**: 661-708. doi: 10.1146/annurev.ento.45.1.661.

EMLEN, D.J.; J. HUNT & L.W. SIMMONS. 2005. Evolution of sexual dimorphism and male dimorphism in the expression of beetle horns: phylogenetic evidence for modularity, evolutionary lability, and constraint. **The American Naturalist 166** (Suppl.): S42-S68.

EMLEN, D.J.; L.C. LAVINE & B. EWEN-CAMPEN. 2007. On the origin and evolutionary diversification of beetle horns. **Proceedings of the National Academy of Sciences 104**: 8661-8668.

GOLDSMITH, S.K. 1985. Male Dimorphism in *Dendrobias mandibularis* Audinet-Serville (Coleoptera: Cerambycidae). **Journal of the Kansas Entomological Society 58**: 534-538.

GOULD, S.J. 1966. Allometry and size in ontogeny and phylogeny. **Biological Reviews 41**: 587-640. doi: 10.1111/j.1469-185X.1966.tb01624.x

HAMMER, O. 2002. Morphometrics – brief notes50p. Available at http://folk.uio.no/ohammer/past/morphometry.pdf [Accessed: February 2014].

HAMMER, O.; D.A.T. HARPER & P.D. RYAN. 2001. Past: Palaeontological Statistics Software Package for Education and Data Analysis. **Palaeontological Electronica 4** (1): 9p. Available online at: http://palaeo-electronica.org/2001_1/past/past.pdf [Accessed: March 2011].

HOLLOWAY, B.A. 1982. **Anthribidae (Insecta: Coleoptera). Fauna of New Zealand 3.** Wellington, DSIR, 269p.

HOWDEN, A.T. 1992. Oviposition Behavior and Associated Morphology of the Neotropical Anthribid *Ptychoderes rugicollis* Jordan (Coleoptera: Anthribidae). **Coleopterists Bulletin 46:** 20-27.

HOWDEN, A.T. 1995. Structures related to oviposition in Curculionoidea. **Memoirs of the Entomological Society of Washington 14:** 53-100.

HUXLEY, J.S. 1932. **Problems of relative growth.** London, Methuen, 276p.

HUXLEY, J.S. 1950. Relative growth and form transformation. **Proceedings of the Royal Society London 137:** 465-469. doi:10.1098/rspb.1950.0055.

KAWANO, K. 2000. Genera and Allometry in the Stag Beetle Family Lucanidae, Coleoptera. **Annals of the Entomological Society of America 93:** 198-207. doi: http://dx.doi.org/10.1603/0013-8746(2000)093[0198:GAAITS]2.0.CO;2.

KAWANO, K. 2006. Sexual Dimorphism and the Making of Oversized Male Characters in Beetles (Coleoptera). **Annals of the Entomological Society of America 99:** 327-341. doi: doi: http://dx.doi.org/10.1603/0013-8746(2006)099[0327:SDATMO]2.0.CO;2.

MATSUO, Y. 2005. Extreme Eye Projection in the Male Weevil *Exechesops leucopis* (Coleoptera: Anthribidae): Its Effects on Intrasexual Behavioral Interferences. **Journal of Insect Behavior 18:** 465-477. doi: 10.1007/s10905-005-5605-y.

MERMUDES, J.R.M. 2002. *Systaltocerus platyrhinus* Labram & Imhoff, 1840: redescrições e considerações sobre a sinonímia com *Homalorhamphus vestitus* Haedo Rossi & Viana, 1957 (Coleoptera, Anthribidae, Anthribinae). **Revista Brasileira de Entomologia 46:** 579-590. doi: http://dx.doi.org/10.1590/S0085-56262002000400013.

MERMUDES, J.R.M. 2005. Revisão sistemática, análise cladística e biogeografia dos gêneros *Tribotropis e Hypselotropis* (Coleoptera, Anthribidae, Anthribinae, Ptychoderini). **Revista Brasileira de Entomologia 49:** 465-511. doi: http://dx.doi.org/10.1590/S0085-56262005000400009.

MERMUDES, J.R.M. & D.S. Napp. 2006. Revision and cladistic analysis of the genus *Ptychoderes* Schoenherr, 1823 (Coleoptera, Anthribidae, Anthribinae, Ptychoderini) **Zootaxa 1182:** 1-130.

MERMUDES, J.R.M. & I. MATTOS. 2010. Description of Males of *Ptychoderes brevis* and *Ptychoderes jekeli*, with a cladistical reanalysis of *Ptychoderes* (Coleoptera: Anthribidae). **Annals of the Entomological Society of America 105:** 523-531. doi: http://dx.doi.org/10.1603/AN10016.

MERMUDES, J.R.M. & J.M.S. RODRIGUES. 2010. Description of two new species of *Hypselotropis* Jekel with a revised key and

phylogenetic reanalysis of the genus (Coleoptera, Anthribidae, Anthribinae). **Zootaxa 2575:** 49-62.

MOCZEK, A.P. 2006. A matter of measurements: challenges and approaches in the comparative analysis of static allometries. **American Naturalist 167:** 606-611.

MOCZEK, A.P. & D.J. EMLEN. 2000. Male horn dimorphism in the scarab beetle, *Onthophagus taurus*: do alternative reproductive tactics favour alternative phenotypes? **Animal Behaviour 59:** 459-466. doi: 10.1006/anbe.1999.1342.

MOCZEK, A.P.; J. HUNT; D.J. EMLEN & L.W. SIMMONS. 2002. Threshold evolution in exotic populations of a polyphenic beetle. **Evolutionary Ecology Research 4:** 587-601.

OKSANEN, J.; F.G. BLANCHET; R.P. LEGENDRE; P.R. MINCHIN; R.B. O'HARA; G.L. SIMPSON; P. SOLYMOS; M.H.H. STEVENS & H. WAGNER. 2013. **Vegan: Community Ecology Package. R package version 2.0-7.** Available online at: http://cran.r-project.org/web/packages/vegan/index.html [Accessed: July 2013]

POSADAS, P.; E. ORTIZ-JAUREGUIZAR & M.E. PÉREZ. 2007. Dimorfismo sexual y variación morfométrica geográfica en *Hybreoleptops aureosignatus* (Insecta: Coleoptera: Curculionidae). **Anales de la Academia *Nacional* de Ciencias Exactas, Físicas y Naturales 59:** 141-150.

R CORE TEAM. 2013. **R: A language and environment for statistical computing.** Vienna, R Foundation for Statistical Computing. Available online at: http://www.R-project.org/ [Accessed: July 2013]

SCHLAGER, S. 2013. **Morpho: Calculations and visualizations related to Geometric Morphometrics.** R package version 0.25. Available at http://sourceforge.net/projects/morpho-rpackage [Accessed: July 2013]

SHIOKAWA, T. & O. IWAHASHI. 2000. Mandible dimorphism in males of a stag beetle, *Prosopocoilus dissimilis okinawanus* (Coleoptera: Lucanidae). ***Applied Entomology* and Zoology 35** (4): 487-494. doi: 10.1303/aez.2000.487.

SLIPINSKI, S.A.; R.A.B. LESCHEN & J.F. LAWRENCE. 2011. Order Coleoptera Linnaeus, 1758. *In*: Z.Q. ZHANG (Ed.). Animal biodiversity: An outline of higher-level classification and survey of taxonomic richness. **Zootaxa 3148:** 203-208.

THOMPSON, G.H. 1963. **Forest Coleoptera of Ghana. Biological notes and host trees.** Oxford, Forestry Memoirs 24, 78p.

TOMKINS, J.L.; J.S. KOTIAHO & N.R. LEBAS. 2005. Matters of scale: Positive allometry and the evolution of male dimorphisms. **American Naturalist 165:** 389-402.

TOMKINS J.L. & A.P. MOCZEK. 2009. Patterns of threshold evolution in polyphenic insects under different developmental models. **Evolution 62:** 459-468. doi: 10.1111/j.1558-5646.2008.00563.x.

VALENTIN, J.L. 2000. **Ecologia numérica: Uma introdução à análise multivariada de dados ecológicos.** Rio de Janeiro, Interciência, 117p.

YOSHITAKE, H. & I. KAWASHIMA. 2004. Sexual Dimorphism and Agonistic Behavior of *Exechesops leucopis* (Jordan) (Coleoptera: Anthribidae: Anthribinae). **The Coleopterists Bulletin 58:** 77-83.

Appendix 1. Mean, standard deviations, and variance measures of morphological characters of males and females of *S. platyrhinus* and *H. prasinata*. The result of the independent T test for sexual dimorphism is also shown. (***) for p ≤ 0.0001.

Measures	S. platyrhinus							H. prasinata						
	Males			Females			T test	Males			Females			T test
	Mean	SD	Variance	Mean	SD	Variance		Mean	SD	Variance	Mean	SD	Variance	
RL1	0.54	0.13	0.02	0.41	0.06	0.00	0.0091	0.45	0.01	0.07	0.44	0.01	0.09	0.4807
RAW	0.21	0.08	0.01	0.23	0.04	0.00	0.3405	0.26	0.12	0.34	0.21	0.01	0.08	0.6784
MW1R	0.03	0.12	0.01	0.16	0.03	0.00	0.0041	–	–	–	–	–	–	–
MW2R	0.30	0.11	0.01	0.23	0.04	0.00	0.1091	–	–	–	–	–	–	–
RBW	0.38	0.10	0.01	0.31	0.05	0.00	0.0688	0.09	0.01	0.08	0.12	0.01	0.08	0.2619
HW	0.30	0.07	0.01	0.32	0.05	0.00	0.4375	0.26	0.00	0.06	0.29	0.01	0.08	0.1792
PL	0.31	0.07	0.00	0.34	0.10	0.01	0.2870	0.55	0.00	0.07	0.55	0.01	0.08	0.7185
PW	0.37	0.07	0.01	0.40	0.04	0.00	0.3047	0.49	0.00	0.07	0.51	0.01	0.09	0.4486
EL	0.65	0.06	0.00	0.68	0.04	0.00	0.3094	0.91	0.00	0.07	0.91	0.01	0.08	0.8098
EW	0.42	0.07	0.01	0.45	0.05	0.00	0.2969	0.58	0.00	0.07	0.60	0.01	0.09	0.3034
MEW	-0.03	0.06	0.00	0.02	0.04	0.00	0.0231	0.08	0.00	0.05	0.09	0.00	0.07	0.6900
DIS	0.11	0.14	0.02	0.05	0.05	0.00	0.2983	0.09	0.01	0.08	0.11	0.01	0.08	0.2525
IEW	0.09	0.09	0.01	0.09	0.06	0.00	0.8413	0.07	0.01	0.08	0.07	0.01	0.08	0.8187
II	-0.56	0.12	0.01	-0.60	0.10	0.01	0.4182	-0.18	0.02	0.13	-0.36	0.01	0.08	***
III	-0.05	0.12	0.01	-0.35	0.11	0.01	***	0.13	0.02	0.13	-0.19	0.01	0.08	***
IV	0.02	0.12	0.01	-0.41	0.04	0.00	***	0.08	0.02	0.15	-0.25	0.01	0.09	***
V	0.06	0.11	0.01	-0.40	0.05	0.00	***	0.14	0.18	0.43	-0.33	0.01	0.09	***
VI	0.04	0.10	0.01	-0.42	0.05	0.00	***	0.09	0.04	0.21	-0.39	0.01	0.08	***
VII	0.01	0.09	0.01	-0.49	0.04	0.00	***	0.09	0.04	0.20	-0.42	0.01	0.08	***
VIII	-0.07	0.09	0.01	-0.49	0.06	0.00	***	0.07	0.03	0.18	-0.41	0.01	0.09	***
IX	-0.23	0.08	0.01	-0.44	0.10	0.01	***	-0.05	0.03	0.19	-0.35	0.01	0.08	***
X	-0.47	0.07	0.01	-0.61	0.10	0.01	0.0002	-0.48	0.01	0.11	-0.61	0.01	0.10	***
XI	-0.28	0.07	0.01	-0.42	0.09	0.01	***	-0.09	0.02	0.16	-0.39	0.02	0.12	***
VL_IV	-0.40	0.09	0.01	-0.34	0.06	0.00	0.0725	-0.22	0.00	0.05	-0.15	0.01	0.08	0.0002
VL_V	-0.36	0.09	0.01	-0.23	0.11	0.01	0.0010	-0.10	0.00	0.06	0.02	0.00	0.07	***

Appendix 2. Mean, standard deviations, and variance measures of the morphology of major males and minor males of *S. platyrhinus*. The result of T test for independent polyphenism in males is also shown. (***) for p ≤ 0.0001.

Measures	Major Males			Minor Males			T test
	Mean	Variance	SD	Mean	Variance	SD	
RL1	4.17	0.43	0.65	2.67	0.45	0.67	***
RAW	1.78	0.05	0.21	1.37	0.04	0.19	0.0002
MW1R	1.17	0.07	0.27	0.99	0.04	0.20	0.1450
MW2R	2.29	0.15	0.38	1.60	0.10	0.31	0.0002
RBW	2.74	0.13	0.36	1.94	0.16	0.39	***
HW	2.20	0.05	0.22	1.71	0.05	0.21	***
PL	2.21	0.05	0.23	1.79	0.07	0.26	0.0002
PW	2.59	0.06	0.24	2.06	0.10	0.32	***
EL	4.88	0.16	0.40	3.92	0.20	0.45	***
EW	2.86	0.08	0.28	2.29	0.14	0.37	***
MEW	0.99	0.01	0.11	0.87	0.01	0.10	0.0109
DIS	1.50	0.16	0.40	1.06	0.07	0.26	0.0108
IEW	1.41	0.03	0.17	1.02	0.03	0.18	***
II	0.31	0.01	0.11	0.24	0.00	0.05	0.0181
III	1.01	0.05	0.23	0.77	0.02	0.14	0.0195

Continues

Appendix 2. Continued.

Measures	Major Males			Minor Males			T test
	Mean	Variance	SD	Mean	Variance	SD	
IV	1.19	0.04	0.19	0.87	0.05	0.23	0.0013
V	1.30	0.02	0.14	0.94	0.05	0.23	***
VI	1.25	0.01	0.09	0.91	0.05	0.23	***
VII	1.15	0.01	0.09	0.86	0.04	0.19	***
VIII	0.96	0.01	0.11	0.73	0.02	0.14	***
IX	0.63	0.01	0.10	0.52	0.01	0.09	0.0129
X	0.36	0.00	0.06	0.32	0.00	0.04	0.2229
XI	0.57	0.01	0.08	0.48	0.01	0.08	0.0132
VL_IV	0.44	0.01	0.07	0.34	0.00	0.05	0.0011
VL_V	0.47	0.01	0.11	0.40	0.01	0.07	0.0460

21

Natural history of *Micrablepharus maximiliani* (Squamata: Gymnophthalmidae) in a Cerrado region of Northeastern Brazil

Francisco Dal Vechio[1,3], Renato Recoder[1], Hussam Zaher[2] & Miguel Trefaut Rodrigues[1]

[1] *Departamento de Zoologia, Instituto de Biociências, Universidade de São Paulo. Rua do Matão, Travessa 14, 321, Cidade Universitária, 05508-090 São Paulo, Brazil.*
[2] *Museu de Zoologia, Universidade de São Paulo. Avenida Nazaré 481, Ipiranga, 04263-000 São Paulo, Brazil.*
[3] *Corresponding author. E-mail: francisco.dalvechio@gmail.com*

ABSTRACT. *Micrablepharus maximiliani* (Reinhardt & Luetken, 1861) is a microteiid lizard widely distributed in the open areas of South America. Little is known about its ecology and reproductive biology. Here, we analyzed aspects of the natural history of a population of *M. maximiliani* from a Cerrado area in the state of Piauí, northeastern Brazil. Our results suggest that the reproductive activity of *M. maximiliani* might be seasonal in the Cerrado, since reproductive females were observed only in the dry season, whereas reproductive males were present in both seasons. Vitellogenic follicles and oviductal eggs were found simultaneously in one female, suggesting that females may produce more than one clutch per season. Sexual dimorphism was observed in body shape, and individuals were mainly restricted to a typical savanna physiognomy. The diet consisted of small arthropods, including spiders, crickets and cockroaches as the most important items.

KEY WORDS. Diet; ecology; lizards; reproduction; sexual dimorphism.

Micrablepharus maximiliani (Reinhardt & Luetken, 1861) is a small gymnophthalmid lizard (about 4 cm of snout-vent length) that is widely distributed throughout open areas of tropical South America (Vanzolini 1988, Ávila-Pires 1995, Rodrigues 1996). In Brazil, it is found in open and semi open habitats, along the Cerrado domain, and in mesic Caatinga habitats (Vanzolini *et al.* 1980, Werneck & Colli 2006, Moura *et al.* 2010). It is also present in the Atlantic Forest of eastern Brazil, associated with coastal restingas, open habitat enclaves, in transitional areas between the Caatinga and Cerrado domains, or in isolated forests known as "Brejos de Altitude" (Freire 1996, Silva *et al.* 2006, Moura *et al.* 2010, Abrantes *et al.* 2011).

This species is a semi-fossorial lizard, often associated with rock outcrops, leaf-litter and bare ground, usually found in sandy-soil habitats, or inside termite mounds (Rodrigues 1996, 2003, Mesquita *et al.* 2006, Werneck *et al.* 2009). Sexual dimorphism is present and only males have femoral pores (Rodrigues 1996). Although commonly sampled in herpetofaunal inventories, little is known about basic aspects of their natural history and there is practically no information available on its reproductive strategy.

Here, we provide for the first time a comprehensive study of the natural history of *M. maximiliani*, presenting information on the reproductive condition, diet, habitat use, and sexual dimorphism of a population from a locality in the northern Cerrado, state of Piauí, Brazil.

MATERIAL AND METHODS

Field work was carried out at Estação Ecológica Uruçuí-Una (EEUU) (08°50'S, 44°10'W), located in the northern extremity of the Brazilian Cerrado, in the southwestern corner of the state of Piauí. The vegetation of the EEUU is classified as woody savanna (Castro 1985). The relief is characterized by dissected plateaus that range from 480 to 620 m a.s.l., separated by valleys. The climate is dry to sub-humid, with high mean annual temperatures around 25°C, and the dry and wet seasons extending from March to November and from December to March, respectively (Castro 1985).

Samplings were conducted along three campaigns, two during the rainy season (February 10th to March 9th 2000 and January 9th to February 1st 2001) and one in the dry season (16-28th July 2000). We sampled the main physiognomic subunits present in the region with the use of pitfall traps with drift fences, complemented by active search. Each sampling unit contained one central and three external 30 L buckets arranged in a Y-shape disposition and joined together by four meters long plastic fences.

Four main physiognomies were sampled: **Cerrado *sensu stricto*** – typical savanna vegetation found around the EEUU headquarters and at the top of Serra Grande plateau, characterized by dense tree and herbaceous layers, but without a closed canopy. The soil is sandy and covered by significant floral diversity; **Gallery forest** – evergreen forests composed by trees up to 15 m high, forming a closed canopy. The soil is less sandy than in the surrounding savannas, and the ground is covered with a dense litter; **Dry forest** – a dense but thin arboreal vegetation with a discontinuous canopy, usually present around rocky outcrops surrounded by savannas; **Hillside forest** – a forest characterized by a closed canopy and sandy soil, with high deposition of organic matter forming a dense litter. Table I shows the total effort and capture rates in each physiognomy. A detailed methodology, map and habitat photographs can be found in Dal Vechio et al. (2013).

A representative number of the captured specimens were euthanized with a lethal injection of anesthetics, fixed in 10% formol, and preserved in 70% alcohol. Voucher specimens are housed at the herpetological collection of the Museu de Zoologia da Universidade de São Paulo (MZUSP), São Paulo, Brazil. The specimens examined were: MZUSP-90226, MZUSP-90236, MZUSP-90237, MZUSP-90242, MZUSP-90243, MZUSP-90246-51, MZUSP-90262, MZUSP-90265, MZUSP-90772-75, MZUSP-90777-81, MZUSP-90784, MZUSP-90873.Ten morphometric measurements were taken using a Mitutoyo digital caliper: snout-vent length (SVL), length between limbs (LBL), head height (HH), head width (HW), head length (HL), femur length (FL), tibia length (TL), foot length (FTL), humeral length (HUL), forearm length (FAL).

To assess the reproductive condition and diet, we dissected the preserved individuals. Their stomachs were removed and their contents were examined under stereomicroscope. Each prey item was identified to the more inclusive group up to the level of Order. We calculated the importance index (Ix), as modified from Howard (1999), which represents the relative importance of each category in relation to the entire diet. We also calculated niche breadth using the inverse of the diversity index of Simpson (Simpson 1949). Diet overlap between the sexes was calculated with the niche overlap index of Pianka (1974). The diet overlap index varies from 0 (no overlap) to 1 (complete overlap).

The reproductive condition was determined by gonad inspection. Females were considered sexually mature when vitellogenic follicles or oviductal eggs were present. Males were considered sexually mature when convoluted epididymides and/or testes hypertrophy were present. We evaluated sexual dimorphism in morphometry through analysis of variance (ANOVA) performed using SPSS v.15.0 with a significance level of 0.05.

RESULTS

A total of 67 individuals of *Microblepharus maximiliani* (Fig. 1) were captured during all three campaigns at the EEUU, 53 individuals were captured in the cerrado *sensu stricto* (Fig. 2), seven in the dry forest, six in the gallery forest and only one in the hillside forest. The capture rate was similar for both seasons, but higher for the cerrado physiognomy (Table I).

Table I. Number of specimens of *M. maximiliani* captured in pitfall traps at Estação Ecológica Uruçuí Una in the wet and dry seasons. The sampling effort expended (in buckets x days) and capture rate (individuals per bucket) in each season are presented.

	N captured		Effort		Capture rate		
	Wet	Dry	Wet	Dry	Wet	Dry	Total
Cerrado	44	9	2840	800	0.0155	0.0113	0.0267
Dry forest	4	3	1080	680	0.0037	0.0044	0.0081
Gallery forest	2	4	1320	720	0.0015	0.0056	0.0071
Hillside forest	1	0	1080	0	0.0009	0	0.0009
Total	51	16	6320	2200	0.0081	0.0073	0.0079

A sample of 24 adults was collected and preserved (12 males and 12 females): four females and six males captured in the dry season, eight females and six males captured in the rainy season. The sexes did not differ in body size (Anova; $F_{1,23}$ = 3.487, p = 0.08). Sexual dimorphism was observed in shape, with females presenting longer LBL (Anova; $F_{1,23}$ = 7.634, p = 0.01), and males presenting longer FL (Anova; $F_{1,23}$ = 4.607, p = 0.04) and HL (Anova; $F_{1,23}$ = 4.396, p = 0.04) (Table II).

Table II. Morphometric variables with the corresponding mean ± standard deviation and level of significance p from analysis of variance (Anova) for sexual dimorphism. Values with significance < 0.05 are presented in bold.

	Males (N = 12) Mean ± SD	Females (N = 12) Mean ± SD	F	p
SVL	37.09 ± 1.83	38.35 ± 1.44	3.487	0.08
LBL	**18.52 ± 1.29**	**19.85 ± 1.24**	**7.634**	**0.01**
HH	4.03 ± 0.33	3.91 ± 0.3	0.843	0.37
HW	5.63 ± 0.44	5.37 ± 0.31	2.772	0.11
HL	**7.84 ± 0.42**	**7.58 ± 0.11**	**4.396**	**0.04**
FL	**5.61 ± 0.38**	**5.3 ± 0.34**	**4.607**	**0.04**
TL	4.7 ± 0.25	4.69 ± 0.35	0.004	0.95
FTL	6.93 ± 0.55	6.85 ± 0.61	0.113	0.74
HUL	3.66 ± 0.43	3.48 ± 0.32	0.205	0.25
FAL	6.92 ± 0.59	6.87 ± 0.46	0.038	0.85

All four females captured during the dry season were reproductive, with three (75%) presenting vitellogenic follicles and one (25%) presenting vitellogenic follicles and two oviductal eggs simultaneously. All eight females captured during the wet season were not reproductive, and presented only translucent follicles. All dissected males presented convoluted epididymides and enlarged testes, and were thus considered reproductive regardless of the season.

Twenty four stomachs were analyzed and 36 items were identified in six prey categories. One item was not identified.

Figures 1-2. (1) Individual of *Microblepharus maximiliani* with autotomized tail, captured in the Estação Ecológica Uruçuí Una; (2) a typical Cerrado habitat where the species was found.

Four stomachs were empty. Prey categories with higher importance index correspond to Araneae (0.48), Orthoptera (0.48) and Blattaria (0.34) (Table III). Niche breadth was 4.26, whereas dietary overlap between sexes was 0.92.

Table III. Number (N), percentage (% N), frequency (F), frequency in percentage (% F) and importance index (Ix) of prey categories in the diet of *Microblepharus maximiliani* from the Estação Ecológica Uruçuí Una.

	N	% N	F	% F	Ix
Orthoptera	11	29.73	9	37.50	0.48
Blattaria	8	21.62	6	25.00	0.34
Coleoptera	3	8.11	3	12.50	0.14
Diptera	2	5.41	2	8.33	0.10
Aranae	11	29.73	9	37.50	0.48
Hymenoptera	1	2.70	1	4.17	0.05
Non-identified arthropod	1	2.70	1	4.17	0.05
Total	37		24		
Niche breadth	4.26				

DISCUSSION

Our results show that *Microblepharus maximiliani* is more commonly found in the cerrado *sensu stricto* physiognomy in the EEUU. Preference for open habitats in this species is well established (Vanzolini 1988, Mesquita *et al.* 2006, Nogueira *et al.* 2009). In fact, even in the Atlantic Forest, *M. maximiliani* is usually found in association with open formations such as restingas (Freire 1996, Silva *et al.* 2006) or forest edges (Rodrigues 1996), being rarely found inside rainforests (Nogueira *et al.* 2009, Moura *et al.* 2010).

Sexual dimorphism has been previously reported for the species, with females lacking femoral pores (Rodrigues 1996). Our data suggest that females and males of *M. maximiliani* may also differ in the proportional lengths of trunk, hind limbs and head. The presence of longer trunks in females may be associated with a fertility advantage conferred by a larger space in the peritoneal cavity for egg development (Olsson *et al.* 2002, Cox *et al.* 2003). A larger trunk length in females is recurrent in many species of Gymnophthalmidae (Vitt 1982, Vitt & Ávila-Pires 1998, Balestrin *et al.* 2010). On the other hand, males presented larger femoral and head lengths than females, a pattern that seems to be common in species of lizards in which the exploration or defense of territories by males, favors success in finding mates (Perry *et al.* 2004, Peterson & Husack 2006, Kaliontzopoulou *et al.* 2010). A similar pattern is also observed in other small-bodied microteiid lizards, in which males present larger hind limbs than females (Vitt 1982). Nevertheless, in the other species of the genus, *Microblepharus atticolus* (Rodrigues 1996), the sexes do not show dimorphism in morphometry (Vieira *et al.* 2000). Curiously, in this species both sexes present femoral pores (Rodrigues 1996).

The Cerrado has two well-defined weather seasons (a dry-winter and a rainy-summer), which favor a seasonal reproductive cycle for most inhabitant lizard species (Colli 1991, Van Sluys 1993, Wiederhecker *et al.* 2002, Colli *et al.* 2003, Mesquita & Colli 2003). Our results suggest that the reproductive activity of females of *M. maximiliani* occurs in the dry season. Thus juveniles would be found from August to November, but sampling throughout the year would be necessary to confirm this pattern. Food supply may not be a limiting factor in the dry season, as it would be advantageous for the offspring spawning at a time with greater prey availability (Telford 1971, Mesquita *et al.* 2006). Nevertheless, the presence of adequate microenvironments for the eggs to develop may also be constrained during the rainy season (TELFORD 1971), as well as opportunities for females to thermoregulate, which may compromise the development of their eggs.

The presence of females bearing vitellogenic follicles and oviductal eggs simultaneously suggests that *M. maximiliani* can produce more than one clutch during the reproductive season. *Vanzosaura rubricauda* (Boulenger, 1902) a related species, also shows multiple clutches of two eggs, with a continuous reproduction cycle in the Caatinga (VITT 1982), but a seasonal one in the Chaco (CRUZ 1994). A fixed clutch size of two eggs is a recurring pattern in many species of Gymnophthalmidae (FITCH 1970, TELFORD 1971, SHERBROOKE 1975, VITT 1982, ÁVILA-PIRES 1995, PIANKA & VITT 2003, BALESTRIN *et al.* 2010).

Our results suggest that specimens of *M. maximiliani* from the EEUU are opportunistic predators that feed on a variety of small arthropods, especially spiders, crickets and cockroaches. Although with a relatively small sample and without considering prey abundance in the habitat, our results are similar to those obtained from specimens of *M. maximiliani* from the Jalapão region, which mainly consumed crickets, spiders, and hemipterans (MESQUITA *et al.* 2006), and from the Paranã River valley, which mainly consumed spiders, hemipterans and crickets (WERNECK *et al.* 2009). The prey items consumed by *M. atticolus* were crickets, hemipterans, and spiders (VIEIRA *et al.* 2000). We observed a large dietary overlap between males and females in our sample, suggesting a lack of food partitioning between the sexes.

ACKNOWLEDGEMENTS

We are indebted to Deocleciano G. Ferreira, Head of the Ibama Regional Office, for his support. We are grateful to Mário de Vivo, Erika Hingst-Zaher, Renato Gregorin, Maria J. de J. Silva, Ana P. Carmignotto, Alexandre Percequillo, Renata C. Amaro, Luis F. Silveira, Andrés C. Mendez, Marcos A.N. de Sousa, Cristiano Nogueira, Dante Pavan, Marianna Dixo, Dalton M. Novaes, Giovanna G. Montingelli, Ricardo A. Fuentes, Rodrigo P. Ribeiro, Guilherme R. Britto, Felipe F. Curcio, and Paula H. Valdujo for their help in the field. We are also indebted to Carolina Mello for support with the collection at the MZUSP, Pedro Dias and Silvio Nihei (IB-USP) for identification of prey items, and Mauro Teixera Jr. for critically reading and commenting on a previous draft of this paper. This research was supported by grants from the Fundação de Amparo à Pesquisa do Estado de São Paulo (FAPESP), Conselho Nacional de Desenvolvimento Científico e Tecnológico (CNPQ), and Fundação Boticário para a Conservação da Natureza.

REFERENCES

ABRANTES, S.H.F.; M.M.R. ABRANTES & A.C.G.P. FALCÃO. 2011. A fauna de lagartos em três brejos de altitude de Pernambuco, nordeste do Brasil. **Revista Nordestina de Zoologia 5:** 23-29.

ÁVILA-PIRES, T.C.S. 1995. Lizards of Brazilian Amazonia (Reptilia: Squamata). **Zoologische Verhandelingen 299:** 1-706.

BALESTRIN, R.L.; L.H. CAPPELLARI & A.B. OUTEIRAL. 2010. Biologia reprodutiva de *Cercosaura schreibersii* (Squamata, Gymnophthalmidae) e *Cnemidophorus lacertoides* (Squamata, Teiidae) no escudo Sul-Riograndense, Brasil. **Biota Neotropica 10:** 131-139. doi: 10.1590/s1676-06032010000100013.

CASTRO, A.A.F.J. 1985. Vegetação e Flora da Estação Ecológica de Uruçuí-Una (Resultados Preliminares), p. 251-261. *In*: Anais do XXXVI Congresso Nacional de Botânica. Curitiba, Sociedade Botânica do Brasil, Ibama, vol. 1.

COLLI, G.R. 1991. Reproductive ecology of *Ameiva ameiva* in the Cerrado of central Brazil. **Copeia 1991:** 1002-1012. doi: 10.2307/1446095.

COLLI, G.R.; D.O. MESQUITA; P.V.V. RODRIGUES & K. KITAYAMA. 2003. The ecology of the gecko *Gymnodactylus geckoides amarali* in a neotropical savanna. **Journal of Herpetology 37:** 694-706. doi: 10.1670/180-02a.

COX, R.M.; S.L. SKELLY & H.B. JOHN-ALDER. 2003. A Comparative test of adaptive hypotheses for size dimorphism in lizards. **Evolution 57:** 1653-1669. doi: 10.1554/02-227.

CRUZ, F.B. 1994. Actividad reproductiva en *Vanzosaura rubricauda* (Sauria: Teiidae) del chaco accidental en Argentina. **Cuaderno de Herpetología 8:** 112-1118.

DAL VECHIO, F.; R.S. RECODER; Z. HUSSAM & M.T. RODRIGUES. 2013. The herpetofauna of the Estação Ecológica de Uruçuí-Una, state of Piauí, Brazil. **Papéis Avulsos de Zoologia 53 (16):** 225-243. doi: 10.1590/s0031-10492013001600001.

FITCH, H.S. 1970. **Reproductive cycles of lizards and snakes.** Lawrence, Museum of Natural History, University of Kansas, 247p.

FREIRE, E.M.X. 1996. Estudo ecológico e zoogeográfico sobre a fauna de lagartos (Sauria) das dunas de Natal, Rio Grande do Norte e da restinga de Ponta de Campina, Cabedelo, Paraíba, Brasil. **Revista Brasileira de Zoologia 13 (4):** 903-921. doi: 10.1590/S0101-81751996000400012.

HOWARD, A.K.; J.D. FORESTER; J.M. RUDER; J.S. PARMELEE & R. POWELL. 1999. Natural History of a Terrestrial Hispaniolan Anole: *Anolis barbouri*. **Journal of Herpetology 33:** 702-706. doi: 10.2307/1565590.

KALIONTZOPOULOU, A.; M.A. CARRETERO & G.A. LLORENTE. 2010. Sexual dimorphism in traits related to locomotion: ontogenetic patterns of variation in *Podarcis* wall lizards. **Biological Journal of the Linnean Society 99:** 530-543. doi: 10.1111/j.1095-8312.2009.01385.x.

MESQUITA, D.O. & G.R. COLLI. 2003. The ecology of *Cnemidophorus ocellifer* (Squamata, Teiidae) in a neotropical savanna. **Journal of Herpetology 37:** 498-509. doi: 10.1670/179-02a.

MESQUITA, D.O.; G.R. COLLI; F.G.R. FRANÇA & L.J. VITT. 2006. Ecology of a Cerrado Lizard Assemblage in the Jalapão Region of Brazil. **Copeia 2006:** 460-471. doi: 10.1643/0045-8511(2006)2006[460:eoacla]2.0.co;2.

MOURA, M.R.; J.S. DAYRELL & V.A. SÃO-PEDRO. 2010. Reptilia, Gymnophtalmidae, *Micrablepharus maximiliani* (Reinhardt and Lutken, 1861): Distribution extension, new state record and geographic distribution map. **Check List 6**: 419-426.

NOGUEIRA, C.; G.R. COLLI & M. MARTINS. 2009. Local richness and distribution of the lizard fauna in natural habitat mosaics of the Brazilian Cerrado. **Austral Ecology 34**: 83-96. doi: 10.1111/j.1442-9993.2008.01887.x.

OLSSON, M.; R. SHINE; E. WAPSTRA; B. UJVARI & T. MADSEN. 2002. Sexual dimorphism in lizard body shape: the roles of sexual selection and fecundity selection. **Evolution 56**: 1538-1542. doi: /10.1111/j.0014-3820.2002.tb01464.x.

PERRY, G.; K. LEVERING; I. GIRARD & T. GARLAND-JR. 2004. Locomotor performance and social dominance in male *Anolis cristatellus*. **Animal Behaviour 67**: 37-47. doi: 10.1016/j.anbehav.2003. 02.003.

PETERSON, C.G. & J.F. HUSACK. 2006. Locomotor performance and sexual selection: individual variation in sprint speed of collared lizards (*Crotaphytus collaris*). **Copeia 2006**: 216-224. 10.1643/0045-8511(2006)6[216:lpassi]2.0.co;2.

PIANKA, E.R. 1974. Niche overlap and diffuse competition. **Proceedings of the National Academy of Sciences 71**: 2141-2145. doi: 10.1073/pnas.71.5.2141.

PIANKA, E.R. & L.J. VITT. 2003. **Lizards: Windows to the Evolution of Diversity.** Berkeley, University of California Press, 333p.

RODRIGUES, M.T. 1996. A New Species of Lizard, Genus *Microblepharus* (Squamata: Gymnophthalmidae), from Brazil. **Herpetologica 52**: 535-541.

RODRIGUES, M.T 2003. Ecologia e Conservação da Caatinga, p. 181-236. *In:* R. LEAL; J.M.C. DA SILVA & M. TABARELLI (Eds). **Herpetofauna da Caatinga.** Recife, Editora UFPE, 804p.

SHERBROOKE, W.C. 1975. Reproductive cycle of a tropical teiid lizard, *Neusticurus ecpleopus* Cope, in Peru. **Biotropropica 7**: 194-207. doi: 10.2307/2989623.

SILVA, S.T.; U.G. SILVA; G.A.B. SENA & F.A.C. NASCIMENTO. 2006. A biodiversidade da Mata Atlântica alagoana: anfíbios e répteis, p. 65-76. *In:* F.B.P. MOURA (Ed.). **A Mata Atlântica em Alagoas.** Maceió, Editora UFAL.

SIMPSON, E.H. 1949. Measurement of diversity. **Nature 163**: 688. doi: 10.1038/163688a0.

TELFORD-JR, S.R. 1971. Reproductive patterns and relative abundance of two microteiid lizard species in Panama. **Copeia 1971**: 670-675. doi: 10.2307/1442635.

VAN SLUYS, M. 1993. The reproductive cycle of *Tropidurus itambere* (Sauria, Tropiduridae) in southeastern Brazil. **Journal of Herpetology 27**: 28-32. doi: 10.2307/1564901.

VANZOLINI, P.E. 1988. Proceedings of a workshop on Neotropical distributional patterns, p. 317-342. *In*: P.E. VANZOLINI & W.R. HEYER (Eds). **Distributional patterns of South American lizards.** Rio de Janeiro, Academia Brasileira de Ciências.

VANZOLINI, P.E; A.M.M. RAMOS-COSTA & L.J. VITT. 1980. **Répteis das Caatingas.** Rio de Janeiro, Academia Brasileira de Ciências, 161p.

VIEIRA, G.H.C.; D.O. MESQUITA; A.K. PÉRES-JR; K. KITAYAMA & G.R. COLLI. 2000. Natural history: *Microblepharus atticolus*. **Herpetological Review 31**: 241-242.

VITT, L.J. 1982. Sexual dimorphism and reproduction in the microteiid lizard, *Gymnophthalmus multiscutatus*. **Journal of Herpetology 16**: 325-329. doi: 10.2307/1563730.

VITT, L.J. & T.C.S. ÁVILA-PIRES. 1998. Ecology of Two Sympatric Species of *Neusticurus* (Sauria: Gymnophthalmidae) in the Western Amazon of Brazil. **Copeia 1998**: 570-582. doi: 10.2307/1447787.

WERNECK, F.P. & G.R. COLLI. 2006. The lizard assemblage from Seasonally Dry Tropical Forest enclaves in the Cerrado biome, Brazil, and its association with the Pleistocenic Arc. **Journal of Biogeography 33**: 1983-1992. doi: 10.1111/j.1365-2699.2006.01553.x.

WERNECK, F.P.; G.R. COLLI & L.J. VITT. 2009. Determinants of assemblage structure in Neotropical dry forest lizards. **Austral Ecology 34**: 97-115. doi: 10.1111/j.1442-9993.2008.01915.x.

WIEDERHECKER, H.C.; A.C.S. PINTO & G.R. COLLI. 2002. Reproductive ecology of *Tropidurus torquatus* (Squamata: Tropiduridae) in the highly seasonal Cerrado biome of central Brazil. **Journal of Herpetology 36**: 82-91. doi: 10.2307/1565806.

22

The feeding habits of the eyespot skate *Atlantoraja cyclophora* (Elasmobranchii: Rajiformes) in Southeastern Brazil

Alessandra da Fonseca Viana[1,2] & Marcelo Vianna[1]

[1] *Laboratório de Biologia e Tecnologia Pesqueira, Instituto de Biologia, Universidade Federal do Rio de Janeiro. Avenida Carlos Chagas Filho 373, Bloco A, 21941-902 Rio de Janeiro, RJ, Brazil.*
[2] *Corresponding author. E-mail: fviana.ale@gmail.com*

ABSTRACT. The stomach contents of the eyespot skate, *Atlantoraja cyclophora* (Regan, 1903), were examined with the goal to provide information about the diet of the species. Samples were collected off the southern coast of Rio de Janeiro, Brazil, near Ilha Grande, between January 2006 and August 2007, at a depth of about 60 m. The diet was analyzed by sex, maturity stages and quarterly to verify differences in the importance of food items. The latter were analyzed by: frequency of occurrence, percentage of weight and in the Alimentary Index. The trophic niche width was determined to assess the degree of specialization in the diet. Additionally, the degree of dietary overlap between males and females; juveniles and adults and periods of the year were defined. A total of 59 individuals of *A. cyclophora* were captured. Females and adults were more abundant. The quarters with the highest concentrations of individuals were in the summer of the Southern Hemisphere: Jan-Feb-Mar 06 and Jan-Feb-Mar 07. Prey items were classed into five main groups: Crustacea, Teleosts, Elasmobranchs, Polychaeta, and Nematoda. The most important groups in the diet of the eyespot skate were Crustacea and Teleosts. The crab *Achelous spinicarpus* (Stimpson, 1871) was the most important item. The value of the niche width was small, indicating that a few food items are important. The comparison of the diet between males and females and juveniles and adults indicates a significant overlap between the sexes and stages of maturity; and according to quarters, the importance of prey groups differed (crustaceans were more important in the quarters of the summer and teleost in Jul-Aug-Sep and Oct-Nov-Dec 06), indicating seasonal differences in diet composition. Three groups with similar diets were formed in the cluster analysis: (Jan-Feb-Mar 06 and 07); (Apr-May-Jun 06 and Jul-Aug-Sep 07); (Jul-Aug-Sep 06 and Oct-Nov-Dec 06).

KEY WORDS. Diet; elasmobranch; trophic niche; Rajidae; Rio de Janeiro.

Dietary analyses may be used to understand variations in the growth, reproduction, migration and the behavioral aspects of food capture. They may result in increased knowledge of resource sharing and competition between organisms (ROSECCHI & NOUAZE 1987), and a broader understanding of trophic ecology, which can be used in ecosystem management (ZAVALA-CAMIN 1996). Fish diet may differ ontogenetically, spatially and seasonally. For instance, juveniles and adults differ in the size of the food they consume, and differences in food availability may affect the diet of broadly distributed species (ZAVALA-CAMIN 1996).

Elasmobranchs play an important role in marine ecosystems, occupying high trophic levels (EBERT & BIZARRO 2007). They play an important role in the energy flow between the benthic and pelagic regions (AGUIAR & VALENTIN 2010). As a result of their place in the food chain, benthonic skates may accumulate contaminants such as mercury, and are therefore good indicators of pollution (LACERDA *et al.* 2000).

Benthonic skates have suffered increased fishing pressure in recent years (MASSA *et al.* 2006). They are also caught very frequently as bycatch. These two factors make fishery the main anthropogenic factor that affects elasmobranch populations. (VOOREN & KLIPPEL 2005). Overfishing of skates and rays can change the abundance and distribution of their populations, and result in proportional growth of populations of species in lower trophic levels (WALKER & HISLOP 1998).

Studies on the diet of elasmobranchs are less frequent than studies involving other marine fishes. According to AGUIAR & VALENTIN (2010), only 44 studies on the alimentary biology and ecology of elasmobranchs have been published in Brazilian journals, and only one of these analyzed the diet of *Atlantoraja cyclophora* (Regan, 1903). Elasmobranchs are carnivorous, eating less variety of prey than the herbivorous or omnivorous teleosts. EBERT & BIZARRO (2007) studied 60 species of Rajiformes and indicated that in the diet of skates, teleosts and decapods are dominant groups. Rajiformes are secondary or tertiary consumers and are classified as benthopelagic or epibenthic predators specialized in marine invertebrates or small crustaceans (AGUIAR & VALENTIN 2010).

The eyespot skate *A. cyclophora* belongs to Rajidae. This oviparous skate with demersal habits is found in the South Atlantic Ocean, from Rio de Janeiro (Brazil) to the south of the Mar del Plata (Argentina). The species is found from 30 to 300 m depth (mainly below 50 m) (GOMES *et al.* 2010), generally in cold waters of the continental shelf and upper continental slope, and does not seem to have a preference for any peculiar granulometry of the sediment. Normally, females of *A. cyclophora* are heavier and wider than males (ODDONE & VOOREN 2004). The species is caught as bycatch in coastal demersal fisheries, and has suffered growing fishing pressure. As a result, it is considered "vulnerable" in the Red List of the International Union for Conservation of Nature (CHEUNG *et al.* 2005). Given the scarcity of studies on *A. cyclophora*, it is the time for scientists to focus their attention on the species (MASSA *et al.* 2006).

The goal of this study was to generate information about the diet of *Atlantoraja cyclophora* in southeastern Brazil, by identifying the main food items to species; and to ascertain differences in the importance of food items according to sex, maturity stages and periods of the year.

MATERIAL AND METHODS

Between January 2006 and August 2007, 14 samples were hauled from the southern coast of Rio de Janeiro, Brazil, near Ilha Grande (Fig. 1). Each haul lasted one hour, and was conducted at about 60 m depth. The boat worked with a bottom-pair trawl. The net had a 20 mm mesh between opposing knots at the body and sleeves and 18 mm in codend.

The area of the study is located near the Ilha Grande Bay (22°50'-23°20'S, 44°00'-44°45'W), in the southern coast of Rio de Janeiro. The sediment is mainly characterized by medium/fine sand, but also silt and clay. In this region three water masses occur: Tropical Water, Coastal Water and South Atlantic Central Water. The Tropical Water (TW) occurs in the upper layer, with high temperatures and salinity (T > 20°C, S > 36.4). The Coastal Water (CW), with high temperature (24°C) and low salinity (34.9), normally occurs in the inner shelf. The South Atlantic Central Water (SACW) has low temperature and salinity (T < 20°C, 34 < S < 36.4) and can emerge when the wind conditions are favorable for upwelling. Therefore, in late spring and summer, this mass occupies the inner shelf, and in winter and autumn it moves offshore.

After collection, the material was chilled and later frozen until processing. Samples of *A. cyclophora* were identified according to GOMES *et al.* (2010), measured (disc width, cm), weighed (total weight, g) and sexed (voucher C.DBAV UERJ1256). Then, the skates were dissected and the maturity stage of each specimen (juvenile or adult) was determined according to a combination of internal and external characteristics (ODDONE & VOOREN 2005). The stomachs were removed, weighed, fixed in 10% formalin and conserved in 70% ethanol. The stomach contents were analyzed and prey items were

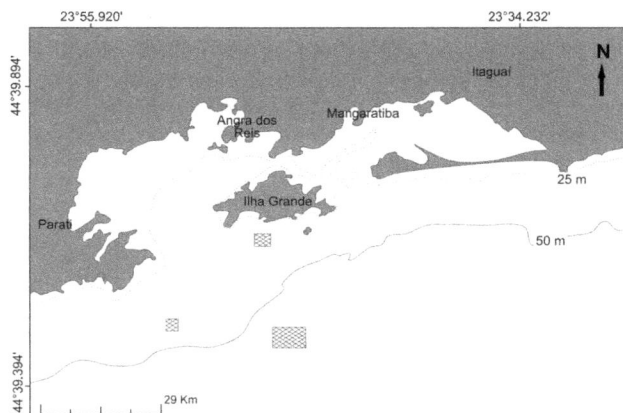

Figure 1. Location of samples taken, between January 2006 and August 2007, near Ilha Grande, on the southern coast of Rio de Janeiro, Brazil.

weighed and identified to the lowest possible taxonomic level. Identification was made with the help of specialists.

A cumulative prey curve (CORTES 1997, FERRY & CAILLIET 1996) was employed to determine whether the sample was large enough to precisely describe the diet. This method plots the cumulative number of randomly pooled stomachs against the cumulative number of prey types. This curve was constructed using the program EstimateSWin820 and fifty randomizations were performed.

Food items were analyzed by Frequency of Occurrence (%FO), expressed as the percentage of the stomachs analyzed containing the prey item; and Percentage Weight (%W), which is the percentage of the weight of the prey item with respect to the total weight of all prey items combined. These measurements were combined in the Alimentary Index (IAi), according to KAWAKAMI & VAZZOLER (1980), and modified to use the Percentage Weight (%W), according to the equation: $\%IAi_1 = ((\%FO_1 \times \%W_1) / \Sigma (\%FO_T \times \%W_T)) \times 100$, where $\%FO_1$ = percentage of stomachs in which determined item occur; $\%W_1$ = percentage of weight of determined item. The trophic niche width was determined according to the Standardized Levins Index (HURLBERT 1978) to assess the degree of diet specialization, according to the following equation: $Bi = [(\Sigma jPij^2)^{-1} - 1] (n-1)^{-1}$, where Pij = proportion of the prey j in the diet of the predator I; n = number of prey categories. In order to provide information about prey importance and feeding strategy, an adaptation of the COSTELLO's (1990) method by AMUNDSEN *et al.* (1996) was employed.

The diet was analyzed by sex, maturity stages of the skates and quarterly. The Trophic Niche Overlap Index by PIANKA (1973), was employed to determine the degree of dietary overlap between males and females and juveniles and adults, according to the following equation: $Oxy = \Sigma (Pxi \times Pyi) / \sqrt{\Sigma(\Sigma Pxi^2 \times \Sigma Pyi^2)}$, where Pxi = proportion of prey i in predator x; Pyi = proportion of prey i in predator y.

In the temporal analyses, six quarters were defined: 1 (Jan-Feb-Mar 06), 2(Apr-May-Jun 06), 3(Jul-Aug-Sep 06), 4(Oct-Nov-Dec 06), 5 (Jan-Feb-Mar 07), 6 (Jul-Aug-Sep/07). The diet was determined for each quarter and a cluster analysis was performed, employing Ward's Method, using the Past Statistic Program to define groups with similar diet.

RESULTS

We analyzed a total of 59 stomachs of *A. cyclophora*; of these, two were empty and were not included in the analyses. Of the remaining 57 stomachs, those from females and adults were more abundant (Table I), with disc width ranging from 14.8 to 47.8 cm (mean = 38.04 ± 8.25; median = 40.3). While Jan-Feb-Mar 06 were the quarters with the highest concentration of skates, Apr-May-Jun 06 and Jul-Aug-Sep 07 (Table I) had the lowest concentrations. The cumulative prey curve (Fig. 2) stabilized after about 40 stomachs, indicating that this number was representative for the sampling area.

Figure 2. Cumulative prey curve in the diet, of *Atlantoraja cyclophora*, with a confidence interval of 95% upper and lower, in southeastern Brazil.

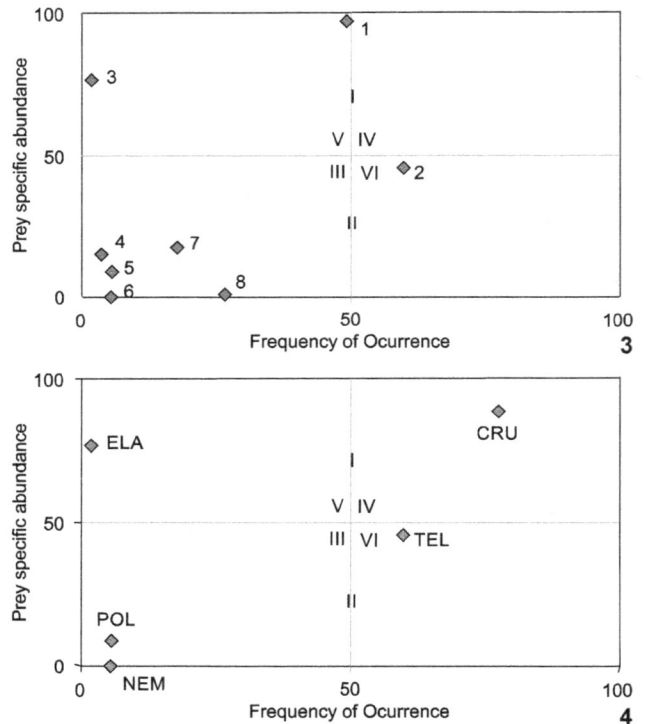

Table I. Sex ratio according to maturity stages and quarters of the year of *Atlantoraja cyclophora*, collected in southern coast of Rio de Janeiro, Brazil.

	Females	Males	Total
Juveniles	10	7	17
Adults	24	16	40
Jan-Feb-Mar 06	14	10	24
Apr-May-Jun 06	0	1	1
Jul-Aug-Sep 06	5	2	7
Oct-Nov-Dec 06	7	6	13
Jan-Feb-Mar 07	8	3	11
Jul-Aug-Sep 07	0	1	1

Prey items were classified into five main groups (Crustacea, Teleosts, Elasmobranchs, Polychaeta, Nematoda) (Table II). The group Crustacea was the most important in the diet (79.4%IAi). Teleosts were also important, since this group has a significant FO%. The crab *Achelous spinicarpus* (Stimpson, 1871) (= *Portunus spinicarpus*) was the most important item overall (53.5%IAi), followed by Portunidae (23.6%IAi). Item fragments of teleosts were important (9.7%IAi), and *Dactylopterus volitans* (Linnaeus, 1758) was a very representative teleost (8.2%IAi). The trophic niche width of *A. cyclophora* was 0.1, indicating that a few food items are very important.

The prey importance and feeding strategy diagrams (Figs 3 and 4) show that no food item was dominant in the diet of *A. cyclophora*. We reached this conclusion because no items occur in area IV of the diagram (Fig. 3). However, two items were the most important, Brachyura and fragments of Teleosts. The other items were rare, because they occur in area III (Fig. 3). When whole groups were taken into consideration, Crustacea were dominant (presence in are IV), but Fishes were

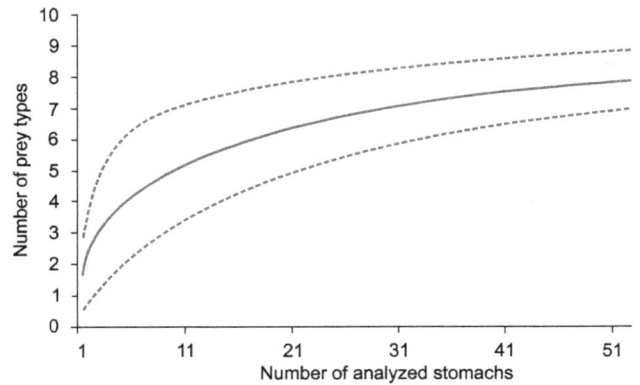

Figures 3-4. Diet of *Atlantoraja cyclophora* analyzed by Costello's method (1990), adapted by Amundsen et al. (1996), in southeastern Brazil. where: feeding strategy – specialist (I) or generalist (II); importance of prey – rare (III) or dominant (IV) and width of trophic niche – high between phenotype (V) or high within phenotype (VI) by food items (3) and by groups (4). Food items: 1) Brachyura, 2) Teleosts fragments, 3) Rajidae, 4) Stomatopoda, 5) Polychaeta, 6) Nematoda, 7) Caridea/Dendobranchiata, 8) Crustacean fragments. Groups: CRU) Crustacea, TEL) Teleost fishes, ELA) Elasmobranch, POL) Polychaeta, NEM) Nematoda.

Table II. Frequency of ocurrence (%FO), percentage weight (%W) and alimentary index (%IAi), of prey items in the total diet of *Atlantoraja cyclophora*, in southeastern Brazil, according to sex (females and males) and stages of maturity.

Prey items	Sex									Stages of maturity					
	Total (n = 57)			Females (n = 34)			Males (n = 23)			Juveniles (n = 17)			Adults (n = 40)		
	% FO	% W	% IAi	% FO	% W	% IAi	% FO	% W	% IAi	%FO	% W	%IAi	%FO	% W	%IAi
Crustacea	77.2	747.0	79.4	79.0	72.4	77.3	78.3	85.9	89.5	82.4	85.5	90.5	80.0	73.8	79.7
Caridea/Dendobranchiata	5.3	0.3	0.1	2.9	0.0	0.0	8.7	1.6	0.9	11.8	2.5	1.6	2.5	0.1	0.0
Caridea	10.5	2.1	1.3	11.8	1.5	0.9	8.7	4.6	2.7	5.9	2.4	0.8	12.5	2.0	1.2
Dendobranchiata	1.8	0.1	0.0				4.3	0.1	0.0				2.5	0.0	0.0
Brachyura	10.5	3.5	2.2	11.8	2.7	1.5	8.7	7.0	4.1	11.8	14.4	9.6	10.0	2.5	1.2
Leucosiidae	1.8	1.5	0.2				4.3	8.9	2.6				2.5	1.7	0.2
Parthenopidae	1.8	0.1	0.0				4.3	0.4	0.1				2.5	0.1	0.0
Portunidae	14.0	27.9	23.6	14.7	25.8	18.2	13.0	37.9	33.5				20.0	30.3	28.8
Achelous spinicarpus	22.8	38.9	53.5	29.4	42.0	59.2	17.4	24.4	28.8	11.8	64.8	43.1	30.0	36.8	52.3
Stomatopoda	3.5	0.1	0.0	5.9	0.1	0.0				5.9	0.1	0.0	2.5	0.1	0.0
Crustacean fragments	26.3	0.3	0.5	23.5	0.1	0.1	30.4	1.1	2.3	41.2	1.4	3.4	20.0	0.2	0.2
Teleosts	59.6	25.1	20.6	61.8	27.3	22.7	56.5	14.0	10.5	64.7	11.0	9.2	57.5	26.2	20.3
Dactylopterus volitans	8.8	15.5	8.2	14.7	18.7	13.2							12.5	16.8	10.0
Teleost fragments	50.9	3.2	9.7	50.0	2.5	6.0	52.2	6.4	22.6	64.7	11.0	40.4	45.0	2.5	5.3
Pleuronectiformes	1.8	1.2	0.1				4.3	7.3	2.1				2.5	1.3	0.2
Symphurus sp.	1.8	5.1	0.5	2.9	6.1	0.9							2.5	5.5	0.7
Polydactylus sp.	1.8	0.1	0.0				4.3	0.3	0.1				2.5	0.1	0.0
Elasmobranchs	1.8	0.2	0.0	2.9	0.3	0.0				5.9	3.0	0.2			
Rajidae	1.8	0.2	0.0	2.9	0.3	0.0				5.9	3.0	1.0			
Polychaeta	5.3	0.0	0.0	5.9	0.0	0.0	4.3	0.0	0.0	11.8	0.3	0.0	2.5	0.0	0.0
Euclymene sp.	1.8	0.0	0.0	2.9	0.0	0.0				5.9	0.1	0.0			
Sthenelais sp.	3.5	0.0	0.0	2.9	0.0	0.0	4.3	0.0	0.0	5.9	0.2	0.1	2.5	0.0	0.0
Nematoda	5.3	0.0	0.0	2.9	0.0	0.0	8.7	0.1	0.0	11.8	0.1	0.0	2.5	0.0	0.0
Nematods	5.3	0.0	0.0	2.9	0.0	0.0	8.7	0.1	0.0	11.8	0.1	0.1	2.5	0.0	0.0

also important (presence in area VI, with a high frequency of occurrence). Polychaeta and Nematoda were rare (Fig. 4).

The value of dietary overlap between females and males was high, 0.93. In both cases Crustaceans were more important (Table II) (77.3% IAi and 89.5% IAi respectively) followed by teleost, which were more heavily represented by females (22.7% IAi and 10.5% IAi respectively). Regarding the items, Portunidae (18.2% IAi for females and 33.5% IAi for males) and *A. spinicarpus* (59.2% IAi for females and 28.8% IAi for males) were the most important for both sexes. Other items like Caridea and crustacean fragments were more represented in the diet of males (2.7 and 2.3% IAi respectively).

The value of dietary overlap between juveniles and adults was also quite high, 0.99. In this case, the crustaceans were also more representative for the both stages of maturity (Table II) (90.5 and 79.7% IAi respectively), followed by teleost (9.2 and 20.3% IAi respectively). The polychaetes and nematodes

were less relevant, but more relevant to juveniles. In both groups *A. spinicarpus* was the most representative item (43.1% IAi for juveniles and 52.3%.IAi for adults).

The importance of prey groups differed according to season (Table III). The cluster analysis (Fig. 5) indicated that three groups were formed: Group 1 (Jan-Feb-Mar 06 and Jan-Feb-Mar 07); Group 2 (Apr-May-Jun 06 and Jul-Aug-Sep 07) and Group 3 (Jul-Aug-Sep 06 and Oct-Nov-Dec 06), indicating seasonal differences in diet composition. The quarters Jan-Feb-Mar 06 and Jan-Feb-Mar 07 had a great importance of crustaceans (97.3 and 98.9% IAi respectively), mainly Portunidae and *A. spinicarpus*. In the quarters of Jul-Aug-Sep 06 and Oct-Nov-Dec 06 teleost were the most representative (57.7 and 63.7% IAi respectively), especially *D. volitans*. The quarters, Apr-May-Jun 06 and Jul-Aug-Sep 07 had predominance of crustaceans (76.2 and 100% IAi respectively), but the number of individuals captured during this period was less representative.

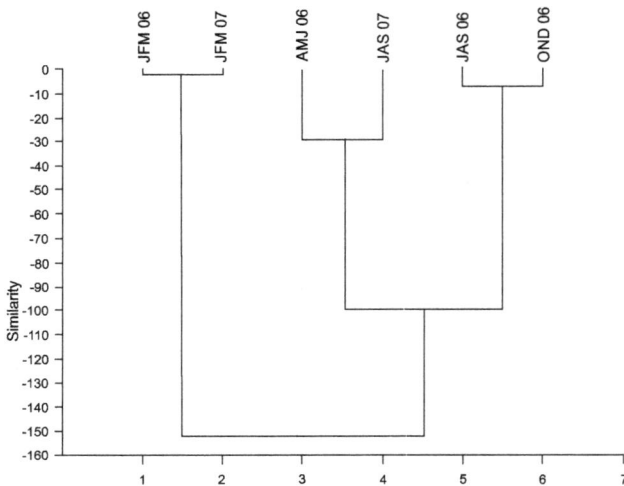

Figure 5. Dendrogram resulting from the cluster analysis, employing Ward's Method, according to quarter, in southeastern Brazil, where JFM correspond to Jan-Feb-Mar; AMJ to Apr-May-Jun, JAS to Jul-Aug-Sep and OND to Oct-Nov-Dec.

DISCUSSION

Females of *A. cyclophora* were larger at onset of sexual maturity than males, which is in agreement with results of other studies on Rajidae (ODDONE & VOOREN 2005, ODDONE *et al.* 2008). Females also reached larger size than males, as observed in other studies with *A. cyclophora* (ODDONE & VOOREN 2004, ODDONE *et al.* 2008). Skates were more abundant in the summer (Jan-Feb-Mar 06 and Jan-Feb-Mar 07). Summer is the season of greatest upwelling of the South Atlantic Central Water (SACW), which generates the oceanographic conditions favored by this species (ODDONE & VOOREN 2004).

Atlantoraja cyclophora feeds on benthic prey, mainly crustaceans and teleost. Crabs, mainly *A. spinicarpus*, were the most important item in the diet of *A.cyclophora*, probably because they are abundant in the area (PIRES 1992). The prey importance and feeding strategy diagrams confirmed these results, showing the dominance of crustaceans in the diet of the skate, followed by teleosts. Our results indicate that the diet of *A.cyclophora* is specialized. Although the Fishes did not have a high value of Alimentary Index (% IA$_I$) and was not a dominant item, this group

Table III. Frequency of Ocurrence (%FO), Percentage Weight (%W) and Alimentary Index (%IAi) of prey items in the diet of *Atlantoraja cyclophora* according to the quarters, in southeastern Brazil.

Prey items	Jan-Mar 06 (n = 24)			Apr-Jun 06 (n = 1)			Jul-Sep 06 (n = 7)			Oct-Dec 06 (n = 13)			Jan-Mar 07 (n = 11)			Jul-Sep 07 (n = 1)		
	% FO	%W	IAi	% FO	%W	IAi	% FO	%W	IAi	% FO	%W	IAi	% FO	%W	IAi	%FO	%W	IAi
Crustacea	83.3	95.6	97.3	100	76.2	76.2	85.7	37.9	42.3	46.2	51.1	36.3	100	97.2	98.9	100	100	100
Caridea/Dendobranchiata	4.2	0.6	0.1	100	76.2	76.2				7.7	0.0	0.0						
Caridea	4.2	0.6	0.1				57.1	6.8	17.9				9.1	0.7	0.2			
Dendobranchiata																100	66.7	66.7
Brachyura	16.7	10.4	5.5							7.7	1.1	0.4	9.1	0.3	0.1			
Leucosiidae										7.7	9.9	3.7						
Parthenopidae										7.7	0.4	0.1						
Portunidae	4.2	11.7	1.5				28.6	30.1	39.8	7.7	14.2	5.3	36.4	50.9	67.1			
Achelous spinicarpus	37.5	72.2	85.7				14.3	1.1	0.7	15.4	23.9	17.7	18.2	45.0	29.7			
Stomatopoda										15.4	0.6	0.5						
Crustacean fragments	29.2	0.2	0.1				14.3	0.0	0.0	7.7	1.0	0.4	45.5	0.3	0.5	100	33.3	33.3
Teleosts	50.0	4.4	2.7	100	23.8	23.8	71.4	62.1	57.7	84.6	48.8	63.7	54.5	1.9	1.0			
Dactylopterus volitans							14.3	36.0	23.8	23.1	40.5	45.1	9.1	0.8	0.2			
Teleost fragments	50.0	4.4	6.9	100	23.8	23.8	28.6	1.0	1.3	69.2	8.0	26.6	45.5	1.1	1.8			
Pleuronectiformes							14.3	4.9	3.3									
Symphurus sp.							14.3	20.2	13.3									
Polydactylus sp.										7.7	0.4	0.1						
Elasmobranchs													9.1	0.8	0.1			
Rajidae													9.1	0.8	0.3			
Polychaeta										7.7	0.0	0.0	18.2	0.1	0.0			
Euclymene sp.										7.7	0.0	0.0						
Sthenelais sp.													18.2	0.1	0.1			
Nematoda	4.2	0.0	0.0							7.7	0.0	0.0	9.1	0.0	0.0			
Nematods	4.2	0.0	0.0							7.7	0.0	0.0	9.1	0.0	0.0			

had a high Frequency of Occurrence (% FO), particularly in teleost fragments. These values are small because of the low% P and prey-specific abundance. Therefore we cannot estimate the true importance of this group. As previously observed by ZAVALA-CAMIN (1996), Fish are digested more rapidly than other groups that have a carapace, such as crustaceans.

A diet based on crustaceans and teleosts was also observed for a congeneric species, and for *A. cyclophora* by other authors (SOARES *et al.* 1992, BIZARRO *et al.* 2007). In a study of the diet of *A. cyclophora* in Ubatuba (23°20'S-24°00'S, 44°30'W-45°30'W), São Paulo, Brazil, the most important items were crustaceans and teleosts, and Brachyura was also relevant (SOARES *et al.* 1992). In the same study, the feeding habits of *Raja castelnaui* (= *A. castelnaui*) were also analyzed and showed greater representation of crustaceans and teleosts. The diets of *Raja binoculata, R. inornata*, and *R. rhina* were studied in California (USA), and crustaceans and fish were the most important groups (BIZARRO *et al.* 2007).

The diets of juveniles and adults overlapped significantly. However, teleost and *A. spinicarpus* are more important in the diet of adults and Caridea/Dendobranchiata, polychaetes and nematodes are more important in the diet of juveniles. The diets of both sexes were also very similar. Nevertheless, teleost and crabs are slightly more frequent in the diet of females, while males tend to eat more shrimp. These small differences can be attributed to differences in the method of catching food. It is difficult for smaller individuals to capture teleosts, which are more agile (VIANNA *et al.* 2000).

The temporal analysis of feeding may have been compromised by the low abundance of individuals at certain times of the year, as for instance in Apr-May-Jun 06 and Jul-Aug-Sep 07. However, analysis of the other quarters indicated that there are two groups with similar diets. The water masses found in the region modify the physical and chemical parameters of the water during the seasons, which suggests a change in the structure and dynamics of the benthic fauna in the region (PIRES 1992). Portunidae and *A. spinicarpus* were the most important groups consumed in Jan-Feb-Mar 06 and Jan-Feb-Mar 07, i.e., the summer of 2006 and the summer of 2007 (when the SACW approaches the coast), which may be explained by the increased availability of *A. spinicarpus* in this region at this time of year (BRAGA *et al.* 2005). At Ubatuba, *A. spinicarpus* is also important in the food of other rajids such as *R. agassizi* and *P. extenta*, especially in the summer, which is related to the local abundance of the species (MUTO *et al.* 2001). Therefore, the period of greatest abundance of *A. spinicarpus* corresponds to the optimal oceanographic conditions for *A. cyclophora*. The upwelling of SACW may explain the importance of this crab in the diet of the skate. Seasonal changes in diet were observed in other studies of rajids (MUTO *et al.* 2001, BORNATOWSKI *et al.* 2010).

Although this study did not analyze a large number of stomachs, the data are important for interpreting the feeding biology of this skate, and to evaluate its responses to environmental conditions and fishing pressures.

ACKNOWLEDGEMENTS

The authors are grateful to the Laboratório de Biologia e Tecnologia Pesqueira group for the help in samples, measurements and dissection of the skates, to Paulo C. Paiva for helping with polychaete identification, and to Tereza C.G. da Silva and Karina A. Keunecke for helping with crustacean identification.

REFERENCES

AGUIAR, A.A. & J.L. VALENTIN. 2010. Biologia e Ecologia Alimentar de Elasmobrânquios no Brasil. **Oecologia Australis 14** (2): 464-489. doi:10.4257/oeco.2010.1402.09

AMUNDSEN, P.A.; H.M. GABLER & F.J. STALDVIK. 1996. A new approach to graphical analysis of feeding strategy from stomach contents data-modification of the Costello (1990) method. **Journal of Fish Biology 48:** 607-614. doi:10.1111/j.1095-8649.1996.tb01455.x

BIZARRO, J.J.; H.J. ROBINSON; C.S. RINEWALT & D.A. EBERT. 2007. Comparative feeding ecology of four sympatric skate species off central California, USA. **Environmental Biology of Fishes 80** (2-3): 197-220. doi:10.1007/s10641-007-9241-6

BORNATOWSKI, H.; M.C. ROBERT & L. COSTA. 2010. Feeding of guitarfish *Rhinobatos percellens* (Walbaum, 1972) (Elasmobranchii, Rhinobatidae), the target of artisanal fishery in Southern Brazil. **Brazilian Journal of Oceanography 58** (1): 45-52. doi:10.1590/S1679-87592010000100005

BRAGA, A.A.; A. FRANSOZO; G. BERTINI & P.B. FUMIS. 2005. Composição e abundância dos caranguejos (Decapoda, Brachyura) nas regiões de Ubatuba e Caraguatatuba, litoral norte paulista, Brasil. **Biota Neotropica 5** (2). doi:10.1590/S1676-06032005000300004

CHEUNG, W.W.L.; T.J. PITCHER & D. PAULY. 2005. A fuzzy logic expert system to estimate intrinsic extinction vulnerabilities of marine fishes to fishing. **Biological Conservation 124:** 97-111. doi:10.1016/j.biocon.2005.01.017

CORTES, E. 1997. A critical review of methods of studying fish feeding based on analysis of stomach contents: application to elasmobranch fishes. **Canadian Journal of Fisheries and Aquatic Sciences 54:** 726-738. doi: 10.1139/f2012-051

COSTELLO, M.J. 1990. Predator feeding strategy and prey importance: a new graphical analysis. **Journal of Fish Biology 36:** 261-263. doi:10.1111/j.1095-8649.1990.tb05601.x

EBERT, D.A & J.J BIZARRO. 2007. Standardized diet compositions and trophic levels of skates (Chondrichthyes: Rajiformes: Rajoidei). **Environmental Biology of Fishes 80:** 221-237. doi:10.1007/s10641-007-9227-4

FERRY, L.A & G.M. CAILLIET. 1996. Sample size and data analysis: are we characterizing and comparing diet properly?, p. 71-80. *In:* D. MACKINLAY & K. SHEARER (Eds). **Feeding Ecology and Nutrition in Fish.** São Francisco, American Fisheries Society.

GOMES, U.L.; C.N. SIGNORI; O.B.F. GADIG & H.R.S. SANTOS. 2010. **Guia para identificação de tubarões e raias do Rio de Janeiro.** Technical Books, Rio de Janeiro. 234p.

HURLBERT, S.H. 1978. The measurement of niche overlaps and some relatives. **Ecology 59:** 67-77. doi:10.2307/1936632

KAWAKAMI, E. & G. VAZZOLER. 1980. Método gráfico e estimativa de índice alimentar aplicado no estudo de alimentação de peixes. **Boletim do Instituto Oceanográfico 29** (2): 205-207. doi: 10.1590/S0373-55241980000200043

LACERDA, L.D.; H.H.M. PARAQUETTI; R.V. MARINS; C.E. REZENDE; I.R. ZALMON; M.P. GOMES & V. FARIAS. 2000. Mercury content in shark species from the south-eastern Brazilian coast. **Revista Brasileira de Biologia 60** (4): 571-576. doi: 10.1590/S0034-71082000000400005

MASSA, A.; N. HOZBOR & C.M. VOOREN. 2006. *Atlantoraja cyclophora.* *In:* IUCN (Ed.). **Red List of Threatened Species.** Version 2009.1. Available online at: http://www.iucnredlist.org [Accessed: 21 july 2009]. doi: 10.1007/978-1-4020-9703-4_8

MUTO, E.Y.; L.S.H. SOARES & R. GOITEIN. 2001. Food resource utilization of the skates *Rioraja agassizi* (Müller & Henle, 1841) and *Psammobatis extenta* (Garman, 1913) on the Continental shelf off Ubatuba, south-eastern Brazil. **Revista Brasileira de Biologia 61** (2): 217-238. doi: 10.1007/978-1-4020-9703-4_8

ODDONE, M.C & C.M. VOOREN. 2004. Distribution, abundance and morphometry of *Atlantoraja cyclophora* (Regan, 1903) (Elasmobranchii: Rajidae) in southern Brazil, Southwestern Atlantic. **Neotropical Ichthyology 2** (3): 137-144. doi: 10.1590/S1679-62252004000300005

ODDONE, M.C. & C.M. VOOREN. 2005. Reproductive biology of *Atlantoraja cyclophora* (Regan 1903) (Elasmobranchii: Rajidae) off southern Brazil. **ICES Journal of Marine Science 62:** 1095-1103.

ODDONE, M.C.; W. NORBIS; P.L. MANCINI & A.F. AMORIM. 2008. Sexual development and reproductive cycle of the Eyespot skate *Atlantoraja cyclophora* (Regan, 1903) (Condrichthyes: Rajidae: Arhynochobatinae), in southeastern Brazil. **Acta Adriatica 49** (1): 73-87. doi: 10.1016/j.icesjms.2005.05.002

PIANKA, E.R. 1973. The structure of lizard communities. **Annual Review of Ecology, Evolution and Systematics 4:** 53-74. doi: 10.1146/annurev.es.04.110173.000413

PIRES, A.M.S. 1992. Structure and dynamics of benthic megafauna on the continental shelf offshore of Ubatuba, southeastern Brazil. **Marine Ecology Progresses Series 86:** 63-76. doi: 10.3354/meps086063

ROSECCHI, E. & Y. NOUAZE. 1987. Comparison de cinq índices alimentaires utilises dans lánalyse des contenus stomacaux. **Revue des Travaux de L'institut des Pêches Maritimes 49** (4): 111-123.

SOARES, L.S.H.; C.L.B. ROSSI-WONGTSCHOWSKI; L.M.C. ALVARES; E.Y. MUTO & M. LOS ANGELES. 1992. Grupos tróficos de peixes demersais da plataforma continental interna de Ubatuba, Brasil. I. Condrichthyes. **Boletim do Instituto Oceanográfico 40** (1/2): doi: 10.1590/S0373-55241992000100006

VIANNA, M.; C.A. ARFELLI & A.F. AMORIM. 2000. Feeding of *Mustelus canis* (Elasmobranchii, Triakidae) caught off south-southeast coast of Brazil. **Boletim do Instituto de Pesca 26** (1): 79-84.

VOOREN, C.M. & S. KLIPPEL. 2005. Diretrizes para a conservação de espécies ameaçadas de elasmobrânquios, p. 213-228. *In:* C.M. VOOREN & S. KLIPPEL (Eds). **Ações para a conservação de tubarões e raias do Brasil.** Porto Alegre, Editora Igaré.

WALKER, P.A & G. HISLOP. 1998. Sensitive skates or resilient rays? Spatial and temporal shifts in ray species composition in the central and north-western North Sea between 1930 and the present day. **ICES Journal of Marine Science 55:** 392-402. doi:10.1006/jmsc.1997.0325

ZAVALA-CAMIN, L.A. 1996. **Introdução aos estudos sobre alimentação natural em peixes.** Maringá, EDUEM, 129p.

23

Morphology and 18S rDNA gene sequence of *Spirostomum minus* and *Spirostomum teres* (Ciliophora: Heterotrichea) from Rio de Janeiro, Brazil

Noemi M. Fernandes[1,2] & Inácio D. da Silva Neto[1]

[1] *Laboratório de Protistologia, Departamento de Zoologia, Universidade Federal do Rio de Janeiro. 21941-590, Rio de Janeiro, Rio de Janeiro, Brazil.*
[2] *Corresponding author. E-mail: noemi.fernandes@yahoo.com.br*

ABSTRACT. Species of *Spirostomum* Ehrenberg, 1838 are widely used as model organisms in ecological studies of environmental impacts and symbioses between ciliates and human pathogenic bacteria. However, the taxonomy of this genus is confused by the superficiality of the morphological descriptions of its included species, and the use of only a few characters for their differentiation. The present study provides details of total infraciliature, nuclear apparatus, morphometric data and 18S rDNA gene sequences of *Spirostomum teres* Claparède & Lachmann, 1858 and *Spirostomum minus* Roux, 1901, isolated from a sewage treatment plant and a freshwater lake in the city of Rio de Janeiro, Brazil, respectively. For the morphological descriptions of *S. teres* and *S. minus*, living cells were observed using bright-field and differential interference contrast (DIC) microscopy, the total infraciliature and nuclear apparatus were revealed by staining with protargol, and ciliary patterns were observed also with scanning electron microscopy (SEM). The complete sequences of the 18S rDNA of *S. teres* and *S. minus* were obtained using *eukaryotic* universal *primers, and then compared with sequences of other species and populations of Spirostomum deposited in the GenBank database. Living S. minus measured* 400-800 µm in length and 55-115 µm in width, with the following characteristics: adoral zone of membranelles approximately 112 µm long; inconspicuous paroral kinety; 30-40 kineties in somatic ciliature; moniliform macronucleus with 9-25 nodes, approximately 12 micronuclei; single and posterior contractile vacuole; and yellow-brown cytoplasm. Living and fully extended *S. teres* measured approximately 250 µm in length and 65 ìm in width, with the following characteristics: adoral zone of membranelles approximately 92 µm long; approximately 30 somatic kineties; compact macronucleus, approximately five micronuclei; macronuclear groove present; single and posterior contractile vacuole; and colorless cytoplasm. Evidence from 18S rDNA sequences confirms the identification of *S. teres* and suggests the existence of cryptic species closely related to *S. minus*. The use of silver impregnation technique (protargol) allowed the observation and description of a greater number of characters in *S. minus* and *S. teres*, thus assisting the research that require identification of these species.

KEY WORDS. Bioindicators; ciliates; cryptic species; heterotrichs; morphology.

Spirostomum Ehrenberg, 1838 are conspicuous ciliates protists that are easily recognized by their large sizes (500-1000 µm) and elongate bodies, being easily confounded with small helminths. The name *Spirostomum* refers to the ability these ciliates have to contract in a spiral mode. This type of contraction is due to the presence of post-ciliary, sub-pellicular fibers that arise on the anterior end and spiral in a counterclockwise direction toward the posterior end of the body (ISHIDA *et al.* 1988).

Spirostomum was mentioned for the first time by EHRENBERG (1838) and morphological studies have been subsequently carried out for some species (PACKARD 1948, FINLEY *et al.* 1964, DANIEL & MATTERN 1965, TUFFRAU 1967, KUDO 1971). However, these studies were based exclusively on observations of specimens *in vivo*. REPAK & ISQUITH (1974) emphasized the difficulty of separating and identifying species of *Spirostomum* owing to the superficial-

ity of their morphological descriptions and the use of only a few characters for species differentiation. Moreover, most of those characters vary significantly within species (REPAK & ISQUITH 1974), making identification even more difficult. For this reason, the taxonomy of *Spirostomum* has become extremely confusing. More detailed morphological and morphometric studies that increase the number of characters available for species identification are needed to solve this problem. Descriptions should include observations on infraciliary patterns and ultrastructural characters.

Currently, species of *Spirostomum* are identified using only morphological features such as the shape and size of the cell, ratio of the total length/length of the adoral zone of membranelles, location of the contractile vacuole, and configuration of the macronucleus (FOISSNER *et al.* 1992, BERGER *et al.* 1997). However, it should be noted that it is easy to misidentify some species of *Spirostomum* (Foissner *et al.* 1992). In recent decades, sequences of the 18S small-subunit rRNA gene have been widely used in the phylogenetic study of ciliates, and also in species identification (Lynn & Small 2002, Gong *et al.* 2007, Li *et al.* 2009). The small-subunit rRNA sequences of European populations of *Spirostomum minus* Roux, 1901and *Spirostomum teres* Claparède & Lachmann, 1858 were obtained by Schmidt *et al.* (2007) and have been used in recent phylogenetic studies of the Heterotrichea by several authors.

Species of *Spirostomum* have been used currently as model organisms in research on human pathogenic bacteria (Fokin *et al.* 2003, 2005). Some species (e.g., *S. minus* and *S. teres*) have also been used in studies of environmental impacts because they are considered good indicators of water quality (Foissner *et al.* 1992, Berger *et al.* 1997, Berger & Foissner 2003) and show sensitivity to certain toxic substances (e.g. heavy metals like nickel, copper, mercury, and zinc; the phenol Na-PCP) (Madoni *et al.* 1992, Madoni 2000). Nevertheless, studies on the morphology and taxonomy of *Spirostomum* species are quite rare (Packard 1948, Finley *et al.* 1964, Daniel & Mattern 1965, Tuffrau 1967, Kudo 1971, Repak & Isquith 1974, Foissner *et al.* 1992).

In the present work, morphological descriptions and morphometric data of *S. teres* and *S. minus* isolated from a sewage treatment plant and a freshwater lake in the municipality of Rio de Janeiro, Brazil, are presented. Descriptions of morphological characters observed on living and protargol-stained organisms, which show details of the nuclear apparatus and infraciliature, are included. Images obtained by electron microscopy (SEM) are presented to complement the descriptions and 3D visualization of characters of the cell surface. The small-subunit 18S rDNA sequences of *S. minus* and *S. teres* from Rio de Janeiro are also presented and characterized.

MATERIAL AND METHODS

Specimens of *Spirostomum teres* and *S. minus* were isolated from samples collected in a sewage treatment plant in Penha, Rio de Janeiro (22°52′S, 43°13′W), and from a freshwater lake located at the Universidade Federal do Rio de Janeiro (22°51′S, 43°13′W), respectively. Samples from both localities were obtained manually using plastic containers and then maintained in the laboratory for several days at room temperature (about 24°C) as a raw culture for studies as proposed by Foissner (1992). Living cells were picked out from cultures and observed using bright-field and differential interference contrast microscopy. The infraciliature and nuclear apparatus were revealed by staining with protargol according to the method of Dieckmann (1995). Specimens were examined at magnifications of 100-1000x. Patterns of somatic and oral ciliature were observed with scanning electron microscopy according to the methodology of Silva-Neto *et al.* (2012). Measurements were made with Image Pro-plus 5.0® (Media Cybernetics, Inc., Bethesda, MD) using an Olympus BX 51 compound microscope.

Slides of protargol-stained *S. minus* and *S. teres* were deposited in collection of the Laboratório de Protistologia (Universidade Federal do Rio de Janeiro), with registry numbers: IBZ-UFRJ0017-X and IBZ-UFRJ0018-X, respectively.

Extraction of genomic DNA from *S. minus* and *S. teres*, PCR amplification of the genes for 18S rRNA, and sequencing of rDNA were performed according to the method described by Schmidt *et al.* (2007). The gene for 18S rRNA was amplified using the *eukaryotic* universal *primers EukA/EukB* (Medlin *et al.* 1988) that extend over the full length of the gene. Sequences of *S. minus* and *S. teres* were edited into contigs using BioEdit (Hall 1999) and deposited at the NCBI/GenBank database with the accession numbers JQ282896 and JQ282897, respectively. These two sequences were compared with the 18S rDNA sequences in the GenBank database of *S. minus* (AM398200), *S. teres* (AM398199) and *Spirostomum ambiguum* Ehrenberg, 1838 (AM398201) obtained by Schmidt *et al.* (2007) using the BLASTN algorithm. The frequencies of individual nucleotides and GC composition of sequences of *S. minus* and *S. teres* from Rio de Janeiro were obtained using CompSeq (Rice *et al.* 2000). The similarity structural matrix and absolute distance matrix between *Spirostomum* species was generated with BioEdit (Hall 1999) and PAUP* 4.0, respectively.

RESULTS

Morphology of *Spirostomum minus* Roux, 1901

Living and fully extended organisms have vermiform, elongated bodies, tapered at the ends and with a total length of 400-800 µm (Fig. 1). Contracted organisms have ellipsoid bodies (Figs 3, 4 and 6), a total length of 175-262 µm (mean = 206 µm) and a width of 55-115 µm. During contraction, specimens of *S. minus* twist their oral and somatic ciliature in a counterclockwise direction (Fig. 5). The adoral zone of membranelles (AZM) was 112µm (64-142 µm) long and extended from the anterior end of the ciliate to a cytostome located approximately at the midpoint of the body (Figs 3, 4 and 6). In contracted specimens, the AZM made a counter-clockwise turn around the body (Figs 3, 4 and 6-8). Inconspicuous paroral kinety located to the left of the AZM (Figs 3 and 4). Somatic ciliature composed of an average of 35 (range 30-40) kineties (Fig. 5). In contracted individuals, torsion of the cell caused somatic kineties to be oriented obliquely in relation to the central axis of the body (Fig. 5). Moniliform macronucleus, with 9-25 nodes connected by nuclear bridges (Fig. 3); approximately 12 micronuclei (mean diameter 1.8 µm) near or overlapping the macronucleus (Fig. 3). Single contractile vacuole located at the posterior end of the body (Figs 1 and 2). Living organisms were yellow-brown in color when observed under low magnification (Fig. 1). This coloration was due to the presence of pale brown cortical granules (approx. 0.5 µm) that were densely packed between the kineties (Fig. 2). Individuals moved by gliding slowly over the substrate or by swimming freely above it (Table I).

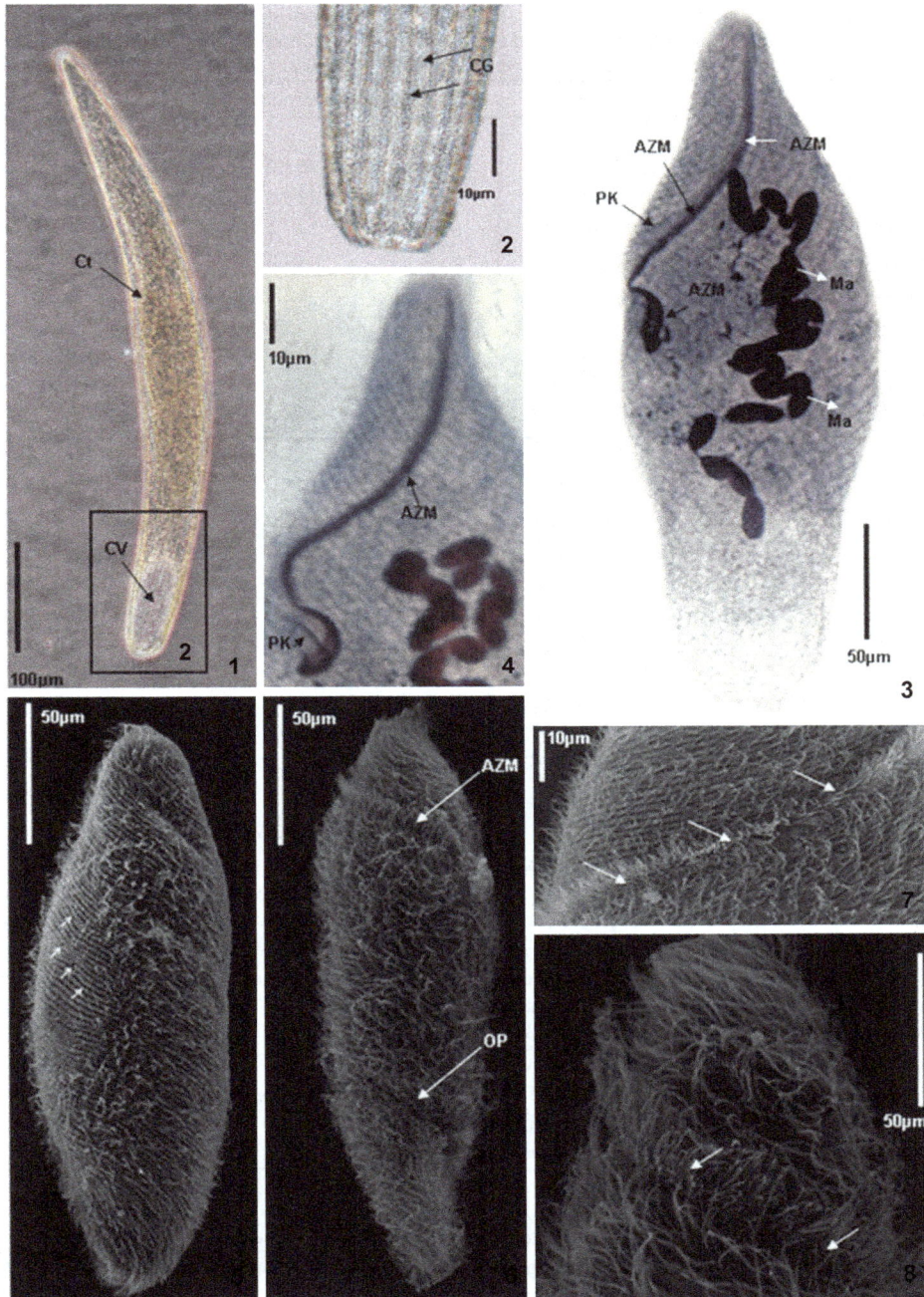

Figures 1-8. Morphology of *Spirostomum minus*: (1) image obtained from phase contrast microscopy of the living organism showing general shape of the cell, cytostome, and single contractile vacuole in posterior end of the body; (2) detail of posterior end of cell showing cortical granules; (3) ventral view of *S. minus* after staining with protargol showing nuclear apparatus, somatic kineties, and oral infraciliature; (4) closeup of oral region showing the arrangement of adoral zone of membranelles and paraoral kinetie; (5) scanning electron micrography showing general shape of the cell in dorsal view and somatic kineties (arrows) obliquely twisted in relation to the central axis by contraction; (6) dividing cells showing morphogenesis of the oral primordial of daughter cells; (7-8) scanning electron micrography showing the general morphology of the oral region and adoral zone of membranelles (arrows). (AZM) Adoral zone of membranelles, (PK) paroral kinety, (CV) contractile vacuole, (Ma) macronucleus, (Ct) cytostome, (CG) cortical granules, (OP) oral primordium.

Table I. Morphometric data of *Spirostomum minus* and *S. teres* from sewage treatment system by activated sludge in Rio de Janeiro. All measures were performed on specimens protargol impregnated and are given in micrometer (μm). (SD) standard deviation, (SE) standard error, (CV) coefficient of variation, (n) number of specimens investigated, (AZM) adoral zone of membranelles.

Characters	Mean	Median	Minimum	Maximum	SD	SE	CV (%)	n
Spirostomum minus								
Body length	206.0	200.0	175.0	262.0	23.5	4.7	0.1	25
Body width	82.0	82.0	55.0	115.0	15.1	3.0	0.1	25
Length of AZM	112.0	111.0	64.0	142.0	19.3	3.8	0.1	25
Total length of macronucleus	232.0	228.0	125.0	362.0	60.9	12.1	0.2	25
Width of macronucleus	8.0	8.0	5.0	12.0	1.8	0.3	0.2	25
Diameter of micronuclei	1.8	1.9	1.3	3.0	0.3	0.0	0.1	25
Number of micronuclei	12.0	8.0	3.0	30.0	3.8	2.7	0.3	25
Number of macronuclear nodes	17.0	17.0	9.0	25.0	4.4	0.8	0.2	25
Number of somatic kineties	35.0	34.0	30.0	40.0	1.1	0.6	0.1	25
Spirostomum teres								
Body length	165.0	167.0	93.0	196.0	25.4	4.6	0.1	30
Body width	66.0	61.0	38.0	96.0	15.5	2.8	0.2	30
Length of AZM	92.0	93.0	56.0	114.0	15.3	2.8	0.1	30
Total length of macronucleus	38.0	38.0	20.0	52.0	6.2	1.1	0.1	30
Width of macronucleus	12.0	13.0	7.0	16.0	2.6	0.4	0.2	30
Diameter of micronuclei	2.0	2.0	1.0	2.9	0.4	0.0	0.2	30
Number of micronuclei	5.0	5.0	1.0	11.0	2.9	0.5	0.5	30
Number of macronuclear nodes	1.0	1.0	1.0	1.0	0.0	0.0	0.0	30
Number of somatic kineties	30.0	40.0	22.0	37.0	5.9	0.1	0.1	30

Morphology of *Spirostomum teres*

Fully extended living organisms had a rounded posterior end and tapered anterior end. They were 150-250 μm in length and approximately 65 ìm in width (Fig. 9). Contracted cells had an ellipsoid body, with slightly tapered ends (Figs 10, 12 and 13). The oral and somatic ciliature of *S. teres* twisted in a counterclockwise direction during contraction (Fig. 10). The adoral zone of membranelles (AZM) was approximately 92 μm (56-114 μm) long and extended from the anterior end of the ciliate to the cytostome, which was located in the anterior third of the body (Figs 10-14). In contracted cells, the AZM made a counterclockwise turn around the body (Figs 10-14). Inconspicuous paroral kineties were located to the left of the AZM (Fig. 14). Somatic ciliature composed of about 30 kineties (Fig. 10). In contracted individuals, torsion of the cell caused somatic kineties to be oriented obliquely in relation to the central axis of the body (Fig. 10). Nuclear apparatus composed of single, compact macronucleus located approximately in the center of the body (Fig. 12 and 13). In some cells, macronuclear grooves that hold micronuclei were observed (Fig. 15). Approximately 5 micronuclei (5 μm in diameter) were dispersed throughout the cytoplasm or overlapping the macronucleus. A single, conspicuous contractile vacuole occupied almost the entire posterior third of the body (Fig. 9). Live cells were colorless (Fig. 9), and moved slowly, gliding over the substrate (Table I).

Molecular characterization

The nucleotide composition and GC content of the 18S rDNA gene sequences of the *S. minus*, *S. teres* and *S. ambiguum* obtained by SCHMIDT *et al.* (2007) and the sequences of *S. minus* and *S. teres* obtained in this study are presented for comparison in Table II. The 18S rDNA sequences confirmed the morphological identification of *S. teres* collected in Brazil. This sequence is composed by 1,558 bp being more similar to that of *S. teres* (98.7%), obtained by SCHMIDT *et al.* (2007), differing from it by 10 nucleotides (Table III).

Table II. Nucleotidic composition and GC content of *S. minus*, *S. teres* and *S. ambiguum* by SCHMIDT *et al.* (2007) and *S. minus* and *S. teres* from Rio de Janeiro. Data obtained from an alignment of 1312 bp. All values are given in percent.

Species	Adenine	Cytosine	Guanine	Thymine	GC content
S. minus	26.91	19.21	27.67	26.22	46.88
S. minus RJ	26.96	19.35	27.65	25.97	46.99
S. teres	26.98	19.28	27.82	25.91	47.10
S. teres RJ	26.84	19.26	27.98	25.93	47.23
S. ambiguum	26.98	19.36	27.52	26.14	46.88

The sequence from *S. minus* obtained in the present study consisted of 1,348 bp and 46.99% of the GC content. This se-

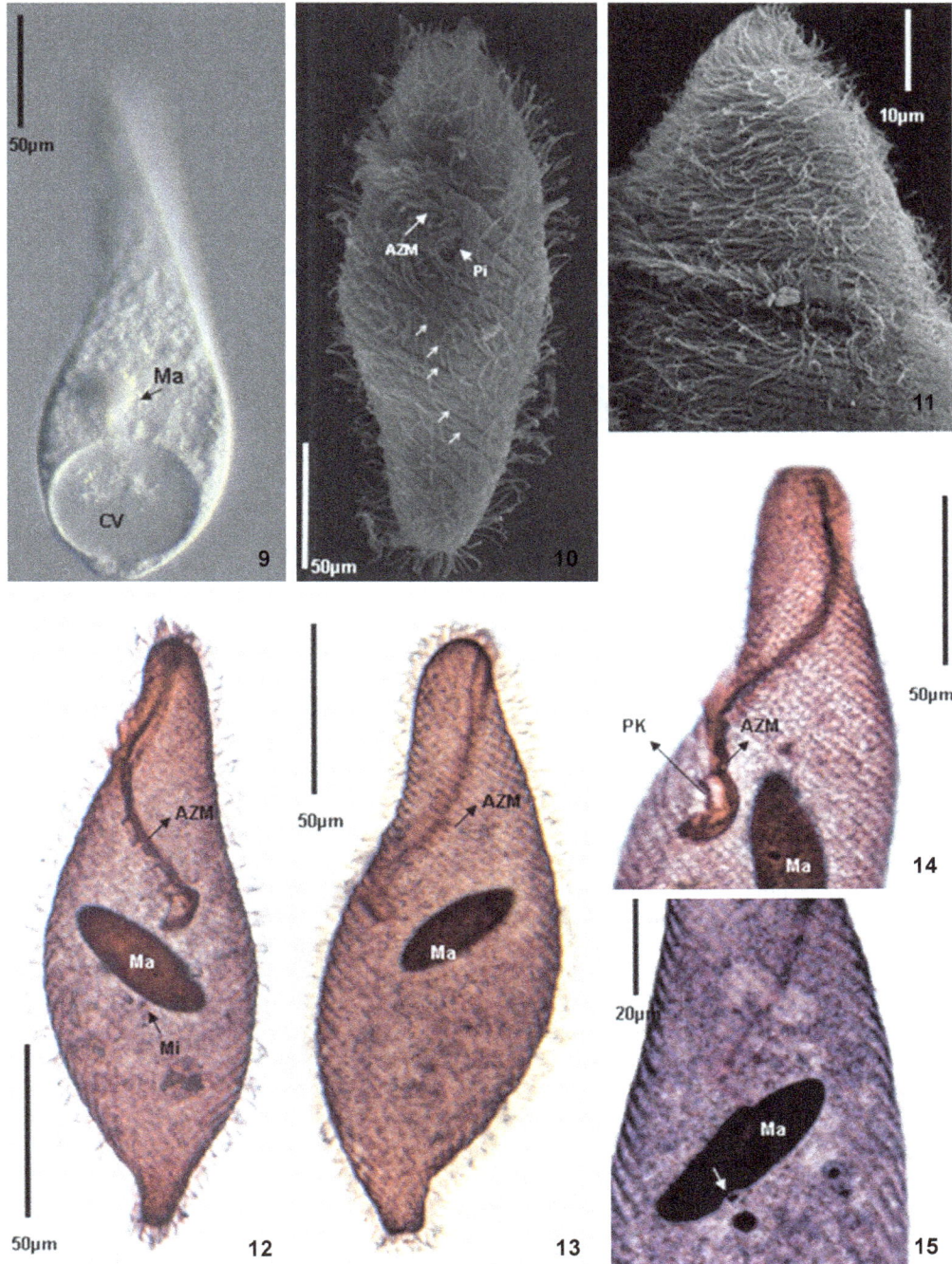

Figures 9-15. Morphology of *Spirostomum teres* (9) image obtained with differential interference contrast (DIC) microscopy of the living organism showing general shape of the cell, compact macronucleus and single contractile vacuole; (10) scanning electron micrography in ventral view showing adoral zone of membranelles, peristomial infundibulum and somatic kineties (arrows) obliquely twisted; (11) closeup of anterior end showing part of the adoral zone of membranelles. (12-15) Protargol-stained cells: (12) ventral view showing adoral zone of membranelles and nuclear apparatus; (13) dorsal view; (14) closeup of oral ciliature showing adoral zone of membranelles and paroral kinety; (15) macronuclear groove and micronuclei (arrow). (AZM) Adoral zone of membranelles, (PK) paroral kinety, (CV) contractile vacuole, (Ma) macronucleus, (Mi) micronuclei, (Pi) peristomial infundibulum.

quence showed 97.3% of similarity to the sequence from *S. minus* obtained by SCHMIDT *et al.* (2007) and 31 different nucleotide sites (Table III). The relatively low percentage of similarity and the high number of nucleotide differences between these two sequences suggest that they are cryptic species. The pairwise distances (absolute distance) and differences between 18S rDNA sequences of *Spirostomum* populations/species are given in Table III.

Table III. The structural similarities (%) of the 18S rDNA gene sequences of *S. minus* and *S. teres* from Rio de Janeiro (RJ), and others species/populations of *Spirostomum* species. sequenced. Absolute distance (number of nucleotidic differences) is shown in parenthesis.

Species	*S. minus*	*S. minus* RJ	*S. ambiguum*	*S. teres*
S. minus	–			
S. minus RJ	0.973 (31)	–		
S. ambiguum	0.982 (23)	0.970 (35)	–	
S. teres	0.978 (28)	0.972 (32)	0.975 (32)	–
S. teres RJ	0.973 (28)	0.964 (36)	0.971 (31)	0.987 (10)

DISCUSSION

Studies of the morphology of species of *Spirostomum* are rare. The most recent and complete work was presented by FOISSNER *et al.* (1992). In it, aspects of the ecology and morphology of *S. ambiguum*, *Spirostomum caudatum* Delphy, 1939, *S. minus* and *S. teres* were described. The reviews of REPAK & ISQUITH (1974) and FOISSNER *et al.* (1992) were used by us as a basis for species identification.

Of the nine species that currently comprise *Spirostomum*, five (including *S. minus*) possess a moniliform macronucleus. These species are distinguished mainly by the shape of the body extremities and the size of the peristome (REPAK & ISQUITH 1974). The other species of moniliform macronucleus differ from *S. minus* in the following characteristics (Table IV): *Spirostomum inflatum* Kahl, 1932 is a marine organism with a wider or "inflated" posterior end, whereas *S. minus* is a freshwater ciliate with a slightly tapered posterior end. Specimens of *Spirostomum loxodes* Stokes, 1885 have a tapered anterior end, with a peristome that occupies one-third of the total length of the body (vs. 1/2 in *S. minus*). *Spirostomum intermedium* Kahl, 1932 has a smaller body (300-400 µm) than other species. The species that most resembles *S. minus* is *S. ambiguum* (Table IV). Those two species are the only ones that have yellow pigments, and the major difference between them is size. Individuals of *S. ambiguum* are larger, reaching a length of up to 4 mm (Table IV), and have a larger peristome and a greater number of macronuclear nodules, somatic kineties, and micronuclei than *S. minus* (Table IV). The strain of *S. minus* described in the present study is very similar to the strains described by REPAK & ISQUITH (1974) and FOISSNER

et al. (1992) (Table IV). All morphological characteristics overlap in these three populations, except the number of somatic kineties, which is greater in the Brazilian population (30-40 vs. 20-24). However, according to REPAK & ISQUITH (1974), the number of somatic kineties is a quite variable characteristic among morphospecies of *Spirostomum*. Thus, based on observation of morphological characters as compared to previous populations described by REPAK & ISQUITH (1974) and FOISSNER *et al.* (1992), and also by comparison with similar species, we conclude that the Brazilian species described is *S. minus*.

Based on molecular data obtained for *S. minus* in the present study and by comparison with 18S rDNA gene sequence of *S. minus* provided by SCHMIDT *et al.* (2007) (single sequence available in GenBank database for this species), we observed a relatively low percentage of similarity (97.3%) between these two populations and 31 different nucleotide sites. This fact suggests the possibility of cryptic species within the *S. minus* complex. However, morphological characters of *S. minus* were not presented by SCHMIDT *et al.* (2007), therefore it is not possible to compare both populations morphologically. There is an evident need for sequencing a greater number of *S. minus* morphotypes in order to ascertain the existence of cryptic species.

Among all species of *Spirostomum*, only *S. teres* and *Spirostomum ephrussi* Delphy, 1939 have a compact macronucleus. These two species also share other similar characteristics such as length, number of kineties and number of micronuclei (Table V). However, there have been no new observations or new reports on *S. ephrussi* since its original description. Thus, based on morphological similarity REPAK & ISQUITH (1974) considered *S. ephrussi* as junior synonym of *S. teres*. The population of *S. teres* described in the present study is very similar to the population studied by FOISSNER *et al.* (1992), but the number of micronuclei is greater in the former (Table V). All other morphological characteristics observed for the Brazilian strain of *S. teres* overlap with the population described previously (Table V) and confirm the identity of the species.

The phase of the life cycle must be taken into account in comparative studies of *Spirostomum* species. PACKARD (1948) performed a morphological study of the nuclear apparatus of *S. teres* and observed a relationship between age since division and number of macronuclear nodules. Specimens that have recently gone through division or conjugation have fewer nodules, and micronuclei with smaller diameters compared with more mature cells. Internal factors (e.g., cytoplasmatic pressure) or external (e.g., chemical composition) can also alter the shape of the macronucleus in *S. teres* (PACKARD 1948). In the present study, the presence of a macronuclear groove housing one micronucleus was observed in some specimens of *S. teres* (Fig. 15). This characteristic was also observed in populations of *S. teres* by PACKARD (1948).

Many authors state in their descriptions that *Spirostomum* species lack a paroral membrane (TUFFRAU 1967, REPAK & ISQUITH 1974, ISHIDA *et al.* 1988). However, FERNANDEZ-LEBORANS (1985)

Table IV. Morphological characteristics of populations of *Spirostomum minus* and *S. ambiguum*.

	S. ambiguum	S. ambiguum	S. minus	S. minus	S. minus
Total length (μm)	1000-4000	1000-4000	500-800	300-800	400-800
Length of peristome	2/3 of body	–	1/2 of body	–	1/2 of body
Color	yellow	yellow	yellow	yellow	yellow
Macronuclear shape	moniliform	moniliform	moniliform	moniliform	moniliform
Number of macronuclear nodes	12-50	10-50	24	8-50	9-25
Number of micronuclei	12-100	–	4-20	–	3-30
Number of kineties	46	70-90	20-24	20-24	30-40
Reference	Repak & Isquith (1974)	Foissner *et al.* (2002)	Repak & Isquith (1974)	Foissner *et al.* (2002)	Present study

Table V. Morphological comparisons between similar species and populations of *Spirostomum teres*.

	S. ephrussi	S. teres	S. teres	S. teres
Total lenght (μm)	450	150-400	150-600	150-250
Lenght of peristoma	3/5 of body	1/2 of body	1/2 of body	1/2 of body
Color	–	colorless	colorless	colorless
Macronuclear shape	compact	ellipsoid	ellipsoid	ellipsoid
Number of macronuclear nodes	1	1	1	1
Number of micronuclei	–	–	1-2	1-11
Number of kineties	–	14-24	25-30	22-37
Reference	Repak & Isquith (1974)	Repak & Isquith (1974)	Foissner *et al.* (2002)	Present study

used silver-staining to confirm the presence of a paroral kinety consisting of a single row of cilia on the right margin of the peristome in *S. teres*. A paroral kinety was visible in cells of *S. minus* and *S. teres* stained with protargol in the present study (Figs 4 and 14), corroborating that observation. This paroral kinety may correspond to the paroral membrane of other heterotrichs, but a morphogenetic study is needed to verify their homology with one another.

Spirostomum species are excellent model organisms and suitable bioindicators for microbiological, ecological, environmental, and ecotoxicological analyses (Madoni *et al.* 1992, Berger *et al.* 1997, Madoni 2000, Berger & Foissner 2003). Fast, accurate characterization of species for environmental analyses requires molecular approaches that can complement or even completely replace traditional morphological methods, which are often resource- and time-consuming. However, the gene coding the 18S rRNA has been sequenced for only three species of *Spirostomum*, which emphasizes the need for molecular characterization of more *Spirostomum* species to be used in studies of environmental impact.

The use of silver-staining allowed the observation and description of a greater number of characters in *S. minus* and *S. teres*. The 18S rDNA sequences confirm the identification of *S. teres* from Brazil and suggest the existence of cryptic species for Brazilian *S. minus* and European population presented by

Schmidt *et al.* (2007). The present study contributed to a better understanding of the morphology of these species, which are widely used as models in ecological studies of environmental impact and in studies of symbiosis between ciliates and human pathogenic bacteria.

ACKNOWLEDGEMENTS

This work was supported by CAPES (Coordenação de Aperfeiçoamento de Pessoal de Nível Superior) and BIOTA-FAPERJ, Proc. E-26/110.022/2011 (Fundação de Amparo à Pesquisa do Estado do Rio de Janeiro). We also thank Maximiliano Dias for collection of samples from a sewage treatment plant.

REFERENCES

Berger, H. & W. Foissner. 2003. Illustrated Guide and Ecological Notes to Ciliate Indicator Species (Protozoa, Ciliophora) in Running Waters, Lakes, and Sewage Plants, p. 273-293. *In*: C. Steinberg; W. Calmano; H. Klapper & R.D. Wilken (Eds). **Handbuch Angewandte Limnologie.** Landsberg, Ecomed Verlag, II+160p.

Berger, H.; W. Foissner & F. Kohmann. 1997. **Bestimmung und Ökologie der Mikrosaprobien nach.** Stuttgart, Gustav Fischer, X+291p.

DANIEL, W.A. & C.F. MATTERN. 1965. Some observations on the structure of the peristomial membranelle of *Spirostomum ambiguum*. **Journal of Protozoology 12** (1): 14-27. doi: 10.1111/j.1550-7408.1965.tb01806.x

DIECKMANN, J. 1995. An improved protargol impregnation for ciliates yielding reproducile results. **European journal of Protistology 31** (1): 372-382.

EHRENBERG, C.G. 1838. **Die Infusionsthierchen als vollkommene Organismen.** Leipzig, Leopold Voss, VIII+547p.

FERNANDEZ-LEBORANS, G. 1985. The kinetosomal composition of the adoral zone of membranelles in *Spirostomum teres* and *S. ambiguum* (Ciliophora: Heterotrichida). **Transactions of the American Microscopical Society 104** (1): 129-133.

FINLEY, H.E.; C.A. BROWN & W.A. DANIEL. 1964. Electron microscopy of the ectoplasm and infraciliature of *Spirostomum ambiguum*. **Journal of Protozoology 11** (2): 264-280. doi: 10.1111/j.1550-7408.1964.tb01754.x

FOISSNER, W. 1992. Estimating the species richness of soil protozoa using the "non-ûooded petri dish method", p. B-10.1-B-10.2. *In*: LEE, J.J. & A.T. SOLDO (Eds). **Protocols in Protozoology.** Lawrence, Allen Press, V+652p.

FOISSNER, W.; H. BERGER & F. KOHMANN. 1992. **Taxonomische und ökologische Revision der Ciliaten des Saprobiesystems Band II: Peritrichia, Heterotrichida, Odontostomatida.** Deggendorf, Landesamtes für Wasserwirtschaft, III+502p.

FOKIN, S.I.; M. SCHWEIKERT; H.D. GÖRTZ & M. FUJISHIMA. 2003. Bacterial endocytobionts of Ciliophora. Diversity and some interactions with the host. **European Journal of Protistology 39** (4): 475-480. doi: 10.1078/0932-4739-00023

FOKIN, S.I.; M.F. SCHWEIKERT; F. BRÜMMER & H.D. GÖRTZ. 2005. *Spirostomum* spp. (Ciliophora, Protista), a suitable system for endocytobiosis research. **Protoplasma 225** (1): 93-102. doi: 10.1007/s00709-004-0078-y

GONG, Y.; Y. YU; F. ZHU & W. FENG. 2007. Molecular phylogeny of *Stentor* (Ciliophora: Heterotrichea) based on small subunit ribosomal RNA sequences. **Journal of Eukaryotic Microbiology 54** (1): 45-48. doi: 10.1111/j.1550-7408.2006.00147.x

HALL, T.A. 1999. BioEdit: a user-friendly biological sequence alignment editor and analysis program for Windows 95/98/NT. **Nucleic Acids Symposium Series 41** (1): 95-98.

ISHIDA, H. & Y. SHIGENAKA. 1988. The cell model contraction in the ciliate *Spirostomum*. **Cell Motility and the Cytoskeleton 9** (3): 278-282. doi: 10.1002/cm.970090310.

KUDO, R.R. 1971. **Protozoology.** Illinois, Springfield, 5th ed., IV+1174p.

LI, L.; Q. ZHANG; X. HU; A. WARREN; K.A.S. AL-RASHEID; A.A. AL-KHEDHEIRY & W. SONG. 2009. A redescription of the marine hypotrichous ciliate, *Nothoholosticha fasciola* (Kahl, 1932) nov gen., nov comb. (Ciliophora: Urostylida) with brief notes on its cellular reorganization and SS rRNA gene sequence. **European Journal of Protistology 45** (3): 237-248. doi: 10.1016/j.ejop.2009.01.004.

LYNN, D.H & E.B. SMALL. 2002. Phylum Ciliophora, p. 371-656. *In*: J.J. LEE; P.C BRADBURY & G.F. LEEDALE (Eds). The Illustrated Guide to the Protozoa. Lawrence, Society of Protozoologists, V+1436p.

MADONI, P. 2000. The acute toxicity of nickel to freshwater ciliates. **Environmental Pollution 109** (1): 53-59. doi: 10.1016/S0269-7491(99)00226-2

MADONI, P.; G. ESTEBAN & G. GORBI. 1992. Acute toxicity of cadmium, copper, mercury, and zinc to ciliates from activated sludge plants. **Bulletin of Environmental Contamination and Toxicology 49** (6): 900-905. doi: 10.1007/BF00203165

MEDLIN, L.; H.J. ELWOOD; S. STICKEL & M.L. SOGIN. 1988. The characterization of enzymatically amplified eukariotic 16S-like rRNA-coding regions. **Gene 71** (1): 491-499. doi: 10.1016/0378-1119(88)90066-2

PACKARD, C.E. 1948. The effects of certain chemicals on the macronucleus of *Spirostomum teres*, with notes on the genus. **Transaction of the American Microscopical Society 67** (3): 275-279.

REPAK, A. & I.R. ISQUITH. 1974. The systematics of the genus *Spirostomum* Ehrenberg, 1838. **Acta Protozoologica 12** (1): 325-333.

RICE, P.; I. LONGDEN & A. BLEASBY. 2000. EMBOSS: The European Molecular Biology Open Software Suite. **Trends in Genetics 16** (6): 276-277. doi:10.1016/S0168-9525(00)02024-2

SCHMIDT, S.L.; T. TREUNER; M. SCHLEGEL & D. BERNHARD. 2007. Multiplex PCR Approach for Species Detection and Differentiation within the Genus *Spirostomum* (Ciliophora, Heterotrichea). **Protist 158** (2): 139-145. doi:10.1016/j.protis.2006.11.005

SILVA-NETO, I.D.; T.S. PAIVA; R.J.P. DIAS; C.J.A. CAMPOS & A.E. MIGOTTO. 2012. Redescription of *Licnophora chattoni* Villeneuve-Brachon, 1939 (Ciliophora, Spirotrichea), associated with *Zyzzyzus warreni* Calder, 1988 (Cnidaria,Hydrozoa). **European Journal of Protistology 48** (1): 48-62. doi: 10.1016/j.ejop.2011.07.004

TUFFRAU, M. 1967. Les structures fibrillaires somatiques et buccales chez les ciliés Hétérotriches. **Protistologica 3** (1): 369-394.

24

Parental care behavior in the Guiana dolphin, *Sotalia guianensis* (Cetacea: Delphinidae), in Ilha Grande Bay, Southeastern Brazil

Rodrigo H. O. Tardin[1,3], Mariana A. Espécie[1], Liliane Lodi[2] & Sheila M. Simão[1]

[1] *Laboratório de Bioacústica e Ecologia de Cetáceos, Departamento de Ciências Ambientais, Instituto de Florestas, Universidade Federal Rural do Rio de Janeiro. Rodovia BR 465, km 7, 23890-000 Seropédica, RJ, Brazil.*
[2] *Projeto Golfinho-Flíper, Instituto Aqualie. Rua Edgard Werneck 428, casa 32, Jacarepaguá, 22763-010 Rio de Janeiro, RJ, Brazil.*
[3] *Corresponding author: E-mail: rhtardin@gmail.com*

ABSTRACT. Parental care is any form of parental behavior that increases offspring fitness. To the authors' knowledge, this study is the first to analyze the intensity of parental care in the Guiana dolphin, *Sotalia guianensis* (van Bénéden, 1864). The objectives of this study are as follows: 1) to quantify the degree of parental care in *S. guianensis* in Ilha Grande Bay, Rio de Janeiro; 2) to investigate the influence of behavioral state and group size on the degree of parental care; and 3) to evaluate the differences between the intensity of parental care provided to calves and juveniles. Our results indicate that the intensity of parental care is high in *S. guianensis* and that care is more intense in larger groups. It is possible that these differences serve to maximize hydrodynamic gains and to minimize risks. Our results suggest that parental care is more intense during travel. A possible reason for this greater intensity is that the feeding dynamics show a more random pattern than other behavioral states. Moreover, the results indicate that calves receive more intense care than juveniles. These results suggest that parent-offspring conflict is possible in the study population.

KEY WORDS. Adult-offspring relationship; parent-offspring conflict.

In mammalian societies, the production and care of the young is a fundamental element because useful models of ecology, social behavior and population dynamics can be developed to incorporate the consequences of parental care (WHITEHEAD & MANN 2000). From this perspective, parental care can be defined as any form of behavior by the parent that appears to enhance the fitness of its offspring (CLUTTON-BROCK *et al.* 2006). Several authors have investigated the amount of time and energy that parents allocate to the care of their offspring (MURRAY *et al.* 2009), whereas others have evaluated how the residual reproductive and survival values may be diminished by intense parental care (PAREDES *et al.* 2005).

In mammals, parental care is associated, in general, with the mother. In this taxon, care is extended beyond gestation. As a result of postgestational parental care, females may lose weight (MILLAR 1978) and modify their behavior (SZABO & DUFFUS 2008). Parental care in mammals is better studied in terrestrial taxa, such as carnivores (e.g., CLUTTON-BROCK *et al.* 2006), but it is poorly understood for marine mammals, especially for species that spend the majority of the time submerged, as is the case for cetaceans.

Cetaceans are long-lived animals and K-strategists. They have extended gestational periods (WHITEHEAD & MANN 2000), and the offspring are known to have an underdeveloped physiology (NOREN 2007) associated with high levels of mortality (WHITEHEAD & MANN 2000). Therefore, offspring often tend to stay with the mother for prolonged periods of time, even if they are nutritionally independent (MANN & SMUTS 1999).

Studies that systematically and quantitatively investigate the mother-offspring relationship in wild cetaceans are scarce. Most of these studies concentrate on accessible species known through long-term research. These species include the North Atlantic right whale, *Eubalaena glacialis* (P.S.L Müller, 1776) (HAMILTON & COOPER 2010); the humpback whale, *Megaptera novaeangliae* (Borowski, 1781) (SZABO & DUFFUS 2008); the Indo-Pacific bottlenose dolphin, *Tursiops aduncus* (Ehrenberg, 1833) (MANN & SMUTS 1999); and the killer whale, *Orcinus orca* (Linnaeus, 1758). Such studies have focused on infant carrying behavior and the associated infant echelon position (MANN & SMUTS 1999), which is reported to provide hydrodynamic (NOREN 2007) and offspring feeding benefits (MANN & SMUTS 1999). Infant carrying behavior is observed in six of 19 orders of eutherian mammals (ROSS 2001) and is the second most costly behavior in mammals (ALTMANN & SAMUELS 1992). This behavior is an evolutionary strategy in species in which the offspring accompany the adults and are subject to environmental challenges after birth (ROSS 2001), as is the case for cetaceans. Parental care appears to differ with offspring age and with other characteristics of the offspring. TRIVERS (1974) hypothesized that parent and offspring would have a conflict of interest over the duration of parental care. For example, the care of juveniles would require an increasing amount of time, and the parents would benefit if they decreased the intensity of care provided to the juveniles to invest in calves. This conflict would be enhanced as a result of the maturational processes affecting the offspring. Parental care of the calves would be intense, but the care of the juveniles would be less intense because the parent would be interested in investing more energy in future offspring.

Despite substantial effort during recent decades, parental care is poorly understood for the "data deficient" (SECCHI 2010) Guiana dolphin, Sotalia guianensis (van Bénéden, 1864) (e.g., MONTEIRO-FILHO et al. 2008, RAUTENBERG & MONTEIRO-FILHO 2008). Sotalia guianensis occurs from central Honduras (CARR & BONDE 2000) to southern Brazil (SIMÕES-LOPES 1988) and displays a multi-male mating system with sperm competition but does not exhibit sexual dimorphism (ROSAS & MONTEIRO-FILHO 2002). The gestational period is estimated to be 12 months, and lactation is estimated to last for 8.7 months. The age of females at sexual maturity may be approximately 5-8 years (ROSAS & MONTEIRO-FILHO 2002), and calving intervals may range between two and three years (SANTOS et al. 2001).

Recently, the Brazilian National Plan for Small Cetaceans (ROCHA-CAMPOS et al. 2011) recommended that behavioral aspects of resident populations of S. guianensis be investigated for the next five years to help re-evaluate the conservation status of these animals. One objective of this study is to quantify the degree of parental care shown by S. guianensis in Ilha Grande Bay. In this study, we test the hypothesis of parental care and evaluate the prediction that a demonstrable degree of parental care would be observed in this species. This expectation is consistent with previous studies of the common bottlenose dolphin, Tursiops truncatus (Montagu, 1821), which reported that offspring show an underdeveloped physiology (NOREN 2007). The second objective of the study is to investigate how the behavioral state and group size influence the intensity of parental care in groups containing both adults and offspring. The working hypothesis is that the intensity of care will vary according to different behavioral states and in groups of different sizes. The associated prediction is that parental care in groups with larger mean sizes will be stronger because mother-offspring dyads can associate with different individuals and therefore experience increased rates of agonistic behavior, especially from males seeking to copulate. An additional prediction is that parental care will be less intense during feeding behavior because the physical proximity of the offspring can constrain the mothers' ability to herd and capture prey. This hypothesis is derived from data on T. truncatus indicating that mothers reduced their speed of movement up to 76% compared to mothers without offspring (NOREN 2007). The third objective of this study is to evaluate differences in the intensity of parental care furnished to calves and juveniles. Based on the hypothesis and predictions that we aim to test, the intensity of care is expected to vary according to maturational processes. According to this expectation, juveniles would receive less parental care. In this study, we propose a novel method of analyzing parental care. This method involves investigations of

behavior that facilitates infant movement and of behavior that protects offspring against agonistic interactions in S. guianensis.

MATERIAL AND METHODS

Ilha Grande Bay (23°02'S, 44°26'W) is located along the southern coast of the state of Rio de Janeiro. Together with Sepetiba Bay, it forms an extensive estuarine system (SIGNORINI 1980). The western part of the bay, where boat trips were conducted as part of this study (23°02'S, 44°26'W), is relatively shallow (<10 m) (NOGARA 2000) and receives organic matter from river drainages and from biomass produced by mangroves (SIGNORINI 1980). This bay receives nutrient-rich sea waters derived from the South Atlantic Central Waters (SACW) (SIGNORINI 1980). The western part is preferentially used by S. guianensis (LODI 2003a) and is surrounded by an outer region ranging in depth between 20 and 40 m (Nautical chart #1633,) with a smaller proportion of islands and rocky coasts than the inner region.

This area contains the largest aggregation observed for this species at a single time was found in this area (LODI & HETZEL 1998), with individuals exhibiting different degrees of residence in the area (ESPÉCIE et al. 2010).

We conducted seven boat trips per season (about six hours of observation effort per trip) for three years (from May 2007 to March 2010) on board a 7.5 m vessel equipped with a YAMAHA® 22 HP inboard engine. When a group of dolphins was sighted, the boat's velocity was reduced and a 15 m distance from the group was maintained. Focal-animal sampling procedures were used. The sampling was continuous, and a SONY DCR-TRV 120® digital handycam was used to collect data. The use of the digital handycam was desirable for recording adult-offspring relationships because spatial and temporal interactions could be observed in slow motion, increasing the quality of the analysis. This advantage is especially important because the time that dolphins spend at the surface, permitting visual observations, can be very brief.

Our definition of a group followed the SMOLKER et al. (1992) chain rule, in which dolphins separated by a distance of 10 m or less were considered to be members of the same group. Distance was measured during the analysis of the video clips. The dolphins' body sizes could be used as proxies for distance to determine whether animals belonged to the same group.

As in certain other delphinids (e.g., T. aduncus), S. guianensis displays fission-fusion social dynamics, with group size and members changing in minutes or seconds (e.g., SANTOS & ROSSO 2008). We believe that it was possible to count given groups twice but that the probability of double counting was reduced because of the fission-fusion dynamics and the size of the population.

We used the category of adults throughout this study, but we believe that the adults observed rearing the offspring were parents because the individuals that are affected by the costs associated with this form of behavior are those that have previously invested a substantial amount of energy to benefit the offspring, as is the case for parental care in mammals (CLUTTON-BROCK 1991). Therefore, we assumed that any adult located less than one adult body length from the offspring was the mother. This assumption may have been a source of bias in the data because allomaternal care might be occurring. However, this approach was used because there is no sexual dimorphism in this species (ROSAS & MONTEIRO-FILHO 2002). The only way to effectively determine the sex of an adult is to perform a genetic analysis of the living animal.

To investigate how the size and behavior of the group influenced parental care, we first defined the category of offspring to include both calves and juveniles. Within this category, we defined individuals as calves (individuals ≤ ½ of adult length) and juveniles (individuals larger than ½ and smaller than ⅔ of adult length) (GEISE et al. 1999), to investigate variation in parental care as a function of the developmental stage of the offspring.

The behavioral states used in this study were based on those defined in KARCZMARSKI et al. (2000): feeding was defined, as an absence of directional movements, accompanied by diving in an asynchronous manner, whereas traveling was defined as the presence of continuous and directional movements.

To investigate if parental care was performed, we developed a new analytical approach. We used two specific categories for analysis; longitudinal and transverse care. Two positions were established for both longitudinal and transverse care.

To characterize longitudinal care, we considered that parental care was evident if an adult positioned itself longitudinally ahead of the offspring. In this study, this position included the infant echelon position commonly described in the cetacean literature (MANN & SMUTS 1999). This position was defined as longitudinal position 1 (Fig. 1). If the adult did not position itself ahead of the offspring, i.e., if it positioned itself longitudinally behind the offspring, no parental care was assumed. This position was defined as longitudinal position 2 (Fig. 2).

To characterize transverse care, we considered that parental care was evident if an adult positioned itself between the boat and the offspring. The boats considered in this context included any boat located near the dolphins, including the research boat. This position was defined as transverse position 1 (Fig. 3). If the adult did not position itself between the boat and the offspring, i.e., the offspring was located between the boat and the adult, no parental care was assumed. This position was defined as transverse position 2 (Fig. 4). Given

Figures 1-4. Positions of adult-offspring dyad of *Sotalia guianensis*, in Ilha Grande Bay, Rio de Janeiro: (1) longitudinal position 1, when the adult located itself longitudinally ahead the offspring; (2) longitudinal position 2, when the adult was locating itself longitudinally behind the offspring; (3) transversal position 1 when the adult positioned itself between the boat and the offspring; (4) transversal position 2, when the adult was not positioning itself between the boat and the offspring. Photos by Mariana Espécie.

these positions for longitudinal and transverse care, we quantified the amount of time that each type of parental behavior was observed. The length of time was chosen as the measure because groups in some situations were observed for prolonged periods (approximately 1-5 minutes), allowing the exact duration of parental care to be quantified.

To evaluate which of the parental care types (longitudinal or transverse) was preferred by adults, we defined four sub-categories, measured in seconds:

Total Care (TC): The adult behaved to protect the offspring both longitudinally and transversely (longitudinal and transverse positions 1). Total Longitudinal Care (TLC): The adult behaved to protect its offspring longitudinally but not transversely (longitudinal position 1 and transverse position 2). Total Transversal Care (TTC): The adult behaved to protect the offspring transversely but not longitudinally (transverse position 1 and longitudinal position 2). Absence of Care (AC): The adult did not protect the offspring longitudinally or transversely (longitudinal and transverse positions 2)

We acknowledge that offspring may take an active role by taking protected positions to increase their own benefits. However, we observed that mothers had an active role in most cases. On many occasions, we observed mothers actively providing care both longitudinally and transversely. For instance, this behavior pattern sometimes occurred if a calf was closer to the boat than its mother and an active change in position was initiated by the mother. Moreover, mothers are not subject to the same physiological and muscular constraints that influence the behavior of the offspring. For this reason, the mothers are best capable of changing their position and have the flexibility necessary to position themselves.

To quantify the tendency of the offspring to be protected by the adult both longitudinally and transversely, we used a Mann-Whitney U test. The same test was used to determine whether group sizes influenced the time taken by adults to protect their offspring both longitudinally and transversely.

To determine whether the behavioral state of the group containing the adult and its offspring influenced longitudinal and transverse care, we used a Kruskal-Wallis test. The variables included in this test were as follows: offspring feeding in longitudinal/transverse position 1 (P1F), offspring feeding in longitudinal/transverse position 2 (P2F), offspring traveling in longitudinal/transverse position 1 (P1T) and offspring traveling in longitudinal/transverse position 2 (P2T). The groups were used as the cases for the test. If we discriminated between offspring in two developmental stages (calves and juveniles), we used a Kruskal-Wallis test to determine whether parental care (both longitudinal and transverse) was performed in degrees that differed between these two developmental stages. We used the following variables in this approach: calves in longitudinal/transverse position 1, calves in longitudinal/transverse position 2, juveniles in longitudinal/transverse position 1 and juveniles in longitudinal/transverse position 1. By quantifying

the four sub-categories created, the same test was used to investigate which type of parental care (longitudinal or transverse) was observed more frequently

RESULTS

We conducted 28 boat trips, which resulted in 100.5 hours of effort and 42.1 hours of direct observation (41.9%). We observed a total of 1,343 groups, including possible double counts (16.7 ± 17.9 individuals, range: 2-200 individuals). Because a total of 1,268 (94.4%) of these groups included offspring, we could conduct an analysis of parental care.

Longitudinal care

Adults located themselves ahead of their offspring more often than expected, and calves received care for a longer total time than juveniles (Table I). In longitudinal care, we observed adults locating themselves ahead of their offspring for a total of 160,830 seconds (86.5%, mean value 57 ± 64 s). Adults located themselves behind their offspring for a total of 25,080 seconds (13.5%, mean value 43 ± 54 s). A Mann-Whitney test showed significant differences between the time offspring were in position 1 and in position 2 ($U = 656.5$, $N_{position\ 1} = 2,933$, $N_{position\ 2} = 584$, $p = 0.000001$) (Table I). If we divided offspring into developmental stages, a Kruskal-Wallis test showed significant differences between the length of time adults spent longitudinally protecting calves and juveniles ($N = 2,656$, $H_3 = 92.1$, $p = 0.0001$) (Table I). A *post hoc* multiple comparison of means did not find significant differences between juveniles in position 1 and calves in position 2 ($p = 0.5$) or between juveniles in position 2 and calves in position 2 ($p = 0.5$).

The mean group size observed for longitudinal position 1 was 27.5 ± 30.1, and that for position 2 was 24.4 ± 34.1. The size of the groups including both adult and offspring differed significantly relative to the duration of longitudinal care (Mann-Whitney test, $U = 740.0$, $N_{position\ 1} = 2,940$, $N_{position\ 2} = 580$, $p = 0.000001$).

The length of time that the adults located themselves in position 1 and in position 2 differed significantly according to the behavioral state (Kruskal-Wallis test, $N = 2,655$, $H_3 = 188.1$, $p = 0.0001$) (Table I). A *post hoc* multiple comparison of means did not find significant differences between offspring traveling in position 2 and offspring feeding in position 1 ($p = 1.0$).

Transverse care

In transverse care, we observed the adults locating themselves between their offspring and the boat more often than expected, and the adults protected calves more often than they protected juveniles (Table I). We observed this position (transverse position 1) for a total of 172,446 seconds (89.1%, mean 60 s ± 67 s). We observed offspring between the boat and the adult (transverse position 2) for a total of 21,054 seconds (10.9%, mean 34 s ± 61 s). A Mann-Whitney test found significant differences between the length of time offspring were in

Table I. Duration in seconds of each category and position of parental care for calves, and juveniles of *Sotalia guianensis* and batched as offspring for feeding and travelling behavior in Ilha Grande Bay, Rio de Janeiro.

Parental care	Developmental Stages		Offspring	
	Calf	Juvenile	Feeding	Travelling
Longitudinal position 1	60.3 ± 70.2	44.4 ± 44.5	47.9 ± 36.6	79.9 ± 98.9
Longitudinal position 2	47.3 ± 64.3	30.5 ± 24.3	37.3 ± 30.1	57.4 ± 80.3
Transversal position 1	62.4 ± 75.1	45.3 ± 44.5	49.2 ± 37.4	81.9 ± 102.2
Transversal position 2	34.5 ± 27.2	27.1 ± 17.6	31.7 ± 22.5	36.9 ± 29.5

position 1 and 2 (U = 710.0, $N_{position\ 1}$ = 2,895, $N_{position\ 2}$ = 619, p = 0.0000001) (Table I). If we divided the offspring into developmental stages, a Kruskal-Wallis test showed significant differences between the length of time adults transversely protected calves and juveniles (N = 2,656; H_3 = 156.9, p = 0.0001) (Table I). A *post hoc* multiple comparison of means did not find significant differences between juveniles in position 2 and calves in position 2 (p = 0.2).

The mean group size was 28.4 ± 31.7 for transverse position 1 and 20.2 ± 25.5 for transverse position 2. The group size showed significant differences relative to the duration of transverse care (Mann-Whitney test, U = 702.0, $N_{position\ 1}$ = 2,904, $N_{position\ 2}$ = 616, p = 0.0000001) (Table I).

The length of time that adults located themselves in transverse position 1 and in transverse position 2 differed significantly according to the behavioral state of the group,(Kruskal-Wallis test, N = 2,633, H_3 = 245.1, p = 0.0001) (Table I). A *post hoc* multiple comparison of means did not find a significant difference between offspring traveling in position 2 and offspring feeding in position 2 (p = 0.5).

Longitudinal vs. transverse care

Using the four previously defined sub-categories, we observed adults positioning themselves both ahead of the offspring and between the offspring and the boat (TC) for a total of 153,384 seconds (mean 60 ± 67 s, range 7-565 s) (79.3%). Adults were observed locating themselves between the boat and the offspring but not ahead of the offspring (TTC) for a total of 19,596 seconds (mean 53 ± 63 s, range 15-361 s) (10.1%); for a total of 15,504 seconds (mean 38 ± 31 s, range 12-252 s) (8.0%), adults were observed locating themselves ahead of the offspring but not between the boat and the offspring (TLC); and for a total of 5,010 seconds (mean 24 ± 15 s, range: 3-134 s) (2.6%), adults were not observed ahead of the offspring or between the offspring and the boat (AC).

DISCUSSION

Offspring care

The shallow and protected waters of the western part of Ilha Grande Bay offer abundant prey (MATSUURA 1978) and may provide an ideal region to raise offspring. The combination of these features can represent a possible explanation for the high concentration of groups containing offspring within Ilha Grande Bay, the highest concentration ever reported within the geographical distribution of the Guiana dolphin (TARDIN *et al.* 2011). Other populations of this species throughout its distribution have been found to include large numbers of groups containing calves. In the Paranaguá estuarine complex (Paraná State), SANTOS *et al.* (2010) reported that 86.4% of all observed groups of Guiana dolphins contained offspring. In Sepetiba Bay, a region adjacent to Ilha Grande Bay where large aggregations of individuals have been observed, NERY *et al.* (2010) reported that 80.3% of all groups observed contained offspring.

Our data indicate that the degree to which offspring receive care is high in this population. Previous work with a closely related species, the common bottlenose dolphin, reported that offspring show a limited pulmonary capacity (NOREN *et al.* 2002), a low concentration of oxygen in the blood (NOREN *et al.* 2002) and small muscle mass (DEAROLF *et al.* 2000) and that for these reasons, parental care is needed to increase offspring fitness.

By positioning themselves ahead of offspring, adults may reduce the thrust required by offspring to swim by almost 60% (WEIHS 2004) because the offspring benefit from the wave pressure created by the large body of the adult. This positional arrangement also reduces the fluke stroke amplitude by up to 24%, as shown by captive studies of *T. truncatus* (NOREN & EDWARDS 2011). In this position, offspring may also have easy access to the adult's mammary glands (MANN & SMUTS 1999), facilitating nursing. Moreover, the location of the offspring behind the adults enables the offspring to observe behaviors that are essential to their survival, especially feeding behavior. In this way, they can learn how to capture prey and can learn about the locations where concentrations of prey occur. These observations indicate a potential opportunity for social learning (BENDER *et al.* 2008).

Studies of different species of cetaceans – e.g., *E. glacialis* (HAMILTON & COOPER 2010), *T. aduncus* (MANN & SMUTS 1999) and *M. novaeangliae* (SZABO & DUFFUS 2008) – have reported that behavior in which (the adults locate themselves ahead of the offspring is a common strategy, suggesting that the benefits received by the offspring are high enough to be maintained. However, the costs imposed on adults as a result of this position may directly affect the fitness of the adults. NOREN (2007), working with captive *T. truncatus*, reported that the costs of this behavior for the adults are similar to those associated with the infant carrying behavior displayed by primates. The speed of adults traveling ahead of their offspring was found to be 76% less than the speed of adults traveling without offspring (Noren 2007). Moreover, the distance to which dolphins could reach when hitting their flukes in the water was decreased by 13% (Noren 2007). Despite the lack of data for Guiana dolphins, we believe that adults face energetic costs similar to those reported for *T. truncatus*. Despite all of these costs, longitudinal care, i.e. adults locating themselves ahead of the offspring, is clearly evident in *S. guianensis*.

Our investigation of the influence of group size on longitudinal care showed that care was more intense if there were more individuals in a group. One hypothesis that we may derive from this outcome is that mothers maintaining offspring in close proximity can directly observe offspring behavior by visually or acoustically monitoring the offspring. This capability appears to be of great importance in large groups in which several individuals may be difficult to detect through direct observation due to the dilution effect. The ability to monitor offspring is especially important during travel to the outer part of the bay, as indicated by evidence of shark bites on a free-ranging *S. guianensis* during our sampling cruises (unpublished data)When traveling in close proximity, offspring may derive greater benefits from social learning. In fact, information about social learning is useful for understanding the behavioral dynamics of certain mammalian populations and may also be useful for conservation purposes (Custance *et al.* 2002).

Our data show that longitudinal care is more intense during traveling than during feeding. During traveling, when the dolphins swam in fixed directions for a prolonged period, a greater amount of time spent in longitudinal care appears to be more beneficial because it allows the offspring to accompany the entire group. This opportunity can be important because the underdeveloped physiology of the offspring does not allow them to travel unaided for long distances (Noren 2007). They gain hydrodynamic benefits by locating themselves behind adults (Weihs 2004). During feeding bouts observed in Ilha Grande Bay, individuals performed movements of short distance and duration to capture prey (Tardin *et al.* 2011). The benefits to be gained from care during feeding appear to be less important than the benefits to be gained while traveling. During feeding, it is more important that the offspring accompany the group to learn the coordinated feeding tactics used to capture prey. This function is facilitated by the ability to behave more freely within the group. Tardin *et al.* (2011), working with the same species in the same study area, reported that offspring were present in 95% of all groups engaged in coordinated feeding behavior.

We observed that adults providing transverse care located themselves between the boat and the offspring more often than would be the case according to a random scenario. This result

may suggest that the benefits provided to the offspring may be high enough to favor the continued performance of this behavior by adults. Transverse care allows offspring to benefit from the dilution effect. The rate of collisions with boats is reduced because another individual is positioned between the potential source of injury (the boat) and the offspring (Turner & Pitcher 1986).

These advantages suggest that parental care yields an increase in offspring fitness, reflecting the offspring's enhanced probability of survival. Indeed, a high rate of mortality is observed in cetaceans at this age (Whitehead & Mann 2000), and collisions with boats may be fatal to the offspring, especially in view of the weaker physiological state of immature animals (Noren *et al.* 2002). However, transverse care increases the rate of collisions between adults and boats, decreasing adult survival as a cost associated with this form of parental care. In fact, boats may be a relevant source of injury in Ilha Grande Bay, where almost 30% of injured dolphins showed marks and scars resulting from collisions with boats (Felipe Torres D'Azeredo, pers. comm.).

Our comparisons involving the sizes of the groups containing both the adult and the offspring were accompanied by the observation that transverse care was more intense in larger groups. This finding may suggest that the close transverse proximity to the adult may minimize agonistic interactions with other individuals, especially in view of the high mortality associated with this age class, as found in *T. aduncus* (Whitehead & Mann 2000). These agonistic interactions between adults and offspring occur because the females are not receptive for mating during the lactation period. If the offspring dies, the female may become able to copulate and produce other offspring (Whitehead & Mann 2000). In a study of *S. guianensis* conducted in Sepetiba Bay, a bay adjacent to the study area, Nery & Simão (2008) observed a case of infanticide, in which adult individuals interacted agonistically with the offspring, causing its death. Such interactions are not unique to this taxon. In mammals, infanticide is also known in langurs, *Semnopithecus entellus* (Dufresne, 1797) (Ren *et al.* 2011). If adults remain in close proximity to their offspring, locating themselves between a possible source of injury (in this case an adult male) and their offspring, the offspring obtains a benefit from this behavior through the dilution effect (Landeau & Terborgh 1986). In this situation, the rate of attacks on the offspring is reduced because the adult positions itself to create a physical barrier that prevents males from reaching the offspring.

We found that the behavioral state of the group significantly influenced the degree of transverse care. Transverse care was more intense during traveling than during feeding. Note that traveling dolphins often swam to outer areas of the bay. The features of these areas differ from those of the protected and shallow inner waters. In the open and deep waters of the outer bay, offspring may encounter predators and, therefore, risk death. Weir *et al.* (2008), working with the Dusky dolphin,

Lagenorhynchus obscurus (Gray, 1828) in New Zealand, reported that groups with calves preferred shallow waters, a possible strategy for avoiding predators. Moreover, feeding behavior involves the other adults in the group in prey capture. During feeding, the other adults are not concerned with the behavior of the mother-offspring dyad. As a result, the offspring are less in need of protection.

Our analysis of the frequencies of parental care strategies (longitudinal or transverse) indicated that Total Care, i.e., adult protection of offspring both longitudinally and transversely, was the most frequent parental care strategy and showed the highest mean duration. However, we observed a preference to protect offspring transversely if Total Care could not be provided: transverse care showed the second highest mean duration. This finding may suggest that adult protection of offspring resulting from the adult's physically locating its body between a source of injuries and the offspring appears to be relatively more important to offspring fitness than longitudinal care. These sources of injuries, such as boat collisions, predators and agonistic interactions, may pose serious threats to offspring survival.

Parent-offspring conflict

Our data indicate that parental care was more intense, both longitudinally and transversely, in calves than in juveniles. The physiological condition of the calves, with their underdeveloped locomotion and their dependence on their mothers for milk, appears to drive the decision to increase the parental care of calves. Juveniles, in contrast, are more capable of locomotion and are nutritionally independent of the adults.

Nevertheless, our data indicate that the parental care of juveniles was still intense. Our results indicate that the adults located themselves ahead of the juveniles and between the juveniles and boats more often than randomly expected. These findings suggest that the juveniles still gained benefits from adult care. However, the intensity of parental care for the juveniles was lower than the intensity of parental care for the calves. MANN & SMUTS (1999) reported that such differences in care were in accordance with maturational processes for *T. aduncus* in Shark Bay, Australia. SZABO & DUFFUS (2008), working with humpback whales in Alaska, reported that the time spent by adults with offspring during intervals of diving became less as the calves grew older.

This paper is the first study to quantify and test hypotheses about parental care in *S. guianensis*. The methodology represents a novel approach for assessing parental care. Replications of this research in other Guiana dolphin populations are needed to improve the knowledge of this phenomenon because, to our knowledge, this report is the first published study to investigate parental care in this species. We found that offspring were protected and that the intensity of parental care varied with group size, behavior and the developmental stage of the offspring. The data provide the first evidence of parent-offspring conflict in this species because the intensity of parental care varied with the age of the offspring. Future studies should investigate the

potential of offspring to play an active role in parental care to increase the benefits that they obtain. Quantifying individual differences and the intensity of investment in parental care will help to broaden the limited knowledge of the behavior of Guiana dolphins because the parent-offspring relationship is an important element of mammalian biology. Behavioral information such as that reported here can aid the conservation of the species because it increases the knowledge of the parent-offspring relationship. This knowledge is useful for understanding social systems (SUTHERLAND 1998). The dolphins may be affected by the increase in boat traffic at Ilha Grande Bay. As a result, they may change their use of the area. The dolphins may move from neighboring islands in shallow waters (LODI 2003), used primarily as a feeding and nursery site (TARDIN *et al.* 2011), to other areas that may pose risks of predation to calves. Intense human activities in the area may affect the energy needed to care for calves and thus the probability of offspring survival. These issues merit consideration and must be recognized by future studies of the area.

ACKNOWLEDGMENTS

We thank Suzanne Beck for initial English revision, Dona Elza, Tico, Gilberto and students of the LBEC for their support. Rodrigo H.O. Tardin is in Programa de Pós Graduação em Ecologia e Evolução, Universidade do Estado do Rio de Janeiro **and Mariana A. Espécie in** Programa de Pós-Graduação em Biologia Animal, Universidade Federal Rural do Rio de Janeiro. Personnel for this study were partially supported by the Fundação de Amparo à Pesquisa do Estado do Rio de Janeiro (FAPERJ) (R.H.O. Tardin, Grant #E-26/151.047/2007 and Grant #E-26/100.866/2011); the Conselho Nacional de Pesquisa e Desenvolvimento (CNPq) (M.A.Espécie, Grant #111555/2008-6), and the Coordenação de Aperfeiçoamento de Pessoal de Nível Superior (CAPES) (R.H.O. Tardin, and M.A.Espécie) and the Cetacean Society International.

REFERENCES

ALTMANN, J. & A. SAMUELS. 1992. Costs of maternal care: infant-carrying baboons. **Behavioral Ecology and Sociobiology** 29 (6): 391-398. **doi:** 10.1007/BF00170168

BENDER, C.E.; D.L. HERZING & D.F. BJORKLUND. 2008. Evidence of teaching in Atlantic spotted dolphins by mother dolphins foraging in the presence of their calves. **Animal Cognition** 12 (1): 43-53. **doi:** 10.1007/s10071-008-0169-9

CARR, T. & R.K. BONDE. 2000. Tucuxi (Sotalia fluviatilis) occurs in Nicaragua, 800km north of its previously known range. **Marine Mammal Science** 16 (2): 447-452.

CLUTTON-BROCK, T. 1991. **The Evolution of Parental Care.** Princeton, Princeton University Press, 368p.

CUSTANCE, D.M.; A. WHITEN & T. FREDMAN. 2002. Social learning and primate reintroduction. **International Journal of Primatology** 23 (3): 479-499. **doi:** 0164-0291/02/0600-0479/0

DEAROLF, J.L.; W.A. MCLELLAN; R.M. DILLAMAN; D.J.R. FRIERSON &
D.A. PABST. 2000. Precocial development of axial locomotor
muscle in bottlenose dolphins (*Tursiops truncatus*). **Journal
of Morphology 244** (3): 203-215. **doi**: 10.1002/(SICI)1097-
4687(200006)244:3<203::AID-JMOR5>3.0.CO;2-V

ESPÉCIE, M.A.; R.H.O. TARDIN & S.M. SIMÃO. 2010. Degrees of
residence of Guiana dolphins (*Sotalia guianensis*) in Ilha Grande Bay, south-eastern Brazil: a preliminary assessment. **Journal
of Marine Biological Association of the United Kingdom
90** (8): 1633-1639. **doi**: 10.1017/S0025315410001256

GEISE, L.; N. GOMES & R. CERQUEIRA. 1999. Behaviour, habitat use
and population size of *Sotalia fluviatilis* (Gervais, 1853)
(Cetacea: Delphinidae) in the Cananéia estuary region, SP,
Brazil. **Revista Brasileira de Biologia 59** (2): 183-194.

HAMILTON, P.K. & L.A. COOPER. 2010. Changes in North Atlantic right
whale (*Eubalaena glacialis*) cow-calf association times and use
of the calving ground: 1993-2005. **Marine Mammal Science
26** (4): 896-916. **doi**: 10.1111/j.1748-7692.2010.00378.x

KARCZMARSKI, L.; V.C. COCKCROFT & A. MCLACHLAN. 2000. Habitat
use and preferences of Indo-Pacific humpback dolphins
Sousa chinensis in Algoa Bay, South Africa. **Marine Mammal
Science 16** (1): 65-79.

LANDEAU, L. & J. TERBORGH. 1986. Oddity and the confusion effect
in predation. *Animal Behavior* **34** (5): 1372-1380.

LODI, L. 2003. Seleção e uso do hábitat pelo boto-cinza, *Sotalia
guianensis* (van Béneden, 1864) (Cetacea: Delphinidae), na Baía
de Paraty, Estado do Rio de Janeiro. **Bioikos 17** (1-2): 5-20.

LODI, L. & B. HETZEL. 1998. Grandes agregações do boto-cinza
(*Sotalia fluviatilis*) na Baía da Ilha Grande, Rio de Janeiro.
Bioikos 12 (2): 26-30.

MANN, J. & B. SMUTS. 1999. Behavioral development in wild
bottlenose dolphin newborns (*Tursiops sp.*). **Behavior 136**
(5): 529-566.

MATSUURA, Y. 1978. **Exploração e avaliação de estoque de peixes pelágicos no sul do Brasil – Projeto integrado para o
uso e exploração racional do ambiente marinho.** São Paulo, Relatório Técnico do Instituto Oceanográfico, 46p.

MILLAR, J.S. 1978. Energetics of reproduction in *Peromyscus
leucopus*: the cost of lactation. **Ecology 59** (8): 1055-1061.

MONTEIRO-FILHO, E.L.A.; M.M.S. NETO & DOMIT, C. 2008. Com-
portamento de infantes, p. 127-138. *In*: E.L.A. MONTEIRO-FI-
LHO & K.D.K.A. MONTEIRO (Eds). **Biologia, Ecologia e Conservação do Boto-Cinza.** São Paulo, Páginas & Letras Editora e Gráfica, 274p.

MURRAY, C.F.; E.V. LONSDORF; L.E. EBERLY & A.E. PUSEY. 2009.
Reproductive energetics in free-living female chimpanzees
(*Pan troglodytes schweinfurthii*). **Behavioral Ecology 20** (6):
1211-1216. **doi**: 10.1093/beheco/arp114

NERY, M.F & S.M. SIMÃO. 2008. Sexual coercion and aggression
towards a newborn calf of marine tucuxi dolphins (*Sotalia
guianensis*). **Marine Mammal Science 25** (2): 450-454.
doi: 10.1111/j.1748-7692.2008.00275.x

NERY, M.F.; S.M. SIMÃO & T.C.L. PEREIRA. 2010. Ecology and behavior

of the estuarine dolphin, *Sotalia guianensis* (Cetacea,
Delphinidae), in Sepetiba Bay, south-eastern Brazil. **Journal
of Ecology and Natural Environment 2** (9): 194-200.

NOGARA, P.J. 2000. **Caracterização dos ambientes marinhos
da Área de Proteção Ambiental de Cairuçu – Município
de Paraty – RJ.** Rio de Janeiro, Technical Report, Fundação
SOS Mata Atlântica, 83p.

NOREN, S.R. 2007. Infant carrying behaviour in dolphins: costly
parental care in an aquatic environment. **Functional Ecology
22** (2): 284-288. **doi**: 10.1111/j.1365-2435.2007.01354.x

NOREN, S.R. & E.F. EDWARDS. 2011. Infant position in mother-calf
dolphin pairs: formation locomotion with hydrodynamic
benefits. **Marine Ecology Progress Series 424**: 229-236.
doi: 10.3354/meps08986

NOREN, S.R.; G. LACAVE; R.S. WELLS & T.M. WILLIAMS. 2002. The
development of blood oxygen stores in bottlenose dolphins
(*Tursiops truncatus*): implications for diving capacity. **Journal of
Zoology 258** (1): 105-113. **doi**: 10.1017/S0952836902001243

PAREDES, R.; I.L. JONES &._D.J. BONESS. 2005. Reduced parental care,
compensatory behaviour and reproductive costs of thick-
billed murres equipped with data loggers. **Animal Behavior
69** (1): 197-208. **doi**: 10.1016/j.anbehav.2003.12.029

RAUTENBERG, M. & E.L.A. MONTEIRO-FILHO. 2008. Cuidado parental,
p. 139-156. *In*: E.L.A. MONTEIRO-FILHO & K.D.K.A. MONTEIRO
(Eds). **Biologia, Ecologia e Conservação do Boto-Cinza.**
São Paulo, Páginas & Letras Editora e Gráfica, 274p.

REN, B.; D. LI; X. HE; J. QIU & M. LI. 2011. Female resistance to
invading males increases infanticide in langurs. **Plos One 6**
(4): 1-4. **doi**: 10.1371/journal.pone.0018971

ROCHA-CAMPOS, C.C.; I.G. CÂMARA & D.J. PRETTO. 2011. **Plano de
Ação Nacional para a Conservação dos Mamíferos Aquáticos: Pequenos Cetáceos.** Brasília, Instituto Chico Mendes
de Conservação da Biodiversidade, Série Espécies Ameaçadas.

ROSAS, F.C.W. & E.L.A. MONTEIRO-FILHO. 2002. Reproduction of the
estuarine dolphin (*Sotalia guianensis*) on the coast of Paraná,
Southern Brazil. **Journal of Mammalogy 83** (2): 506-515.
doi: 10.1644/1545-1542(2002)083<0507:ROTEDS>2.0.CO;2

ROSS, C. 2001. Park or ride? Evolution of infant carrying in
primates. **International Journal of Primatology 22** (5): 749-
771. **doi**: 10.1023/A:1012065332758

SANTOS, M.C.O. & S. ROSSO. 2008. Social organization of marine
tucuxi dolphins, *Sotalia guianensis*, in the Cananéia Estuary
of Southeastern Brazil. **Journal of Mammalogy 89** (2): 347-
355.

SANTOS, M.C.O.; L. BARÃO-ACUÑA & S. ROSSO. 2001. Insights on
site fidelity and calving intervals of the marine tucuxi
dolphin (*Sotalia fluviatilis*) in south-eastern Brazil. **Journal
of Marine Biological Association of the United Kingdom
81** (6): 1049-1052.

SANTOS, M.C.O.; J.E.F. OSHIMA; E.S. PACÍFICO & E. SILVA. 2010. Group
size and composition of Guiana dolphins (*Sotalia guianensis*)
(van Bénèden, 1864) in the paranaguá estuarine complex,
Brazil. **Revista Brasileira de Biologia 70** (1): 111-120.

SECCHI, E. 2010. *Sotalia guianensis*. *In*: IUCN 2012 (Ed.). **IUCN Red List of Threatened Species. Version 2012.1.** Available online at: http://www.iucnredlist.org [Acessed: 05.VII.2012].

SIGNORINI, S.R. 1980. A study of the circulation in bay of Ilha Grande and bay of Sepetiba. Part I, an assessment to the tidally and wind-driven circulation using a finite element numerical model. **Boletim do Instituto Oceanográfico 29** (1): 41-55.

SIMÕES-LOPES, P.C. 1988. Ocorrência de uma população de *Sotalia fluviatilis* Gervais, 1853, (Cetacea: Delphinidae) no Limite Sul da sua distribuição, Santa Catarina, Brasil. **Biotemas 1** (1): 57-62.

SMOLKER, R.A.; A.F. RICHARDS; R.C. CONNOR & J.W. PEPPER. 1992. Sex differences in patterns of association among Indian Ocean bottlenose dolphins. **Behavior 123** (1-2): 38-69.

SUTHERLAND, W.J. 1998. The importance of behavioural studies in conservation biology. **Animal Behaviour 56** (4): 801-809.

SZABO, A. & D. DUFFUS. 2008. Mother-offspring association in the humpback whale, *Megaptera novaeangliae*: following behaviour in an aquatic mammal. **Animal Behavior 75** (3): 1085-1092. DOI:10.1016/j.anbehav.2007.08.019

TARDIN, R.H.O.; M.A. ESPÉCIE; M.F. NERY; F.T. D'AZEREDO & S.M. SIMÃO. 2011. Coordinated feeding tactics of the Guiana dolphin, *Sotalia guianensis* (Cetacea: Delphinidae), in Ilha Grande Bay, Rio de Janeiro, Brazil. **Zoologia 28** (3): 291-296. **DOI:** 10.1590/S1984-46702011000300002

TRIVERS, R. 1974. Parent-offspring conflict. **American Zoologist 14** (1): 249-264.

TURNER, G.F. & T.J. PRICHTER. 1986. Attack abatement: A model for group protection by combined avoidance and dilution. **American Naturalist 128**: 228-240.

WEIR, J.S.; N.M.T. DUPREY & B WÜRSIG. 2008. Dusky dolphin (*Lagenorhynchus obscurus*) subgroup distribution: are shallow waters a refuge for nursery groups? **Canadian Journal of Zoology 86** (11): 1225-1234.

WEIHS, D. 2004. The hydrodynamics of dolphin drafting. **Journal of Biology 3** (2): 1-23.

WHITEHEAD, H. & J. MANN. 2000. Female reproductive strategies of cetaceans: Life histories and calf care, p. 219-246. *In*: J. MANN; R.C. CONNOR; P. TYACK & H. WHITEHEAD (Eds). **Cetacean Societies: Field studies of dolphins and whales.** Chicago, University of Chicago Press, 448p.

PERMISSIONS

LIST OF CONTRIBUTORS

Cecília P. Alves-Costa and Adriana Ayub
Laboratório de Ecologia e Restauração da Biodiversidade (LERBIO), Departamento de Botânica, Universidade Federal de Pernambuco. Avenida Prof. Moraes Rego 1235, Cidade Universitária, 50670-901 Recife, PE, Brazil

Marco A.R. Mello
Departamento de Biologia Geral, Instituto de Ciências Biológicas, Universidade Federal de Minas Gerais. Avenida Antônio Carlos 6627, Pampulha, Caixa Postal 486, 31270-910 Belo Horizonte, MG, Brazil

Raissa Sarmento
Laboratório de Ecologia e Restauração da Biodiversidade (LERBIO), Departamento de Botânica, Universidade Federal de Pernambuco. Avenida Prof. Moraes Rego 1235, Cidade Universitária, 50670-901 Recife, PE, Brazil

Austin L. Hughes
Department of Biological Sciences, University of South Carolina, Columbia SC 29205 USA

Bruno V.B. Rodrigues, Alexandre B. Bonaldo and Regiane Saturnino
Laboratório de Aracnologia, Coordenação de Zoologia, Museu Paraense Emílio Goeldi. Avenida Perimetral 1901, Terra Firme, 66077-830 Belém, Pará, Brazil

Luíza Loebens, Rafael Lazzari and Silvio T. da Costa
Departamento de Zootecnia e Ciências Biológicas, Centro de Educação Norte do Rio Grande do Sul, Universidade Federal de Santa Maria. 98300-000 Palmeira das Missões, RS, Brazil

Luciane T. Gressler and Fernando J. Sutili
Programa de Pós-graduação em Farmacologia, Universidade Federal de Santa Maria. 97105-900 Santa Maria, RS, Brazil

Bernardo Baldisserotto
Departamento de Fisiologia e Farmacologia, Universidade Federal de Santa Maria. 97105-900 Santa Maria, RS, Brazil

Diane Nava, Rozane M. Restello and Luiz U. Hepp
Programa de Pós-graduação em Ecologia, Universidade Regional Integrada do Alto Uruguai e das Missões. Avenida Sete de Setembro 1621, 99709-910 Erechim, RS, Brazil

Qi Liu and Li-Biao Zhang
Guangdong Public Laboratory of Wild Animal Conservation and Utilization & Guangdong Key Laboratory of Integrated Pest Management in Agriculture, Guangdong Entomological Institute, Guangzhou 510260, China

Fu-Min Wang
Guangdong Provincial Wildlife Rescue Center, Guangzhou 510520, China

Li Wei
College of Ecology, Lishui University, Lishui 323000, China

Manuel Pedraza
Programa de Pós-Graduação, Museu de Zoologia, Universidade de São Paulo. Avenida. Nazaré 481, Ipiranga, 04263-000 São Paulo, SP, Brazil

José Eduardo Martinelli-Filho
Faculdade de Oceanografia, Instituto de Geociências da Universidade Federal do Pará. Campus Universitário do Guamá, 66075-110 Belém, PA. Brazil

Célio Magalhães
Instituto Nacional de Pesquisas da Amazônia. Caixa Postal 2223, 69080-971 Manaus, AM, Brazil

Diego Dias da Silva
Programa de Pós-graduação em Sistemática e Evolução, Centro de Biociências, Universidade Federal do Rio Grande do Norte. Campus Universitário Lagoa Nova, 59072-970 Natal, RN, Brazil

Bruno Cavalcante Bellini
Programa de Pós-graduação em Sistemática e Evolução, Centro de Biociências, Universidade Federal do Rio Grande do Norte. Campus Universitário Lagoa Nova, 59072-970 Natal, RN, Brazil

Departamento de Botânica, Ecologia e Zoologia, Centro de Biociências, Universidade Federal do Rio Grande do Norte. 59072-970 Natal, RN, Brazil

Lucas Rodriguez Forti
Programa de Pós-Graduação Interunidades em Ecologia Aplicada, Escola Superior de Agricultura Luiz de Queiroz, Universidade de São Paulo. Avenida Centenário 303, 13400-970 Piracicaba, SP, Brazil

Rafael Márquez
Fonoteca Zoológica, Departamento de Biodiversidad y Biología Evolutiva, Museo Nacional de Ciencias Naturales. José Gutiérrez Abascal 2, Madrid, Spain

Jaime Bertoluci
Departamento de Ciências Biológicas, Escola Superior de Agricultura Luiz de Queiroz, Universidade de São Paulo. Avenida Pádua Dias 11, 13418-900 Piracicaba, SP, Brazil

Denis Rafael Pedroso
Laboratório de Aracnologia, Museu Nacional, Universidade Federal do Rio de Janeiro, Brazil

Renner Luiz Cerqueira Baptista
Laboratório de Diversidade de Aracnídeos, Instituto de Biologia, Universidade Federal do Rio de Janeiro, Brazil

Rogério Bertani
Laboratório Especial de Ecologia e Evolução, Instituto Butantan. Avenida Vital Brazil 1500, 05503-900 São Paulo, SP, Brazil

Aline S. Maciel, Thereza de A. Garbelotto, Ingrid C. Winter, Talita Roell and Luiz A. Campos
Departamento de Zoologia, Universidade Federal do Rio Grande do Sul. Avenida Bento Gonçalves 9500, Agronomia, 91501-970 Porto Alegre, RS, Brazil

Alex José de Almeida, Melina Maciel F. Freitas and Sônia A. Talamoni
Programa de Pós-graduação em Zoologia de Vertebrados, Pontifícia Universidade Católica de Minas Gerais. Avenida Dom José Gaspar 500, 30535-610 Belo Horizonte, MG, Brasil

Marcela A. Souza
Laboratório de Ecologia de Peixes, Universidade Federal de Lavras. Campus Universitário, 37200-000 Lavras, MG, Brazil

Daniela C. Fagundes
Programa de Pós-Graduação em Ecologia Aplicada, Universidade Federal de Lavras. Campus Universitário, 37200-000 Lavras, MG, Brazil

Cecília G. Leal
Programa de Pós-Graduação em Ecologia Aplicada, Universidade Federal de Lavras. Campus Universitário, 37200-000 Lavras, MG, Brazil
Departamento de Biologia, Universidade Federal de Lavras. Campus Universitário, 37200-000 Lavras, MG, Brazil

Paulo S. Pompeu
Departamento de Biologia, Universidade Federal de Lavras. Campus Universitário, 37200-000 Lavras, MG, Brazil

Fabio Cleisto Alda Dossi and Fernando Luis Cônsoli
Laboratório de Interações em Insetos, Departamento de Entomologia e Acarologia, Escola Superior de Agricultura "Luiz de Queiroz", Universidade de São Paulo, Avenida Pádua Dias 11, 13418-900 Piracicaba, SP, Brazil

Amilcar Brum Barbosa and Sonia Barbosa dos Santos
Laboratório de Malacologia Límnica e Terrestre, Departamento de Zoologia, Instituto de Biologia Roberto Alcantara Gomes, Universidade do Estado do Rio de Janeiro. Rua São Francisco Xavier 524, PHLC sala 525-2, 20550-900 Rio de Janeiro, RJ, Brazil

Ludmilla M.S. Aguiar and Ricardo B. Machado
Departamento de Zoologia, Instituto de Ciências Biológicas, Universidade de Brasília, Campus Darcy Ribeiro, Asa Norte, 70910-900 Brasília, DF, Brazil

Enrico Bernard
Departamento de Zoologia, Universidade Federal de Pernambuco. Rua Nelson Chaves, Cidade Universitária, 50670-420 Recife, PE, Brazil

Débora de S. Silva-Camacho, Joaquim N. de S. Santos, Rafaela de S. Gomes and Francisco G. Araújo
Laboratório de Ecologia de Peixes, Universidade Federal Rural do Rio de Janeiro. Antiga Rodovia Rio-SP km 47, 23851-970 Seropédica, RJ, Brazil

Tatiana P. Portella
Laboratório de Dinâmica Evolutiva e Sistemas Complexos, Departamento de Zoologia, Universidade Federal do Paraná. Caixa Postal 19020, 81531-990 Curitiba, PR, Brazil

Diego R. Bilski
Programa de Pós-Graduação em Ecologia e Conservação, Universidade Federal do Paraná

Fernando C. Passos
Laboratório de Biodiversidade, Conservação e Ecologia de Animais Silvestres, Departamento de Zoologia, Universidade Federal do Paraná. Caixa Postal 19020, 81531-990 Curitiba, PR, Brazil

Tatiana P. Portella
Laboratório de Dinâmica Evolutiva e Sistemas Complexos, Departamento de Zoologia, Universidade Federal do Paraná. Caixa Postal 19020, 81531-990 Curitiba, PR, Brazil
Programa de Pós-Graduação em Ecologia e Conservação, Universidade Federal do Paraná

Fernanda M. de Souza, Eliandro R. Gilbert, Maurício G. de Camargo and Wagner W. Pieper
Centro de Estudos do Mar, Universidade Federal do Paraná. Caixa Postal 50002, 83255-976 Pontal do Paraná, PR, Brazil

Ingrid Mattos and José Ricardo M. Mermudes
Laboratório de Entomologia, Departamento de Zoologia, Universidade Federal do Rio de Janeiro. Caixa Postal 68044, 21941-971 Rio de Janeiro, RJ, Brazil

Mauricio O. Moura
Departamento de Zoologia, Universidade Federal do Paraná. Caixa Postal 19020, 81531-980 Curitiba, PR, Brazil

Francisco Dal Vechio, Renato Recoder and Miguel Trefaut Rodrigues
Departamento de Zoologia, Instituto de Biociências, Universidade de São Paulo. Rua do Matão, Travessa 14, 321, Cidade Universitária, 05508-090 São Paulo, Brazil

Hussam Zaher
Museu de Zoologia, Universidade de São Paulo. Avenida Nazaré 481, Ipiranga, 04263-000 São Paulo, Brazil

Alessandra da Fonseca Viana and Marcelo Vianna
Laboratório de Biologia e Tecnologia Pesqueira, Instituto de Biologia, Universidade Federal do Rio de Janeiro. Avenida Carlos Chagas Filho 373, Bloco A, 21941-902 Rio de Janeiro, RJ, Brazil

Noemi M. Fernandes and Inácio D. da Silva Neto
Laboratório de Protistologia, Departamento de Zoologia, Universidade Federal do Rio de Janeiro. 21941-590, Rio de Janeiro, Rio de Janeiro, Brazil

Rodrigo H. O. Tardin, Mariana A. Espécie and Sheila M. Simão
Laboratório de Bioacústica e Ecologia de Cetáceos, Departamento de Ciências Ambientais, Instituto de Florestas, Universidade Federal Rural do Rio de Janeiro. Rodovia BR 465, km 7, 23890-000 Seropédica, RJ, Brazil

Liliane Lodi
Projeto Golfinho-Flíper, Instituto Aqualie. Rua Edgard Werneck 428, casa 32, Jacarepaguá, 22763-010 Rio de Janeiro, RJ, Brazil

Index

www.ingramcontent.com/pod-product-compliance
Lightning Source LLC
Chambersburg PA
CBHW082023190326
41458CB00010B/3254